ELEMENT	SYMBOL	ATOMIC NUMBER	ATOMIC MASS, amu	ORBITALS			MELTING POINT, °C	DENSITY (SOLID), Mg/m³ (=g/cm³)	CRYSTAL STRUCTURE, 20° C	APPROX. ATOMIC RADIUS nm†	VALENCE (MOST COMMON)	APPROX. IONIC RADIUS, nm‡
				3d	4s	4p						
Iron	Fe	26	55.85	Ar + 6	2		1538	7.87	bcc	0.1241	2+	0.074
									fcc	0.1269	3+	0.064
Cobalt	Co	27	58.93	Ar + 7	2		1495	8.83	hcp	0.125	2+	0.072
Nickel	Ni	28	58.71	Ar + 8	2		1455	8.90	fcc	0.1246	2+	0.069
Copper	Cu	29	63.54	Ar + 10	1		1084	8.93	fcc	0.1278	1+	0.096
Zinc	Zn	30	65.38	Ar + 10	2		420	7.13	hcp	0.139	2+	0.074
Germanium	Ge	32	72.59	Ar + 10	2	2	937	5.32	§	0.1224	4+	—
Arsenic	As	33	74.92	Ar + 10	2	3	816	5.78	—	0.125	3+	—
Krypton	Kr	36	83.80	Ar + 10	2	6	−157	—	fcc	0.201	inert	—
				4d	5s	5p						
Silver	Ag	47	107.87	Kr + 10	1		961.9	10.5	fcc	0.1444	1+	0.126
Tin	Sn	50	118.69	Kr + 10	2	2	232	7.17	bct	0.1509	4+	0.071
Antimony	Sb	51	121.75	Kr + 10	2	3	630.7	6.7	—	0.1452	5+	—
Iodine	I	53	126.9	Kr + 10	2	5	114	4.93	ortho	0.135	1−	0.220
Xenon	Xe	54	131.3	Kr + 10	2	6	−112	2.7	fcc	0.221	inert	—
				4f	5d	6s						
Cesium	Cs	55	132.9	Xe +		1	28.6	1.9	bcc	0.265	1+	0.167
Tungsten	W	74	183.9	Xe + 14	4	2	3410	19.25	bcc	0.1367	4+	0.070
Gold	Au	79	197.0	Xe + 14	10	1	1064.4	19.3	fcc	0.1441	1+	0.137
Mercury	Hg	80	200.6	Xe + 14	10	2	−38.86	—	—	0.155	2+	0.110
Lead	Pb	82	207.2	Hg + $6p^2$			327.4	11.38	fcc	0.1750	2+	0.120
Uranium	U	92	238.0	Rn + $5f^3$	6d	$7s^2$	1133	19.05	—	0.138	4+	0.097

* From various sources, including the ASM *Handbooks.*

† One half of the closest approach of two atoms in the elemental solid. For noncubic structures, the average interatomic distance is given; e.g., in hcp, the atom is slightly ellipsoidal.

‡ Radii for CN = 6; otherwise, $0.97 \, R_{CN=8} \approx R_{CN=6} \approx 1.1 \, R_{CN=4}$. Patterned after Ahrens.

§ Diamond cubic.

Sixth Edition

ELEMENTS OF MATERIALS SCIENCE AND ENGINEERING

Sixth Edition

STRUCTURE · PROPERTIES · PERFORMANCE

ELEMENTS OF MATERIALS SCIENCE AND ENGINEERING

Lawrence H. Van Vlack

UNIVERSITY OF MICHIGAN

Ann Arbor, Michigan

Addison-Wesley Publishing Company

Reading, Massachusetts • Menlo Park, California • New York
Don Mills, Ontario • Wokingham, England • Amsterdam
Bonn • Sydney • Singapore • Tokyo • Madrid • San Juan

This book is in the
Addison-Wesley Series in Metallurgy and Materials Engineering

Consulting Editors
Morris Cohen
Merton C. Flemings

Sponsoring Editor: Don Fowley
Production Supervisor: Bette J. Aaronson
Copy Editor: Lyn Dupré
Text Designer: Deborah Schneck
Technical Art Consultant: Joseph Vetere
Illustrators: Dick Morton and C & C Associates, Inc.
Cover Designer: Marshall Henrichs
Production Coordinators: Helen Wythe and Sheila Bendikian
Manufacturing Supervisor: Hugh Crawford

Library of Congress Cataloging-in-Publication Data

Van Vlack, Lawrence H.
　　Elements of materials science and engineering / by Lawrence H. Van
　　Vlack. — 6th ed.
　　　　p. cm.

　　Includes index.
　　ISBN 0-201-09314-6
　　1. Materials.　2. Solids.　I. Title.
TA403.V35 1989
620.1′1 — dc19 　　　　　　　　　　　　　　　　88-15383
　　　　　　　　　　　　　　　　　　　　　　　　　　CIP

Reprinted with corrections February, 1990

To Today's Students, who will be Tomorrow's Engineers

PREFACE

No course in science or engineering may remain static. Not only does technology advance and scientific understanding increase, the academic framework undergoes changes. Thus, periodic revisions are desirable in an effort to optimize the value of a textbook for students who will be tomorrow's engineers.

Developments such as the high-temperature superconductors are exciting, and the scientific data such as that obtained from tunnelling electron microscopes provide new insights. During the last decade, however, the evolving structure of the academic environment probably has had a more direct impact on introductory materials courses within the engineering curricula. Whereas academic departments will continue to have specialists in ceramics, in polymers, and in metals, the trend has been toward departments of materials science and engineering. The reasons for this merging are many; for example, fracture toughness applies equally to the failure of metals, of ceramics, and of polymers, as well as to hybrid composites. Likewise, graduate students working with polyblends give cognizance to phase immiscibilities and to the microstructure/property relationships utilized by ceramists and metallurgists. Particulate processing is no longer restricted to ceramics, nor are engineering designs using magnets limited to metallic materials.

In view of these changes, the majority of the current generation of instructors can easily extend the topics of crystals from single-component metals to binary ceramic compounds, and even introduce simple molecular crystals when they teach an introductory materials course. Likewise, although reaction rates may differ, the same principles hold for the phase relationships of ceramics and polymers as they do for metals. Today's instructor easily handles these topics generically for the several types of materials.

The major modification to this edition has been in the attention to the commonality found within the materials field, in which structures and properties are considered generically for all materials rather than categorically by material classes—metals, polymers, ceramics, and semiconducters. The three photos present on the cover and chapters of this sixth edition are symbolic of this generic view; each chosen to pictorially demonstrate the connection between structure, properties, and performance.

Overview

Chapter 1 remains as an introduction to the topic of materials, since undergraduate students generally relate to their product without giving thought to the materials within them. Chapter 2 reviews the necessary chemistry from the students' previous general chemistry courses, but in doing so extends the topics of bonding and atomic coordination. The topics of Chapters 3, 4, and 5 are common to all materials—crystal structure, disorder in solids, and phase relationships, respectively. Included for the first time in Chapter 5 are several molecular phase diagrams, chosen to emphasize immiscibility, which is pertinent to the more recently developed polyblends.

Chapter 6 combines and extends the subject of reaction rates, while Chapter 7 does the same for an introduction to microstructure. Although the three principal classes of materials have distinct differences with respect to these two topics, the bases of the differences are instructive to the subject; for example, the crystallization rates of metallic, silicate, and polymeric materials.

Chapters 8, 9, and 10 focus on the mechanical behavior of solids. In sequence, they consider deformation, strengthening, and the characteristics of polymers and composites. Chapters 11, 12, and 13 look at the electromagnetic behavior of solids—conductivity, magnetic, and the dielectric and optical, respectively. These six chapters are written to give the instructor options regarding the topics to be selected, depending on the available time and curricular requirements.

The final chapter (14) addresses performance in service, particularly for severe conditions in which corrosion, fatigue, heat, or radiation may alter the structure and hence the properties of materials.

Pedagogy

Teaching aids within the text include not only the Summary at the end of each chapter, but also nearly 175 Examples in which a procedure is outlined before the calculations are made. Wherever appropriate, followup comments supplement the calculations. Practice Problems at the end of each chapter offer a trial run for the student, and answers to these several hundred problems are available at the end of the text. A new end-of-chapter feature is the inclusion of Test Problems. Of the nearly 400 such problems throughout the book, the majority either are new to the text or are significantly modified from those in previous editions. More than 400 Terms and Concepts are defined in a glossary.

Supplementary Material

A Study Guide that accompanies the text is available for students' use with this edition. It provides both a means for self-instruction when desired, or facilitates self-help for a lagging student. The Study Guide contains Quiz Samples (and their answers) as well as expanded solutions to the Practice Problems of the text. Also included are the more widely used study sets that appeared in *Study Aids for Introductory Materials Courses.* These visual aids, revised as necessary, have proven their merit in previous years in not having to delay class progress for those individuals who are not immediately clear on crystal structures, phase diagrams, diffusion, or other basic concepts.

Acknowledgments

It is with regret that I can not cite each and every individual who has contributed to the updating of this text. The list would include literally hundreds of students at The University of Michigan who have given feedback on assigned topics and study problems. The critical comments and suggestions of my academic colleagues in Ann Arbor have been most welcome and helpful. Likewise, academic associates in other materials science and engineering departments deserve recognition for both letters and personal discussions in regard to content and possible improvements.

The role of Professors Morris Cohen (Massachusetts Institute of Technology), Richard Porter (North Carolina State), and Ronald Gibala (University of Michigan) should be specifically acknowledged. Each critiqued the contents of this new edition in detail and provided an assurance of appropriateness of the changes. On the publishing side, I want to thank Don Fowley, Bette Aaronson, and all of the Addison-Wesley personnel for their attention to the multitude of editing and production details that lead to a quality product.

Finally, and most importantly, none of the revision efforts would have been possible without my wife Fran's patience and tolerance during the recent months.

Ann Arbor, Michigan   L.H.V.V.

CONTENTS

Chapter Six
REACTION RATES   **187**

Chapter Seven
MICROSTRUCTURES   **215**

FOREWORD

MATERIALS SCIENCE AND ENGINEERING—A PERSPECTIVE

The history of human civilization and social development is strongly intertwined with the pervasive role of *materials*—namely, the substances that are accessible to mankind and can be processed to exhibit the desired properties for making things. This connection between the human race and its materials has expanded enormously over the ages until, now, billions of tons of raw materials are taken annually from nature to be manipulated step by step into innumerable products and systems for societal purpose. We have come to recognize materials as one of the primary resources of mankind, ranking with living space, food, energy, information, and manpower itself. Indeed, the vast field of materials, both natural and manmade, epitomizes the intimate interdependence between society and nature. It is no wonder, then, that many aspects of science and engineering are directed presently to the understanding and utility of materials. The very definition of engineering adopted by the Accreditation Board for Engineering and Technology illuminates the central position of materials: "Engineering is the profession in which a knowledge of the mathematical and natural sciences gained by study, experience, and practice is applied with judgment to develop ways to utilize, economically, the *materials* and *forces* of nature for the benefit of mankind" (emphasis added). In a real sense, virtually every modern technology is materials-limited at the present time with respect to performance, reliability, or cost.

In the light of this background, perhaps it is somewhat surprising to find that the field of materials science and engineering (MSE) came into intellectual focus as an identifiable body of inquiry and endeavor only about 35 years ago. The essential framework of MSE is composed of the scientific and practical interrelationships at play among the processing, structure, properties, and performance of all classes of materials that are potentially useful to society. This linkage is illustrated in Fig. I. A special feature of this interplay lies in the countercurrent flows and synergistic intermixing of scientific knowledge gained from research and empirical knowledge gained from experience. It is an operational mode that has been functioning successfully in metallurgy—the science and engineering of the metallic state—for well over a century, beginning with the disclosure of microstructure and its significance in the dynamics of Fig. I. Accordingly, the discipline of metallurgy offered a logical, time-tested "role model" for the younger but more comprehensive arena of MSE, a multidiscipline that is basically capable of encompassing all classes of materials, including metals, in a new unity.

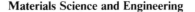

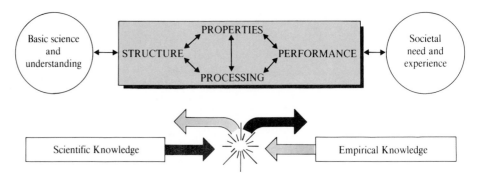

FIG. I

A model of materials science and engineering as an interactive information-transfer network, linking the interrelationships among the structure, properties, processing, and performance of materials, and highlighting the countercurrent flows of scientific and empirical knowledge.

This book, *Elements of Materials Science and Engineering,* is the sixth in a series of texts that have pioneered in the educational approach to MSE, and have literally brought the evolving concept of MSE to over one million students around the world. Nevertheless, the thrust of the Sixth Edition constitutes an intrinsic departure in that it places accent mainly on the *generic* phenomena and behavioral characteristics of materials, across the board, rather than on the separate classes of materials. This pedagogical change reflects the growing coherence and overall importance of MSE, and thereby establishes a sound foundation for later courses dealing in greater detail with specific kinds of materials. While earlier texts have tended to feature the individual classes of materials—metallic, ceramic, polymeric, electronic, or other engineering materials, the Sixth Edition represents a definite advance in providing a fresh access to modern MSE, now portrayed as an integrated field instead of merely the sum of its parts.

In recent years, MSE has attracted particular attention because of the dramatic discovery of unusual materials, the ingenious development of novel processing methods, and the significant improvement of mechanical, chemical, electrical, optical, and magnetic properties. An instructive case in point is presented in Fig. II, which charts the progress achieved in the strength of permanent magnets since the beginning of the century, but especially during the past few decades, culminating in the intermetallic alloys based on $Nd_2Fe_{14}B$. With careful processing, the latter materials have sufficient magnetic strength to permit sharply enhanced regimes of miniaturization and efficiency in the design of electric motors and electronic devices. This materials development, covering the entire spectrum from

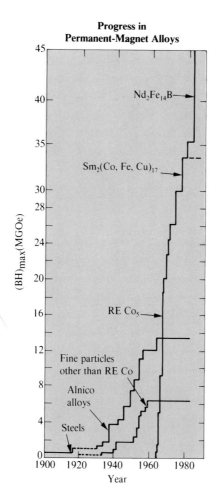

**Progress in
Permanent-Magnet Alloys**

FIG. II

Advances in the strength of permanent-magnet materials during the 20th century, based on the energy product $(BH)_{max}$ as a figure of merit. (From National Materials Advisory Board Report NMAB-426 on *Magnetic Materials,* 1985.) The striking $Nd_2Fe_{14}B$ development was published in 1984 in the United States and Japan.

fundamental research to industrial production, provides a nice example of the MSE system at work; it is a manifestation of all the mutual interchanges illustrated by the progressively reinforcing information-transfer network of Fig. I.

Overall then, MSE operates to (a) generate new knowledge toward a deeper understanding of materials as a substantive component of nature, and then to (b) harness this knowledge effectively in the service of mankind.

Cambridge, Massachusetts   **Morris Cohen**

Chapter 1

INTRODUCTION TO MATERIALS SCIENCE AND ENGINEERING

Materials have always been an integral part of human culture and civilization; for example, we designate periods in the past as the Stone, Bronze, and Iron Ages. Today's advanced technologies rely on sophisticated materials—all of them use devices, products, and systems that consist of materials. The engineer's expertise lies in *adapting materials and energy to society's needs.*

The theme of this text is that *the properties of materials depend on their internal structure.* In turn, the properties influence the *performance* of a material, both during manufacture and during service. To change the performance of a material, we must modify its internal structure. Conversely, if service conditions alter a material's structure, the engineer must anticipate that changes will occur in properties and performance.

1-1

MATERIALS AND CIVILIZATION

Among the various attributes of the human species is the ability to make things. Of course, all objects, tools, components, and engineering systems require the use of materials to meet their purposes. The role of materials has been sufficiently important in the progress of civilization that the anthropologists and historians have identified early cultures by the most significant material used then — consider the *Stone,* the *Bronze,* and the *Iron* Ages of the past. These days, however, we are not limited to one predominant material. Rather, we have a multitude of sophisticated materials — plastics, silicon, titanium, high-technology ceramics, optical fibers, and so on (Fig. 1–1.1). We can refer to our modern society as the age of technology.

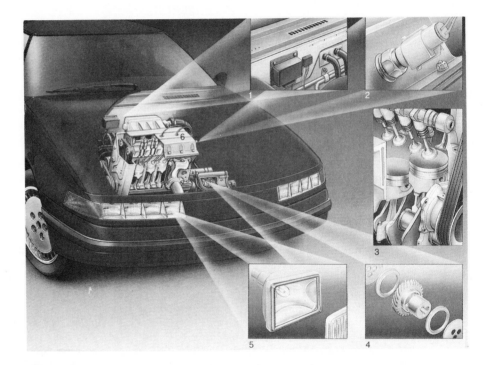

FIG. 1-1.1

Materials System. An automobile is a system of materials, each with suitable characteristics for processing, appropriate shapes for assembly, and specific properties for service. Availability, economic feasibility, safety, and aesthetic qualities also are required.
(Xydar®, courtesy of Amoco Performance Products, Inc.)

With improvements in materials made by the artisans and technologists across history, people could make improved products. Our clothing protected us more efficiently from the elements or was more attractive; our tools relieved our toil; our homes were more comfortable; our weapons became more sophisticated as we tried to obtain security against our opponent's advanced weapons; our vehicles achieved longer ranges and quicker transport and gave us more ready access to food, supplies, and enjoyment. Not only were functional products improved with the adaptation of materials, but also the arts and crafts were influenced by the artisans' abilities to work (process) these materials into objects that other people desired.

In early civilizations, as today, materials affected a broad spectrum of human activities. During the Stone Age, the sites of the better flint deposits became the sites of villages and the crossroads of primitive barter. In ensuing centuries, the requirements for materials to meet societal needs brought forth exploration and political downfalls, as well as wars. The transition from the Bronze Age to the Iron Age, which started some 3500 years ago, introduced additional nontechnical changes. Iron was a "democratic material" in these early civilizations, because its widespread availability affected the common person in every clan through the introduction of implements, tools, and utensils. No longer were the benefits of available metals limited to the affluent. Probably the most significant feature of this new democratic material was that the number of users exploded. More crafts-people were required in the materials trades. Advances now occurred in centuries and decades, rather than in millennia. This acceleration continues to the present day.

Closer to modern times, we can cite the American and British railroad systems as examples of the symbiosis of materials technology and socioeconomic changes. As early as 1830, people attempted to use steam power for land transportation. Rails were necessary for several obvious reasons; however, rails at that time were merely straps of soft wrought iron nailed to planks. An economical metal with the required characteristics of strength and durability was not available. Were it not for the improvements made by Kelly (U.S.) and by Bessemer (U.K.) in steel production, the railroad system could not have developed to open the West of the United States and to industrialize England. Conversely, were it not for the industrial and agricultural demand for transportation, the incentives, the capital, and the technology for the manufacture of steel would have been missing.

A current example of the interplay of technological and nontechnical areas involves silicon. This material has introduced a multibillion-dollar industry. It has facilitated communication over all distances, from hearing aids to extraterrestrial telemetry. Our day-to-day life is being altered via the availability of in-home entertainment such as videocassettes, and by the introduction of computers inexpensive enough that individual users can afford them. Changes are not solely technical.

1-2

MATERIALS AND ENGINEERING

Many prospective engineering students have asked, "What do engineers do?" The simplest, most general answer is

Engineers adapt materials and energy for society's needs.

More specifically, engineers design products and systems, make them, and monitor their usage. Every product is made with materials; and energy is involved in production, in use, and even in communication. This association with design, production, and usage is the reason that engineering students take a course in materials as part of their undergraduate curricula.

The various engineering disciplines focus on their own subset of products and systems. As examples, the mechanical engineer (M.E.) gives attention to automobiles, power-generating equipment, and other mechanical products. The naval architect considers ships for marine transport. The electrical engineer (E.E.) constructs circuits ranging from megawatt distribution systems to nanowatt calculator signals (Fig. 1–2.1). The petroleum refinery is designed by the chemical engi-

(a)   (b)

FIG. 1–2.1

Electrical Materials. (a) Power distribution system. (L. H. Van Vlack, *A Textbook of Materials Technology,* Addison-Wesley, Reading, Mass., with permission.)
(b) Internal structure (circuit) of hand calculator. (Courtesy of Eastman Kodak Company.) All devices, products, and systems require materials, which must be specified by the design, quality-controlled in production, and monitored in service.

neer (Ch.E.), the bridge girder by the civil engineer (C.E.), and the space shuttle by the aerospace engineer. All these products must be made out of materials. The materials that are specified must have the correct properties—first for production, and subsequently for service. They must not fail in use, producing a liability. Costs of the processed materials must be acceptable, and the choice of materials must be compatible with the environmental standards—from raw-material sources, through manufacturing steps and product usage, to eventual discard. This means that newer technical fields, such as environmental engineering and public-policy engineering, also must give consideration to materials.

The age of technology is also an age of materials barriers. Nowhere is this more dramatically revealed than in energy conversion. Coal-to-watts conversion becomes more efficient at higher steam temperatures. Thus, we are not surprised that, as soon as materials were developed for a 400° C (750° F) power plant, design engineers asked for materials capable of 450° C (840 °F). That is now available; as expected, designs now call for 540° C (1000° F), or more for steam generation.

In the other thermal direction, superconductivity was until recently limited to temperatures less than 20K (−425° F). New ceramic materials have been identified that possess superconductivity above the temperature of liquid N_2 (> 78K, or −320° F). Thus, a new world of opportunities has opened. There remain technical barriers to be solved, however, because the brittle ceramic oxides must be processed into wire or similar current paths without breaking (Fig. 1–2.2).

Less heralded, there has been equally important, steady progress through this

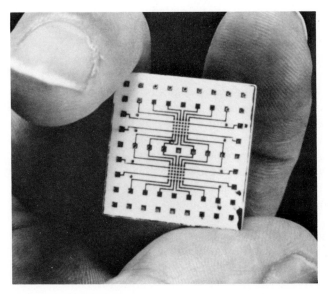

FIG. 1–2.2

Superconductor Circuit. Being ceramics, the new +78K superconductors are brittle. Until processes are developed to shape these materials into wires, processing will be limited to plasma spraying and similar techniques. (Courtesy of International Business Machines Corporation.)

FIG. 1–2.3

Materials Progress (Permanent Magnets). With more magnetic permanence (see BH_{max} in Chapter 12), significant changes can be made in engineering designs. For example, with 300 kJ/m³ available, the size of an electric motor can be cut in half. This development offers the design engineer many new opportunities. Although trends such as those shown here cannot be expected to continue without a limit, every material is a candidate for improvement within the professional lives of present-day engineering students. (cf. Fig. II of the Foreword.)

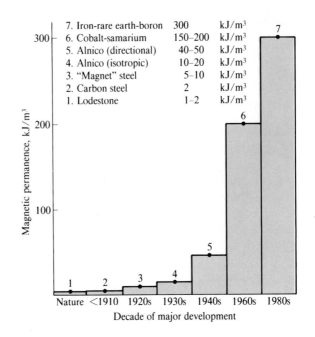

7. Iron-rare earth-boron	300	kJ/m³
6. Cobalt-samarium	150–200	kJ/m³
5. Alnico (directional)	40–50	kJ/m³
4. Alnico (isotropic)	10–20	kJ/m³
3. "Magnet" steel	5–10	kJ/m³
2. Carbon steel	2	kJ/m³
1. Lodestone	1–2	kJ/m³

Magnetic permanence, kJ/m³

Decade of major development

Nature <1910 1920s 1930s 1940s 1960s 1980s

century on the capabilities of permanent magnets (Fig. 1–2.3). Currently, materials are being developed that could permit the design of 1-horse power (hp) motors not much larger than a human fist. As with superconductors, these new materials can lead to new devices, machines, and systems that involve many engineers. Examples of the role of materials in technology are numerous. The advancement of technology will always be closely associated with the development of new materials and their processing.

Materials substitution is another arena of engineering activity. The substitution of aluminum for steel to develop lighter weight, more fuel-efficient cars is not automatic. Aluminum and steel are not processed identically, so, to make the switch, production plants must purchase different capital equipment. The two materials have different rigidities, which influences the technical design. The substitution of stainless steel for regular steel to grant longer life to bridges or ocean-going boats could appear to be rational, until the engineer compares costs, fabricability, and sources of our raw materials with their associated international complications. The role of materials in modern society is manifold.

Some engineers will not encounter materials in their first assignment after graduating, and probably a few of those engineers will not advance to more responsible positions where they consider materials; however, sooner or later the majority of engineers will be involved in materials selection or performance, either directly or with colleagues in their technical organizations. Materials (and energy) are seldom avoided by today's engineer.

Engineering Properties of Materials

Materials may be characterized by their properties. They may be strong or ductile, and they may have a high electrical resistivity. Conversely, they may be brittle, or soft, or they may have good conductivity. The engineer needs to have available quantitative values for these properties, so that comparisons of merit are available for product design and materials selection. We shall obtain these values in later chapters; in the meantime, it will be helpful for you to know the engineering definitions of a few of these characteristics (Table 1–2.1), so that you will understand the examples of engineering applications of materials.

The resistance of a wire in a circuit is a function of the size of the wire as well as of the material:

$$R = \rho \left(\frac{L}{A} \right). \qquad \text{(1–2.1)}$$

The resistance, R, increases with increased length, L, and reduced cross-sectional area, A. The proportionality constant is the *electrical resistivity,* denoted by the Greek letter rho, ρ, and is a property of the material. The dimensional unit for resistivity is an ohm·m.

Electrical conductivity, denoted by the Greek letter sigma, σ, is the current density, J, in an electric field, $\mathscr{E}$. It is also the reciprocal of the resistivity:

$$\sigma = \frac{J}{\mathscr{E}} = \frac{1}{\rho}. \qquad \text{(1–2.2)}$$

TABLE 1–2.1 Selected Engineering Properties

PROPERTY	SYMBOL	DEFINITION	UNITS
Conductivity (electrical)	σ	Charge flux along a voltage gradient	ohm$^{-1}\cdot$m^{-1}
Conductivity (thermal)	k	Energy flux along a thermal gradient	W/m·° C [(Btu/ft²·s)/(° F/in.)]
Resistivity	ρ	Reciprocal of electrical conductivity	ohm·m
Stress	s	Force per unit area	N/m², or MPa [psi]
Strength	S	Critical stress for failure	N/m², or MPa [psi]
Strain	e	Dimensional response to stress	fraction, or %
Ductility	#	Plastic strain prior to fracture	fraction, or %
Elastic modulus	E	Ratio of stress to elastic strain	N/m², or MPa [psi]
Hardness	Hdn	Resistance to penetration	#
Toughness	#	Energy absorbed prior to fracture	#

[ ] English units
More than one procedure is used. See Chapter 8.

Thus, it has $\text{ohm}^{-1} \cdot \text{m}^{-1}$ as its most commonly used dimensional units. However, some people use mhos/m. As yet, the International System of Units (SI) notation units, siemans/m, has not been widely adopted.

In SI units, *thermal conductivity, k,* is expressed as power per unit area (watts/m^2) along a temperature gradient (K/m). These units reduce to $\text{W/m} \cdot \text{K}$. We do not use the reciprocal of thermal conductivity (which would be a thermal resistivity) as a labeled property of a material.

Stress, s, is a measure of force per unit area, F/A. Thus, it has the units of newtons/m^2, or pascals. The levels of stress encountered in engineering applications are commonly in the megapascal range; hence MPa units are widely used.* *Strength, S,* is the critical stress to initiate failure. Thus, it is carefully monitored for structural materials.

Strain, e, is the dimensional response to stress (it is *not* a synonym for stress). It is expressed as a fraction (or as a percent), and is therefore dimensionless, $\Delta L/L_O$. Strain is positive under tensile stresses and negative when compressive stresses are applied. Strain may be reversible *(elastic)* or permanent *(plastic)*. *Ductility* is the plastic strain that accompanies fracture; it is critical in deformation processing, such as forging and rolling.

An *elastic modulus, E,* is the ratio of stress to elastic strain:

$$E = s/e_{\text{el}} = \frac{(F/A)}{(\Delta L/L_O)}. \qquad (1\text{-}2.3)$$

It is the material property that contributes to the rigidity of a product or structure.

Hardness is the resistance of a material to penetration. It is easy to measure, and, since it parallels the strength of ductile materials, it is a widely used specification. (It is not as good an index of strength of nonductile (brittle) materials.)

Finally, *toughness* is a measure of the energy absorbed prior to fracture. Thus, it is as important in engineering design as are strength, rigidity, and other properties.

Example 1-2.1

A copper wire has a diameter of 0.9 mm. (a) What is the resistance of a 30-cm wire? (b) How many watts are expended if 1.5 volts dc are applied across 30 m of this wire?

Procedure Part (a) relates resistance to resistivity. Therefore, use data from Appendix C ($\rho = 17$ ohm $\cdot$ nm), and Eq. (1-2.1). Part (b) draws on elementary science, from which watts = volts $\times$ amps, ($P = EI = E^2/R$), and volts = amps $\times$ ohms, ($E = IR$).

---

* This text will include certain English units parenthetically—particularly those pertaining to mechanical properties. Fortunately, the major companies in the United States are making a shift to SI units. Many senior engineers however, possess a wealth of technical experience within the nonmetric framework. Therefore, it is often desirable for us to be "bilingual" in respect to units, if we are to capitalize on their expertise.

Calculation

(a) $R = \dfrac{(17 \times 10^{-9} \text{ ohm} \cdot \text{m})(0.3 \text{ m})}{(\pi/4)(9 \times 10^{-4} \text{ m})^2} = 0.008 \text{ ohm}$

(b) $P = \dfrac{E^2}{R} = \dfrac{E^2 A}{\rho L} = \dfrac{(1.5 \text{ V})^2 (\pi/4)(9 \times 10^{-4} \text{ m})^2}{(17 \times 10^{-9} \text{ ohm} \cdot \text{m})(30 \text{ m})} = 2.8 \text{ watts}$

Example 1-2.2

Which part has the greater stress: (a) a rectangular aluminum bar of 24.6 mm × 30.7 mm (0.97 in. × 1.21 in.) cross-section, under a load of 7640 kg, and therefore a force of 75,000 N (16,800 lb$_f$); or (b) a round steel bar whose cross-sectional diameter is 12.8 mm (0.505 in.), under a 5000-kg (11,000-lb) load?

Procedure Stress is force per unit area. Also, $f = ma$, with a being $g = 9.8 \text{ m/s}^2$.

Units $\dfrac{\text{newtons}}{(\text{m})(\text{m})} = \text{pascals}$ *Units* $\dfrac{\text{pounds}}{(\text{in.})(\text{in.})} = \text{psi}$

Calculation

(a) $\dfrac{(7640)(9.8)}{(0.0246)(0.0307)} = 100 \text{ MPa}$ (a) $\dfrac{16,800}{(0.97)(1.21)} = 14,300 \text{ psi}$

(b) $\dfrac{(5000)(9.8)}{(\pi/4)(0.0128)^2} = 380 \text{ MPa}$ (b) $\dfrac{11,000}{(\pi/4)(0.505)^2} = 55,000 \text{ psi}$

Example 1-2.3

If the average modulus of elasticity of the steel used is 205,000 MPa (30,000,000 psi), by how much will a wire 2.5 mm (0.1 in.) in diameter and 3 m (10 ft) long be extended when it supports a load of 500 kg (1100 lb$_f$ and 4900 N)?

Procedure Modulus of elasticity = stress/strain; or, $e = s/E$.

Units $\text{m/m} = \dfrac{\text{N/m}^2}{\text{pascals}}$ $\text{in./in.} = \dfrac{\text{lb/in}^2}{\text{psi}}$

Calculation

$\text{strain} = \dfrac{4900/(\pi/4)(0.0025)^2}{205,000 \times 10^6}$ $\text{strain} = \dfrac{1100/(\pi/4)(0.1)^2}{30,000,000}$

$= 0.005 \text{ m/m}$ $= 0.005 \text{ in./in.}$

$\text{extension} = (0.005 \text{ m/m})(3 \text{ m})$ $\text{extension} = (0.005 \text{ in./in.})(120 \text{ in.})$

$= 15 \text{ mm}$ $= 0.6 \text{ in.}$

1-3
STRUCTURE ⇔
PROPERTIES ⇔
PERFORMANCE

Every known type of material must be considered for use by the engineer. Admittedly, some materials are not used widely because of poor availability, initial properties, service performance, or high cost. Others, such as iron, steel, paper, concrete, water, and vinyl plastics, find extensive uses. Of course, some of these categories can be subdivided. For example, there may be as many as 2000 types of steel, 5000 types of plastics, and 10,000 kinds of glass. The engineer certainly has a large number of choices for the product being designed.

Add to the many types of materials the capability of modifying properties of materials, as in the case of gears. Gears must be weak and soft to be machinable during manufacture, but strong and wear-resistant for service in the power train of automobiles, or of the heavy construction equipment. Thus, the task of knowing the properties and behavior of all types of available materials becomes enormous. In addition, several hundred new varieties of materials appear on the market each month. This means that individual engineers cannot hope to be familiar with the properties and performance of all types of materials in their numerous forms. However, the engineer can learn principles for guidance in the selection and application of materials.

Internal Structure and Properties

Since it is obviously impossible for the engineer or scientist to have detailed knowledge of the many thousands of materials already available, as well as to keep abreast of new developments, she must have a firm grasp of the underlying principles that govern the properties of *all* materials. The principle that is of most value to engineers and scientists is that *properties of a material originate from the internal structures of that material.* As an analogy, the operation of a television set or other electronic product (Figs. 1-2.1b, and 1-3.1) depends on the components, devices, and circuits within that product. Anyone can twirl knobs, but the electronic technician must understand the internal circuits if he is to repair a television set efficiently; and the electrical engineer and the physicist must know the characteristics of each circuit element if they are to design or improve the performance of the final product.

The *internal structures* of materials comprise atoms associated with their neighbors in crystals, molecules, and microstructures. In the following chapters, we shall devote much attention to these structures, because technical persons must understand structures to produce and use materials, just as mechanical engineers must understand the operation of an internal-combustion engine to design or improve a car for the demands of the next decade.

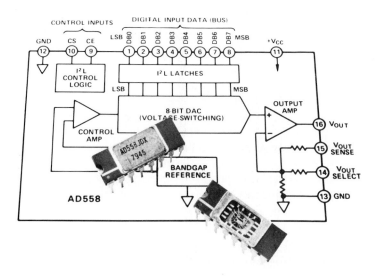

FIG. 1-3.1

Digital-Analog Converter. The performance of this microprocessor device depends on its internal circuits (see Fig. 1-2.1b). Likewise, the performance of a material depends on the structure of its internal components. We shall see that these internal arrangements involve electrons, atoms, crystals, and microstructures. The engineer can select and modify these internal structures, just as a circuit designer can select and modify internal circuits. (Courtesy of Analog Devices and L. H. Van Vlack, *Materials for Engineers: Concepts and Applications,* Addison-Wesley, Reading, Mass., with permission.)

Properties and Processing

Materials must be processed to meet the specifications that the engineer determines for the product she is designing. The engineer calls this processing of materials *production.* The most familiar processing steps of production simply change the shape of the materials by machining or forging. Of course, properties influence ease of processing. Extremely hard materials immediately destroy the edge of a cutting tool, and soft materials such as lead can "gum-up" saw blades, grinding wheels, and other tools. Likewise, strong materials are not amenable to plastic deformation, particularly if they are also nonductile (brittle). For example, it would be prohibitively expensive to produce sheet steels for most car fenders with anything other than the softest types of steel.

Processing commonly involves more than simply changing the shape of the material by machining or by plastic deformation. Not uncommonly, the manufacturing process changes the properties of a material. For example, a wire is strengthened and hardened if it is drawn through a die to decrease its diameter. Typically, this hardening is not desired in a copper wire to be used as an electrical conductor; conversely, the engineer depends on this strengthening that develops during processing the steel wire used in a steel-belted, radial tire. Whether or not we desire property modification, we should expect it whenever the manufacturing process changes the internal structure of the material. The internal structure of a material is altered when that material is deformed; hence, there is a change in properties.

Thermal processing may also affect the internal structure of a material. Such processing includes annealing, quenching from elevated temperatures, and a number of other heat treatments. We must understand the nature of the resulting structural changes, so that, as engineers, we can specify appropriate processing steps.

Properties and Service Behavior

A material in the completed product possesses a set of properties—strength, hardness, conductivity, density, color, and so on—chosen to meet the design requirements. It will retain these properties indefinitely, *provided* there is no change in the internal structure of the material. If the product encounters a service condition that alters the internal structure, however, we must expect the properties and behavior of the material to change accordingly. For example, rubber gradually hardens when exposed to light and the air, aluminum cannot be used in many locations of a supersonic plane, a metal can fatigue under cyclic loading; a drill of ordinary steel cannot cut as fast as a drill of high-speed steel; a magnet loses its polarity in an *rf* field, and a semiconductor can be damaged by nuclear radiation. The list is endless. The engineer must consider not only the initial demand but also those service conditions that will alter the internal structure and hence the properties of a material.

The Engineering Approach

The engineer must first understand the underlying principles of a particular assignment, then develop a solution. As we have emphasized, since the solution to optimum materials selection for product and system *performance* involves the *properties* of the materials, the engineer must understand the internal *structures* that govern those properties.

Figure 1–3.2 associates these three factors for materials selection. This text will focus on the structure ⇔ property relationships. You will gather detailed information on property ⇔ performance relationships when you take sequel courses in the various engineering curricula. In this text, we shall examine numerous examples in which properties are important for performances, during both the raw material-to-product processing and the subsequent service life.

During processing

STRUCTURE ⬌ PROPERTIES ⬌ PERFORMANCE

During service

FIG. 1–3.2

The Materials Spectrum. The structure must be controlled to ensure desired properties, which in turn influence the performance, both during the processing steps of production and in service. Conversely, a processing or service situation that changes the properties does so because the internal structure has been altered.

1-4
TYPES OF MATERIALS

It is convenient to classify materials into various types that have common characteristics. One way to group materials, based on atomic bonds and structures, is to distinguish *metals, polymers,* and *ceramics.* This categorization relates closely to processing. We can also group materials on the basis of properties such as *mechanical, electrical,* or *optical.* Further, we can subdivide the groups. For example, electrical materials are commonly identified as *conductors, semiconductors,* and *insulators.*

Metals, Polymers, and Ceramics

Metals are characterized by their high thermal and electrical conductivities. They are opaque, and usually they can be polished to a high luster (Fig. 1-4.1). Commonly, but not always, they are relatively heavy and deformable.

What accounts for metallic characteristics? The simplest answer is that metals owe their behavior to the fact that the valence electrons are not bound, but rather

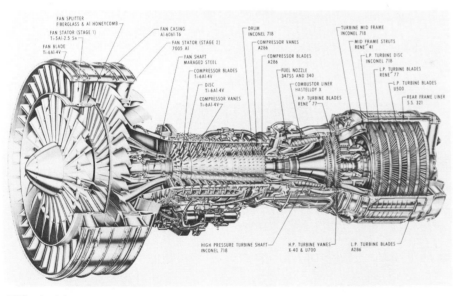

FIG. 1-4.1

Metals in a Jet Engine. The jet engine is a complex system designed for converting fuel energy into motion. Here a major goal is to improve the ratio of thrust to weight. Each of the designated materials is selected and processed to serve an assigned function in harmony with all the other operating materials. (From *Materials and Man's Needs,* Committee on the Survey of Materials Science and Engineering, National Academy of Sciences, Washington, DC, 1974; courtesy of the General Electric Company.)

can leave their "parent" atoms. (Conversely, electrons are not free to roam to the same extent in polymers and ceramics.) Since some of the electrons are independent in metals, they can quickly transfer an electric charge and thermal energy. The opacity and reflectivity of a metal arise from the response of these unbound electrons to electromagnetic vibrations at light frequencies. These properties are another result of the partial independence of some of the electrons from their parent atoms.

Polymers (commonly called plastics, Fig. 1–4.2) are noted for their low density and their use as insulators, both thermal and electrical. They are poor reflectors of light, tending to be transparent or translucent (at least in thin sections). Finally, some of them are flexible and subject to deformation. This latter characteristic is used in manufacturing.

Unlike metals, which have migrant electrons, the *nonmetallic elements* of the upper-right corner of the periodic table (Fig. 2–1.1) have an *affinity* to *attract* or *share additional electrons.* Each electron becomes associated with a specific atom (or pair of atoms). Thus, in plastics, we find only limited electrical and thermal conductivity, because all the thermal energy must be transferred from hot to cold regions by atomic vibrations, a much slower process than is the electronic transport of energy that takes place in metals. Furthermore, the less mobile electrons in plastics are more able to adjust their vibrations to those of light and therefore do not absorb light rays.

Materials that contain *only* nonmetallic elements share electrons to build up large molecules, often called *macromolecules.* We shall see in Section 2–3 that

FIG. 1–4.2

Polymers (Plastics) for an Automobile Liftgate. Use of polymers is expanding, because these materials have a high strength-to-weight ratio, and commonly can be processed simply. (Reprinted with permission from *Journal of Metals,* Vol. 39, No. 4, 1987, a publication of The Metallurgical Society, Warrendale, Pennsylvania 15086 USA.)

FIG. 1–4.3

Advanced Ceramics (Al_2O_3). Shown (clockwise) are ultrapure alumina powder, a medical device that replaces a human hip joint, a tundish nozzle for continuous casting of steel, a silicidized resistor, and a ceramic multilayer substrate for electronics. (Reprinted with permission from *Journal of Metals,* Vol. 39, No. 4, 1987, a publication of The Metallurgical Society, Warrendale, Pennsylvania 15086 USA.)

these large molecules contain many repeating units, or *mers,* from which we get the word *polymers.*

Ceramics, in simplest terms, are *compounds that contain metallic and nonmetallic elements.* There are many examples of ceramic materials, ranging from the cement of concrete (and even the rocks themselves), to glass, to electrical insulators, and to permanent magnets, to name but a few (Fig. 1–4.3).

Each of these materials is relatively hard and brittle. Indeed, hardness and brittleness are general attributes of ceramics, as is higher resistance than that of either metals or polymers to high temperatures and to severe environments. The basis for these characteristics is again the electronic behavior of the constituent atoms. Consistent with their natural tendencies, the metallic elements release their outermost electrons and give these electrons to the nonmetallic atoms, which retain them. The result is that these electrons are immobilized, so the typical ceramic material is a good insulator, both electrically and thermally.*

Equally important, the positive metallic ions (atoms that have lost electrons) and the negative nonmetallic ions (atoms that have gained electrons) develop

---

* Unexpectedly, ceramics have become the focus of research on the new superconductors.

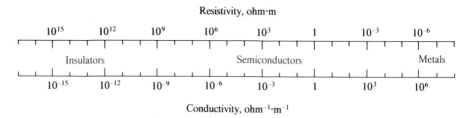

FIG. 1–4.4

Conductivities and Resistivities ($\sigma = 1/\rho$). These are properties of materials. Resistance is a function of size and shape in addition to resistivity.

strong attractions for each other. Each *cation* (positive) surrounds itself with *anions* (negative). Considerable energy (and therefore considerable force) is usually required to separate the two. It is not surprising that ceramic materials tend to be hard (mechanically resistant), refractory (thermally resistant), and inert (chemically resistant).

Conductors, Semiconductors, and Insulators

A wide spectrum of conductivities is encountered in engineering materials (Fig. 1–4.4). Conductivities range from $\sim 10^{-15}$ ohm$^{-1} \cdot$m^{-1} for certain polymers to $\sim 10^8$ ohm$^{-1} \cdot$m^{-1} for copper and silver; superconductors have infinite conductivity.

All metals are *conductors,* because their valence electrons are able to move freely. In contrast, those polymers and ceramics that have tightly bound electrons are *insulators.* For years, the design engineer selected materials for electrical circuits primarily on the basis of insulating or conducting characteristics. Materials with conductivity in the range between 10^{-9} and 10^4 ohm$^{-1} \cdot$m^{-1} were largely ignored. Of course, the advent of the transistor and related devices changed this neglect — nowadays, the *semiconductor* midrange is fundamental to high technology. Silicon, which has a conductivity of 10^{-3} ohm$^{-1} \cdot$m^{-1}, is paramount. Furthermore, many ceramics and a number of polymers display semiconducting behavior. We will consider semiconduction in considerable detail in Chapter 11.

S U M M A R Y

1. Materials have been dominant forces in the cultures of advancing *civilizations.* In fact, we speak of the Stone, Bronze, and Iron Ages. Furthermore, materials were critical for the Industrial Revolution, and they play a major role in the emergence of our current "high-technology" society.

2. The products and systems of all engineering descriptions must be made of materials. Every engineering advance encounters materials barriers. Sooner or later, the majority of engineering graduates become involved with materials selection or performance in their technical products and systems. Principal engineering properties of materials include *strength,* the limiting *stress; elastic modulus,* the ratio of stress to *elastic strain; ductility,* the *plastic strain* before fracture; and *conductivity* and *resistivity* (the reciprocal of conductivity).

3. A *structure–property–performance* relationship provides a systematic approach to materials selection and behavior. Just as the internal circuit arrangement dictates the capabilities of an electrical product, the *internal structure* of a material controls that material's properties and service performance.

4. It is convenient to categorize materials as metals, polymers, and ceramics. *Metallic* elements readily lose or release electrons. *Nonmetallic* elements accept or share electrons and thus produce macromolecular polymers. *Ceramic* materials are compounds of metallic and nonmetallic elements.

We also can group materials as *conductors.* Conductors are predominately metals. Their valence electrons are delocalized; in contrast, *insulators* have tightly bound electrons, which are characteristic of many polymers and ceramics. *Semiconductors* are relatively new to technology; they feature elemental silicon, but also include a growing number of ceramic materials and a few polymers.

KEY WORDS*

Ceramics
Conductivity
Ductility
Engineering
Materials (engineering)
Metals

Modulus of elasticity (elastic modulus, Young's modulus)
Performance (materials)
Plastics
Polymers
Property

Resistivity (ρ)
Strain (e)
Strength (S)
Stress (s)
Toughness

---

* See *Terms and Concepts* for definitions.

PRACTICE PROBLEMS*

1–P21 What is the stress on a 1-mm (0.04-in.) diameter wire that supports a 3-kg (6.6-lb) load?

1–P22 When the stress on the wire of Problem 1–P21 is 37 MPa (5300 psi), the elastic strain is 0.00054. What is the elastic modulus?

1–P23 A wire is 36 ft (11 m) long. It has been stretched $1\frac{3}{8}$ in. (35 mm). What stress has been developed if the elastic modulus is 10,000,000 psi (70 GPa)?

1–P24 Identify the number of significant figures that you should read from your calculator for Problems 1–P21, 1–P22, and 1–P23.

1–P25 Brass has a resistivity of 62×10^{-9} ohm·m. (a) What is the end-to-end resistance of a brass strip 5 cm long by 5 mm wide by 0.5 mm thick? (b) What is its conductivity?

1–P26 Compare and contrast *strength, ductility,* and *toughness.*

1–P27 Compare and contrast *resistance, resistivity,* and *conductivity.* What are the units of each?

1–P41 (a) Cite metal parts that are used in a modern kitchen stove. What property is pertinent to each application? (b) Repeat part (a) for ceramics. (c) Repeat part (a) for polymers.

---

* These *Practice Problems* (1) use equations directly (plug and chug), (2) refer to specific statements in the text, or (3) "mimic" the examples at the end of the sections. You should try them for practice before attempting the *Test Problems* that follow. All *Practice Problems* have answers on the pages at the end of the text. If you need them, the detailed solutions are given in the *Study Guide to Elements of Materials Science and Engineering.* The *Study Guide* also contains Quiz Samples.

TEST PROBLEMS*

121 What is the elastic strain in a brass rod that is stressed 49 MPa (7100 psi)? (Elastic moduli and other properties of common metals are given in Appendix C.)

122 (a) An iron rod 0.50-in. (12.7-mm) in diameter supports a load elastically of 1540 lb (700 kg, and therefore a force of 6860 N). What stress is placed on the rod? (b) How much will the rod be strained by that load?

123 A metal rod should not be stressed to more than 49 MPa (7100 psi) in tension. What diameter is required if it is to carry a load of 1200 kg (2640 lb)?

124 What is the elastic strain in a monel rod that is stressed 49 MPa (7100 psi)? (Elastic moduli and other properties of common metals are given in Appendix C.)

125 (a) A copper rod 0.50-in. (12.7-mm) in diameter supports a load of 1540 lb (700 kg, and therefore a force of 6860 N). What stress is placed on the rod? (b) If the rod of part (a) has a modulus of elasticity of 16,000,000 psi (110 GPa), how much will the rod be strained elastically by that load?

126 A 1-mm (0.04-in.) copper wire that is 380 mm long supports a load of 5 kg elastically. The temperature drops from 10° C to 5° C. By how much must the load be altered to return the wire to its initial length?

127 What is the difference in the end-to-end resistance of a 2-mm (0.08-in.) aluminum wire that is 30.5 m (100 ft) in length and a copper wire with the same dimensions?

128 Examine the cord of a household appliance, such as a toaster or coffee maker. List the materials used and cite the probable reasons for their selection.

131 (a) Dismantle a cheap pen. List the materials that are used. (b) For each material, cite service conditions that the engineer had to consider during materials selection.

132 (a) Examine your car (or one owned by a friend). List all the materials that you can identify. (Do *not* list names of parts.) Compare your list with the lists composed by your classmates. (b) Discuss the attributes that were important in the selection of each material you listed. (c) Suggest plausible substitutes for each material. (d) Examine the list you composed for part (c). Can you explain why each substitute was *not* used?

133 A car fender could be made out of armor plate so that it would not crumple in a collision. Give several reasons why the common fender is not designed in this manner.

141 Examine an incandescent light bulb closely. (a) Which *types* of materials do you see? (b) What thermal and electrical characteristics are required of each material that is present?

142 Cite three critical materials requirements in your engineering field (M.E., C.E., E.E., and so on).

143 Examine a piece of wood. Describe some of its structural characteristics. How does the wood's structure affect the properties of this material?

---

* The preceding set of *Practice Problems* gave you a chance to learn how you can plug equations or establish procedures. The *Test Problems* require you to do more. As with all engineering calculations, it is necessary for you (1) to establish a procedure, (2) to identify and obtain required information, and then (3) to proceed toward a solution on the basis of accumulated knowledge.

In general, data will not be cited in the problems if they are contained in related tables or in the appendices.

Chapter 2

ATOMIC BONDING AND COORDINATION

Interatomic bonds exist in all solids. They provide strength and related electrical and thermal properties. For example, strong bonds lead to high melting temperatures, high moduli of elasticity, shorter interatomic distances, and lower thermal-expansion coefficients, as well as contributing to greater hardness and strength.

Different bonding patterns lead to molecular structures or to extended, three-dimensional structures. To visualize these structures, we need to examine the role of the valence electrons on the primary bonds—*ionic, covalent,* and *metallic*—in sufficient detail that we can anticipate the electrons' effects on interatomic distances and atomic coordination.

Finally, we shall consider generalizations that permit us to make our initial correlations between structures and properties.

2–1
INDIVIDUAL ATOMS AND IONS

The atom is the basic unit of internal structure for our studies of materials. The initial concepts involving individual atoms are familiar to most of you. They include *atomic number, atomic mass,* and the relationships of the *periodic table.* In addition, we will want to pay attention to the bonding forces between atoms. Understanding these forces will permit us to make generalizations regarding poly-atomic structures and their associated properties.

Chemistry Principles

Since atoms are extremely small in terms of our day-to-day concepts of mass, it is convenient to use the *atomic mass unit,* amu, as the basis for many calculations. The amu is defined as one-twelfth of the atomic mass of carbon-12, the most common isotope of carbon. There are $0.6022 \ldots \times 10^{24}$ amu per g. We will use this conversion factor (called *Avogadro's number,* AN) in various ways. Since natural carbon contains approximately 1 percent C^{13}, along with 98.9 percent C^{12}, the average atomic mass of a carbon atom is $12.011 \ldots$ amu. This is the value presented in the *periodic table* (Fig. 2–1.1) and in tables of selected elements (Appendix B). Those atomic masses you will encounter most commonly are shown in the table below. We can round off the values in this table, as indicated, for all but the most precise calculations.

The *atomic number* indicates the number of electrons associated with each neutral atom (and the number of protons in the nucleus). Each element is unique with respect to its atomic number. Appendix B lists selected elements, from hydrogen, with an atomic number of 1, to uranium (92). It is the electrons—particularly the outermost ones—that affect most properties of interest to engineers: (1) they determine the chemical properties; (2) they establish the nature of the interatomic bonding, and therefore the mechanical and strength characteristics; (3) they control the size of the atom and affect the electrical conductivity of materials; and (4) they influence the optical characteristics. Consequently, we shall pay specific attention to the distribution and energy levels of the electrons around the nucleus of the atom.

	PRECISION VALUES	FOR MOST CALCULATIONS
H	$1.0079 \ldots$ amu (or 1.0079 g/(0.602×10^{24} atoms))	1 amu (or 1 g/mol)
C	$12.011 \ldots$	12
O	$15.9994 \ldots$	16
Cl	$35.453 \ldots$	35.5
Fe	$55.847 \ldots$	55.8

Metals — Nonmetals

1A	IIA	IIIB	IVB	VB	VIB	VIIB	VIII	VIII	VIII	IB	IIB	IIIA	IVA	Va	VIA	VIIA	0
1 **H** 1.00797																1 **H** 1.0079	2 **He** 4.0026
3 **Li** 6.939	4 **Be** 9.012											5 **B** 10.811	6 **C** 12.011	7 **N** 14.007	8 **O** 15.9994	9 **F** 18.998	10 **Ne** 20.183
11 **Na** 22.990	12 **Mg** 24.312											13 **Al** 26.98	14 **Si** 28.086	15 **P** 30.97	16 **S** 32.064	17 **Cl** 35.453	18 **Ar** 39.95
19 **K** 39.102	20 **Ca** 40.08	21 **Sc** 44.96	22 **Ti** 47.90	23 **V** 50.94	24 **Cr** 52.00	25 **Mn** 54.94	26 **Fe** 55.85	27 **Co** 58.93	28 **Ni** 58.71	29 **Cu** 63.54	30 **Zn** 65.37	31 **Ga** 69.72	32 **Ge** 72.59	33 **As** 74.92	34 **Se** 78.96	35 **Br** 79.91	36 **Kr** 83.80
37 **Rb** 85.47	38 **Sr** 87.62	39 **Y** 88.91	40 **Zr** 91.22	41 **Nb** 92.91	42 **Mo** 95.94	43 **Tc** 99	44 **Ru** 101.07	45 **Rh** 102.91	46 **Pd** 106.4	47 **Ag** 107.87	48 **Cd** 112.40	49 **In** 114.82	50 **Sn** 118.69	51 **Sb** 121.75	52 **Te** 127.60	53 **I** 126.90	54 **Xe** 131.30
55 **Cs** 132.91	56 **Ba** 137.34	57–71 **La** series*	72 **Hf** 178.49	73 **Ta** 180.95	74 **W** 183.85	75 **Re** 186.2	76 **Os** 190.2	77 **Ir** 192.2	78 **Pt** 195.1	79 **Au** 196.97	80 **Hg** 200.59	81 **Tl** 204.37	82 **Pb** 207.19	83 **Bi** 208.98	84 **Po** 210	85 **At** 210	86 **Rn** 222
87 **Fr** 223	88 **Ra** 226	89– **Ac** series†															

58 **Ce** 140.12	59 **Pr** 140.91	60 **Nd** 144.24	61 **Pm** 145	62 **Sm** 150.35	63 **Eu** 151.96	64 **Gd** 157.25	65 **Tb** 158.92	66 **Dy** 162.50	67 **Ho** 164.93	68 **Er** 167.26	69 **Tm** 168.93	70 **Yb** 173.04	71 **Lu** 174.97
90 **Th** 232.04	91 **Pa** 231	92 **U** 238.03	93 **Np** 237	94 **Pu** 242	95 **Am** 243	96 **Cm** 247	97 **Bk** 249	98 **Cf** 251	99 **Es** 254	100 **Fm** 253	101 **Md** 256	102 **No** 253	103 **Lw** 257

FIG. 2–1.1

Periodic Table of the Elements, Showing the Atomic Number and Atomic Mass (in amu). There are 0.602×10^{24} amu per g; therefore the atomic masses are g per 0.602×10^{24} atoms. Metals readily release their outermost electrons. Nonmetals readily accept or share additional electrons.

The *periodicity of elements* is emphasized in chemistry courses. We shall not repeat those characteristics here except to observe that the periodic table (Fig. 2–1.1) arranges the atoms by sequentially higher atomic numbers, such that the vertical columns, called *groups,* possess atoms of similar chemical and electronic characteristics. In brief, those elements at the far left of the periodic table are readily ionized to give positive ions, called *cations.* Those in the upper-right corner of the periodic table more readily share or accept electrons—they are *electronegative.*

Electrons

Since electrons are components of all atoms, their negative electrical charge is commonly regarded as unity. In physical units, this charge is equal to 0.16×10^{-18} A·s per electron (or 0.16×10^{-18} coul/electron).

The electrons that accompany an atom are subject to rigorous rules of behavior because they have the characteristics of standing waves during their movements in the neighborhood of the atomic nucleus. We can summarize several features of electron behavior. With individual atoms, electrons have specific energy states, called *orbitals.* As shown in Fig. 2–1.2, the available electron energy states around a hydrogen atom can be identified definitely.* To us, the important consequence is that there are large ranges of intermediate energies *not* available for the electrons. The values in these ranges are forbidden because the corresponding frequencies do not permit standing waves. Unless excited by external means, the one electron of a hydrogen atom will occupy the lowest orbital of Fig. 2–1.2.

Figure 2–1.3 shows schematically the energies of the lowest orbitals for sodium. Each orbital can contain no more than two electrons, which must be of opposite spins. Again, there are forbidden energy gaps between the orbitals that are unavailable for electron occupancy.

In our considerations, the topmost occupied orbital will have special significance, since it contains the *valence electrons.* These electrons may be removed by a relatively small electric field, to give us the positive *cations* we mentioned. The energy requirements are called the *ionization energies.* In Section 2–2, we shall see that these outermost or valence electrons are *delocalized* in metallic solids, and are free to move throughout the metal rather than remaining bound to individual atoms. This provides the basis for electrical and thermal conductivity.

When the valence orbitals are not filled, the atom may accept a limited number of extra electrons within these unfilled energy states, to become a negative ion, or *anion.* These electronegative atoms with unfilled valence orbitals may also share electrons. This ability to share electrons is important in covalent bonding; we shall review it in the next section.

---

* This identification is made by spectrographic experiments.

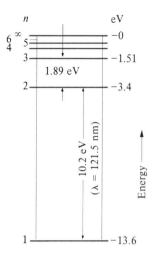

FIG. 2–1.2

Energy Levels for Electrons (Hydrogen). The electron of hydrogen normally resides in the lowest energy level. (At this level, it would take 13.6 eV, or 2.2×10^{-18} J, to separate the electron from the nucleus.) Electrons can be given additional energy, but only at specific levels. Gaps exist between these levels—the values in these gaps are forbidden energies.

FIG. 2–1.3

Energy Levels for Electrons (Sodium). Since a sodium atom possesses 11 electrons, and only two electrons can occupy each level (orbital), several orbitals must be occupied. Gaps exist between these orbitals. It takes 5.1 eV (0.82×10^{-18} J) to remove the uppermost (valence) electron from sodium.

Example 2–1.1

Sterling silver contains approximately 7.5 w/o* copper and 92.5 w/o silver. What are the a/o copper and a/o silver?

Procedure Select a mass basis and calculate the number of atoms of each type of element present. Atomic masses are given in Fig. 2–1.1 and Appendix B.

Calculation Basis: 10,000 amu alloy = 9250 amu Ag + 750 amu Cu.

Ag: 9250 amu Ag/(107.87 amu Ag/atom) = 85.75 atoms = 88 a/o

Cu: 750 amu Cu/(63.54 amu Cu/atom) = 11.80 atoms = 12 a/o

total atoms = 97.55

---

* Weight percent, w/o; atom percent, a/o; linear percent, l/o; volume percent, v/o; mole percent, m/o; and so on. In condensed phases *(solids and liquids),* weight percent is implied unless specifically stated otherwise. In *gases,* v/o or m/o are implied unless specifically stated otherwise.

Example 2–1.2

The mass of a small diamond is 3.1 mg. (a) How many C^{13} atoms are present if carbon contains 1.1 a/o of that isotope? (b) What is the weight percent of that isotope?

Procedure (a) Determine the total number of carbon atoms present from the atomic masses; then 1.1 percent (on an atomic basis). (b) From the numbers of atoms in part (a) and the atomic masses, calculate the mass (or weight) fraction.

Calculation

(a) $$\frac{0.0031 \text{ g}}{(12.011 \text{ g}/0.6022 \times 10^{24} \text{ atoms})} = 1.55 \times 10^{20} \text{ C atoms}$$

$$(1.55 \times 10^{20})(0.011) = 1.7 \times 10^{18} \text{ C}^{13} \text{ atoms}$$

(b) Basis: 3.1 mg.

$$\frac{\text{Mass}_{13}}{\text{Mass}_{\text{total}}} = \frac{(1.7 \times 10^{18})(13 \text{ amu})}{(1.55 \times 10^{20})(12.011 \text{ amu})} = 1.2 \text{ w/o}$$

Comment Since the mass of the C^{13} is greater than that of the average atom, the w/o will be greater than the a/o.

Example 2–1.3

Ten g of nickel are to be electroplated on a steel surface with an area of 0.8953 m². The electrolyte contains Ni^{2+} ions. (a) How thick will the nickel plate be? (b) What amperage is required if the plating is to be accomplished in 50 minutes?

Procedures (a) Set up your own equation based on $\rho = m/V = m/At$. (b) Determine the number of nickel atoms. Two electrons are required per Ni^{2+}, each with 0.16×10^{-18} A·s (or 0.16×10^{-18} C). The (amp)(sec) product is obtained from these.

Calculation

(a) $10 \text{ g}/(8.9 \times 10^6 \text{g/m}^3)(0.8953 \text{ m}^2) = 1.25 \times 10^{-6} \text{ m}$ $\hspace{2em}$ (or 1.25 μm)

(b) $$\left[\frac{10 \text{ g Ni}}{58.71 \text{ g Ni}/0.6 \times 10^{24} \text{ atoms}}\right]\left[\frac{(2 \text{ el/atom})(0.16 \times 10^{-18} \text{ A·s/el})}{(3000 \text{ s})}\right] = 10.9 \text{ amp}$$

Comment By now, you should be alert to the data available in the Appendices.

2–2
MOLECULES

The more common examples of molecules include compounds such as H_2O, CO_2, CCl_4, O_2, N_2, and HNO_3. Other small molecules that will be important to us are shown in Fig. 2–2.1. Within each of these molecules, the atoms are held together

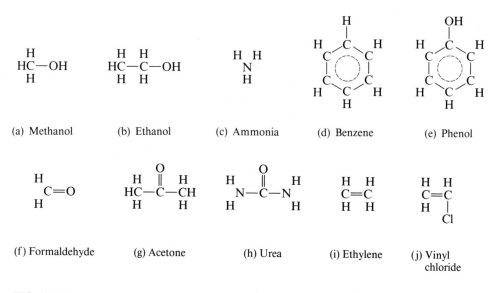

FIG. 2-2.1

Small Molecules. Each carbon atom is surrounded by four bonds, each nitrogen by three, each oxygen by two, and each hydrogen and chlorine by one.

by strong *intra*molecular attractive forces that produce *covalent bonds.* Unlike the strong forces that hold the atoms *within* the molecule, the *inter*molecular bonds *between* molecules are weak; consequently, each molecule is free to act more or less independently. These observations are borne out by the following facts: (1) each of these molecular compounds has a low melting and a low boiling temperature compared with other materials, (2) the molecular solids are soft because the molecules can slide past each other with small stress applications, and (3) the molecules remain intact in liquids and gases.

These molecules are comparatively small; other molecules have large numbers of atoms. For example, pentatriacontane (shown in Fig. 2-2.2c) has over 100

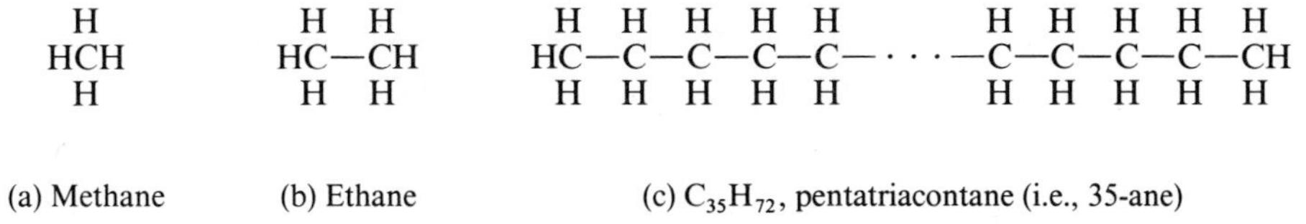

```
  H            H  H         H  H  H  H  H              H  H  H  H  H
 HCH         HC—CH        HC—C—C—C—C—· · · ·—C—C—C—C—CH
  H            H  H         H  H  H  H  H              H  H  H  H  H

(a) Methane   (b) Ethane          (c) C₃₅H₇₂, pentatriacontane (i.e., 35-ane)
```

(c) $C_{35}H_{72}$, pentatriacontane (i.e., 35-ane)

FIG. 2-2.2

Examples of molecules. Molecules are discrete groups of atoms. Primary bonds hold together the atoms within the molecule. Weaker, secondary forces attract molecules to each other.

atoms, and some molecules contain many thousand. Whether the molecule is small like CH_4 or much larger than the one shown in Fig. 2–2.2(c), the distinction between the stronger *intra*molecular and the weaker *inter*molecular bonds still holds.

Other materials such as metals, MgO, SiO_2, and phenol-formaldehyde plastics have continuing three-dimensional structures of strong bonds. The difference between the structures of molecular materials and those with strong (primary) bonds continuing in all three dimensions produces major differences in properties. We shall consider these differences in subsequent chapters.

Covalent Bonds

Covalent bonds are *stereospecific;* that is, each bond is between a specific pair of atoms. The pair of atoms share a pair of electrons (of opposite magnetic spins). We represent these bonds in sketches in several ways: (1) as bond lines — for example, C—C, (2) as pairs of dots (Fig. 2–2.3a), or (3) as regions of high electron probability (Fig. 2–2.3b). The latter sketch is the more appropriate representation, since the electrons behave in a wavelike fashion and their positions cannot be defined precisely.*

Each atom of a material is coordinated with its neighbors. The thermal vibrations on one atom influence the adjacent atoms; the displacement of one atom by mechanical forces, or by an electric field, leads to adjustments of the neighboring atoms. The number of coordinating neighbors that each atom has is important. The carbon atoms of diamond in Fig. 2–2.3(b) each have four neighbors. Thus, we say that those atoms have a *coordination number* of four (CN = 4). Four is the maximum coordination number possessed by covalently bonded carbon. (It is possible to have a coordination number that is less than the maximum; for example, the carbons in ethylene (Fig. 2–2.1i) are bonded to three neighbors.) The maximum coordination numbers of other covalently bonded atoms are indicated in Table 2–2.1.

Typically, covalent bonds are very strong. The hardness of diamond is a result of the fact that each carbon atom is covalently bonded with four neighboring atoms, and each neighbor is bonded with an equal number of atoms to form a rigid three-dimensional structure (Fig. 2–2.3). Likewise, the *intra*molecular bonds within methane, CH_4, polyethylene, and other polymers arise from these covalent bonds of shared electrons, thus producing stable molecules.

---

* It is impossible to pinpoint an electron's location exactly, because of the *uncertainty principle.* However, the probability of an electron's position is greatest along the line between the two atoms (see the shaded areas of Figs. 2–2.3b and 2–2.4).

C **:** C **:** C **:** C

..   **..**   **..**   **..**

C **:** C **:** C **:** C

..   **..**   **..**   **..**

C **:** C **:** C **:** C

..   **..**   **..**   **..**

C **:** C **:** C **:** C

(a)

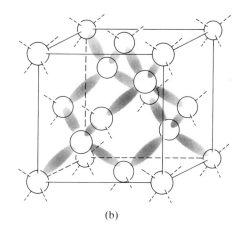

(b)

FIG. 2-2.3

Diamond Structure of Carbon. The strength of the covalent bonds accounts for the great hardness of diamond. (a) Two-dimensional representation with a pair of electrons shared by neighboring atoms. (b) Three-dimensional representation with the bonds shown as the region of high electron probability (shaded).

Bond Lengths and Energies

The strength of bonds between atoms in a molecule, of course, depends on the kind of atoms and the other neighboring bonds. Table 2-2.2 is a compilation of bond lengths and energies for those atom couples most commonly encountered in molecular structures. The energy reported is the amount required to break 1 mole (Avogadro's number) of bonds. For example, 370,000 joules of energy are required to break 0.602×10^{24} C—C bonds, or $370,000/(0.602 \times 10^{24})$ joules per bond. This same amount of energy is released (-0.61×10^{-18} J/bond) if one of these C—C bonds is formed. Only the sign is changed.

TABLE 2-2.1 Covalent Bonds per Atom*

H	1	S	2
F	1	N	3
Cl	1	C	4
O	2	Si	4

* The value also represents the maximum coordination number.

TABLE 2-2.2 Bond Energies and Lengths

BOND	BOND ENERGY* kJ/mol†	BOND ENERGY* kcal/mol†	BOND LENGTH, nm
C—C	370‡	88‡	0.154
C=C	680	162	0.13
C≡C	890	213	0.12
C—H	435	104	0.11
C—N	305	73	0.15
C—O	360	86	0.14
C=O	535	128	0.12
C—F	450	108	0.14
C—Cl	340	81	0.18
O—H	500	119	0.10
O—O	220	52	0.15
O—Si	375	90	0.16
N—H	430	103	0.10
N—O	250	60	0.12
F—F	160	38	0.14
H—H	435	104	0.074

* Values are approximate. They vary with the type of neighboring bonds. For example, methane (CH_4) has the value shown for its C—H bond; however, the C—H bond energy is about 5 percent less in CH_3Cl, and 15 percent less in $CHCl_3$.

† Energies per 0.602×10^{24} bonds.

‡ All values are negative for forming bonds (energy is released), and are positive for breaking bonds (energy is required).

Bond Angles

The chemist recognizes hybrid orbitals in certain covalent compounds, where the s and p orbitals are combined. The most important hybrid for us to review is the sp^3 orbital. Four equal orbitals are formed, instead of distinct $2s$ and $2p$ orbitals occurring in individual atoms (e.g., the individual sodium of Fig. 2-1.3). Methane (CH_4; Fig. 2-2.4), carbon tetrachloride (CCl_4), and diamond (Fig. 2-2.3) provide examples of sp^3 orbitals that connect four identical atoms to the central carbon. Therefore, we find the atoms equally spaced around the central carbon at 109.5° from each other.* Geometrically, this is equivalent to placing the carbon at a cube

---

* The angle 109.5° is a time-averaged value. Any particular H—C—H angle in CH_4 will vary rapidly as a result of thermal vibrations.

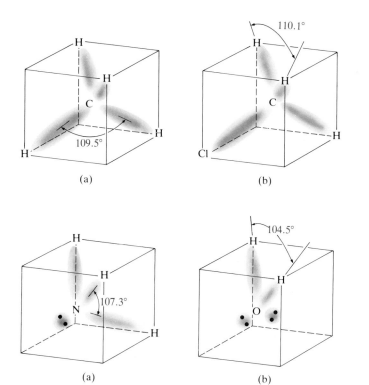

FIG. 2-2.4

Bond Angles. (a) Methane, CH_4, is symmetrical with each of the six angles equal to 109.5°. (b) Chloromethane, CH_3Cl, is distorted.

FIG. 2-2.5

Bond Angles. (a) Ammonia, NH_3, and (b) water have angles between the 109.5° of Fig. 2-2.4(a) and 90°. Ammonia has one lone-pair of electrons; water has two.

center and pointing the orbitals toward four of the eight corners (Fig. 2-2.4a). However, if the orbitals do not bond identical atoms to the central carbon, these time-averaged angles are distorted slightly, as shown for CH_3Cl (Fig. 2-2.4b).

Greater distortions occur in hybrid orbitals when some of the electrons occur as *lone pairs* rather than in the covalent bond. This distortion is particularly evident in NH_3 and H_2O (Fig. 2-2.5), where 107.3° and 104.5° are the time-averaged values for H—N—H and H—O—H, respectively.

One of the bond angles most frequently encountered in the study of materials is the C—C—C angle of the hydrocarbon chains (Fig. 2-2.6). This value will differ

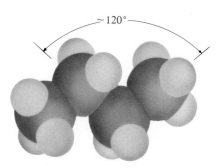

FIG. 2-2.6

Bond Angles (Butane). Although we commonly draw straight chains (Figs. 2-2.2c and 2-3.1b), there is a C—C—C bond angle of about 120°.

slightly, depending on whether hydrogen or some other side radical is present, but we may assume for our purposes that the C—C—C angle is close to 120°.

Delocalized Electrons

Most valence electrons within molecules enter into covalent bonds such as those in Figs. 2–2.3(b), 2–2.4, and 2–2.5. These electrons are shared by specific atoms. The chemist calls these covalent bonds sigma bonds, denoted by the Greek letter σ.

Not all valence electrons are localized so specifically, as is illustrated by benzene, which is shown in Fig. 2–2.7. The classical representation is the one shown in part (a) of that figure. A better representation is the one shown in part (b), because the six electrons that are not included in the σ bonds develop standing waves across the whole molecule. These electrons, which are called pi (denoted by the Greek letter π) electrons, respond to an electric field by moving their center of oscillation toward the side of the molecule that is nearer the positive electrode. However, they cannot leave the molecule (except under unusually catastrophic conditions). There are as many wave patterns for these delocalized electrons *as there are carbon atoms* in the ring (in this case, six). The concept of delocalized electrons will also be useful to us when we consider the conductivity of metals in Chapter 11.

Example 2–2.1

How much energy is required, +, (or released, −) if 2.6 kg of acetylene, C_2H_2, react with hydrogen to produce ethylene, C_2H_4?

Procedure For each molecule of C_2H_4 produced, one H—H bond and one triple carbon bond, C≡C, are eliminated. Conversely, a double carbon bond, C=C, and two C—H bonds, are formed. The energies involved are +435, +890, −680 and 2(−435) kJ/mol, respectively. A mole includes 0.6×10^{24} bonds.

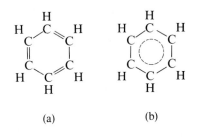

(a)   (b)

FIG. 2–2.7

Delocalized Electrons. (a) Benzene ring (simplified). (b) Benzene ring (preferred). The σ orbitals between the carbon atoms are stereospecific. Electrons in the other orbitals (π) can move from one side of the molecule to the other in response to internal and external electrical fields; they are delocalized.

Calculation

$$\frac{+435 + 890 \text{ kJ}}{0.6 \times 10^{24} \text{ bonds}} + \frac{-680 + 2(-435) \text{ kJ}}{0.6 \times 10^{24} \text{ bonds}} = -375 \times 10^{-21} \text{ J/C}_2\text{H}_2$$

$$(2600 \text{ g})(0.6 \times 10^{24} \text{ amu/g})/(26 \text{ amu/C}_2\text{H}_2) = 60 \times 10^{24} \text{ C}_2\text{H}_2$$

$$(-375 \times 10^{-21} \text{ J/C}_2\text{H}_2)(60 \times 10^{24} \text{ C}_2\text{H}_2) = -22.5 \times 10^6 \text{ J}$$

or

$$(-22.5 \times 10^6 \text{ J})(0.239 \text{ cal/J}) = -5.4 \times 10^6 \text{ cal}$$

Comment The value is negative, so energy is released; thus, once initiated, the reaction could proceed spontaneously.

Example 2-2.2

The structure of naphthalene is a double benzene ring with two carbons of each ring in common. (a) How many delocalized electrons are there? (b) How many wave patterns (electron states) are available to the delocalized electrons?

Procedure Sketch the molecule. Determine the formula. Account for the valence electrons.

Answer See Fig. 2-2.8.

  (a) Formula: $C_{10}H_8$.
        Valence electron balance
            Total valence electrons: $(4/C)(10 \text{ C}) + (1/H)(8H) = 48$
            Electrons in bonds: C—H $= 16$
                                                          C—C (σ) $= 22$
                                                          Remainder (delocalized) $= 10$
  (b) Electron states = number of carbon atoms = 10

Additional Information Each state can contain two electrons (of opposite spin). Therefore, a total of 20 delocalized electrons could be accommodated (were it not for a resulting imbalance of charges in the molecule).

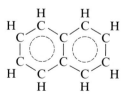

FIG. 2-2.8

Naphthalene (Example 2-2.2). A carbon is located at each hexagon corner. This leaves 10 delocalized electrons.

2-3
MACROMOLECULES (POLYMERS)

As sketched in Fig. 2–2.2(c), molecules may contain many atoms with strong intramolecular bonds. Materials such as polyethylene typically contain thousands. Thus, the term *macromolecules* (large molecules) is appropriate. These molecules provide the basis for many of our *plastics* and for related products.

Nature produces numerous macromolecules. *Cellulose* is one such material; it is present in wood, cotton, and most fibrous plants. Industrially made macromolecules include polyethylene, polyvinyl chloride, and polystyrene. Each of these molecules may be viewed as a series of repeating units, called *mers* (Fig. 2–3.1b). Macromolecules are commonly called *polymers* (many units). The polymer of Fig. 2–3.1(b) was produced by the joining of many single units (*monomers*, Fig. 2–3.1a) into a molecular chain. In this example, the monomer was vinyl chloride, C_2H_3Cl (Fig. 2–2.1j), and the polymer is polyvinyl chloride.

Linear Molecules

The polymer chain of Fig. 2–3.1(b) is *linear*. There is no theoretical limit to its length. Strong covalent bonds exist along its length. Bonds of equivalent strength do not join adjacent molecules. Numerous commercial polymers are linear — these include the polyvinyls and the polyesters, as well as other families that are less familar.

The polyvinyls are a family of polymers that have C_2H_3R mers:

$$\left[\begin{array}{c} H \;\; H \\ -C-C- \\ H \;\; | \\ R \end{array}\right]_n \tag{2-3.1}$$

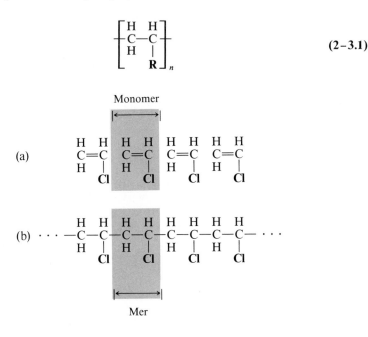

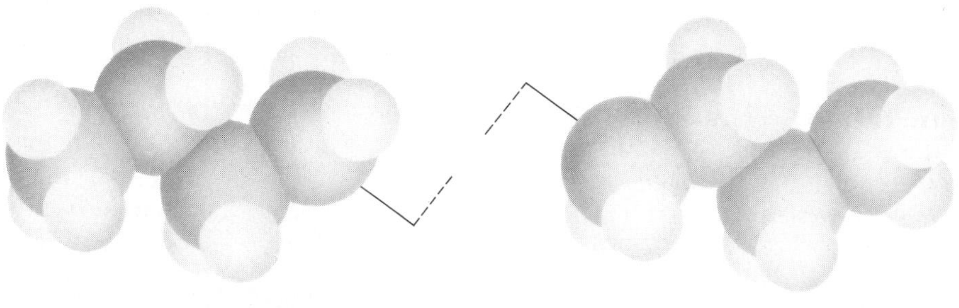

FIG. 2–3.2

Linear Polymer (Polyethylene). The molecular length varies from molecule to molecule. Typically, these molecules contain hundreds of carbon atoms.

where *n* is the *degree of polymerization*—that is, the number of mers per molecule. The **R** is a chlorine atom (Cl) in *polyvinyl chloride* (Fig. 2–3.1b). In *polystyrene,* **R** is a benzene ring, whereas in *polyethylene,* it is simply another hydrogen atom (Fig. 2–3.2). Table 2–3.1 summarizes the more common vinyl polymers, as well as several vinylidene polymers:

$$\left[\begin{array}{c} \quad\; \mathbf{R'} \\ \mathrm{H} \;\; | \\ -\mathrm{C}-\mathrm{C}- \\ \mathrm{H} \;\; | \\ \quad\; \mathbf{R''} \end{array} \right]_n \qquad\qquad (2\text{–}3.2)$$

The mers of polyesters, polyurethanes, and similar materials, are somewhat more complex, and commonly are the product of two precursors. Like the vinyls, however, they have a chain of covalently bonded atoms that forms a "backbone" structure. The monomers of all these linear polymers must be *bifunctional;* that is, they must be able to react with *two* adjacent molecules (Fig. 2–3.1).*

Secondary Bonds

Covalent bonds are very strong. Therefore, we call them *primary* bonds. In contrast, we call the weak intermolecular bonds between the molecules *secondary* bonds. Nonetheless, intermolecular bonds are important.

The forces of attraction that produce secondary bonds exist because of the local electric fields within and around uncharged atoms. The simplest (and weakest) of

---

* As an analogy, consider a railroad freight car with couplers on both ends. In contrast, a typical highway semitrailer is "monofunctional," since it can connect at one end only.

TABLE 2–3.1 Structures of Selected Vinyl Polymers

POLYMER	1	2	3	4	INTERNATIONAL ABBREVIATION
Polyvinyls (general)	H	H	H	—R	
Polyethylene	H	H	H	—H	PE
Polyvinyl chloride	H	H	H	—Cl	PVC
Polyvinyl alcohol	H	H	H	—OH	PVAl
Polystyrene	H	H	H	⬡	PS
Polypropylene	H	H	H	—CH_3	PP
Polyvinyl acetate	H	H	H	—$OCCH_3$ (O double bond)	PVAc
Polyacrylonitrile	H	H	H	—CN	PAN
Polyvinylidenes (general)	H	H	R′	R″	
Polyvinylidene chloride	H	H	—Cl	—Cl	PVDC
Polymethyl methacrylate	H	H	—CH_3	—$COCH_3$ (O double bond)	PMMA
Polyisobutylene	H	H	—CH_3	—CH_3	PIB
Polytetrafluoroethylene	F	F	F	F	PTFE

these secondary bonds arises from electron oscillations within the atom or molecule. In brief, the center of negative charges for the electrons is displaced momentarily ($< 10^{-15}$ s), but repeatedly, from the center of positive charges that accompanies the protons in the nuclei. These oscillations produce small electric *dipoles,* where the positive side of one atom or molecule is attracted to the negative side of an adjacent atom or molecule. The resulting dipoles provide the only forces of attraction in many of our ambient vapors and gases. Only at very low temperatures, where the thermal energy is nearly absent, are these attractive forces able to condense and solidify these materials. For example, the melting temperatures, T_m, of Ne, Kr, H_2, O_2, and CH_4 are $-248°$ C, $-157°$ C, $-259°$ C, $-218°$ C, and $-185°$ C, respectively.

Another type of secondary bond exists with asymmetric molecules. Chloromethane, CH_3Cl (Fig. 2–2.4b), is an example. The 17 electrons of chlorine locate the negative center of charge away from the center of molecular mass. Likewise, the hydrogen nuclei (protons) at the ends of the other three bonds cause the

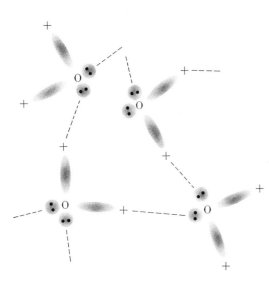

FIG. 2–3.3

Hydrogen Bridge (in Water). The hydrogen at the end of the orbital is an exposed proton (+). It is attracted to the electron lone-pairs of the adjacent water molecules (Fig. 2–2.5b). The hydrogen bridge gives water the highest boiling of any material with a low molecular weight (18 amu).

positive center of charge to locate in the opposite side of the center of mass. The result is an electric dipole, and the molecule is intrinsically *polar.* In general, the *dipole moments, p,* of polar molecules, which are measured by Qd (the product of charge, Q, and displacement, d), are greater than are those for comparable values in atoms and in nonpolar molecules. Thus, materials possessing polar molecules remain solid to relatively high temperatures, before the secondary forces are over-come to produce melting and vaporization.

The third type of secondary bond is the *hydrogen bridge.* It is categorized separately because it produces the strongest of these secondary forces of attraction. Covalently bonded hydrogen atoms—for example, C—H and O—H, expose bare protons on the end of the bonds. Those protons can easily be attracted to negative charges of other molecules, because the protons are not shielded by electrons. Likewise, the lone-pairs of electrons such as exist in H_2O and NH_3 are unshielded (Figs. 2–2.5a and b). Figure 2–3.3 schematically shows the resulting attractions between molecules in water.

The properties of water are influenced significantly by the hydrogen bridge. Although the molecule is established by the covalency of H—O—H, each H_2O molecule is independent except for these hydrogen bridges. The bridges are of sufficient strength, however, that water has the highest melting point of any mole-cule of its size (18 amu and $T_m = 0°$ C). Likewise, its heat of vaporization, $(\Delta H_v)_{100° C}$, is very high (540 cal/g, or 2250 J/g). The freezing expansion of water also can be related to the stereospecific nature of this type of bonding in solid ice.

Linear molecules are *thermoplastic.* By that, we mean that they soften on heating, and reharden on cooling.* The explanation is straightfoward. Relatively

---

* In a technical sense, melting is not involved in this softening, because melting is the loss of crystallin-ity (Chapter 3).

weak intermolecular forces along the molecular chain hold the molecules to-gether. As the temperature is increased, shear stresses can break these weaker secondary bonds and permit the molecules to move by each other; expectedly, continued increases in temperature accelerate the flow, because the secondary bonding becomes still less effective against the added thermal energy. Conversely, cooling decreases the flow rate until these materials reharden.

Consequences of thermoplasticity are important in manufacturing, because the heated (and softened) polymers can be injected into a mold that has cavities of desired shapes. The mold cools the polymer to produce a commercial product (Fig. 2–3.4a). Of course, there is also a "bad news" side to this characteristic. Thermoplastics cannot be used where service conditions exceed the softening temperature (Fig. 2–3.4b).

Example 2–3.1

Teflon, with a mer of C_2F_4 (Table 2–3.1), polymerizes to polytetrafluoroethylene, $(C_2F_4)_n$. What is the degree of polymerization if the mass of the molecule is 33,000 amu (or 33,000 g/mol)? (b) How many molecules are there per gram?

Procedure (a) We want repeat units per molecule; therefore, we need the mer weight. We consider C_2F_4 to be the repeat unit rather than CF_2, since CF_2 cannot be stable. (The carbon must have four bonds.) (b) There are 33,000 g per Avogadro's number of molecules.

Solution

(a) The mer mass is $(4)(19) + (2)(12) = 100$ amu:

   (33,000 amu/molecule)/(100 amu/mer) = 330 mers/molecule.

(b) $(0.6 \times 10^{24}$ molecules)/(33,000 g) $= 1.8 \times 10^{19}$/g

Comment Teflon is a trade name; PTFE (for polytetrafluoroethylene) is generic.

FIG. 2–3.4

Thermoplasticity (Polyvinyl Spoon for Ice Cream). (a) The original spoon. The design requirements are: simple, fast production (by injection molding), and use with ice-cream sundaes at an ice-cream shop. (b) The spoon after it has been used to stir hot coffee. Thermoplastic softening has deformed the utensil. Was the spoon poorly designed?

Example 2-3.2

How much energy is given off when 70 g of ethylene, C_2H_4, react to give polyethylene?

Procedure With each added C_2H_4 molecule, one $C=C$ is eliminated, and two $C-C$ bonds are formed. From Table 2-2.2, these changes involve $+680$ kJ per mole and $2(-370$ kJ/mol$)$ of energy, respectively ($+$, for energy to break the bond; $-$, for energy released as a bond is formed). A mole includes 0.602×10^{24} bonds (Fig. 2-3.1 and Example 2-2.1).

Calculation

$$\frac{+680,000 \text{ J}}{0.602 \times 10^{24} \text{ molecules}} - \frac{2(370,000 \text{ J})}{0.602 \times 10^{24} \text{ molecules}} = -9.96 \times 10^{-20} \text{ J}/C_2H_4$$

$$70 \text{ g } (0.602 \times 10^{24} \text{ amu/g})/(28 \text{ amu}/C_2H_4) = 1.5 \times 10^{24} \text{ } C_2H_4$$

$$(-9.96 \times 10^{-20} \text{ J}/C_2H_4)(1.5 \times 10^{24} \text{ } C_2H_4) = -150,000 \text{ J}$$

or

$$-150,000 \text{ J } (0.239 \text{ cal/J}) = -36 \text{ kcal}$$

Comment This reaction is the basic reaction for making large vinyl-type molecules that are used in plastics.

2-4
THREE-DIMENSIONAL BONDING

The molecules described in the preceding sections do not possess structures with primary bonds in three dimensions. Although materials with linear (and planar) molecules may have many favorable properties, those mechanical properties that are related to strength are improved if the primary bonds develop in all three dimensions.

Network Structures

Polymerization can lead to a *network* structure if some of the units are *polyfunctional;* that is, if the precursor molecules can react with three or more adjacent molecules. Our prototype for network structures will be phenol-formaldehyde (PF), which was one of the first synthetic polymers, and was marketed under the trade name Bakelite.

The atomic arrangements within formaldehyde (CH_2O) and phenol (C_6H_5OH) are shown in Fig. 2-4.1(a). At room temperature, formaldehyde is a gas; phenol is a low-melting solid. The polymerization that results from the reaction between these two compounds is shown schematically in Fig. 2-4.1(b). The formaldehyde has supplied a CH_2 "bridge" between the benzene rings of two phenols. In the process, a hydrogen is stripped from each of the two phenols, and an oxygen is

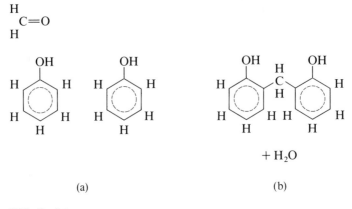

(a)   (b)

FIG. 2–4.1

Phenol–Formaldehyde Reaction. The phenols (C_6H_5OH) contribute hydrogen and the formaldehyde (CH_2O) contributes oxygen to produce water as a by product. The two rings are joined by a —CH_2— bridge.

removed from the formaldehyde. The two hydrogens and the oxygen form water, which can volatilize and escape. The reaction of Fig. 2–4.1 can occur at several points around the phenol molecule.* As a result of this polyfunctionality, a molecular network is formed, rather than a simple linear chain (Fig. 2–4.2).

In production, a partially polymerized mixture is molded at an elevated temperature, where the reaction is completed to give a rigid three-dimensional network. Thus, the heating produces a "set" in the material. In contrast to the earlier *thermoplastic* products, PF is a *thermosetting* product. Reheating will not soften the three-dimensional network significantly.

Network structures also are present in silicate glasses. Fused silica (SiO_2), for example, has silicon atoms that are joined to four adjacent oxygen atoms, which in turn bridge between two silicons. This structure is shown in a three-dimensional sketch in Fig. 2–4.3. The resulting network is very stable, and does not begin to soften until 1200 to 1500° C (depending on the time and forces that are involved). There is considerable open space within the network's structure. This open space can accommodate intense thermal vibrations with very little change in volume. As a result, fused silica is valued for its low thermal-expansion coefficient.

Ionic Bonding

Positively charged cations and negatively charged anions have mutual attraction for each other. Conversely, ions of like charges possess a mutual repulsion. The

* Three is the normal maximum, because there simply is not enough space to attach more than three CH_2 bridges. The number is limted by *stereohindrance.*

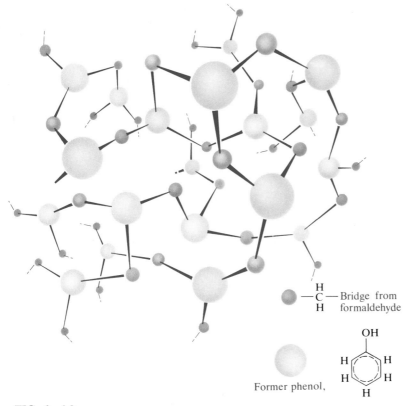

—C— Bridge from
formaldehyde

Former phenol,

FIG. 2-4.2

Network Structure of Polyfunctional Units. Deformation does not occur as readily as in
linear polymers, which are composed of bifunctional units (see Fig. 2-4.1).

coulombic forces of attraction (unlike charges) and repulsion (like charges) can
produce polyatomic structures, as shown in the two-dimensional sketch in Fig.
2-4.4. Negative ions are coordinated directly with positive ions (and vice versa).
The anion-to-anion distances and the cation-to-cation distances are greater.
Therefore, the attractive forces predominate, and unlimited numbers of ions can
be bonded to produce a solid material with primary bonds in three dimensions.

The attractive and repulsive coulombic forces, F_C, are inversely proportional,
k_0, to the square of the separation distance, x, as follows:

$$F_C = k_0(Z_1 q)(Z_2 q)/x^2 \qquad (2-4.1)$$

Here, q is the electron charge of 0.16×10^{-18} A·s, and Z_1 and Z_2 are the ionic
valences.* In NaCl, for example, $Z_1 = +1$ and $Z_2 = -1$; in MgO, they are $+2$ and

---

* The proportionality term, k_0, is $1/4\pi\epsilon_0$, or 9×10^{-9} V·m/A·s.

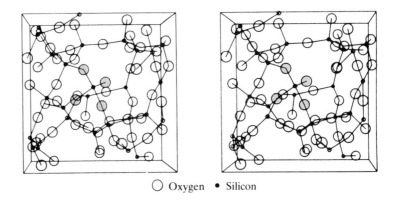

○ Oxygen • Silicon

FIG. 2–4.3

A Stereographic Sketch of Silica Glass at 20° C. The figure should
be viewed with a suitable stereo-viewer, or by controlled eye focus.
The simplest way to view this sketch in stereo is to position the
book at a natural reading distance. Close your eyes, and focus them
straight ahead. Then open your eyes. Each eye should focus on a
separate sketch. A cardboard partition between the two drawings
may help. As with the crystalline SiO_2, each silicon is among four
oxygens, and each oxygen joins (bridges) two tetrahedral units. (By
permission of T. F. Soules, General Electric Co., Cleveland.)

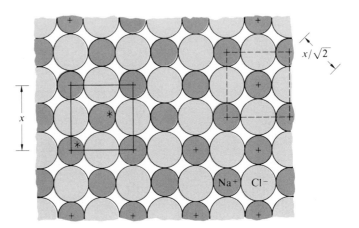

FIG. 2–4.4

NaCl (2-Dimensional). The ionic bonds between the positive, Na^+,
and negative, Cl^-, ions extend in all directions. There are no
isolated NaCl molecular pairs. (See Fig. 3–1.1 for a three-
dimensional sketch of NaCl.)

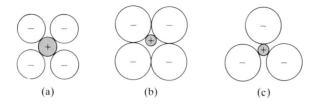

FIG. 2-4.5

Ionic Coordination (Two-Dimensional). (a) Coordination with $r/R > 0.41$. The smaller cation is coordinated with four anions (CN = 4 for two-dimensions). (b) Coordination with $r/R < 0.4$. The positive ion does not have maximum contact with all four neighboring negative ions. Likewise, there is repulsion between the contacting anions. (c) When $r/R < 0.4$, then CN = 3 is favored (two-dimensions).

-2. Hence, the interatomic bonding is much greater in MgO than in NaCl. Melting temperatures, T_m, of 2800° C and 800° C, respectively, support this conclusion.

Ionic Coordination Numbers

Not uncommonly, covalently bonded solids are loosely packed, and possess considerable free space. We could see this free space in the stereo figure of fused silica (Fig. 2-4.3). Likewise, the center of the diamond structure (Fig. 2-2.3b) has an open space large enough for an additional atom. In general, ionic solids are close packed, with less free space, because coulombic attractions are *omnidirectional.* Positive ions, for example, will attract as many negative ions as space will permit (assuming the charge balance is maintained).

Positive ions are generally smaller than negative ions are, because they have been stripped of their valence electrons, whereas negative ions have accepted extra electrons.* As shown in the simplified two-dimensional sketch of Fig. 2-4.5(a), each positive ion is coordinated with four neighbors. $(CN_{2-D} = 4.)$ Unlike neighbors are in direct contact; like neighbors are not. With larger negative ions (or smaller cations), as shown in Fig. 2-4.5(b), there is not direct contact with all the nearby unlike ions. Furthermore, the negative ions come closer together and introduce repulsion. This is an unstable situation that does not facilitate bonding. For ions of these relative sizes, a more stable coordination is shown in Fig. 2-4.5(c), where the coordination number is three (in this two-dimensional example). The maximum coordination number is dictated by the ratio of the radii, r/R, where r is the radius of the smaller ion (almost always positive), and R is the radius of the larger ion.

---

* There are some exceptions. For example, the alkali halide, CsF, has atomic numbers of 55 and 19, respectively, to provide a very large cation and one of the smallest anions.

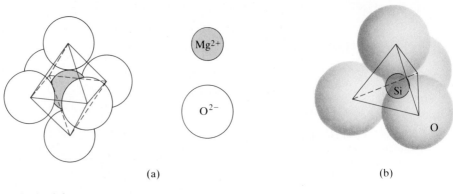

(a)   (b)

FIG. 2–4.6

Coordination Numbers for Ionic Bonding (3-Dimensional). (a) A maximum of six
oxygen ions (O^{2-}) can surround each magnesium ion (Mg^{2+}). (b) The coordination
number of Si^{4+} among O^{2-} is only 4, because the ion-size ratio is less than 0.4 (Table 2–4.1).

Figure 2–4.5 shows two-dimensional sketches. The same concept is applicable
to three dimensions; a coordination number (CN) of six is shown in Fig. 2–4.6(a).
The calculation of the minimum r/R ratio for CN = 6 is obtained in Example
2–4.2 from the sketch in Fig. 2–4.7(a), where the r/R ratio permits anion contact
and the beginning of strong repulsion between the larger negative ions. Other
limiting r/R ratios are given in Table 2–4.1. As with covalent bonding, these limits
are for *maximum coordination numbers,* and thus represent minimum radii
ratios. It is not uncommon to have fewer than the maximum number of neighbors,
especially when it is necessary to maintain a charge balance.

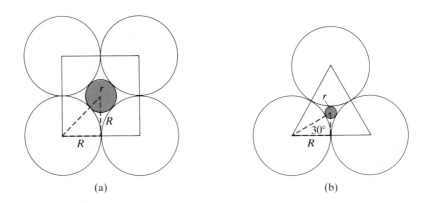

(a)   (b)

FIG. 2–4.7

Coordination Calculations. (a) Minimum r/R for six-fold coordination. (b) Minimum
r/R for three-fold coordination. (Compare with Example 2–4.2 and Fig. 2–4.6).

TABLE 2-4.1 Coordination Numbers versus Minimum Radii Ratios

COORDINATION NUMBER	RADII RATIOS, r/R*	COORDINATION GEOMETRY
3-fold	≥ 0.15	
4	≥ 0.22	
6	≥ 0.41	
8	≥ 0.73	
12	1.0	—

* r—smaller radius; R—larger radius.

Metallic Bonding

Metals can be characterized by their high electrical conductivity. Thus, neither covalent bonding (shared electron pairs) nor ionic bonding (unlike charges) are realized, because both types of bonding localize the valence electrons and preclude conduction. Strong bonding does occur in metals, however, as is evident from the everyday engineering applications of steel, aluminum, and so on. Another bonding type must exist. Although a precise explanation of the bonding of metals is beyond the scope of this text, we can make some observations that will be helpful as we examine the relationships between metallic structures and their properties.

Recall from the discussion of Fig. 2-2.7 that the π electrons in benzene are delocalized. That is, they are not shared covalently by a specific pair of atoms; rather, they develop standing waves across the whole C_6H_6 molecule. There were as many wave patterns for these delocalized electrons as there are carbon atoms in the benzene ring. These delocalized electrons can respond to an electric field by centering their movements toward the side of the molecule nearer the positive electrode; however, they cannot leave the molecule.

The valence electrons of metals also are delocalized. They move in a wavelike pattern through the total metal. Furthermore, there are as many wave patterns as there are atoms in the contiguous metal (a terrific number!). An electrical field will shift the electrons toward the positive electrode, and, if an external connection is made, conductivity occurs.

The simplest way to describe metallic bonding is to view the metal as containing a periodic structure of positive ions surrounded by a "sea" of delocalized electrons (negative). The attraction between the two provides the bond. (We will look more closely at energies of the delocalized electrons in Chapter 11, when we consider conductivity).

The coordination number of atoms in typical metals can be as high as 12 (Fig. 2–4.8), because a pure metal has atoms of only one size; thus $r/R = 1.0$. This high coordination number leads to efficient packing, which is realized by 70 to 80 percent of all pure metals. Approximately 40 percent of the pure metals solidify with CN = 8. (The total exceeds 100 percent, because metals such as iron change from CN = 8 to CN = 12 on cooling from elevated temperatures.)

Example 2–4.1

A plastics molding company buys a phenol-formaldehyde raw material that is only two-thirds polymerized; that is, there is an average of only two $-CH_2-$ bridges joining each phenol, rather than the maximum three. (See the footnote on p. 38.) (a) How many g of additional formaldehyde are required per kg of the above raw material to complete the network formation (that is, to make the phenols fully trifunctional)? (b) How many g of water will be formed in this thermosetting step?

Procedure The raw material corresponds to a linear polymer with the following structure and bridging reaction.

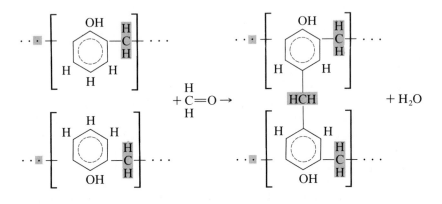

Calculation (Using g/mol)

Raw material   formaldehyde   product   water
$$2[6(12) + 3 + 16 + 1) + (12 + 2)] + (12 + 16 + 2) \rightarrow \quad 224 \quad + \quad 18$$

(a) $x/1 \text{ kg} = 30/212; x = 0.142 \text{ kg}$ (or 142 g)

(b) $y/1 \text{ kg} = 18/212; y = 85 \text{ g}$

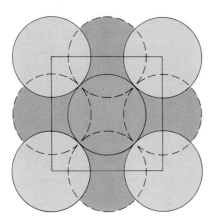

FIG. 2-4.8

Coordination of 12. When all the atoms are the same size, it is possible for each atom to have to have 12 immediate neighbors. Solid circles: four neighbors in the same plane as the central atom. Dashed circles: four neighbors above, *and* four neighbors below. Each neighbor also will be coordinated with 12 neighbors.

Comments The heated raw material is thermoplastic during molding, since it is essentially linear and lacks a network structure. While hot and still under pressure, the described "curing" reaction gives a "set" to the product. The material is thermosetting, because it develops a network structure. It will not soften on reheating.

Example 2-4.2

Show the origin of 0.41 as the minimum ratio for a coordination number of 6.

Procedure The minimum ratio of possible sizes to permit a coordination number of 6 is sketched in Fig. 2-4.7(a). From Fig. 2-4.6(a), note that the fifth and sixth ions sit above and below the center ion of Fig. 2-4.7(a).

Solution

$$(r + R)^2 = R^2 + R^2, \qquad r = \sqrt{2}\,R - R, \qquad \text{and } \frac{r}{R} = 0.41$$

2-5
INTERATOMIC DISTANCES

Attractive forces pull atoms together into solids (and into liquids). Equation (2-4.1) indicated that the attraction becomes stronger as the interatomic distance is reduced. For a given pair of atoms,

$$F_C \propto x^{-2} \qquad\qquad\qquad\qquad (2\text{-}5.1)$$

This may be plotted in Fig. 2-5.1(a) as a hyperbola that would pull the atoms (or ions) into coincidence at $x = 0$. However, there is a limit. The minimum interatomic distance (center-to-center) is approximately 0.2 nm in most solids, and is

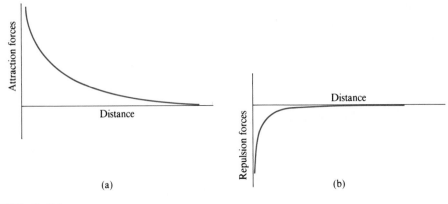

(a)   (b)

FIG. 2–5.1

Interatomic Forces. (a) Attraction (+) forces. Coulombic forces are inversely proportional to the square of the distance. (b) Repulsion (−) forces. These forces arise at close range from the mutual repulsion of electrons, becoming significant at <1 nm.

very specific for a given material. In the polyethylene of Fig. 2–3.2, for example, the distance between carbon nuclei is 0.1544 nm. In iron, the center-to-center distance at 20° C is 0.2482 nm. At the same temperature, the ion centers in MgO are separated by 0.21056 nm. (Note the precision!) If Eqs. (2–4.1 and 2–5.1) apply, why do the atoms abruptly halt at these distances?

Electronic Repulsion

Very strong, short-range repulsive forces enter the picture when the atoms approach within a nanometer of each other. These forces develop because each atom is accompanied by numerous electrons — subvalence as well as valence ones. When atoms are brought into close proximity, there is mutual *electronic repulsion.* The repulsion forces, F_R, are inversely proportional to a higher power of the separation distance than are the attractive forces of Eq. (2–5.1):

$$F_R = -b/x^n. \tag{2–5.2}$$

Here, b, is the proportionality constant, and n may be as high as 9 or 10. The means that these electronic repulsive forces (negative) operate at a much closer range than do the coulombic forces (Fig. 2–5.1b). Thus, the coulombic forces predominate at greater distances of atomic separation, and the repulsive forces predominate at closer interatomic spacings (Fig. 2–5.2a). The equilibrium spacing, $o—x'$, is a natural result when

$$F_C + F_R = 0 \tag{2–5.3}$$

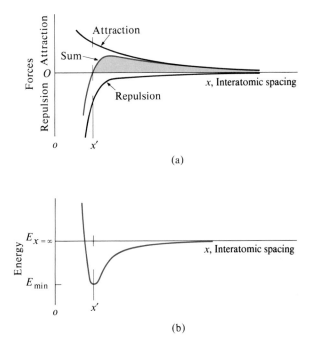

(a)

(b)

FIG. 2–5.2

Interatomic Distances. (a) The equilibrium distance o—x' is that distance at which the net coulombic attractive forces are equal to the electronic repulsive forces. (b) The lowest potential energy occurs when o—x' is the interatomic distance. Since $E = \int F\, dx$, the shaded area of (a) equals the depth of the energy well in (b).

A tension force is required to overcome the predominant forces of attraction if the spacing is to be increased. Conversely, a compressive force has to be applied to push the atoms closer together against the rapidly increasing electronic repulsion.

The equilibrium spacing is a very specific distance for a given pair of atoms, or ions. It can be measured to five significant figures by X-ray diffraction (Chapter 3), if temperature and other factors are controlled. It takes a large force to stretch or compress that distance as much as 1 percent. (Based on Young's modulus, a stress of 2000 MPa (300,000 psi) is required for iron.) It is for this reason that *hard balls* provide a usable model for atoms for many purposes where strength or atom arrangements are considered.*

---

* The hard-ball model is not suitable for all explanations of atomic behavior. For example, a neutron (which does not have a charge) can travel through the space among the atoms without being affected by the electronic repulsive forces just described. Likewise, atomic nuclei can be vibrated vigorously by increased thermal energy, with only a small expansion of the *average* interatomic spacing. Finally, by a momentary distortion of their electrical fields, atoms can move past one another in a crowded solid. (See diffusion in Chapter 6.)

Bonding Energy

The sum of the above two forces provides us with a basis for bonding energies (Fig. 2–5.2b). Since the product of force and distance is energy,

$$E = \int_{\infty}^{x} (F_C + F_R)\, dx \qquad (2-5.4)$$

We will use infinite atomic separation as our energy reference, $E_{x=\infty} = 0$. As the atoms come together, energy is *released* in an amount equal to the shaded area of Fig. 2–5.2(a). The amount of energy released is shown in Fig. 2–5.2(b). Note, however, that at o—x', where $F = 0 = dE/dx$, there is a minimum of energy because energy would have to be supplied to force the atoms still closer together. The depth of this *energy well*, $E_{x=\infty} - E_{min}$, represents the bonding energy, because that much energy would be released as two atoms are brought together at $0°$ K (or be required for complete separation). Table 2–2.2 lists such values for covalent bonds.

The schematic representation of Fig. 2–5.2 can be useful to us for qualitative concepts. For example, the equilibrium distance, x', applies at $0°$ K. However, heating adds kinetic energy, as shown in Fig. 2–5.3. With that energy, x can vary from $<x'$ to $>x'$. However, since the energy well is asymmetrical, the mean value of x is increased. With heating, the average interatomic distance increases, and this produces a thermal expansion.

In subsequent chapters, we shall examine other properties arising from the relationships in Fig. 2–5.2.

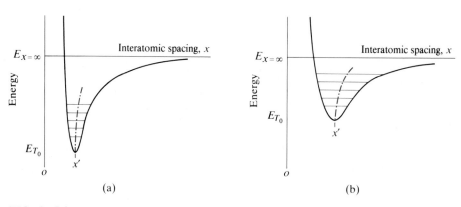

(a)   (b)

FIG. 2–5.3

Energy and Expansion. (a) Strongly bonded solid. (b) Weakly bonded solid. With equal additions of thermal energy, above absolute zero, T_0, the mean interatomic spacing changes less in a material with a deeper energy well (see Fig. 2–5.2b). The expansion becomes more pronounced at higher temperatures (higher energy).

Atomic and Ionic Radii

The equilibrium distance between the centers of two neighboring atoms may be considered to be the sums of their radii (Fig. 2–5.4). In metallic iron, this mean distance is 0.2482 nm at room temperature. Since both atoms are the same, the radius is 0.1241 nm (or 1.241 Å).

Several factors can change the distance between atom centers. The first is temperature, as we mentioned. Ionic valence is a second factor that influences the interatomic spacing. The ferrous iron ion (Fe^{2+}) has a radius of 0.074 nm, which is smaller than that of the metallic iron atom (Table 2–5.1 and Appendix B*). Since the two outer valence electrons of the iron ion have been removed (Fig. 2–5.5), the remaining 24 electrons are pulled in closer to the nucleus, which still maintains a positive charge of 26. A further reduction in interatomic spacing is observed when another electron is removed to produce the ferric ion (Fe^{3+}). The radius of this ion is 0.064 nm, or only about one-half that of metallic iron (and the volume reduces to one-eighth).

A negative ion is larger than its corresponding atom. Since there are more electrons surrounding the nucleus than there are protons in the nucleus, any added electrons are not as closely attracted to the nucleus as are the original electrons.

A third factor affecting the size of the atom or ion is the number of adjacent atoms. An iron atom has a radius of 0.1241 nm when it is in contact with eight adjacent iron atoms, which is the normal arrangement at room temperature. If the atoms are rearranged to place this one iron atom in contact with 12 other iron atoms, the radius of each atom is increased slightly, to ~0.127 nm. With a larger number of adjacent atoms, there is more electronic repulsion from neighboring atoms, and consequently the interatomic distances are increased (Table 2–5.1).

We can apply a rule of thumb for the comparative radii of ions, since the radii are ~10% larger for CN = 6 than for CN = 4. Conversely, the radii are ~3% smaller for CN = 6 than for CN = 8. (See the footnotes to Table 2–5.1 and to Appendix B).

We generally do not speak of atomic radii in covalently bonded materials because the electron distributions may be far from spherical (Fig. 2–2.4a). Furthermore, the limiting factor in atomic coordination is not the atom size, but rather the number of electron pairs obtainable. Even so, we can make some comparisons of interatomic distances (bond lengths) when we look at Table 2–5.1. In ethane with a single C—C bond, this nucleus-to-nucleus distance is 0.154 nm as compared with 0.13 nm for a C=C bond and 0.12 nm for the C≡C bond. This change is to be expected, since the *intra*molecular bonding energies should be greater with the multiple bonds (Table 2–2.2).

---

* The metallic radii used in this book are from the ASM *Metals Handbook*. The ionic radii are patterned after Ahrens.

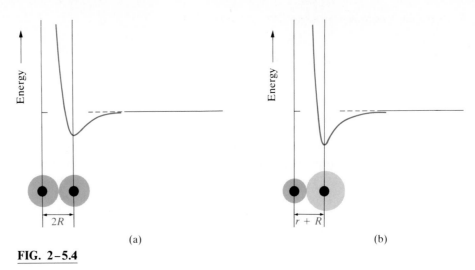

(a)   (b)

FIG. 2–5.4

Bond Lengths. The distance of minimum energy between two adjacent atoms is the bond length. It is equal to the sum of the two radii. (a) In a pure metal, all atoms have the same radius. (b) In an ionic solid, the radii are different because the two adjacent ions are never identical.

TABLE 2–5.1 Selected Atomic Radii

ELEMENT	METALLIC ATOMS		IONS			COVALENT BONDS	
	CN*	RADIUS, nm	VALENCE	CN*	RADIUS, nm†	½ BOND LENGTH,‡ nm	
Carbon						Single	0.077
						Double	0.065
						Triple	0.06
Silicon			4+	6	0.042	Single	0.117
			4+	4	0.038		
Oxygen			2−	8	0.144	Single	0.075
			2−	6	0.140	Double	0.065
			2−	4	0.127		
			2−	2	~0.114		
Chlorine			1−	8	0.187	Single	0.099
			1−	6	0.181		
Sodium	8	0.1857	1+	6	0.097		
Magnesium	12	0.161	2+	6	0.066		
Aluminum	12	0.1431	3+	6	0.051		
			3+	4	0.046		
Iron	8	0.1241	2+	6	0.074		
	12	~0.127	3+	6	0.064		
Copper	12	0.1278	1+	6	0.096		

* CN = coordination number—that is, the number of immediate neighbors. For ions, $1.1 R_{CN=4} \approx R_{CN=6} \approx 0.97 R_{CN=8}$.

† These values vary slightly with the system used. Patterned after Ahrens.

‡ From Table 2–2.2.

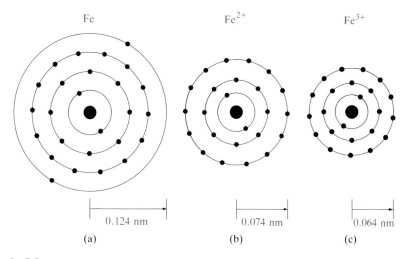

Fe Fe^{2+} Fe^{3+}

0.124 nm   0.074 nm   0.064 nm

(a)   (b)   (c)

FIG. 2-5.5

Atom and Ion Sizes (Schematic). (a) Both iron atoms and iron ions have the same number of protons (26). (b) If two electrons are removed, the remaining 24 electrons and adjacent negative ions are pulled closer to the 26-proton nucleus. (c) A ferric ion holds its 23 electrons still closer to the nucleus.

2-6
GENERALIZATIONS BASED ON ATOMIC BONDING

Several engineering properties may now be associated *qualitatively* with the atomic relationships discussed in this chapter. In subsequent chapters, we shall elaborate on the generalizations presented here.

Density is controlled by atomic weight, atomic radius, and coordination number. The last is a significant factor, because it controls the atomic packing (Chapter 3).

Melting

Melting and *boiling temperatures* can be correlated with the depth of the energy well shown in Fig. 2-5.2(b). Atoms have minimum energy (at the bottom of the well) at a temperature of absolute zero. Increased temperatures raise the energy until the atoms are able to separate themselves from one another.

Hardness

Strength is influenced by the height of the total force or sum curve of Fig. 2-5.2(a). That force, when related to the cross-sectional area, gives the stress required to separate atoms. (As we shall see in Section 8-3, materials can deform through a process other than direct separation of the atoms. However, the amount of stress required to deform them is still governed by the interatomic forces.) Also, since larger interatomic forces of attraction imply deeper energy wells, we observe that materials with high melting points are the *harder* materials; diamond, Al_2O_3, and TiC are examples. In contrast, in materials with weaker bonds, there is a correlation between softness and low melting point; lead, plastics, ice, and grease are examples. Apparent exceptions to these generalizations can arise when more than one type of bond is present, as in graphite and polyethylene.

Elasticity

The *modulus of elasticity* (stress/strain) is related to the slope of the sum curve of Fig. 2–5.2(a), where the net force is zero, because dF/dx relates force to distance. Since both the melting temperature, T_m, and the elastic modulus, E, relate to the bonding energy, they are also correlated (Fig. 2–6.1).

Thermal Expansion

Thermal expansions of materials with comparable atomic packing vary inversely with the melting temperatures of those materials. This indirect relationship exists because the higher-melting-point materials have deeper and therefore more symmetrical energy wells (see Fig. 2–5.3). Thus the mean interatomic distances of more strongly bonded materials increase less with a given change in thermal energy. Examples of several metals are shown in the following table (see also Fig. 2–6.2).

FIG. 2–6.1

Bond Strength (Metals). A stronger bond requires (1) a higher temperature to break free the atoms from their positions within solids, and (2) higher stresses to produce a given strain. Hence, there is a correlation between melting temperature, T_m, and the modulus of elasticity, E.

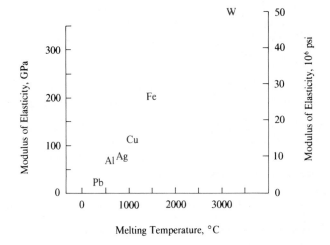

METAL	MELTING TEMPERATURE	COEFFICIENT OF LINEAR EXPANSION
Hg	$-39°$ C	40×10^{-6} m/m·° C
Pb	$327°$ C	29×10^{-6} m/m·° C
Al	$660°$ C	22×10^{-6} m/m·° C
Cu	$1084°$ C	17×10^{-6} m/m·° C
Fe	$1538°$ C	12×10^{-6} m/m·° C
W	$3410°$ C	4.2×10^{-6} m/m·° C

Conductivity of Metals

Electrical conductivity is dependent on the nature of the atomic bonds. Both ionically and covalently bonded materials are poor conductors, because electrons are not free to leave their host atoms. On the other hand, the delocalized electrons of metals move easily along a potential gradient. We shall consider semiconductors in Chapter 11; here, we note that the conductivity of these materials is controlled by the freedom of movement of their electrons.

Thermal conductivity is high in materials with metallic bonds, because delocalized electrons are efficient carriers of thermal as well as electrical energy (Section 11–2).

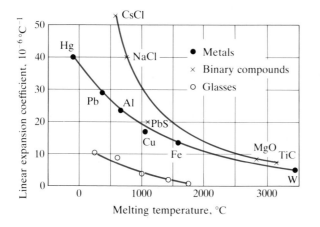

FIG. 2–6.2

Melting Temperatures and Expansion Coefficients (20° C). The more strongly bonded, higher-melting-point materials have lower expansion coefficients (see Fig. 2–6.1). Comparisons must be made among materials with *comparable* structures.

SUMMARY

1. We must use atomic weights in our study of materials. The *atomic mass unit,* amu, is useful. It is defined as one-twelfth of the mass of the carbon-12 isotope. It is also equal to 1 gram divided by Avogadro's number — 0.6022×10^{24}. Thus, *atomic weights* may be expressed either in amu or in g/mol. Electrons have the characteristics of standing waves in individual atoms. The resulting *orbitals* possess unique energies. Intervening energies are precluded, and large forbidden energy gaps exist.

2. *Molecules* are groupings of coordinated atoms that possess strong bonds among themselves *(intramolecular),* but weak *intermolecular* bonds. *Bond lengths* and *bond energies* can be measured. Energy must be supplied (+) to break bonds; it is released (−) when atoms are joined. In general, and with other factors equal, stronger bonds are shorter. Covalent σ bonds are *stereospecific;* that is, they bond two specific atoms. However, π electrons are *delocalized* and establish standing waves across the total molecule. We will also encounter delocalized electrons in metals (Chapter 11).

3. Macromolecules contain hundreds (and thousands) of atoms. *Bifunctional monomers* can polymerize into long *linear* molecules. Typical of these are the polyvinyls, where the mer, or repeatable unit is $+C_2H_3R+$. The type of vinyl depends on the nature of **R**. It is a —**Cl** in *polyvinyl chloride;* a benzene radical, —**C₆H₅**, in *polystyrene;* an —**OH** in *polyvinyl alcohol;* and so on (Table 2–3.1). *Secondary* bonds are relatively weak when compared to *primary* bonds. They arise from electric dipoles that develop in atoms and molecules. These weaker bonds account for the *thermoplasticity* of linear polymers.

4. Polyfunctional monomers can produce *network structures,* which have three-dimensional rigidity.

Also, since coulombic attractions are omnidirectional, ionic materials develop primary bonds in three dimensions. The resulting structures involve the coordination of neighboring ions. The *radii ratio, r/R,* dictates the maximum coordination number, CN, of ionic materials (Table 2–4.1). The valence electrons of metals are delocalized, which leads to high electrical and thermal conductivities.

5. The distances between atoms can be measured precisely in most materials. These *interatomic distances* are established as a balance between the attractive forces that pull the atoms together and the repulsive forces that arise when too many electrons move into close proximity. The equilibrium spacing for the attractive and repulsive forces is also the spacing of minimum energy, so an *energy well* can be indicated. The *atomic radius* is one-half of the interatomic distance between like atoms. *Ionic radii* are less readily determined, since the interatomic distance involves the sum of two different radii. Radii are a function of temperature, valence, and the coordination number. (For ions, $1.1\,R_{CN=4} \approx R_{CN=6} \approx 0.97\,R_{CN=8}$.)

6. It is possible to make some initial generalizations based on atomic coordinations. *Density* is controlled by atomic weight, atomic radius, and coordination number. *Melting temperatures* are correlated with the depth of the energy well. Materials that have higher interatomic forces of attraction commonly possess greater *strengths, hardnesses,* and *moduli of elasticity.* Other factors being equal, materials with strong bonds and high melting temperatures have low *thermal expansion coefficients.* Finally, *electrical* and *thermal conductivities* are greater in those materials with delocalized valence electrons.

KEY WORDS*

Anion	Atomic weight	Bond energy
Atomic mass unit (amu)	Avogadro's number (AN)	Bond length
Atomic number	Bifunctional	Cation
Atomic radius (elements)	Bond angle	Compound

---

* See *Terms and Concepts* for definitions.

Coordination number (CN)
Coulombic force
Covalent bond
Degree of polymerization (n)
Delocalized electron
Dipole
Dipole moment (p)
Electron charge (q)
Electronic repulsion
Energy well
Hydrogen bond (bridge)
Ion

Ionic bond
Ionic radius
Lone-pair
Macromolecules
Mer $+$ $+$
Metallic bond
Mole
Molecule
Molecular weight
Monomer
Network structure
Orbital

Periodic table
Polar group
Polyfunctional
Polymer
Primary bond
Secondary bond
Stereospecific
Thermal expansion coefficient (α)
Thermoplastic
Thermosetting
Valence electrons
Vinyl compounds (CH_3R)

P R A C T I C E   P R O B L E M S *

2–P11 (a) What is the mass of a magnesium atom in grams? (b) The density of magnesium is 1.74 Mg/m³ (1.74 g/cm³); how many atoms are there per mm³?

2–P12 A copper wire weighs 5.248 g, is 2.15 mm in diameter, and is 162 mm long. (a) How many atoms are present per mm³? (b) What is the wire's density?

2–P13 (a) How many silver atoms are there per gram? (b) What is the volume of a silver wire containing 10^{21} atoms? (Obtain the density from one of the appendices.)

2–P14 Chromium is to be plated onto a steel surface (1610 mm²) until it is 7.5 μm thick. (a) How much chromium (Cr^{2+}) is required? (b) How many amperes will be required to do this in 5 minutes? (Obtain the density from one of the appendices.)

2–P15 Distinguish between atomic number and atomic weight.

2–P21 Determine the molecular weight for each of the molecules of Fig. 2–2.1.

2–P22 A small diamond has 10^{20} atoms. How many covalent bonds does it have?

2–P31 A common polymer has C_2H_3Cl as a mer (Table 2–3.1). It has an average mass of 35,000 amu per molecule. (a) What is its mer mass? (b) What is its degree of polymerization?

2–P32 Using the data of Section 2–2, what is the net energy change as 18 grams of vinyl alcohol polymerize to polyvinyl alcohol?

2–P33 There are 9.2×10^{18} molecules per g of polystyrene. (a) What is the average molecular size? (b) What is the degree of polymerization?

2–P34 (a) How many $C{=}C$ bonds are eliminated per mer during polymerization of polyvinyl chloride? (b) How many additional $C{-}C$ bonds are formed?

2–P35 How much energy is released per mer in Problem 2–P34?

2–P36 Show how the bonds are altered for the polymerization of propylene to polypropylene.

2–P41 The radius of a K^+ ion is 0.133 nm when CN = 6. (a) What is it when CN = 4? (b) What is it when CN = 8?

2–P42 All the ions in CsI have CN = 8. What is the anticipated center-to-center distance between Cs^+ and the I^- ions?

2–P43 (a) From Appendix B, cite three divalent cations that can have CN = 6 with Se^{2-} ($R_{CN=6}$ = 0.191 nm), but not CN = 8. (b) Cite two divalent ions that can have CN = 8 with F^-.

---

* See the footnote with the *Practice Problems* of Chapter 1.

2–P44 Why can a neutron move readily through materials, even though the atoms "touch?"

2–P45 Strontium is not listed in Appendix B. On the basis of the periodic table (Fig. 2–1.1) and of other data in that appendix, predict the radius of the Sr^{2+} ion.

2–P61 Estimate the linear thermal expansion coefficient, α_L, of molybdenum, which has a melting temperature of 2880° C.

2–P62 Estimate the elastic modulus of molybdenum based on Fig. 2–6.1.

T E S T   P R O B L E M S *

211 A bronze bell contains 10 w/o tin and 90 w/o copper. What are the atom percents of each element?

212 Using data from the appendixes, determine the number of (a) iron atoms per cm³; (b) aluminum atoms per cm³; (c) chromium atoms per cm³.

213 (a) Cr_2O_3 has a density of 5.2 Mg/m³ (5.2 g/cm³). How many atoms are present per mm³? (b) How many atoms are present per gram?

214 A compound has the formula Cu_3Au. What is the weight percent of each element?

215 A flashlight bulb has a resistance of 8 ohms when it is used in a 6-volt flashlight. How many electrons move through the filament per minute?

216 An electroplated surface of sterling silver (92.5 w/o Ag; 7.5 w/o Cu) is to be plated on a brass spoon (80 Cu–20 Zn). How many amperes are required to plate 1 mg/sec? (Ag^+, Zn^{2+}, and Cu^{2+}.)

217 Explain why electrons can have only selected energy levels within atoms, and why all other levels are forbidden.

218 Indicate the orbital arrangements for a single atom (a) of chlorine; (b) of potassium. (See Fig. 2–1.3.)

221 An organic compound contains 40 w/o carbon, 6.7 w/o hydrogen, and 53.3 w/o oxygen. Name a possible compound.

222 Refer to Fig. 2–2.4(b). Methyl iodide (CH_3I) has a similar structure; however, the H—C—H angles are equal to 111.4°. What is the H—C—I angle in CH_3I? (This is a problem in trigonometry, but it will illustrate the distortion in a polar molecule.)

223 The energy of the C≡N bond is not listed in Table 2–2.2. However, since the energy of the C—N bond is 305 kJ/mol, should the energy of the C≡N bond be 915 kJ/mol, greater than that, or less than that?

231 (a) How many C=C bonds are eliminated per mer during the polymerization of butadiene? (b) How many additional C—C bonds are formed?

232 Sketch the structures of three vinyl monomers. Repeat for the mers of the same three vinyl polymers.

233 Refer to Problem 231. How much energy is released when 7 g of butadiene are polymerized?

234 Engineers who design polymerization processes must give attention to heat removal. Why?

235 By a sketch, show why the
$$-\overset{\displaystyle Cl}{\underset{\displaystyle H}{C}}-$$
portion of polyvinyl chloride is a polar group.

241 Show the origin of "0.73" in Table 2–4.1.

242 (a) What is the radius of the smallest cation that can have an eight-fold coordination with S^{2-} ions without distortion? (b) What is it for one with a six-fold coordination?

---

* See the footnote with the *Test Problems* of Chapter 1.

243 Look ahead to Fig. 3–1.1. (a) What is the coordination number for each Na^+ ion? (b) What is it for each Cl^- ion? (Assume that the structure continues with the same pattern beyond the sketch.)

244 Show the origin of "0.22" in Table 2–4.1. (*Hint:* Place the four large ions of Fig. 2–4.6(b) at the four corners of a cube, and place the small ion at the cube center.) (See Fig. 2–2.4a.)

245 (a) Use data from the appendices to determine the mass of a single gold atom. (b) How many atoms are there per mm^3 of gold? (c) Based on its density, what is the volume of a bead of gold that contains 10^{21} atoms? (d) Assume the gold atoms are spheres ($R_{Au} = 0.1441$ nm), and ignore the spaces among them. What is the volume occupied by the 10^{21} atoms? (e) What volume percent of the space is occupied?

246 Change the gold in Problem 245 to aluminum ($R_{Al} = 0.1431$ nm), and to calcium ($R_{Ca} = 0.197$ nm), and answer the same questions. In addition, (f) point out the major similarity between your answers to the two problems.

247 (a) Repeat Problem 245 (parts c, d, and e) for iron ($R_{Fe} = 0.1241$ nm). (b) Repeat Problem 245, (parts c, d, and e) for tungsten ($R_W = 0.137$ nm). (f) Point out the major similarity between your answers for iron and tungsten.

248 Why is there a lower limit, but not an upper limit, on the radii ratios for $CN = 6$?

251 A cubic volume of CaO that is 0.478 nm along each edge contains four Ca^{2+} ions and four O^{2-} ions. What is the density of CaO?

252 Six O^{2-} ions with the radii shown in Table 2–5.1 surround a Ca^+ ion ($r_{CN=6} = 0.099$ nm). Consider the ions to be hard balls. What is the distance between the "surfaces" of the O^{2-} ions?

253 Six Cl^- ions surround a K^+ ion ($r_{CN=6} = 0.133$ nm). What is the distance between the "surfaces" of the Cl spheres based on the data of Table 2–5.1?

254 The hard-ball model is widely used for metal atoms and ions, but is not applicable for molecules. Why?

261 Gold and nickel are among the elements listed in Appendix B, but not among the materials in Appendix C. Provide an estimate for the thermal expansion coefficient and elastic modulus of each metal.

262 Plot α_L versus E data for unalloyed metals in Appendix C. Add other metals to this list from data in a handbook. Explain your results in your own words.

Chapter 3

CRYSTALS (ATOMIC ORDER)

Now that we have considered atom-to-atom bonding, our next step along the structural scale is to look at the *long-range patterns of atomic order.* Our task is relatively simple in crystalline solids, because unit cells are formed that repeat in each of the three dimensions. Each unit cell has all the geometric characteristics of the total crystal.

In this chapter, we shall look specifically at the atomic arrangements in a few of the more simple structures (bcc, fcc, and hcp) and establish a credibility for their existence through density calculations.

You should become familiar with the notations for unit-cell locations, crystal directions, and crystal planes, because we shall use them subsequently to relate the crystal structure to the properties and behavior of materials.

3-1
CRYSTALLINE PHASES

In the context of materials science and engineering, a *phase* is that part of a material that is distinct from other parts *in structure or composition.* Consider "ice-water." Although they have the same composition, water is a fluid liquid, whereas ice is a crystalline solid. The *phase boundary* between the two locates a discontinuity in the structure; they are *separate phases.* Likewise, both common salt and a saltwater brine contain NaCl, but the two are different phases—the discontinuity at their mutual boundary marks a change in both composition and structure. Now consider a 50–50 combination of water and alcohol. The two are mutually soluble (or *miscible*), so there is only one phase. However, a similar 50–50 mixture of water and oil is *immiscible;* the latter pair comprises two separate phases with a compositional discontinuity at the phase boundary.

Crystals

Essentially all metals, a major fraction of ceramic materials, and certain polymers crystallize when they solidify. To many people, the word *crystal* implies a faceted, transparent, sometimes precious material. Crystalline phases, however, have a more basic characteristic, one that we must consider if we are to understand the internal structures of metals and other materials.

Crystals possess a *periodicity* that produces *long-range order.* By this we mean that the local atomic arrangement is repeated at regular intervals millions of times in the three dimensions of space.

The order found in crystals can be illustrated on a local scale by the atomic coordinations sketched in Fig. 3–1.1: (1) each Na^+ ion has only Cl^- ions as immediate neighbors, and each Cl^- ion has only Na^+ ions as immediate neighbors; (2) the distance between immediate neighbors in NaCl is fixed—that is, $(r_{Na^+} + R_{Cl^-})$ is always equal to 0.097 nm plus 0.181 nm, or 0.278 nm; (3) neighbors of any individual ion are always found in the identical directions, as are neighbors for other corresponding ions.

Although all these local relationships are important, it is of greater significance that an extension of these atomic (or ionic) coordinations in three dimensions produces the characteristic long-range periodicity. This extension is indicated by the line and dot sketch in Fig. 3–1.1, which suggests infinite extrapolation. The atoms (or ions) of a small volume called the *unit cell* (u.c.) are repeated at specific intervals. All unit cells in a crystal are identical. If we describe one, we have described all. For us, this will simplify our later analyses and descriptions of internal structures.

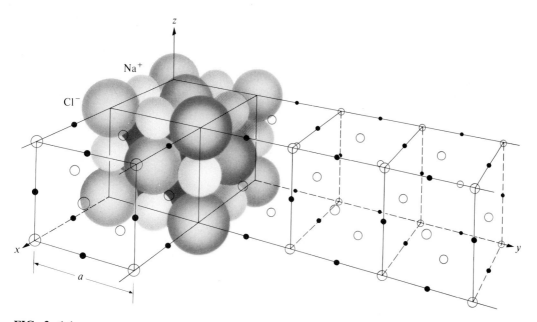

FIG. 3-1.1

Crystal Structure (NaCl). The atomic (or ionic) coordination is repeated to give a long-range periodicity. In this particular structure, the centers of all cube faces duplicate the pattern at the cube corners. (There is also an Na^+ ion at the center of each cell.)

Crystal Systems

Three-dimensional periodicity, which is characteristic of crystals, can be realized by several different geometries. The unit cell in Fig. 3-1.1 is cubic; the three dimensions are equal and are at right angles. Such crystals are said to belong to the cubic system.

Before we look at other crystal systems, we should select a frame of reference. By convention, we orient the x, y, and z axes with the origin in the lower-left rear corner. The axial angles are labeled with the Greek letters alpha (α), beta (β), and gamma (γ), as shown in Fig. 3-1.2(a). Also, by convention, we label the unit-cell dimensions as a, b, and c, respectively, for the three axial directions (Fig. 3-1.2b). If necessary, to facilitate calculations, we can change the origin and orientation. This change should be indicated to the reader.

Variations in the axial angles and in the relative size of a, b, and c dimensions lead to seven (and only seven) crystal systems. These are presented in Table 3-1.1.

We shall encounter the *cubic system* (which has the greatest symmetry) most often. Sketched in Fig. 3-1.3 are the unit cells of the *tetragonal,* the *orthorhombic,* and the *hexagonal* systems—these systems also will be important to us.

FIG. 3–1.2

**Crystal Axes and Unit-Cell
Dimensions.** The conventional
orientation and labels are
sketched. If necessary, the point
of origin and the orientation can
be changed to facilitate a
particular problem; however,
your reader should be advised of
the change.

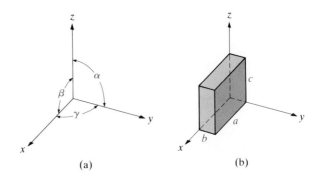

(a)   (b)

TABLE 3–1.1 Crystal Systems (See Fig. 3–1.2)

SYSTEM	AXES	AXIAL ANGLES
Cubic	$a = b = c$	$\alpha = \beta = \gamma = 90°$
Tetragonal	$a = b \neq c$	$\alpha = \beta = \gamma = 90°$
Orthorhombic	$a \neq b \neq c$	$\alpha = \beta = \gamma = 90°$
Monoclinic	$a \neq b \neq c$	$\alpha = \gamma = 90° \neq \beta$
Triclinic	$a \neq b \neq c$	$\alpha \neq \beta \neq \gamma \neq 90°$
Hexagonal	$a = a \neq c$	$\alpha = \beta = 90°; \gamma = 120°$
Rhombohedral	$a = b = c$	$\alpha = \beta = \gamma \neq 90°$

Lattices

As summarized in Table 3–1.1, we can divide space into seven space-filling
systems. Consistent with the seven systems are 14 patterns of points, called Bravais
lattices (Fig. 3–1.4). Three of these are in the cubic system: *simple* cubic (sc),
body-centered cubic (bcc), and *face-centered* cubic (fcc). We shall consider them
numerous times as we proceed through this text. Therefore, let us look at them
more closely.

When viewed in abstract, the lattices in Fig. 3–1.4 define a periodic repetition
of points. Each lattice point has surroundings identical to those of all other lattice
points. The distances to neighboring points, and the directions to neighboring
atoms, are replicated time after time. In simple cubic lattices, the replication
occurs in only the three orthogonal directions of the cubic axes. In body-centered
cubic lattices, the replication also is found in the center of each unit cell. In
face-centered lattices, there is replication at the center of each cube face as well as at
the corners (but there is no duplication at the cube center).

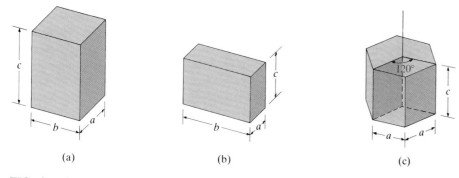

FIG. 3-1.3

Noncubic Unit Cells. (a) Tetragonal: $a = b \neq c$; angles $= 90°$. (b) Orthorhombic: $a \neq b \neq c$; angles $= 90°$. (c) Hexagonal: $a = a \neq c$; angles $= 90°$ and $120°$.

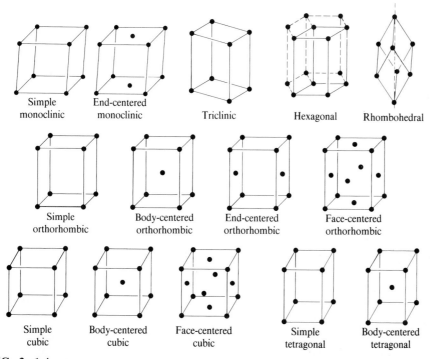

FIG. 3-1.4

Space Lattices. These 14 *Bravais lattices* continue in three dimensions. Each indicated point has identical surroundings. Compare with Table 3-1.1.

FIG. 3–1.5

Face-Centered Cubic (fcc) Lattice (Methane, CH_4). Each lattice point contains a molecule of five atoms. Methane solidifies at $-183°$ C (90 K). Between 20 K and 90 K, the molecules can rotate in their lattice sites. Below 20 K, the molecules have identical alignments, as shown here.

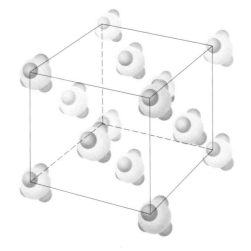

We can "hang" atoms, molecules (Fig. 3–1.5), or other atom combinations on the lattice points, as we shall do in the next section. Admittedly, when we do this, we introduce greater complexity. However, the unit cell still provides us with the structural module for the phase.

Example 3–1.1

The unit cell of chromium is cubic and contains two atoms. Use the data of Appendix B to determine the dimension of the chromium unit cell.

Solution The density of chromium is 7.19 Mg/m³.

$$\text{Mass per unit cell} = 2 \text{ Cr}(52.0 \text{ g})/(0.602 \times 10^{24} \text{ Cr})$$
$$= 172.76 \times 10^{-24} \text{ g}$$
$$\text{Volume} = a^3 = (172.76 \times 10^{-24} \text{ g})/(7.19 \times 10^6 \text{ g/m}^3)$$
$$= 24 \times 10^{-30} \text{ m}^3$$
$$a = 0.2884 \times 10^{-9} \text{ m} \qquad\qquad \text{(or 0.2884 nm)}$$

Comment This same value is obtained experimentally by X-ray diffraction (Section 3–8).

3–2
CUBIC STRUCTURES

Body-Centered Cubic Metals

Iron crystallizes in the cubic system. At room temperature, there is an iron atom at each corner of the unit cell, and another atom at the body center of the unit cell (Fig. 3–2.1). Iron is the most common metal with this body-centered cubic (bcc)

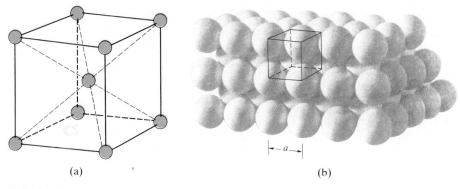

(a) (b)

FIG. 3-2.1

Body-Centered Cubic (bcc) Structure of a Metal. (a) Schematic view showing the
location of atom centers. (b) Model made from hard balls. (G. R. Fitterer. Reproduced
by permission from B. Rogers, *The Nature of Metals,* 2d ed., American Society for
Metals, and the Iowa State University Press, Chapter 3.)

structure, but it is not the only one. Chromium and tungsten, among other metals
listed in Appendix B, also have bcc metal structures.

Every iron atom in this bcc metal structure is surrounded by eight adjacent iron
atoms, whether the atom is located at a corner or at the center of the unit cell.
Therefore, every atom has the same surroundings (Fig. 3-2.1a). There are two
metal atoms per bcc unit cell. One atom is at the center of the cube and eight
octants are located at the eight corners (Fig. 3-2.2).

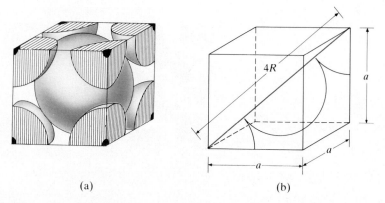

(a) (b)

FIG. 3-2.2

Body-Centered Cubic (bcc) Metal Structure. This structure has two metal atoms per
unit cell, and an atomic packing factor of 0.68. Since atoms are in "contact" along the
body diagonals, Eq. (3-2.1) applies.

The materials with a bcc metal structure have atom contact along the body diagonal (b.d.) of the unit cell. Thus, from Fig. 3–2.2(b), we write

$$(\text{b.d.})_{\text{bcc metal}} = 4R = a_{\text{bcc metal}} \sqrt{3} \qquad (3-2.1a)$$

or

$$a_{\text{bcc metal}} = 4R/\sqrt{3} \qquad (3-2.1b)$$

where a is the *lattice constant*.

We can develop the concept of the atomic *packing factor* (PF) of a bcc metal by assuming spherical atoms (hard-ball model) and calculating the volume fraction of the unit cell that is occupied by the atoms:

$$\text{packing factor} = \frac{\text{volume of atoms}}{\text{volume of unit cell}} \qquad (3-2.2)$$

There are two atoms per unit cell in a bcc metal, and we are assuming spherical atoms, so

$$\text{PF} = \frac{2[4\pi R^3/3]}{a^3} = \frac{2[4\pi R^3/3]}{[4R/\sqrt{3}]^3} = 0.68$$

Face-Centered Cubic Metals

The atomic arrangements in copper (Fig. 3–2.3) are not the same as those in iron, although both are cubic. In addition to an atom at the corner of each unit cell of copper, there is one at the center of each face, but none at the center of the cube.

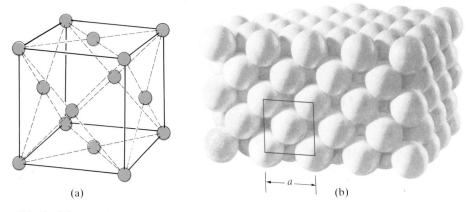

(a)   (b)

FIG. 3–2.3

Face-Centered Cubic (fcc) Metal Structure. (a) Schematic view showing location of atom centers. (b) Model made from hard balls. (G. R. Fitterer. Reproduced by permission from B. Rogers, *The Nature of Metals,* 2d ed., American Society for Metals, and the Iowa State University Press, Chapter 3.)

This face-centered cubic (fcc) structure is as common among metals as is the bcc structure. Aluminum, copper, lead, silver, and nickel possess this structure (as does iron at elevated temperatures).

The fcc metal structure has four atoms per unit cell. The eight corner octants total one atom, and each of the six face-centered atoms add one-half of an atom for a net count of four atoms per unit cell. Since the atoms make contact along the face diagonal (f.d.), we write

$$(f.d.)_{fcc\,metal} = 4R = a_{fcc\,metal}\sqrt{2} \qquad (3-2.3a)$$

or, for the lattice constant,

$$a_{fcc\,metal} = 4R/\sqrt{2} \qquad (3-2.3b)$$

As shown in Example 3–2.1, the packing factor for an fcc metal is 0.74, which is greater than the packing factor of 0.68 for a bcc metal. This difference is to be expected, since each atom in a bcc metal has only eight neighbors. Each atom in an fcc metal has 12 neighbors. You can check this number in Fig. 3–2.4; the front face-centered atom has four adjacent neighbors, four neighbors in contact on the back side, and four additional neighbors (that are not shown) sitting in front.

Other Face-Centered Cubic Structures

The *metal structure* of Figs. 3–2.3 and 3–2.4 is not the only fcc structure. There are several others. Figure 3–1.5 showed the *molecular structure* of solid methane, CH_4, with its five atoms occupying the fcc positions as a molecular unit.

The *NaCl structure* of Fig. 3–1.1 also is fcc. The centers of every face are identical in all respects with the corners. In the compound NaCl, where unlike atoms "touch," the dimensions of the unit cell are obtained from the sum of the two radii:

$$a_{fcc\,NaCl} = 2(r_{Na^+} + R_{Cl^-}) \qquad (3-2.4)$$

This relationship is apparent in Figure 3–2.5. (Note in Fig. 2–4.4 that there is a gap between the Cl^- ions. The Cl^- ions do not touch. Thus, Eq. (3–2.3b) does *not* apply.)

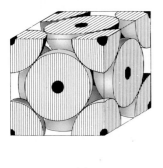

(a)

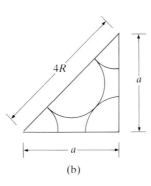

(b)

FIG. 3–2.4

Face-Centered Cubic (fcc) Metal Structure. This structure has four metal atoms per unit cell, and an atomic packing factor of 0.74. Atoms are in "contact" along the face diagonals; therefore, Eq. (3–2.3) applies. Note, however, that Eq. (3–2.3) does *not* apply to other fcc structures (e.g., those of Fig. 3–2.5).

FIG. 3–2.5

AX Structure (NaCl-Type). Compare this figure with Fig. 3–1.1. Each A atom is among six X atoms (and each X atom is among six A atoms). There are four of these 6-f interstitial sites per unit cell. All are occupied. The lattice constant *a* is equal to (2r + 2R).

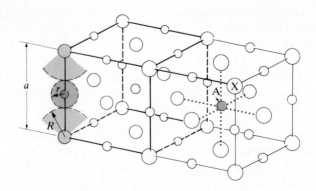

The *diamond-cubic structure* is also fcc. Figure 3–2.6(a) is a redrawn sketch of Fig. 2–2.3(b). Both show covalent bonds rather than spherical atoms or ions in contact; however, that does not matter for our purpose, which is to observe that the centers of all faces are identical in all respects to the corners. There are, however, eight atoms per unit cell for an average of two atoms per fcc position. Experiments show that the C–C interatomic distance is 0.154 nm (center-to-center) in this structure. A close examination of Fig. 3–2.6(b) reveals that the body diagonal is four times this value. Thus,

$$(\text{b.d.})_{\text{diamond}} = 4(\text{C–C}) = (a_{\text{diamond}}) \sqrt{3} \qquad (3\text{–}2.5\text{a})$$

or

$$a_{\text{diamond}} = 4(0.154 \text{ nm})/\sqrt{3} \qquad (3\text{–}2.5\text{b})$$

to give a value of 0.356 nm for a lattice constant, *a*.

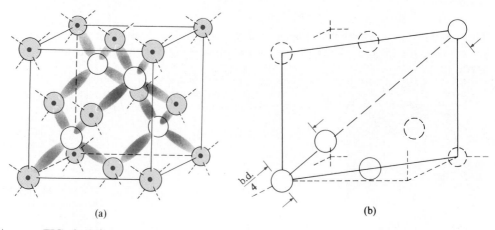

(a)   (b)

FIG. 3–2.6

Diamond-Cubic Structure. (a) Face-centered cubic (fcc). The centers of all six faces are identical to the cube corners. (b) Body diagonal = 4(bond length) = $a\sqrt{3}$.

Silicon, widely used in semiconductors, has a diamond-cubic structure. Likewise, gallium arsenide (GaAs) has a closely related fcc structure in which the atoms of Fig. 3–2.6(a) alternate between gallium and arsenic (Chapter 11).

Simple Cubic Structures

The word *simple* may appear contradictory in the phrase *simple cubic structure,* because such a structure may be more complex than are some bcc and fcc structures. In the crystallographic context, *simple* means a crystal structure with repetition at *only* full unit-cell increments. Recall that a bcc structure is replicated at the center of the unit cell as well as at the corners. Likewise, an fcc structure is replicated at the face centers as well as at the corners. A simple cubic (sc) cell lacks repetition at both of these centered locations.

The alkali halide, CsCl, is sc (Fig. 3–2.7a). Chlorine ions, Cl^-, are located at cell corners; Cs^+ ions are located at cell centers. But this compound is *not bcc,* because the cell center possesses an ion different from that at the cell corners! An alloy of $\sim$ 50-percent copper and $\sim$ 50-percent zinc, called *β'-brass,* also is sc. The choice of location of the unit-cell corner is arbitrary—we can place it at the zinc atom (Fig. 3–2.8a) or at the copper atom (Fig. 3–2.8b).

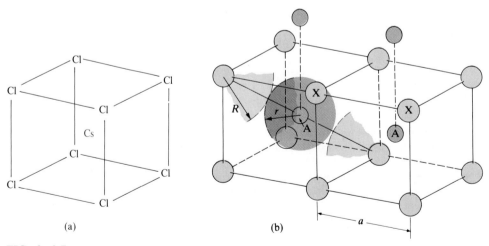

(a)   (b)

FIG. 3–2.7

AX Structure (CsCl-Type). The A atom, or ion, sits in the interstitial site among eight X atoms (8-f sites). These $\frac{1}{2},\frac{1}{2},\frac{1}{2}$ locations are occupied on all unit cells. (See Fig. 3–2.8.) Also note that X atoms sit among eight A atoms. From Eq. (3–2.6), the lattice constant a equals $2(r + R)/\sqrt{3}$.

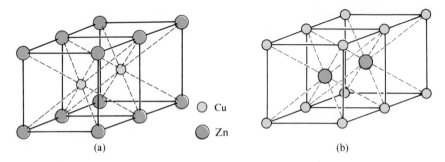

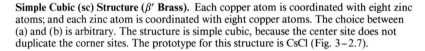

(a)   (b)

FIG. 3–2.8

Simple Cubic (sc) Structure (β' Brass). Each copper atom is coordinated with eight zinc atoms; and each zinc atom is coordinated with eight copper atoms. The choice between (a) and (b) is arbitrary. The structure is simple cubic, because the center site does not duplicate the corner sites. The prototype for this structure is CsCl (Fig. 3–2.7).

These sc structures are *binary;* that is, each possesses two *components.* Therefore, a unit-cell dimension must recognize two radii. From Fig. 3–2.7(b),

$$a_{\text{CsCl-type}} = 2(r_{\text{Cs}^+} + R_{\text{Cl}^-})/\sqrt{3} \qquad (3–2.6)$$

The coordination number is eight for the CsCl-type of sc structure. Therefore, the radii ratio must be greater than the 0.73 that is shown in Table 2–4.1. This constraint precludes MgO from developing this sc structure, since $r_{\text{Mg}^{2+}}/R_{\text{O}^{2-}} = 0.47$, according to the radii data in Appendix B.

Calcium titanate, $CaTiO_3$, is a third compound that has an sc structure. In its cubic form, it has Ca^{2+} ions at each corner of the unit cell, a Ti^{4+} ion at the center, and O^{2-} ions at the centers of each of the six faces (Fig. 3–2.9). It is neither bcc nor fcc, since these centered locations do not replicate the cell corners. (Note from Fig. 3–2.9b that we have a choice of location for the origin of the unit cell.)

Example 3–2.1

Calculate (a) the atomic packing factor of an fcc metal (Fig. 3–2.4); (b) the ionic packing factor of fcc NaCl (Figs. 3–1.1 and 3–2.5).

Procedure Observe in Fig. 3–2.4(a) that there are four atoms per unit cell in an fcc metal structure. The corresponding modification of Fig. 3–1.1 would show four Na^+ and four Cl^- ions per unit cell. Note, however, that the lattice constants relate to the radii differently in the fcc metal structure (Eq. 3–2.3) and in the fcc NaCl structure (Eq. 3–2.4).

Calculation

(a) From Eqs. (3–2.2) and (3–2.3),

$$\text{PF} = \frac{4(4\pi R^3/3)}{a^3} = \frac{16\pi R^3(2\sqrt{2})}{(3)(64R^3)} = 0.74$$

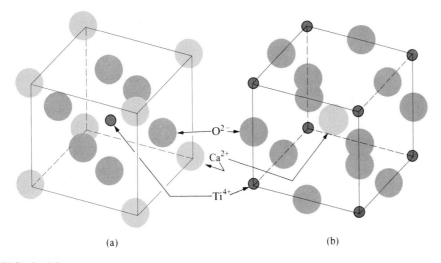

(a)   (b)

FIG. 3-2.9

Cubic CaTiO$_3$. (a) This simple-cubic (sc) structure has a Ti^{4+} ion in the center of the cube, Ca^{2+} ions at the corners, and O^{2-} ions at the center of each face. (b) Alternate presentation. The two are identical, except for our arbitrary choice of origins.

(b) From Eqs. (3-2.2) and (3-2.4), Fig. 3-2.5, and from the radii data in Appendix B,

$$PF = \frac{4(4\pi r^3/3) + 4(4\pi R^3/3)}{(2r + 2R)^3} = \frac{16\pi(0.097^3 + 0.181^3)}{3(8)(0.097 + 0.181)^3} = 0.67$$

Comments It is apparent from this example that the packing factor is independent of atom size, if only one size is present. In contrast, the relative sizes do affect the packing factor when more than one type of atom is present. The fcc metal structure has the highest packing factor (0.74) that is possible for a pure metal, and thus this structure also could be called a *cubic close-packed* (ccp) structure. As we might expect, many metals have this structure. We shall see in the next section that a hexagonal close-packed metal structure also has a packing factor of 0.74.

Example 3-2.2

Copper has an fcc metal structure with an atom radius of 0.1278 nm. Calculate its density and check this value with the density listed in Appendix B.

Procedure Since $\rho = m/V$, we must obtain the mass per unit cell and the volume of each unit cell. The former is calculated from the mass of four atoms (Fig. 3-2.4); the latter requires the calculation of the lattice constant, a, using Eq. (3-2.3) for an fcc metal structure.

Calculation

$$a = \frac{4}{\sqrt{2}}\,(0.1278\ \text{nm}) = 0.3615\ \text{nm}$$

$$\text{density} = \frac{\text{mass/unit cell}}{\text{volume/unit cell}} \qquad\qquad (3\text{–}2.6a)$$

$$= \frac{(\text{atoms/unit cell})(\text{g/atom})}{(\text{lattice constant})^3} \qquad\qquad (3\text{–}2.6b)$$

$$\text{density} = \frac{4[63.5/(0.602 \times 10^{24})]}{(0.3615 \times 10^{-9}\ \text{m})^3} = 8.93\ \text{Mg/m}^3 \qquad\qquad (\text{or } 8.93\ \text{g/cm}^3)$$

The experimental value listed in Appendix B is 8.93 Mg/m³.

Example 3–2.3

Calculate the volume of the unit cell of LiF, which has the same fcc structure as does NaCl (Fig. 3–1.1).

Procedure Although LiF has an fcc structure, we cannot use the geometry of Fig. 3–2.4(b), since the fluoride ions do not touch, as did the metal atoms. Rather, according to Eq. (3–2.4), a is twice the sum of r_{Li^+} and R_{F^-}. (Check Fig. 3–2.5 again.)

Calculation From Appendix B,

$$a = 2(0.068 + 0.133)\ \text{nm}$$

$$a^3 = 0.065\ \text{nm}^3 \qquad\qquad (\text{or } 65 \times 10^{-30}\ \text{m}^3)$$

Comment We use ionic radii, since LiF is an ionic compound. Furthermore, each ion has six neighbors; therefore, the radii of Appendix B do not have to be modified.

Example 3–2.4

Cesium iodide (CsI) has the structure shown in Fig. 3–2.8. What is its packing factor if the radii are 0.172 nm and 0.227 nm, respectively?

Procedure The body diagonal of the unit cell is equal to $(2r + 2R)$. Each unit cell has 1 Cs$^+$ and 1 Cl$^-$ ion.

Calculation

$$a = [2(0.172 + 0.227)]/\sqrt{3} = 0.461\ \text{nm}$$

$$\text{packing factor} = (4\pi/3)(0.172^3 + 0.227^3)/(0.461)^3 = 0.72$$

Comments The packing factor is greater than the one we obtained in the calculation following Eq. (3–2.2) because a bcc metal does not contain a second (different) atom in the center.

These radii differ from those of Appendix B by a factor of 0.97, because the data in the appendix are for CN = 6. Here, CN = 8. (See Section 2–5.)

3-3

NONCUBIC STRUCTURES

The sketches of Figs. 3–3.1(a) and 3–3.1(b) are two representations of *hexagonal* unit cells, and are therefore noncubic. The angles within the base are 120° (and 60°). The rhombic representation of part (b) is more fundamental, because it is the smallest repeating volume; however, the hexagonal representation, which actually contains three of the rhombic cells, is used more often because it shows the hexagonal (six-fold) symmetry characteristics that give the name to this unit cell.

Hexagonal Close-Packed Metals

The hexagonal structure formed by several metals, including magnesium, titanium, and zinc, is shown in Fig. 3–3.2. This hexagonal structure is called the *hexagonal close-packed* (hcp) metal structure. It is characterized by the fact that each atom in one layer is located directly above or below interstices among three

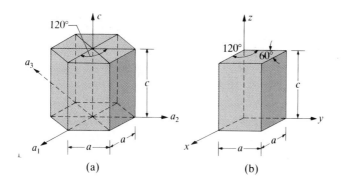

(a)   (b)

FIG. 3–3.1

Hexagonal Cells. (a) Hexagonal representation (four axes). (b) Rhombic representation (three axes). The two representations are equivalent geometrically, with $a \neq c$, $\gamma = 120°$, and $\alpha = \beta = 90°$; however, $V_a = 3 V_b$.

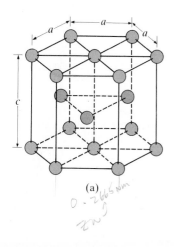

(a)

(b)

FIG. 3–3.2

Hexagonal Close-Packed (hcp) Metal Structure. (a) Schematic view showing the location of atom centers. (b) Model made from hard balls.

atoms in the adjacent layers. Consequently, each atom touches three atoms in the layer below its plane, six atoms in its own plane, and three atoms in the layer above for CN = 12. There is an average of six atoms per unit cell in the hcp metal structure of Fig. 3–3.2 (or two per unit cell, if we use the related rhombic representation).

Since both hcp and fcc metals have a coordination number of 12, they have the same atomic packing factors, 0.74.

Other Noncubic Structures

The vast majority of the crystals that we shall encounter here will be either cubic or hexagonal. There are, however, many noncubic crystals; in fact, these crystals outnumber all the bcc, fcc, and hcp materials combined. We shall look at only a few for illustrative purposes.

Iodine, I_2, belongs to the orthorhombic system (Fig. 3–3.3). From Table 3–1.1, the unit cell has orthogonal axes (90°); but none of the three unit-cell dimensions is equal to another. This crystalline asymmetry occurs because the covalently bonded I_2 molecule is not spherically symmetric.

Polyethylene also develops orthorhombic crystals. The structure is more complex, however, because the molecules are large and linear (Fig. 3–3.4). Polyethylene does not crystallize easily. The large molecules become entangled and thus too unwieldly to be organized readily into a structure with long-range order. One technical advance in polymer processing has been to develop the capability to crystallize these big molecules more readily. The result has been the improvement of a variety of properties of interest to the engineer (Section 7–5).

Graphite is hexagonal (Fig. 3–3.5), but is markedly different from other hexagonal crystals. Graphite contains layers in which carbon atoms are held together tightly with σ bonds (Section 2–2). As with benzene (Fig. 2–2.7), the remaining valence electrons are delocalized and are capable of moving readily through the two-dimensional structure. Meanwhile, the adjacent graphite layers have only secondary attractions to each other. These layers may be considered to be planar

FIG. 3–3.3

Molecular Crystal (Iodine). The molecule of I_2 acts as a unit in the repetitive crystal structure. This lattice is *simple orthorhombic* because $a \neq b \neq c$, and the face-centered positions are *not* identical to the corner positions. (The molecules are oriented differently.) The unit cell axes have 90° angles.

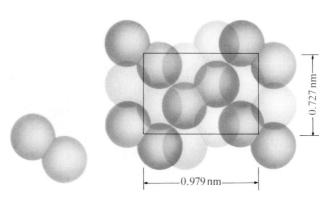

0.727 nm

0.979 nm

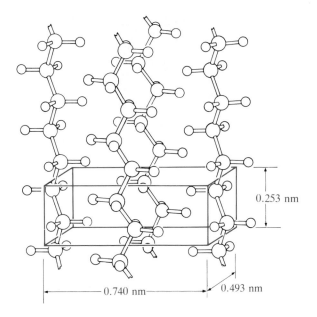

FIG. 3–3.4

Molecular Crystal (Polyethylene). The chains are aligned longitudinally. The unit cell is orthorhombic with 90° angles. (M. Gordon, *High Polymers,* Iliffe, and Addison-Wesley. After C. W. Bunn, *Chemical Crystallography,* Oxford.)

0.253 nm

0.740 nm

0.493 nm

macromolecules, which leads to some unusual properties. First, graphite has good electrical and thermal conductivities in *two dimensions,* but poor conductivities in the third dimension. The material is *anisotropic;* that is it has a directional variation in structure and properties. Second, the strong σ bonds within the layers retain this crystalline solid to more than 2200° C. At the same time, the weak bonds between planes permit graphite layers to slide by one another when sheared. Thus, graphite has lubricating characteristics.

Iron carbide, sometimes called *cementite,* has the chemical formula of Fe_3C. This does not mean that it forms discrete molecules, as does iodine; rather, it simply means that the crystal lattice contains the two elements in a fixed 3-to-1 ratio of iron to carbon. Iron carbide is orthorhombic, with 12 iron atoms and 4

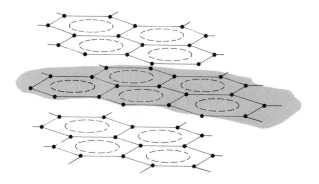

FIG. 3–3.5

Delocalized Electrons in Graphite Layers. Each layer contains "multiple benzene rings" (Fig. 2–2.7). The conductivity is more than 100 times greater in the parallel direction than it is in the perpendicular direction.

carbon atoms in each unit cell. The small carbon atoms are in the interstices among the larger iron atoms. This binary compound is found as a minor phase in almost all steels. Although minor, it greatly affects the properties of steel because Fe_3C is much harder and more rigid than are the other phases in steel (Chapter 9).

Barium titanate, $BaTiO_3$, is the final noncubic phase that we shall consider. It is *tetragonal,* giving its unit cell a square base, $a = a = 0.398$ nm (Fig. 3–3.6), but the cell height, c, is different. In this case, $c = 0.403$ nm; but in other tetragonal cells, c may be smaller than a. An examination of the structure in Fig. 3–3.6 reveals that there are five atoms per unit cell; neither the centers of the face nor the cell center match the cell corners. Therefore, the structure is *simple tetragonal* (st).

Example 3–3.1

The atomic packing factor of magnesium, like that of all hcp metals, is 0.74. What is the volume of magnesium's unit cell, which is shown in Fig. 3–3.2(a)?

Procedures From Appendix B, magnesium has $\rho = 1.74$ Mg/m³ (or 1.74 g/cm³); its atomic mass = 24.31 amu, and $R_{Mg} = 0.161$ nm. From Fig. 3–3.2(a), $12/6 + 2/2 + 3 = 6$ atoms/unit cell.

Two approaches are possible. (a) Since the packing factor is given as 0.74, calculate the volume of six atoms per unit cell. (b) Since $\rho = m/V$, calculate the mass of six atoms per unit cell, then the volume per unit cell.

Calculation

(a) $V_{u.c.} = 6$ atoms $(4\pi/3)(0.161$ nm$)^3/0.74$

   $= 0.14$ nm³   (or 1.4×10^{-28} m³)

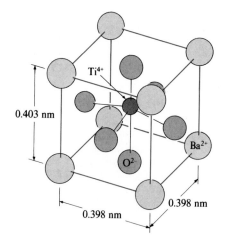

FIG. 3–3.6

Tetragonal BaTiO₃. The base of the unit cell is square; but unlike in CaTiO₃ (Fig. 3–2.9), $c \neq a$. Also, the Ti⁴⁺ and O²⁻ ions are not symmetrically located with respect to the corner Ba²⁺ ions. Above 120° C, BaTiO₃ changes from this structure to that of CaTiO₃.

(b) $m_{u.c.} = 6$ atoms $(24.31 \text{ g})/(0.602 \times 10^{24}) = 2.42 \times 10^{-22}$ g

$$V_{u.c.} = \frac{m}{\rho} = (2.42 \times 10^{-22} \text{ g})/(1.74 \times 10^6 \text{ g/m}^3)$$

$$= 1.4 \times 10^{-28} \text{ m}^3 \qquad\qquad\qquad (\text{or } 0.14 \text{ nm}^3)$$

Comments The average radius, $\overline{R}$, of the magnesium atom is 0.161 nm. However, X-ray diffraction data show that the Mg atoms are compressed almost 1 percent to become oblate spheroids. (See the Comments in Example 3–3.2 and the footnote to Appendix B.)

Example 3–3.2

Assume spherical atoms. What is the c/a ratio of an hcp metal?

Derivation Refer to Fig. 3–3.2(a); consider the three central atoms plus the one at the center of the top. This is an equilateral tetrahedron with edges $a = 2R$. From geometry,

$$h = a\sqrt{2/3}$$
$$c = 2h = 2a\sqrt{2/3} = 1.63a \qquad\qquad\qquad (3\text{–}3.1)$$

Comments The c/a ratios of hcp metals depart somewhat from this figure: Mg, 1.62; Ti, 1.59; Zn, 1.85. This means we must envision magnesium and titanium atoms as slightly compressed spheres, and zinc atoms as elongated spheroids.

Example 3–3.3

The unit-cell volume of hcp titanium at 20° C is 0.106 nm³, as sketched in Fig. 3–3.2(a). The c/a ratio is 1.59. (a) What are the values of c and a? (b) What is the radius of the Ti in a direction that lies in the base of the unit cell?

Procedure Work from a, the dimension of the bottom edge; then calculate the unit-cell volume from the height ($= 1.59\ a$) and the basal area, A.

Calculation

(a) $\qquad A = 6(1/2)(a)(a \sin 60°) = 2.60\ a^2$

$\qquad$ volume $= (1.59a)(2.60a^2) = 4.13a^3 = 0.106$ nm³

$\qquad\qquad a^3 = 0.02566$ nm³; $\quad$ and $\quad a = 0.2950$ nm

$\qquad\qquad c = 1.59(0.295 \text{ nm}) = 0.469$ nm

(b) $\qquad a = 2R_{base}; \therefore R_{base} = 0.1475$ nm

Comments The average radius is 0.146 nm (Appendix B). The atoms of titanium (and the unit cell) are slightly compressed in the c direction. (See Comments with Example 3–3.2.)

3–4

POLYMORPHISM

Diamond ↔ Graphite

We are aware that carbon can exist in two forms—as diamond and as graphite. These two phases are called *polymorphs* (meaning, multiple forms), and they are distinctly different. Diamond is very hard, transparent, and an electrical insulator (Fig. 11–1.1). Graphite has lubricating characteristics, and conducts an electric current ($\sigma = 10^5$ ohm$^{-1}\cdot$m^{-1}; >20 orders of magnitude more than diamond). These contrasts arise from differences in bonding, and therefore differences in structure. Diamond has a three-dimensional structure of covalent bonds (Fig. 3–2.6). Graphite has π electrons that possess mobility in the crystal layers of a two-dimensional structure (Fig. 3–3.5).

Diamond will change to graphite if it is heated sufficiently for the thermal energy to break the bonds and coordinate the atoms with new neighbors. The reaction is reversible:

$$C_{diamond} \xleftrightarrow{\text{pressure heat}} C_{graphite} \qquad (3–4.1)$$

However, the graphite-to-diamond reaction must be accompanied by very high pressures (at high temperatures). The role of pressure is expected, since 200 milligrams (1 carat) of diamond occupies only 60 mm^3, whereas the same mass of graphite occupies 90 mm^3.

Body-Centered Cubic Iron ↔ Face-Centered Cubic Iron

Our principle example of polymorphism in metals will be iron, because our ability to heat-treat steel and to modify its properties stems from the fact that, as iron is heated, it changes from the bcc to the fcc metal structure at 912° C, and back to the bcc at 1400° C.

At room temperature, bcc iron has a coordination number of 8, an atomic packing factor of 0.68, and an atomic radius of 0.1241 nm. When pure iron

FIG. 3–4.1

Thermal Expansion of Iron. Because iron changes from bcc, to fcc between 912° C and 1400° C, it contracts in that temperature range. The fcc polymorph has a higher packing factor than does the bcc polymorph—0.74 versus 0.68. Its atoms also have larger radii. (From L. H. Van Vlack, *Materials for Engineers: Concepts and Applications*, Addison-Wesley, Reading, Mass., Fig. 2–6.1, with permission.)

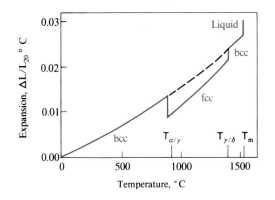

changes to fcc at 912° C, its coordination number is 12, its atomic packing factor is 0.74, and its atomic radius is 0.129 nm. [At 912° C (1673° F) the atomic radius of bcc iron, due to thermal expansion, is 0.126 nm.]

The reactions,

$$Fe_{bcc} \xleftrightarrow{912° C} Fe_{fcc} \xleftrightarrow{1400° C} Fe_{bcc} \qquad (3-4.2)$$

involve volume changes. As calculated in Example 3–4.1, there is a contraction of 1.4 v/o (−0.47 l/o) when the temperature is raised past 912° C (Fig. 3–4.1). Note that there are two factors influencing this volume change. First, the radius increases from 0.126 nm to 0.129 nm due to the larger coordination number of the fcc atoms (Section 2–5). Counteracting and exceeding that effect is the decrease in volume as the atomic packing factor changes. Reactions (3–4.2) reverse during cooling, so expansion occurs when the fcc phase changes back to bcc.

$ZrO_{2_{cubic}} \rightarrow ZrO_{2_{tetragonal}} \rightarrow ZrO_{2_{monoclinic}}$

Zirconia (ZrO_2) has not been a well-known ceramic until recently. Its presence can toughen an alumina (Al_2O_3) product significantly. This toughness can be achieved because zirconia has three (and more) polymorphs. Its high-temperature polymorph has the CaF_2-fcc structure shown in Fig. 3–4.2. On cooling, the atoms shift to make a tetragonal unit cell. The cell retains a square base, but changes its height slightly. The next step,

$$ZrO_{2_{tetragonal}} \rightarrow ZrO_{2_{monoclinic}} \qquad (3-4.3)$$

is more significant. This change to the monoclinic polymorph (Table 3–1.1) involves dimensional strain that consumes energy before fracture, thus toughening the ceramic (Section 9–8).

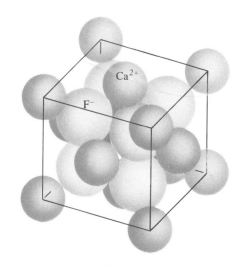

FIG. 3–4.2

The CaF_2-Type Structure. The lattice is face-centered cubic (fcc). Zirconia (ZrO_2) has this structure at high temperatures, but changes to tetragonal and monoclinic at lower temperatures. The structure also is that of the nuclear fuel, UO_2. (The unoccupied positions at the center of the cube and at the mid-edges can accommodate the fission products that accumulate.)

Example 3–4.1

Iron changes from the bcc to the fcc metal structure at 912° C (1673° F). At this temperature, the atomic radii of the iron atoms in the two structures are 0.126 nm and 0.129 nm, respectively.

(a) What is the percent of volume change, v/o, as the structure changes?

(b) What is the percent of linear change, l/o?

(*Note:* As indicated in Section 2–5 and Table 2–5.1, the higher the coordination number, the larger the radius.)

Procedure Use a common basis for the two structures. The simplest is four iron atoms; therefore, *two* unit cells of bcc iron and *one* unit cell of fcc iron. Also $a_{bcc} \sqrt{3} = 4R_{bcc}$, and $a_{fcc} \sqrt{2} = 4R_{fcc}$.

Calculation

(a) $2V_{bcc} = 2a_{bcc}^3 = 2[4(0.126 \text{ nm})/\sqrt{3}]^3 = 0.0493 \text{ nm}^3$

$V_{fcc} = a_{fcc}^3 = [4(0.129 \text{ nm})/\sqrt{2}]^3 = 0.0486 \text{ nm}^3$

$\dfrac{\Delta V}{V} = \dfrac{0.0486 - 0.0493}{0.0493} = -0.014$ (or − 1.4 v/o change)

(b) $(1 + \Delta L/L)^3 = 1 + \Delta V/V$

$\Delta L/L = \sqrt[3]{1 - 0.014} - 1 = -0.0047$ (or − 0.47 l/o change)

Comment Iron expands thermally until it reaches 912° C, at which point it shrinks abruptly; further heating continues the expansion (Fig. 3–4.1).

Example 3–4.2

The densities of ice and water at 0° C are 0.915 and 0.9995 Mg/m³ (or g/cm³), respectively. What is the percent volume expansion during the freezing of water?

Procedure Determine the volume of 1 Mg of each.

volume ice $= 1.093 \text{ m}^3$

volume liquid $= 1.0005 \text{ m}^3$

Calculation

$\Delta V/V = (1.093 \text{ m}^3 - 1.0005 \text{ m}^3)/1.0005 \text{ m}^3$

$= +0.092$ (or 9.2 v/o)

Comments We are familiar with the major changes that occur during freezing and thawing. In principle, polymorphic changes are similar; that is, there is a change in structure within the solid. This change brings about changes in volume, density, and almost every other physical property.

3–5
UNIT-CELL GEOMETRY

The general convention that we shall follow is to orient a crystal such that its x axis points toward us, its y axis points to our right as we face it, and its z axis points upward. Thus, by convention, the origin is at the lower-left rear corner of the unit cell.* The opposing directions are negative.

Points Within Unit Cells

Every point within a unit cell can be identified in terms of the coefficients along the three coordinate axes. Thus, the origin is $0,0,0$. Since the center of the unit cell is at $\frac{a}{2}, \frac{b}{2}, \frac{c}{2}$, the indices of that location are $\frac{1}{2}, \frac{1}{2}, \frac{1}{2}$. These *coefficients of locations* are always expressed in unit-cell dimensions. Thus, the far corner of the unit cell is $1,1,1$, regardless of the crystal system—cubic, tetragonal, orthorhombic, and so on.

A translation from any selected site within a unit cell by an integer multiple of lattice constants ($a, b,$ or c) leads to an identical position in another unit cell. Thus, in the two-dimensional lattice of Fig. 3–5.1, the two points labeled with an asterisk (*) are separated by translations of $3b$ (parallel to y) and $2c$ (parallel to z). This example is obviously noncubic (or nonsquare). However, integer multiples lead to replication in all crystal systems.

Replication also is realized by $\pm\frac{a}{2}, \pm\frac{b}{2}, \pm\frac{c}{2}$ translations in body-centered unit cells (or into adjacent unit cells). Thus, a point with indices of $0.3, 0.4, 0.2$ duplicates the site at $0.8, 0.9, 0.7$; or one at $0.8, -0.1, 0.7$ in the next unit cell to the left. (See Fig. 3–5.2a.)

In face-centered unit cells, replication of points accompanies translations of $\pm\frac{a}{2}, \pm\frac{b}{2}, 0$; of $\pm\frac{a}{2}, 0, \pm\frac{c}{2}$; and of $0, \pm\frac{b}{2}, \pm\frac{c}{2}$ (in addition to the $\pm a, \pm b, \pm c$

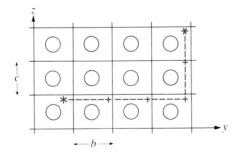

FIG. 3–5.1

Unit Translations. A translation of an integer number of unit dimensions—i.e., lattice constants—leads to a site that is identical in every respect to the original site.

---

* We are not locked into this convention if there is reason to change. However, we shall assume this convention applies in our discussions *unless* informed otherwise. Conversely, your answers should comply with the convention *unless* you advise your reader of the reorientation that you have used.

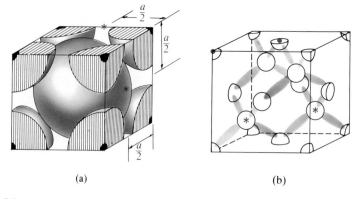

(a)   (b)

FIG. 3–5.2

Unit-Cell Translation. (a) Translation in bcc (metal). A translation of $\pm\frac{a}{2},\pm\frac{a}{2},\pm\frac{a}{2}$ from any point leads to an identical site in *body-centered cubic;* for example, $*$ to $*$ (from $0,\frac{1}{2},1$ to $\frac{1}{2},1,\frac{1}{2}$). (b) Translation in fcc (diamond). A translation of $\mp\frac{a}{2},\pm\frac{a}{2},0$ leads to identical sites in *face-centered cubic.*

translations). As a result, the $\frac{3}{4},\frac{1}{4},\frac{1}{4}$ site is identical to the $\frac{1}{4},\frac{3}{4},\frac{1}{4}$ site in fcc. You can see this identity in the diamond-cubic structure of Fig. 3–5.2(b) with the locations of the two asterisks.

Example 3–5.1

(a) Sketch a noncubic unit cell. Show the locations of points that have the following coefficients:

$0,0,0$ $0,0,\frac{1}{2}$ $\frac{1}{2},\frac{1}{2},\frac{1}{2}$ $\frac{1}{2},\frac{1}{2},0$ $1,1,0$ $1,1,1$ $1,1,2$

(b) Assume the cell is orthorhombic (all axial angles = 90°), and that $a = 0.270$ nm, $b = 0.403$ nm, and $c = 0.363$ nm. What is the translation distance t between points at $0,-1,0$ and at $1,1,2$?

Solution

(a) See Fig. 3–6.1(a).

(b) Since the axial angles are 90°,

$$t = \sqrt{[(1 - 0)(0.270 \text{ nm})]^2 + [(1 - (-1))(0.403 \text{ nm})]^2 + [(2 - 0)(0.363 \text{ nm})]^2}$$
$$= 1.118 \text{ nm}$$

Comment This procedure for locating points within unit cells does not limit us to the reference unit cell.

Example 3–5.2

Consider the following four points within a unit cell:

$\frac{1}{4},\frac{1}{4},\frac{3}{4}$ $\frac{1}{8},\frac{1}{8},\frac{1}{8}$ $\frac{7}{8},\frac{3}{4},\frac{1}{5}$ 0.53, 0.25, 0.17

(a) Identify a second identical site *within* the reference unit cell (bcc) for each of the four points.

(b) Identify three identical sites *within* an fcc unit cell for *each* of the four points.

Answers

(a) The translations in body-centered cells are $\pm 0.5a$, $\pm 0.5a$, $\pm 0.5a$. Therefore.

For $\frac{1}{4},\frac{1}{4},\frac{3}{4}$: $(0.25 + 0.5), (0.25 + 0.5), (0.75 - 0.5) = \frac{3}{4},\frac{3}{4},\frac{1}{4}$

For $\frac{1}{8},\frac{1}{8},\frac{1}{8}$: $(0.125 + 0.5), (0.125 + 0.5), (0.125 + 0.5) = \frac{5}{8},\frac{5}{8},\frac{5}{8}$

For $\frac{7}{8},\frac{3}{4},\frac{1}{5}$: $(0.875 - 0.5), (0.75 - 0.5), (0.20 + 0.5) = \frac{3}{8},\frac{1}{4},0.7$

For 0.53, 0.25, 0.17: $(0.53 - 0.5), (0.25 + 0.5), (0.17 + 0.5) = 0.03, 0.75, 0.67$

(b) The translations in face-centered cells are permutations of $\pm 0.5a$, $\pm 0.5a$, and 0.

For $\frac{1}{4},\frac{1}{4},\frac{3}{4}$:	$\frac{3}{4},\frac{3}{4},\frac{3}{4}$	$\frac{3}{4},\frac{1}{4},\frac{1}{4}$	$\frac{1}{4},\frac{3}{4},\frac{1}{4}$
For $\frac{1}{8},\frac{1}{8},\frac{1}{8}$:	$\frac{5}{8},\frac{5}{8},\frac{1}{8}$	$\frac{5}{8},\frac{1}{8},\frac{5}{8}$	$\frac{1}{8},\frac{5}{8},\frac{5}{8}$
For $\frac{7}{8},\frac{3}{4},\frac{1}{5}$:	$\frac{3}{8},\frac{1}{4},\frac{1}{5}$	$\frac{3}{8},\frac{3}{4},0.7$	$\frac{7}{8},\frac{1}{4},0.7$
For 0.53, 0.25, 0.17:	0.03, 0.75, 0.17	0.03, 0.25, 0.67	0.53, 0.75, 0.67

Comment These translations within body-centered and face-centered cells are independent of whether the crystal is cubic, orthorhombic, or tetragonal.

Example 3–5.3

Calculate the distances in NaCl between the center of a sodium ion and the center of

(a) Its nearest neighbor.

(b) Its nearest positive ion.

(c) Its second nearest Cl^- ion.

(d) Its third nearest Cl^- ion.

(e) The nearest site that is identical.

Procedure Refer to Fig. 3–1.1. From Appendix B, the radii of Na^+ and Cl^- are 0.097 nm and 0.181 nm, respectively (since each has CN = 6).

Calculations

$a = 2(0.097 + 0.181 \text{ nm}) = 0.556 \text{ nm}$

(a) Distance $= a/2 = 0.278$ nm.

(b) Distance $= \sqrt{(a/2)^2 + (a/2)^2} = 0.393$ nm.

(c) Distance $= \sqrt{(a/2)^2 + (a/2)^2 + (a/2)^2} = 0.482$ nm.

(d) Distance $= \sqrt{(a/2)^2 + a^2} = 0.622$ nm.

(e) Same as part (b), since the nearest Na^+ ions are at identical positions. The translation is $\frac{a}{2}, \frac{a}{2}, 0$.

3-6
CRYSTAL DIRECTIONS

When we correlate various properties with crystal structures in subsequent chapters, it will be necessary to identify specific crystal directions, because many properties are directional. For example, the elastic modulus of bcc iron is greater parallel to the body diagonal than it is to the cube edge. Conversely, the magnetic permeability of iron is greatest in a direction parallel to the edge of the unit cell.

Indices of Directions

All parallel directions use the same label or index. Therefore, to label a direction, pick the parallel line that passes through the origin, which is $0,0,0$. The direction is labeled by the coefficient of a point on that line. However, since there is an infinite number of points on any line, we specifically choose the point with the lowest set of integers (Fig. 3–6.1). Thus, the [111] direction passes from $0,0,0$ through $1,1,1$. Note, however, that this direction also passes through $\frac{1}{2}, \frac{1}{2}, \frac{1}{2}$ (and $2,2,2$). Likewise, the [112] passes through $\frac{1}{2}, \frac{1}{2}, 1$; but, for simplicity's sake, we use the integer notation. Observe that we enclose the direction indices in square brackets $[uvw]$, and use the letters u, v, and w for the coefficients arising from the three principal directions — x, y, and z, respectively. Parallel directions always have the same indices. Finally, note that we may have negative coefficients, which we designate with an overbar; a $[11\bar{1}]$ direction will have a component in the minus-z direction. (We do *not* use commas between the indices.)

Angles Between Directions

In certain calculations (e.g., resolved shear stresses), it will be necessary to calculate the angle between two different crystal directions. For most of the calculations we shall encounter, we can do so by simple inspection. Thus, in Fig. 3–6.1, the angle between [110] and [112] directions (i.e., $[110] \not{\times} [112]$) is arctan $2c/\sqrt{a^2 + b^2}$. If that unit cell had been cubic rather than orthorhombic, such that $a = b = c$, the angle would have been arctan $2a/a\sqrt{2}$, or arccos $a\sqrt{2}/a\sqrt{6}$. In fact, with cubic

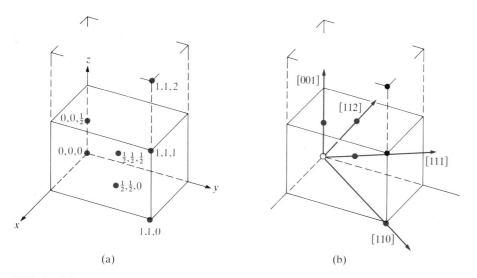

FIG. 3–6.1

Orthorhombic Unit Cell. (a) Indices of points. (b) Indices of directions. As stated in Section 3–5, the origin is commonly but not necessarily located at the lower-left, rear corner. Likewise, as discussed in this section, we use square brackets as closures to indicate crystal directions, $[uvw]$. We will be using parentheses (hkl) to enclose indices for crystal planes (Section 3–7). Commas are omitted within all closures (and closures are not used for the indices of points).

crystals (*only*), we can determine cos $[uvw]$ ∡ $[u'v'w']$ by the *dot product.* This latter procedure will be useful to us, since most of our calculations in this introductory text will involve these symmetric cubic crystals.

Repetition Distances

The repetition distance is a vector between *identical positions* within a crystal. It differs from direction to direction and from structure to structure. For example, in the [111] direction of a bcc metal, there is repetition every $2R$ (which is $a\sqrt{3}/2$). The repetition distance is $a\sqrt{2}$ in the [110] direction of bcc, but $a/\sqrt{2}$ in fcc. You can check these values in Figs. 3–2.1 and 3–2.3.

Conversely, the reciprocals of these repetition distances are the *linear densities.* Thus, in the [110] direction of aluminum, which is fcc with a equal to 0.405 nm, the repetition is one lattice point per $[a/\sqrt{2}]$, which inverts to $\sqrt{2}/(0.405 \times 10^{-6}$ mm$)$, which is equal to 3.5×10^{6}/mm for a linear density.

As we shall observe in Chapter 8, deformation occurs most readily in those directions with the shortest repetition distance (or with the greatest linear density of atoms).

Families of Directions

In the cubic crystal, the following directions are identical except for our arbitrary choice of the x, y, and z labels on the axes:

$$\begin{matrix} [111] & [11\bar{1}] & [1\bar{1}1] & [\bar{1}11] \\ {[\overline{111}]} & [\overline{11}1] & [\overline{1}1\bar{1}] & [1\overline{11}] \end{matrix} = \langle 111 \rangle \qquad (3-6.1)$$

Any directional property* will be identical for these four opposing pairs.† Therefore, it is convenient to identify this *family of directions* as $\langle 111 \rangle$ rather than writing the eight separate indices. Note that the closure symbols are angle brackets $\langle \ \rangle$; again, no commas are used.

Example 3-6.1

All parallel directions possess the same directional indices. Sketch three lines in the $[1\bar{2}0]$ direction that pass through locations (a) 0,0,0; (b) 0,1,0; and (c) $\frac{1}{2}$,1,1.

Sketch   See Fig. 3-6.2(a).

Example 3-6.2

The $\langle 101 \rangle$ family of directions includes what individual directions (a) in a cubic crystal, and (b) in a tetragonal crystal?

Answers

(a) In cubic, $a = b = c$. Therefore,

$[110]$, $[101]$, $[011]$, $[1\bar{1}0]$, $[10\bar{1}]$, $[01\bar{1}]$

(b) In tetragonal cells $a = b \neq c$ (Fig. 3-6.2b). Therefore, only the u and v indices of $\langle uvw \rangle$ are interchangeable; the w index is not:

$[101]$, $[011]$, $[\bar{1}01]$, $[0\bar{1}1]$

Comments   The $[\bar{1}01]$ and the $[10\bar{1}]$ directions are commonly considered to be the two senses of one direction, and not two separate directions. However, if desired, we could list the negative indices of each of the directions shown here, and double the number.

Example 3-6.3

(a) What is the angle between the $[111]$ and $[001]$ directions in a *cubic* crystal?

(b) $[111] \not\triangleleft [\overline{11}1]$?

---

* For example, the modulus of elasticity, magnetic permeability, index of refraction.

† See the comment with Example 3-6.2.

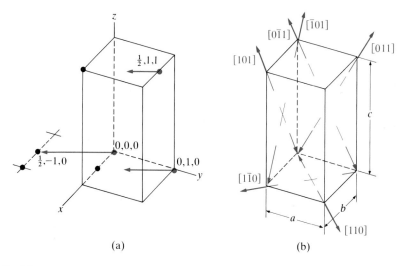

(a)   (b)

FIG. 3-6.2

Crystal Directions. (a) $[1\bar{2}0]$ (see Example 3-6.1). (b) $\langle 101 \rangle$ (see Example 3-6.2b).

Solution From the *cubic* analog of Fig. 3-6.1(b),

    (a) $\cos [111] \not< [001] = a/a\sqrt{3}$

        $[111] \not< [001] = 54.75°$

  (b) Observe that $[001]$ bisects $[111] \not< [\bar{1}\bar{1}1]$; therefore,

       $[111] \not< [\bar{1}\bar{1}1] = 2(54.75°) = 109.5°$

    or, by the dot product (*since the crystal is cubic*):

      $\cos [111] \not< [\bar{1}\bar{1}1] = -\frac{1}{3}$,

        $[111] \not< [\bar{1}\bar{1}1] = 109.5°$

Comment Compare the results here with those in Fig. 2-2.4(a), where all four of the bond angles are 109.5°.

Example 3-6.4

The lattice constant a is 0.357 nm for diamond, which is cubic. (a) What is the linear density of lattice points in the $[\bar{1}11]$ direction? (b) What is the linear density of atoms?

Observations Refer to Fig. 3-2.6(b). Locate the origin at the front, lower-left corner of the unit cell. From there, the $[\bar{1}11]$ direction passes through the rear, upper-right corner. En route, it passes through the center of a carbon atom at the quarter point of the body diagonal. Based on the relocated origin, the repeating distance is from 0,0,0 to $-1,1,1$, or $a\sqrt{3}$. A trip along this route produces *two* atoms for every $a\sqrt{3}$ step.

Calculations $a\sqrt{3} = 0.357\sqrt{3} = 0.618$ nm

(a) linear density $= \dfrac{1}{(0.618 \times 10^{-6}\text{ mm})} = 1.6 \times 10^{6}/\text{mm}$

(b) $\dfrac{\text{two atoms}}{(0.618 \times 10^{-6}\text{ mm})} = \dfrac{3.2 \times 10^{6}}{\text{mm}}$

Comment   Based on our relocated origin, the atom (identified by the leftmost asterisk in Fig. 3–5.2b) is at $-\frac{1}{4}, \frac{1}{4}, \frac{1}{4}$. You can check this location by relating the atom to its four neighbors at

$0,0,0;$ $-\frac{1}{2},0,\frac{1}{2};$ $-\frac{1}{2},\frac{1}{2},0;$ $0,\frac{1}{2},\frac{1}{2}$

3–7

CRYSTAL PLANES

A crystal contains planes of atoms; these planes influence the properties and behavior of a material. Thus, it will be advantageous to identify various planes within crystals.

The lattice planes most readily visualized are those that outline the unit cell, but there are many other planes. The more important planes for our purposes are those sketched in Figs. 3–7.1, 3–7.2, and 3–7.3. These are labeled (010), (110), and (1̄11), respectively, where the numbers within the parentheses (*hkl*) are called the *Miller indices.*

Miller Indices

We can use the plane with the darker tint in Fig. 3–7.4 to explain how (*hkl*) numbers are obtained. The plane intercepts the *x*, *y*, and *z* axes at **1***a*, **1***b*, and **0.5***c*.

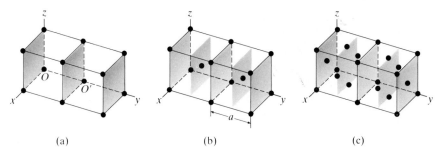

(a)   (b)   (c)

FIG. 3–7.1

(010) Planes in Cubic Structures. (a) Simple cubic (sc). (b) Body-centered cubic (bcc). (c) Face-centered cubic (fcc). (Note that the (020) planes included for bcc and fcc are comparable to (010) planes when they are extended beyond the sketch.)

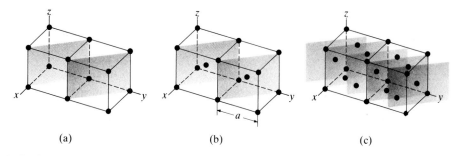

(a)   (b)   (c)

FIG. 3–7.2

(110) Planes in Cubic Structures. (a) Simple cubic (sc). (b) Body-centered cubic (bcc). (c) Face-centered cubic (fcc). (The (220) planes included for fcc are comparable to (110) planes.)

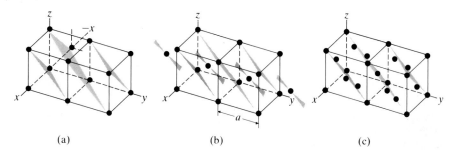

(a)   (b)   (c)

FIG. 3–7.3

($\bar{1}$11) Planes in Cubic Structures. (a) Simple cubic (sc). (b) Body-centered cubic (bcc). (c) Face-centered cubic (fcc). Negative intercepts are indicated by bars above the index. (The ($\bar{2}$22) planes included for bcc are comparable ($\bar{1}$11) planes.)

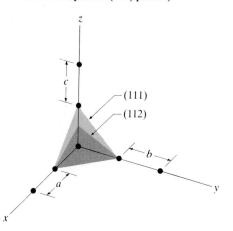

FIG. 3–7.4

Miller Indices. The (112) plane cuts the three axes at 1, 1, and $\frac{1}{2}$ unit distances.

The Miller indices are simply the reciprocals of these intercepts: (112). The plane of lighter tint in Fig. 3–7.4 is the (111) plane, since it intercepts the axes at $1a$, $1b$, and $1c$. Returning to the earlier figures, we have

FIGURE	PLANE	INTERCEPTS	MILLER INDICES
3–7.1(a)	middle	∞a, $1b$, ∞c	(010)
3–7.2(a)	left	$1a$, $1b$, ∞c	(110)
3–7.3(a)	middle	$-1a$, $1b$, $1c$	($\bar{1}11$)

Note that a *minus intercept* is handled readily with an overbar. Furthermore, observe that we use parentheses (*hkl*) to denote planes (and no commas), to avoid confusion with individual directions that were denoted in Section 3–6 with square brackets: [*uvw*]. All parallel planes are identified with the same indices.

We can now give a more rigorous definition of *Miller indices.* They are *the **reciprocals** of the three intercepts that the plane makes with the axes, cleared of fractions and of common multipliers.*

Planes containing an axis, or passing through the origin, would appear to present a problem because their intercepts are not uniquely definable. However, since all parallel planes possess the same indices, we can handle the problem readily by shifting the origin. Consider Fig. 3–7.1(a) where all three shaded planes possess the same geometric features and should be labeled identically. With the origin at O, the middle plane has (010) as its indices. The right plane is calculated to be $(0\frac{1}{2}0)$, but cleared of fractions, as required, it becomes (010). That is, $(0\frac{1}{2}0) \times 2 = (010)$. The left plane cannot be calculated with O as an origin. However, shift the origin to O'. The left plane is now $(0\bar{1}0)$. However, $(0\bar{1}0) \times (-1) = (010)$. This is permissible, because a multiplication simply produces a parallel plane. Multiplication by a negative number transposes the plane to the other side of the origin.

Families of Planes

Depending on the crystal system, two or more planes may belong to the same family of planes. In the cubic system, an example of multiple planes includes the following, which constitute a *family of planes,* or *form:*

$$\begin{array}{ccc} (100) & (010) & (001) \\ (\bar{1}00) & (0\bar{1}0) & (00\bar{1}) \end{array} = \{100\}* \qquad\qquad (3\text{--}7.1)$$

The collective notation for a family of planes is {*hkl*}. Figure 3–1.1 indicates for us the form of the {100} family that has the six planes listed. Each face is identical

---

* These six planes of the {100} family enclose space; hence the term *form.* However, it is permissible to state that the {100} family has only three members—(100), (010), and (001)—since pairs like (100) and ($\bar{1}$00) are parallel, but simply are on the opposite side of the origin. Likewise, the {111} family may be identified with either four or eight planes. (See Example 3–7.7.)

except for the consequences of our arbitrary choice of axis labels and directions. You should verify that the {111} family includes eight planes, and the {110} family includes 12 planes, when all of the permutations and combinations of individual planes are included.

Indices for Planes in Hexagonal Crystals (*hkil*)

The three Miller indices (*hkl*) can describe all possible planes through any crystal. In hexagonal systems, however, it is commonly useful to establish four axes, three of them coplanar (Fig. 3–3.1a). This leads to four intercepts and (*hkil*) indices.* The fourth index *i* is an additional index that mathematically is related to the sum of the first two:

$$h + k = -i \qquad (3-7.2)$$

These optional (*hkil*) indices are generally favored because they reveal hexagonal symmetry more clearly. Although partially redundant, they are used almost exclusively, in preference to the equivalent (*hkl*) indices, in technical papers.

Intersection of Planes

The two planes shown in Fig. 3–7.4 intersect along a line. Based on the previous section, an inspection of that figure tells us that the planes meet along the [1$\bar{1}$0] direction (or the [$\bar{1}$10], which has the opposite sense).

The line of intersection also has a directional index equal to the *cross-product* of the two sets of Miller indices:

$$(hkl) \times (h'k'l') = [uvw] \qquad (3-7.3)$$

Thus, $(111) \times (112) = [1\bar{1}0]$. This identity lets us quickly determine the lines of intersection of higher-order planes that do not readily lend themselves to inspection.

Directions within Planes

Every plane contains an infinite number of directions. (Of course, there also is an uncountable number of directions that do not lie within a given plane.) The coincidence of a direction and a plane is important in the analysis of plastic deformation (Section 8–3). In copper, for example, slip occurs most readily in the ⟨1$\bar{1}$0⟩ directions of the {111} planes.

Whereas we can visualize readily the [1$\bar{1}$0] direction on a (111) plane (Fig. 3–7.4), we can handle more complex orientations with *dot products*. The dot product is equal to zero when the direction lies in the plane:

$$\frac{hu + kv + lw}{\sqrt{u^2 + v^2 + w^2}\sqrt{h^2 + k^2 + l^2}} = 0 \qquad (3-7.4)$$

---

* Called *Miller–Bravais* indices.

Thus, a quick determination of $hu + kv + lw$ lets us know that $[33\bar{4}]$ lies in the $(1\bar{1}0)$ plane, and that $[33\bar{4}]$ does not lie in (110).

Example 3–7.1

Sketch a (111) plane through a unit cell of a simple tetragonal crystal having a c/a ratio of 0.62.

Solution Figure 3–7.5 shows this plane (shaded). The (111) plane cuts the three axes at unit distances. However, the unit distance along the z axis is shorter than is the unit distances on the x and y axes.

Example 3–7.2

Sketch two planes with (122) indices on the axes of Fig. 3–7.5.

Solution The intercepts will be the reciprocals of (122)—$1a$, $0.5b$, and $0.5c$—as shown by the inner set of dashed lines. A plane with intercepts—$2a$, $1b$, and $1c$—is parallel and therefore has the same indices (outer set of dashed lines).

Example 3–7.3

(a) How many atoms per mm² are there on the (100) planes of lead (fcc)? (b) On the (111) planes?

Procedure Sketch the planes in two-dimensions, showing the nuclei.

Solution Pb radius = 0.1750 nm (from Appendix B),

$$a_{Pb} = \frac{4R}{\sqrt{2}} = \frac{4(0.1750 \text{ nm})}{1.414} = 0.495 \text{ nm}$$

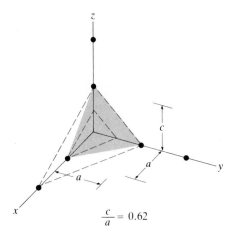

FIG. 3–7.5

Noncubic Intercepts (Tetragonal Structure). The shaded (111) plane cuts the three axes of any crystal at equal unit distances. However, since c may not equal a, the actual intercepting distances are different. (The dashed lines refer to Example 3–7.2.)

$$\frac{c}{a} = 0.62$$

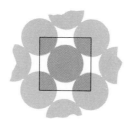

FIG. 3–7.6

Sketch for Example 3–7.3(a). A (100) plane in an fcc metal structure has two atoms per a^2.

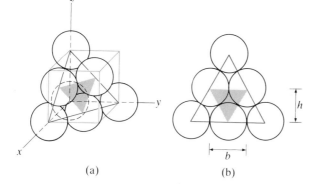

FIG. 3–7.7

Sketch for Example 3–7.3(b). (a) Exposed (111) plane. (b) Rotation for measurement.

(a)   (b)

(a) Figure 3–7.6 shows that the (100) plane contains two atoms per unit-cell face.

$$(100): \quad \frac{atoms}{mm^2} = \frac{2 \text{ atoms}}{(0.495 \times 10^{-6} \text{ mm})^2} = 8.2 \times 10^{12} \text{ atoms/mm}^2$$

(b) Figure 3–7.7(b) shows that the (111) plane contains three one-sixth atoms in the triangular shaded area. That area is $bh/2 = R^2\sqrt{3}$.

$$(111): \quad \frac{atoms}{nm^2} = \frac{\frac{3}{6}}{\frac{1}{2}bh} = \frac{\frac{3}{6}}{\frac{1}{2}(2)(0.1750 \text{ nm})(\sqrt{3})(0.1750 \text{ nm})}$$
$$= 9.4 \text{ atoms/nm}^2 = 9.4 \times 10^{12} \text{ atoms/mm}^2$$

Comment   Determine the atom nuclei per unit area. Since there is one nucleus per atom, you will have calculated the number of atoms per unit area.

Example 3–7.4

A plane includes points at 0,0,0 and $\frac{1}{2},\frac{1}{4},0$ and $\frac{1}{2},0,\frac{1}{2}$. What are its Miller indices?

Solution   Make a sketch (Fig. 3–7.8).

FIG. 3–7.8

The ($\bar{1}$21) Plane (see Example 3–7.4).

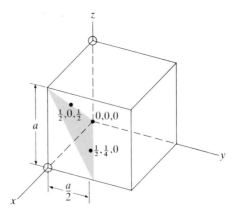

Since the plane passes through the origin, shift the origin, for example, 1 unit in the x direction. The intercepts are now $-1a$, **0.5**a, and $+1a$; therefore, ($\bar{1}$21).

Comment Had we shifted the origin 1 unit upward, the intercepts would have been $+1$, $-\frac{1}{2}$, -1 to give us ($12\bar{1}$). This plane is parallel to ($\bar{1}21$), and therefore equivalent.

Example 3–7.5

What direction is the line of intersection of the (111) and ($1\bar{1}1$) planes?

Answer Refer to Fig. 3–7.9: [$10\bar{1}$] (or [$\bar{1}01$]).

Comment For our purposes, the line of intersection is most readily obtained by inspection. However, it also can be obtained as a cross-product of the indices of the two planes.

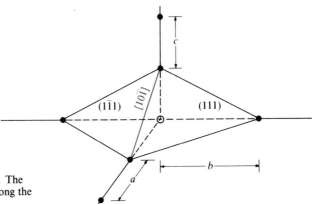

FIG. 3–7.9

Intersection of Planes (Example 3–7.4). The two planes, (111) and ($1\bar{1}1$), intersect along the [$10\bar{1}$] direction.

Example 3-7.6

Refer to Fig. 3-3.1(a). Provide an index for the plane facing the reader.

Answer

Axis:	a_1	a_2	a_3	c
Intercepts:	**1**	∞	**-1**	∞

Therefore, the (*hkil*) indices are ($10\bar{1}0$), since we use reciprocals.

Comment If we had used the alternate presentation of Fig. 3-3.1(b), the (*hkl*) indices would be (100). The additional index, *i*, is related to *h* and *k* by Eq. (3-7.2).

Example 3-7.7

List the individual planes that belong to the {111} family in cubic crystals.

Answer

$$(111) \quad (\bar{1}11) \quad (1\bar{1}1) \quad (11\bar{1})$$
$$(\bar{1}\bar{1}\bar{1}) \quad (1\bar{1}\bar{1}) \quad (\bar{1}1\bar{1}) \quad (\bar{1}\bar{1}1)$$

Comment In reality, there are only four planes (rather than eight) because those of the second set are parallel to those of the first set. However, unlike the negative directions (Example 3-6.2), we commonly list the redundant planes because the eight planes (in this case) enclose a volume and produce a crystal *form*. The form for {111} is a bipyramid.

3-8
X-RAY DIFFRACTION

Excellent experimental verification for the crystal structures that we have been discussing is available through x-ray diffraction. When these high-frequency electromagnetic waves are selected to have a wavelength slightly greater than the *interplanar* spacings of crystals, they are diffracted according to very exacting physical laws. The angles of diffraction let us decipher crystal structures with a high degree of accuracy. In turn, we can readily determine the interplanar spacings (and therefore atomic radii) in metals to four significant figures, and even more precisely, if necessary. Let us first examine the spacings between planes—then, we shall turn to diffraction.

Interplanar Spacings

Recall, from Section 3-7, that all parallel planes bear the same (*hkl*) notation. Thus, the several (110) planes of Fig. 3-7.2 have still another (110) plane that passes directly through the origin. As a result, if we measure a *perpendicular* distance from the origin to the next adjacent (110) plane, we have measured the interplanar distance, d. We will observe that, in the simple cubic structures of Figs. 3–7.1(a), 3–7.2(a), and 3–7.3(a), the interplanar distances are $a, (a\sqrt{2})/2$, and $(a\sqrt{3})/3$ for $d_{010}, d_{110}, d_{\bar{1}11}$, respectively. That is, there is one spacing per cell edge a for d_{010}; two spacings per face diagonal, $a\sqrt{2}$, for d_{110}; and three spacings per body diagonal, $a\sqrt{3}$, for $d_{\bar{1}11}$. We can formulate a general rule for d-spacings in *cubic* crystals:

$$d_{hkl} = \frac{a}{\sqrt{h^2 + k^2 + l^2}} \qquad (3-8.1)$$

where a is the lattice constant and h, k, and l are the indices of the planes.* The interplanar spacings for noncubic crystals may be expressed with equations that are related to Eq. (3–8.1), but take into account the variables of Table 3–1.1. We shall not consider diffraction of noncubic crystals.

Bragg's Law

When x-rays encounter a crystalline material, they are diffracted by the planes of the atoms (or ions) within the crystal. The *diffraction angle,* denoted by the Greek letter theta (θ), depends on the wavelength, denoted by the Greek letter lambda (λ), of the x-rays and the distance d between the planes:

$$n\lambda = 2d \sin \theta \qquad (3-8.2)$$

Consider the parallel planes of atoms in Fig. 3–8.1, from which the wave is diffracted. The waves may be "reflected" from the atom at H or H' and remain in phase at K. However, x-rays are reflected not only from the surface plane, but also from the adjacent subsurface planes. If these reflections are to remain in phase and be *coherent,* the distance $MH''P$ must be equal to one or more integer wavelengths of the rays. The value n of Eq. (3–8.2) is the integer number of waves that occur in the distance $MH''P$.

Diffraction Analyses

The most common procedure for making x-ray diffraction analyses uses very fine powder of the material in question. This powder is mixed with a plastic cement and is formed into a very thin filament that is placed at the center of a circular camera (Fig. 3–8.2). A collimated beam of x-rays is directed at the powder. Since

---

* The reciprocal nature of Miller indices permits this type of simplified calculation. Likewise, the cross-product of the Miller indices for two intersecting planes gives the direction indices for the line of intersection. Finally, in order for a direction to lie in a plane, the dot product of the indices for that direction and for the plane must be equal to zero. In brief, there is a purpose in using the "downside-up" Miller indices.

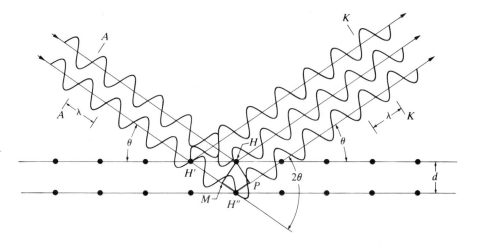

FIG. 3-8.1

X-Ray Diffraction. $MH''P$, which is $2d \sin \theta$, 2θ must equal the wavelength λ to keep the emerging rays in phase.

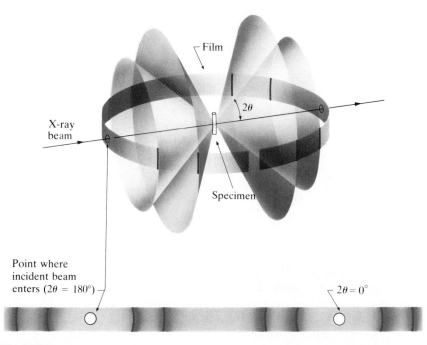

FIG. 3-8.2

The Exposure of X-Ray Diffraction Patterns. Angle 2θ is precisely fixed by the lattice spacing d and the wavelength λ as shown in Eq. (3-8.2). Every cone of reflection is recorded in two places on the strip of film. (B. D. Cullity, *Elements of X-ray Diffraction*, 1st ed., Addison-Wesley.)

there is a very large number of powder particles with essentially all possible orientations, the diffracted beam emerges as a cone of radiation at an angle 2θ from the initial beam. (Observe in Fig. 3–8.1 that the diffracted beam is 2θ away from the initial beam.)

The diffraction cone exposes the film strip in the camera at two places; each is 2θ from the straight-through exit port. There is a separate cone (or pair of *diffraction lines*) for each d_{hkl} value of interplanar spacings. Thus, the diffraction lines may be measured and the *d*-spacings calculated from Eq. (3–8.2). All fcc metals will have a similar set of diffraction lines, but with differing 2θ values, since they have different lattice constants; for example, $a_{Cu} = 0.3615$ nm; $a_{Al} = 0.4049$ nm; $a_{Pb} = 0.495$ nm, and so on. Thus, we can differentiate among various fcc metals.

In Fig. 3–8.3, we see x-ray diffraction films for copper (fcc), tungsten (bcc), and zinc (hcp). It is immediately apparent that the sequences of diffraction lines are different for the three types of crystals. Since we lack space here to explain these differences, we shall simply observe that, with different "fingerprints," it is possible not only to determine the size of the lattice constants with utmost precision, but also to identify the crystal lattice. X-ray diffraction is an extremely powerful tool in the study of the internal structure of materials.

The diffraction patterns of Fig. 3–8.3 were recorded on photographic film. Current practice is to automate the process by using a Geiger counter to detect the diffracted rays. The radiation counts are recorded on a strip chart (Fig. 3–8.4).

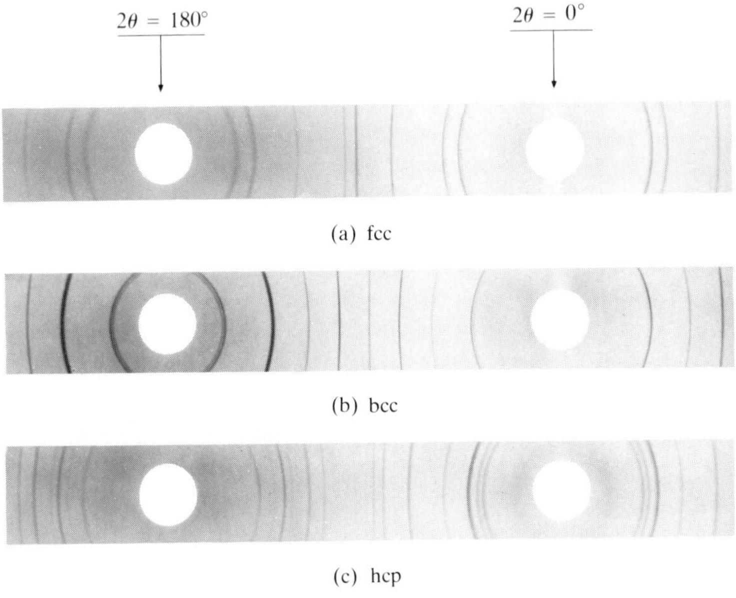

FIG. 3–8.3

X-Ray Diffraction Patterns. Patterns are shown for (a) copper, fcc, (b) tungsten, bcc, and (c) zinc, hcp. The crystal structure and the lattice constants may be calculated from patterns such as these. (B. D. Cullity, *Elements of X-ray Diffraction.* 2d ed., Addison-Wesley.)

$2\theta = 180°$ $2\theta = 0°$

(a) fcc

(b) bcc

(c) hcp

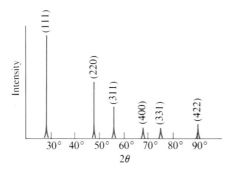

FIG. 3-8.4

Diffraction Pattern of Silicon (Geiger Chart). A Geiger counter measures the intensity of the diffracted x-rays as it moves through the 2θ circle. Each plane appears at a specific angle related to the interplanar spacing (Eq. 3-8.2).

This procedure has a second advantage of quantifying the intensity of each diffraction line by the height of the diffraction peak. This quantification adds significantly to the versatility of diffraction procedures for analyzing the composition and structures of engineering materials.

Example 3-8.1

X-rays of an unknown wavelength are diffracted 43.4° by copper (fcc), whose lattice constant a is 0.3615 nm. Separate determinations indicate that this diffraction line for copper is the first-order ($n = 1$) line for d_{111}.

(a) What is the wavelength of the x-rays?

(b) The same x-rays are used to analyze tungsten (bcc). What is the angle, 2θ, for the second-order ($n = 2$) diffraction lines of the d_{010} spacings?

Solution

(a) Since $2\theta = 43.4°$, using Eqs. (3-8.1) and (3-8.2),

$$(1)\lambda = 2\left[\frac{0.3615 \text{ nm}}{\sqrt{1^2 + 1^2 + 1^2}}\right] \sin 21.7°$$

$$= 0.1543 \text{ nm}$$

(b) From Appendix B,

$$R_w = 0.1367 \text{ nm}$$

From Eq. (3-2.1),

$$a_w = 4(0.1367 \text{ nm})/\sqrt{3} = 0.3157 \text{ nm}$$

From Eqs. (3-8.1) and (3-8.2),

$$\sin \theta = 2(0.154 \text{ nm})(\sqrt{0 + 1 + 0})/(2)(0.3157 \text{ nm}) = 0.488$$

$$2\theta = 58.4°$$

Comments The second-order diffraction of d_{010} is equivalent to the first-order diffraction of a d_{020} spacing—that is, to the perpendicular distance from the origin to a plane that cuts the x, y, and z axes at ∞, $\frac{1}{2}$, and ∞, respectively. You can check this equivalence using Eqs. (3–8.1) and (3–8.2).

The d_{hkl}-spacings for all planes of the same form are equal. For example, $d_{100} = d_{010} = d_{001}$ in the cubic system.

S U M M A R Y

In Chapter 2, we considered the coordination of atoms with their neighbors. In this chapter, we carried our understanding of internal structure one step further by looking at longer-range geometric order. Since all atoms of one kind have similar coordination requirements, it is not surprising that we find the same patterns repeated throughout the material.

1. Crystals are solids with a long-range periodicity. Crystals can be categorized into seven *crystal systems* based on the angles between the reference axes, and on the periodicity of the pattern. Structures that are body-centered and face-centered cubic will receive much, but not all, of our attention.

2. In *body-centered cubic* (bcc) structures, the pattern repeats itself by translations of $\pm\frac{a}{2}$, $\pm\frac{a}{2}$, $\pm\frac{a}{2}$. In *face-centered cubic* (fcc) structures, any two of these three translations produce replication. Packing factors and densities can be calculated from the atom radius in metal structures where a single type of atom is present. We examined several fcc structures; prime among these were the *fcc metal structure,* the *NaCl structure,* and the *diamond-cubic structure.*

3. Nearly 30 percent of the metallic elements form a *hexagonal close-packed* (hcp) structure. Like fcc metals, the packing factor for hcp metals is 0.74, and CN = 12.

4. Iodine and polyethylene form noncubic crystals because their molecules are not spherically symmetric. Graphite has a planar crystal structure that leads to anisotropic properties. Iron and carbon form ortho-rhombic Fe_3C, an important minor phase that plays a significant role in the properties of almost all steels.

5. *Points* within unit cells are indexed by their axial

coefficients. For example, $\frac{1}{2},\frac{1}{2},1$ is the center of the top face of a unit cell, regardless of the crystal system.

6. Many properties vary with *crystal direction.* Directions are indexed as a line passing through the origin and point, u, v, w. Such a direction is identified as $[uvw]$. The values of u, v, and w should be integers with lowest common multiples. Thus, a line from 0,0,0, to the center of the top face is labeled [112] since it extends to the point: 1,1,2. All parallel directions possess the same indices. Families of directions $\langle uvw \rangle$ include all directions that are identical except for our arbitrary choice of coordinate references.

7. *Crystal planes* are labeled by *Miller indices.* The latter are the *reciprocals* of the three intercepts that the plane makes with the three axes, cleared of fractions and of common multipliers. For example, a (120) plane cuts the x and y axes at a and $b/2$, respectively, but is parallel with the z axis (cuts it at ∞). Parallel planes carry the same indices.

We shall pay the greatest, but not exclusive, attention to the (100), (110), and (111) planes; therefore, you should be sure you know that

(i)   Any plane of the {100} family parallels two of the coordinate axes and cuts the third

(ii) Any plane of the {110} family parallels one axis and cuts the other two with equal intercept coefficients

(iii) The planes of the {111} family cut all three axes with equal intercept coefficients

8. Crystal structures are determined by *x-ray diffraction.* The *interplanar spacings* can be calculated to four and more significant figures by measuring the diffraction angles. These are the bases for determining interatomic distances and calculating atomic radii.

The *Study Guide* that accompanies this text contains a study set on "Crystal Directions and Planes." The step-by-step approach used there may be advantageous for some students. In addition, there is a study set on "Hexagonal Indices," which the instructor can use to extend the topic beyond the text.

KEY WORDS*

Anisotropic	Diffraction (x-ray)	Miller indices (*hkl*)
Atomic radius (*R*)	Face-centered cubic (fcc)	NaCl structure
Body-centered cubic (bcc)	Family of directions ⟨*uvw*⟩	Orthorhombic
Bragg's law	Family of planes {*hkl*}	Planar density
Crystal	Hexagonal close-packed (hcp)	Polymorphism
Crystal direction [*uvw*]	Interplanar spacing (*d$_{hkl}$*)	Repetition distance
Crystal plane (*hkl*)	Lattice	Simple cubic (sc)
Crystal system	Lattice constant	Tetragonal
Diamond-cubic structure	Linear density	Translation
Diffraction lines	Long-range order	Unit cell

* See *Terms and Concepts* at end of text for definitions.

PRACTICE PROBLEMS*

3–P11 The unit cell of iron is cubic, with $a = 0.287$ nm. From the density, calculate how many atoms there are per unit cell.

3–P12 There are two atoms per unit cell in titanium. What is the volume of the unit cell?

3–P13 A piece of aluminum foil is 0.08 mm thick and 670 mm² in area. (a) Its unit cells are cubic with $a = 0.4049$ nm. How many unit cells are there in the foil? (b) What is the mass of each unit cell if the density is 2.70 Mg/m³ ($=2.70$ g/cm³)?

3–P21 Calculate the atomic packing factor of vanadium (bcc, with $a = 0.3039$ nm). (Refer to your answer for Problem 247)

3–P22 The volume of the unit cell of chromium in Example 3–1.1 is 24×10^{-30} m³, or 0.024 nm³. Based on the data used in that example problem, plus PF$_{bcc\,metal} = 0.68$, calculate the radius of the chromium atom to verify the value shown in Appendix B.

3–P23 From Fig. 3–2.4, show that CN = 12 for an fcc metal.

3–P24 (a) How many atoms are there per mm³ in solid barium? (b) What is the atomic packing factor? (c) It is cubic. What is its structure? (Atomic number = 56; atomic mass = 137.3 amu; atomic radius = 0.22 nm; ionic radius = 0.143 nm; density = 3.5 Mg/m³.)

3–P25 CsCl has the sc structure of Cl⁻ ions with Cs⁺ ions in the 8-f sites. (a) The radii are 0.187 nm and 0.172 nm, respectively, for CN = 8; what is the packing factor? (b) What would this factor be if r/R were 0.73?

3–P26 The intermetallic compound FeTi has the CsCl-type structure with $a = 0.308$ nm. Calculate its density.

3–P27 X-ray data show that the unit-cell dimensions of cubic MgO are 0.412 nm. This material has a den-

* These *Practice Problems* (1) use equations directly (plug and chug), (2) refer to specific statements in the text, or (3) "mimic" the examples at the end of the sections. You should try them for practice before attempting the *Test Problems* that follow. All Practice Problems have answers on the pages at the end of the text. If you need them, the detailed solutions are given in the *Study Guide to Elements of Materials Science and Engineering*. The *Study Guide* also contains Quiz Samples.

sity of 3.83 Mg/m³. How many Mg^{2+} ions and O^{2-} ions are there per unit cell?

3–P28 Bunsenite (NiO) has an fcc structure of O^{2-} ions with Ni^{2+} ions in all the 6-f sites. (a) The radii are 0.140 nm and 0.069 nm, respectively; what is the packing factor? (b) What would this factor be if r/R were 0.41?

3–P29 Lithium fluoride, LiF, has a density of 2.6 Mg/m³ and the NaCl structure. Use these data to calculate the unit-cell size, a, and compare it with the value you get from the ionic radii.

3–P31 From the data for beryllium (hcp) in Appendix B, calculate the volume of the unit cell (Fig. 3–3.2).

3–P32 What is the ratio of volumes for the two presentations of the hexagonal unit cell in Fig. 3–3.1?

3–P41 The lattice constant, a, for diamond (Fig. 3–2.6) is 0.357 nm. What percent volume change occurs when it transforms to graphite ($\rho = 2.25$ Mg/m³, or 2.25 g/cm³)?

3–P42 The volume of a unit cell of bcc iron is 0.02464 nm³ at 912° C. The volume of a unit cell of fcc iron is 0.0486 nm³ at the same temperature. What is the percent change in density as the iron transforms from bcc to fcc?

3–P43 MnS has three polymorphs. One of these is (a) the NaCl-type structure (Figs. 3–1.1 and 3–2.5). A second is the ZnS-type, which is like diamond (Fig. 3–2.6), *except* that cations and anions alternate positions (see Problem 329). What percent volume change occurs when the second type (ZnS) changes to the first type (NaCl)? (See Appendix B for radii where CN = 6, and the footnote explaining what to do when CN ≠ 6.)

3–P44 Diamond requires higher pressure to be formed than does graphite (e.g., diamond may be formed very deep in the earth, where pressure is high). What does this suggest about the relative densities of diamond and graphite?

3–P45 From Fig. 3–2.3, show that fcc iron could be categorized as body-centered tetragonal (bct) with a c/a ratio of 1.414. Further, the unit cell would be smaller, with only two atoms instead of four. Why do we use fcc rather than bct?

3–P51 Refer to Fig. 3–2.1. We can identify the structure of a bcc metal by placing atoms at only two locations. What are they?

3–P52 Refer to Fig. 3–2.3. We can identify the structure of an fcc metal by placing atoms at only four locations. What are they?

3–P53 Copper is fcc and has a lattice constant of 0.3615 nm. What is the distance between the 0,1,0 and $\frac{1}{2},0,\frac{1}{2}$ locations?

3–P54 MnS has the same structure as NaCl (Fig. 3–1.1). Its lattice constant is ~0.53 nm. What is the center-to-center distance from the Mn^{2+} ion at $0,0,\frac{1}{2}$ and (a) its nearest Mn^{2+} neighbor? (b) its second-nearest Mn^{2+} neighbor?

3–P55 When a copper atom is located at the origin of an fcc unit cell, a small interstitial hole is centered at $\frac{3}{4},\frac{1}{4},\frac{1}{4}$. Where are there other equivalent holes within the same unit cell?

3–P56 Repeat Problem 3–P55, but start with $\frac{1}{4},\frac{3}{4},\frac{3}{4}$ rather than $\frac{3}{4},\frac{1}{4},\frac{1}{4}$.

3–P57 Compare the size and "shape" of the holes that are centered at (a) $\frac{1}{2},0,0$ and (b) $\frac{1}{4},\frac{1}{4},\frac{1}{4}$ of an fcc metal.

3–P61 (a) A line in the [221] direction passes through the origin. Where does it leave the reference unit cell? (b) Another line in a parallel [221] direction leaves the reference unit cell at 1,1,1. Where did it enter the reference unit cell?

3–P62 A line in the [111] direction passes through location $\frac{1}{2},0,\frac{1}{2}$. What are two other locations along its path?

3–P63 (a) In a cubic crystal, what is the tangent of the angle between the [100] direction and the [211] direction? (b) What is it between the [011] direction and the [111] direction?

3–P64 (a) In a cubic crystal, what is cos [113] ⋞ [110]? (b) What is sin [010] ⋞ [122]?

3–P65 (a) What is the center-to-center spacing of atoms in the ⟨110⟩ directions of copper (fcc, a = 0.361 nm)? (b) What is it in the ⟨110⟩ directions of iron (bcc, a = 0.285 nm)?

3–P66 (a) What is the center-to-center spacing of atoms in the [121] direction of copper? (b) What is it in the [121] direction of iron? (See Problem 3–P65 for lattice constants.)

3–P67 What are the several directions in the $\langle 012 \rangle$ family of a cubic crystal? (We usually consider $[uvw]$ and $[\overline{uvw}]$ to be of the same direction, but with an opposite sense; that is, $[\overline{uvw}] = -[uvw]$.)

3–P70 A plane intercepts the crystal axes at $a = 0.5$ and $b = 0.75$. It is parallel to the z axis. What are the Miller indices?

3–P71 A plane intercepts the crystal axes at $a = 1$, $b = 2$ and $c = 1$. What are the Miller indices?

3–P72 What are the indices for a plane with intercepts at $a = 1$, $b = -\frac{3}{2}$, and $c = \frac{2}{3}$?

3–P73 (a) How many atoms are there per mm² in the (100) plane of copper? (b) How many are there in the (110) plane? (c) How many are there in the (111) plane?

3–P74 (a) What is the line of intersection between the (1$\overline{1}$0) and (1$\overline{1}$2) planes of a cubic crystal? (b) What is it between these planes of a tetragonal crystal?

3–P75 (a) What is the line of intersection between the

(112) plane and the (100) plane? (b) What is it between the (112) plane and the (1$\overline{1}$0) plane?

3–P76 What are the $(hkil)$ indices of a plane in hexagonal crystal that intercepts axes at $a_1 = 1$, $a_2 = 1$, and $c = 0.5$?

3–P77 (a) List the planes that belong to the {100} family in tetragonal crystals. (b) List those that belong to the {001} family. (c) Identify those that are positive and negative pairs.

3–P78 (a) What $\langle 110 \rangle$ directions lie in the (11$\overline{1}$) plane of copper? (b) What $\langle 110 \rangle$ directions lie in the (1$\overline{1}$1) plane?

3–P79 (a) What $\langle 111 \rangle$ directions lie in the (101) plane of iron? (b) What $\langle 111 \rangle$ directions lie in the (1$\overline{1}$0) plane?

3–P81 The lattice constant for a unit cell of lead is 0.4950 nm. (a) What is d_{220}? (b) What is d_{111}? (c) What is d_{200}?

3–P82 A sodium chloride crystal is used to measure the wavelength λ of some x-rays. The diffraction angle 2θ is $27°30'$ for the d_{111} spacing of the chloride ions. (a) What is the wavelength? (The lattice constant is 0.563 nm.) (b) What would have been the value of 2θ if λ has been 0.058 nm?

TEST PROBLEMS *

311 There are six atoms per hexagonal unit cell of zinc. What is the volume of the unit cell of zinc?

312 Tin is tetragonal with $c/a = 0.546$. There are four atoms per unit cell. (a) What is the unit-cell volume? (b) What are the lattice dimensions?

313 (a) The corner site of a cubic unit cell is associated with how many unit cells? (b) The corner site of a rectangular area of a plane belongs to how many "unit areas"?

314 Among three-dimensional crystals (Fig. 3–1.4), we find bcc and bct (i.e., body-centered tetragonal). There is an fcc, but no fct. Explain why.

321 Calcium is fcc with a density of 1.55 Mg/m³ (= 1.55 g/cm³). (a) What is the volume per unit cell based on the density? (b) From your answer in part (a), calculate the radius of the calcium atom. Check this value with the radius listed in Appendix B.

---

* The preceding set of *Practice Problems* gave you a chance to learn how you can plug equations or establish procedures. The *Test Problems* require you to do more. As with all engineering calculations, it is necessary for you (1) to establish a procedure, (2) to identify and obtain required information, and then (3) to proceed toward a solution on the basis of accumulated knowledge.

In general, data will not be cited in the problems if they are contained in related tables or in the appendices.

322 Titanium is bcc at high temperatures and its atomic radius is 0.145 nm. (a) How large is the edge of the unit cell? (b) Calculate the density. Compare your answer with the data in the Appendix B. Explain any difference.

323 Strontium is metallic fcc and its atomic radius is 0.215 nm. (a) What is the volume of its unit cell? (b) How large is the "hole" in the center of the unit cell? (c) What other locations have similar-sized holes?

324 Niobium is bcc with a density of 8.57 Mg/m³ (= 8.57 g/cm³). (a) Calculate the center-to-center distance between closest atoms. (b) What is the edge dimension of the unit cell? (Its atomic number is 41 and its atomic weight is 92.91 amu.)

325 Refer to Fig. 3–1.1. KCl has the same structure. (a) What is the distance between the centers of the *closest* Cl^- ions? (b) What is the distance from the center of the K^+ ion to the center of the *second-closest* Cl^- ion? (c) What is the distance between the "surfaces" of the closest Cl^- ions? (d) What is this distance between the "surfaces" of the closest K^+ ions?

326 (a) How many atoms are there per mm³ in solid molybdenum? (b) What is the atomic packing factor? (c) This material is cubic. What is its structure? (Atomic number = 42; atomic mass = 95.94 amu; atomic radius = 0.137 nm; ionic radius = 0.068 nm; density = 10.1 Mg/m³.)

327 CuZn (called β-brass) has the same structure as CsCl (Fig. 3–2.8). However, it is a metallic compound rather than an ionic compound. Calculate its density.

328 Nickel oxide (NiO) has the same structure as NaCl (Fig. 3–1.1). (a) What is its lattice constant, *a*? (b) What is its density? (*Note:* NiO is an ionic compound; therefore, you must use the ionic radii rather than the atomic radii.)

329 ZnS has a density of 4.1 Mg/m³. It has the structure of diamond (Fig. 3–2.6), *except* Zn^{2+} and S^{2-} alternate positions. Based on this information, what is the spacing between the centers of these two ions?

331 Zinc has an hcp structure. The height of the unit cell is 0.494 nm. The centers of the atoms in the base of the unit cell are 0.2665 nm apart. (a) How many atoms are there per hexagonal unit cell?

Explain your reasoning. (b) What is the volume of the hexagonal unit cell? (c) Is the calculated density greater or less than the actual density of 7.135 Mg/m³?

332 The atomic mass of cadmium is 112.4 amu, and the average atomic radius of this hcp metal is 0.156 nm. (a) What is the volume associated with each atom (sphere + interstitial space)? (b) From your answer in part (a), calculate the density.

333 Magnesium is hcp with nearly spherical atoms having a radius of ~0.161 nm. Its *c/a* ratio is 1.62. (a) Calculate the volume of the unit cell from these data. (b) Check your answer by using it to calculate the density of magnesium.

341 Zirconium is bcc in its high-temperature form. The radius increases 1.5 percent when the bcc changes to hcp during cooling. What is the percentage volume change? (Recall that there will be a change in atomic packing factor.)

342 Metallic tin has a tetragonal structure with a = 0.5820 nm and c = 0.3175 nm, and with four atoms per unit cell. Another form of tin (gray) has the cubic structure of diamond (Fig. 2–2.3b) with a = 0.649 nm. What is the volume change as tin transforms from gray to metallic?

343 Estimate the densities of the two polymorphs of MnS in Problem 3–P43.

344 The density of 1.5 g/cm³ (or 1.5 Mg/m³) is given for NH_4Cl in the *Chemical Handbook.* X-ray files state that there are two polymorphs for NH_4Cl; one has an NaCl-type structure with a = 0.726 nm, the other has a CsCl-type structure with a = 0.387 nm. The density value is for which polymorph? (The NH_4^+ ion occupies the crystal lattice as a unit.)

345 It is assumed that, at high pressures, NaCl can be forced into a CsCl-type structure. What would be the percent volume change? (See the footnote of Appendix B for differences in radii with CN = 6 and CN = 8.)

346 Cobalt changes during heating from hcp to fcc at 417° C. How will its volume be altered when (a) compared with iron, which transforms on heating from bcc to fcc (Example 3–4.1), and (b) compared with zirconium, which has two polymorphs (see Problem 341)?

351 Each atom in cesium has eight nearest neighbors. (a) How many second-nearest neighbors are there? (b) If a_{bcc} for cesium is 0.612 nm, what are the center-to-center distances to the second-nearest neighbors?

352 Compare the sizes and "shapes" of the holes that are centered at $\frac{1}{2},\frac{1}{4},0$ and at $\frac{1}{2},\frac{1}{2},0$ of a bcc metal.

353 Refer to Problem 3–P55 and to Fig. 3–2.3. Observe that locations $\frac{1}{4},\frac{1}{4},\frac{1}{4}$ and $\frac{3}{4},\frac{3}{4},\frac{3}{4}$ both lie among four immediate neighbors. Are they equivalent locations? In other words, why was $\frac{1}{4},\frac{1}{4},\frac{1}{4}$ not among the answers for Problem 3–P55, whereas it was in Problem 3–P56?

354 White tin, which is the normal metallic tin, is body-centered tetragonal, with atoms at

$$0,0,0 \qquad \frac{1}{2},\frac{1}{2},\frac{1}{2} \qquad \frac{1}{2},0,\frac{1}{4} \qquad 0,\frac{1}{2},\frac{3}{4}$$

Sketch a unit cell. (a) How many atoms are there per unit cell? (b) What is the shortest interatomic distance? $a = 0.5820$ nm; $c = 0.3175$ nm.

361 Draw a line from $\frac{1}{2},\frac{1}{2},0$ to the center of the *next* unit cell, which could be indexed $\frac{3}{2},\frac{1}{2},\frac{1}{2}$. What is the direction?

362 A line is drawn through a tetragonal unit cell from location $0,\frac{1}{2},0$ to $\frac{3}{4},\frac{1}{8},\frac{3}{4}$. What is the index of that direction?

363 What is the angle between [111] and [1$\bar{1}$1] in a cubic crystal? (*Hint:* [101] bisects this angle.)

364 (a) What is the angle between [101] and [$\bar{1}$01] in a cubic crystal? (b) What is it in a tetragonal crystal where $c/a = 0.55$?

365 Draw a line from a point at $\frac{1}{4},\frac{3}{4},\frac{1}{4}$ to the center of the next unit cell in the rear ($-x$ direction). (a) What is the [uvw] direction between these two points? (b) The unit cell is tetragonal with $a = 0.31$ nm and $c = 0.33$ nm. What is the distance between the two points in part (a)?

366 Draw a line from the $\frac{1}{2},0,\frac{1}{2}$ location of aluminum through the center of the base of next unit cell (x direction). (a) What is the direction? (b) What is the linear density of atoms in that direction? (c) On a sketch, show a parallel direction that passes through the 0,0,1 location. Indicate two other unit-cell locations through which this parallel line passes.

367 (a) How many directions are there in the $\langle 100 \rangle$ family of a cubic crystal? (b) How many are there in the $\langle 110 \rangle$ family? (c) How many are there in the $\langle 111 \rangle$ family? (d) Repeat the $\langle 100 \rangle$ family for a tetragonal crystal.

368 (a) What is the shortest center-to-center distance between atoms in the [211] direction of copper? (b) What is it in the [112] direction? (c) Cite four other directions in the $\langle 121 \rangle$ family of a cubic crystal.

369 Three parallel [210] lines pass through the following three locations: $\frac{1}{2},0,0$; 1,1,0; and $\frac{1}{2},\frac{1}{2},\frac{1}{2}$. (a) Where does each enter and leave the unit cell? (b) A [211] line through 1,1,0 also passes through —, —, —, and —, —, —.

371 What are the axial intercepts for a (111) plane that passes through the center of the unit cell, at $\frac{1}{2},\frac{1}{2},\frac{1}{2}$?

372 A (221) plane contains points at 0,0,0 and $-1,1,0$. What are other crystal locations (of many points) that lie in this plane?

373 A plane includes points at 0,0,0 and 1,0,1 and $1,\frac{1}{2},\frac{1}{2}$. What are its Miller indices?

374 (a) How many atoms are there per mm² on the (111) plane of tungsten? (b) How many are there on the (210) plane of silver?

375 (a) Sketch a unit cell of aluminum (fcc) and shade the (111) plane. (b) What is the planar density of atoms on the (111) plane? (c) There are 6×10^{22} Al atoms per cm³ (see Problem 212b). How far apart are the adjacent (111) planes?

376 (a) Sketch a unit cell of chromium (bcc) and locate the atoms on the (110) plane. (b) What is the planar density of atoms? (c) There are 8.3×10^{22} Cr atoms per cm³ (see Problem 212c). How far apart are the adjacent (110) planes?

377 (a) How many atoms are there per mm² on the (110) plane of diamond? (b) How many are there on the (111) plane? (See Fig. 3–2.6; $a = 0.357$ nm.)

378 The [111] direction is perpendicular to the (111) plane in a *cubic* crystal, but not in a tetragonal crystal. Show why.

379 Which of the atoms sketched in Fig. 3–3.2(a) lie in the (10$\bar{1}$1) plane that intercepts the vertical axis at $c = 1$? (Be very careful. It is easy to mislead yourself in answering this question.)

381 Use x-rays with $\lambda = 0.058$ nm to calculate d_{200} for aluminum. The diffraction angle, 2θ, is $16.47°$. What is the lattice constant? ($n = 1$)

382 Which will have the largest interplanar spacing in chromium (bcc with $R = 0.1249$ nm), d_{200}, d_{110}, or d_{111}?

383 The first line (lowest 2θ) of Fig. 3–8.3(b) is for the d_{110}-spacing in tungsten. (a) Determine the diffraction angle, 2θ, graphically. (b) For tungsten, $R = 0.1367$ nm. What wavelength was used? (c) The second diffraction line is for which d_{hkl} spacing? ($n = 1$.)

384 The first line (lowest 2θ) of Fig. 3–8.3(a) is for the d_{111}-spacing of copper. (a) What is its 2θ value? (b) The radius of the copper atom is 0.1278 nm. Calculate the wavelength of the x-rays that were used. (c) The second diffraction line is for which d_{hkl} spacing?

Chapter 4

DISORDER IN SOLID PHASES

Nature is not perfect; inherently, some disorder is present. Thus, the structures just described in Chapter 3 are subject to exceptions. Often these exceptions are minor—maybe one atom out of 10^{10} is out of place. Even so, they can become important. Other imperfections are broadly distributed through the bulk of the material. Disorder accounts for the behavior of semiconductors, for the ductility of metals, for the color of sapphire; it is the rule within polymers and glasses. Disorder also permits the movement of atoms during heat treating, so that new structures and enhanced properties may be realized.

Structural imperfections may occur at a point, be linear, occur as boundaries, or be bulk in three dimensions. Compositional imperfections may involve substitutional atoms, or atoms located interstitially.

4–1

IMPERFECTIONS IN
CRYSTALLINE SOLIDS

In the previous chapter, we focused on the long-range order that is found in crystals. We saw that the unit cell is a module that is repeated identically. Not only does this structure simplify our analyses of materials and their properties, but also it gives us an assurance of the constancy of nature.

As we said, however, the world is not perfect. Crystals sometimes contain atoms that are in the "wrong" place — or there may be missing atoms, or foreign atoms. There may be bulk displacements if part of the crystal is sheared with respect to the rest of the crystal. Also, two adjacent crystals within a solid will have an atomic mismatch along their interface. Then there are the solids, called glasses, which lack crystallinity. Furthermore, materials are not pure when we examine them with sufficiently sensitive analytical tools. Finally, we intentionally combine materials to produce alloys with improved properties.

In terms of numbers, the imperfections just cited may involve a very small fraction of the atoms. For example, one boron atom among 100 billion unit cells of silicon doubles the conductivity of otherwise pure silicon (Chapter 11). Also, the strength of steel (such as in a car fender) is less than 1 percent of that calculated from bonding forces, because of the dislocation lines that develop in the crystals during processing. These dislocations involve only a minute fraction of all the atoms that are present.

Although we would like to think of the crystal world as being perfect and obedient, we cannot ignore the disorder that is present. In this chapter, we focus first on geometric imperfections, then on the introduction of foreign atoms (impurities). In many situations, we can use these irregularities advantageously to develop useful and desired properties.

When imperfections such as atom *vacancies* involve the absence of one or a few atoms, we call them *point defects*. Other imperfections may be linear through the crystal; hence, we use the term *line defects*. Imperfections may also be two dimensional, involving either external *surfaces,* or internal *boundaries.* Finally, bulk three-dimensional disorder is present in some materials, as we shall discuss in Section 4–2.

Point Defects

The simplest point defect is a *vacancy,* □, which involves a missing atom (Fig. 4–1.1) within a crystal. Such defects can be a result of imperfect packing during the original crystallization, or they may arise from thermal vibrations of the atoms at elevated temperatures (Section 6–4), because as thermal energy is increased there is an increased probability that individual atoms will jump out of their

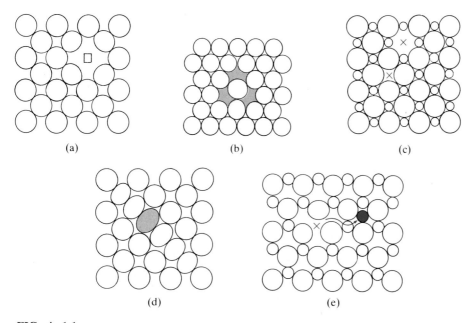

(a)   (b)   (c)

(d)   (e)

FIG. 4-1.1

Point Defects. (a) Vacancy, □. (b) Di-vacancy (two missing atoms). (c) Ion-pair vacancy (Schottky defect). (d) Interstitialcy. (e) Displaced ion (Frenkel defect).

positions of lowest energy. Vacancies may be single, as shown in Fig. 4-1.1(a), or two or more of them may condense into a di-vacancy (Fig. 4-1.1b) or a tri-vacancy.

Ion-pair vacancies (called Schottky imperfections) are found in compounds that must maintain a charge balance (Fig. 4-1.1c). They involve vacancies of pairs of ions of opposite charges. Ion-pair vacancies, like single vacancies, facilitate atomic diffusion (Section 6-5).

An extra atom may be lodged within a crystal structure, particularly if the atomic packing factor is low. Such an imperfection, called an *interstitialcy,* produces atomic distortion (Fig. 4-1.1d).

A *displaced ion* from the lattice into an interstitial site (Fig. 4-1.1e) is called a *Frenkel defect.* Close-packed structures have fewer interstitialcies and displaced ions than vacancies, because additional energy is required to force the atoms into the interstitial positions.

Line Defects (Dislocations)

The most common type of line defect within a crystal is a dislocation. An *edge dislocation* (⊥) is shown in Fig. 4-1.2. It may be described as an edge of an extra plane of atoms within a crystal structure. Zones of compression and of tension

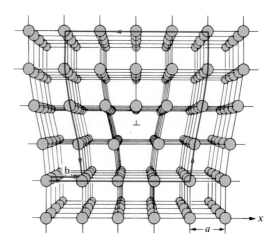

FIG. 4–1.2

Edge Dislocation, ⊥. A linear defect occurs at the edge of an extra plane of atoms. The slip vector **b** is the resulting displacement. (Guy and Hren, *Elements of Physical Metallurgy,* Addison-Wesley.)

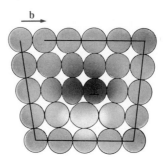

FIG. 4–1.3

Dislocation Energy. Atoms are under compression (darker) and tension (light) adjacent to the dislocation.

accompany an edge dislocation (Fig. 4–1.3), so there is a net increase in energy along the dislocation line. The displacement distance for atoms around the dislocation is called the *slip vector,* **b**.* This vector is at right angles to the edge dislocation line.

A *screw dislocation* (§) is like a spiral ramp with an imperfection line down its axis (Fig. 4–1.4). Its slip vector is parallel to the defect line. Shear stresses are associated with the atoms adjacent to the screw dislocation; therefore, extra energy is involved here, as it is in the previously cited edge dislocations.

---

* Also called a *Burgers vector.*

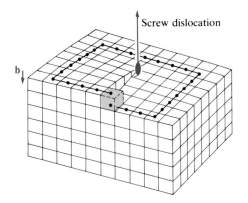

FIG. 4–1.4

Screw Dislocation (Unit Cells Shown). The slip vector **b** is parallel to the linear defect.

Dislocations of both types may originate during crystallization. Edge dislocations, for example, arise when there is a slight mismatch in the orientation of adjacent parts of the growing crystal, such that an extra row of atoms is introduced or eliminated. As shown in Fig. 4–1.4, a screw dislocation allows easy crystal growth, because additional atoms and unit cells can be added to the "step" of the screw. Thus, the term *screw* is apt, because the step swings around the axis as growth proceeds.

Dislocations more commonly originate during deformation. We see this in Fig. 4–1.5, where shear is shown to introduce both edge dislocations and screw dislocations. Both lead to the same final displacement and are in fact related through the dislocation line that connects them.

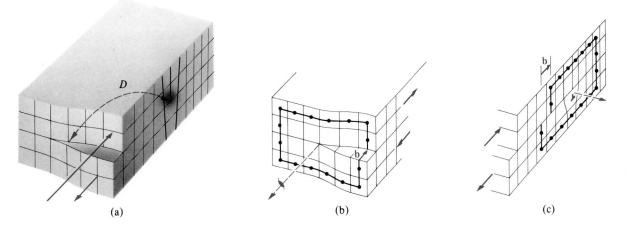

(a)   (b)   (c)

FIG. 4–1.5

Dislocation Formation by Shear. (a) The *dislocation line, D,* expands through the crystal until displacement is complete. (b) This defect forms a screw dislocation where the line is parallel to the shear direction. (c) The linear defect is an edge dislocation where the line is perpendicular to the shear direction.

Surfaces

Crystalline imperfections may extend in two dimensions as a boundary. The most obvious boundary is the external *surface.* Although we can visualize a surface as simply a terminus of the crystal structure, we should quickly appreciate that atomic coordination at the surface is not fully comparable to the atoms within a crystal. The surface atoms have neighbors on only one side (Fig. 4–1.6); therefore they have higher energy and are less firmly bonded than the internal atoms. We can rationalize this energy by looking at Fig. 2–5.2 and by noting that, if additional atoms were to be deposited onto the surface atoms, energy would be released just as it was for the combination of two individual atoms. We find our best visible evidence of this *surface energy* in the case of liquid drops that have spherical shape to minimize the surface area (and therefore the surface energy) per unit volume. Surface adsorption provides additional evidence of the energy differential at the surface.

Grains and Grain Boundaries

Although a material such as copper in an electric wire may contain only one phase—that is, only one structure (fcc)—it contains many crystals of various orientations. These individual crystals are called *grains.* The shape of a grain in a solid usually is controlled by the presence of surrounding grains. Within any particular grain, all of the unit cells are arranged with one orientation and one pattern. However, at the *grain boundary* between two adjacent grains, there is a transition zone that is not aligned with either grain (Fig. 4–1.7).

Although we cannot see the individual atoms illustrated in Fig. 4–1.7, we can quite readily locate grain boundaries in a material under a microscope, if the material has been treated by *etching.* First the material is smoothly polished so that a plane, mirrorlike surface is obtained, and then it is chemically attacked for a short time. The atoms along the region of mismatch between one grain and the next will dissolve more readily than will other atoms, and they will leave a line that can be seen with the microscope (Fig. 4–1.8); the etched grain boundary does not act as a perfect mirror, as does the remainder of the grain (Fig. 4–1.9).

We may consider the grain boundary to be two-dimensional, although it may be curved, and actually it has a finite thickness of 1 or 2 atomic distances. The

Surface

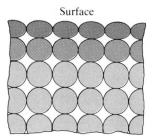

FIG. 4–1.6

Surface Atoms (Schematic). These atoms are not entirely surrounded by others, so they possess more energy than do the internal atoms.

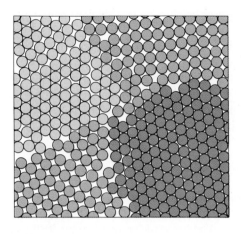

FIG. 4-1.7

Grain Boundaries. Note the disorder at the boundary. (Reproduced by permission from Clyde W. Mason, *Introduction to Physical Metallurgy,* American Society for Metals, Chapter 3.)

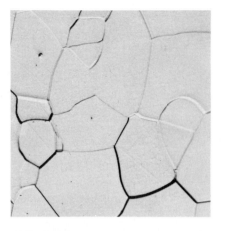

FIG. 4-1.8

Grains and Grain Boundaries (Iron, $\times$500). Each grain is a single crystal. The boundaries between grains are surfaces of mismatch. (See Fig. 4-1.7.) (Courtesy U.S. Steel Corp.)

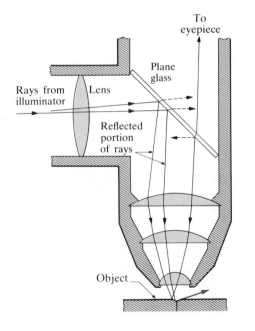

FIG. 4-1.9

Grain-Boundary Observation. The metal has been polished and etched. The corroded boundary does not reflect light through the microscope; thus it appears as a dark line in photomicrographs. (Reproduced by permission from B. Rogers, *The Nature of Metals,* 2d ed., American Society for Metals, and Iowa State University Press, Chapter 2.)

mismatch of the orientation of adjacent grains produces a less efficient packing of the atoms along the boundary. Thus, the atoms along the boundary have a higher *energy* than do those within the grains. This disparity accounts for the more rapid etching along the boundaries. The higher energy of the boundary atom also is important for the nucleation of polymorphic phase changes (Section 3–4). The lower atomic packing along the boundary favors atomic diffusion (Section 6–5), and the mismatch between adjacent grains modifies the progression of dislocation movements (Fig. 4–1.5). Thus, the grain boundary modifies plastic strain of a material.

Example 4–1.1

Accurate measurements can be made to four significant figures of the density of aluminum. When cooled rapidly from 650° C, $\rho_{Al} = 2.698$ Mg/m³. Compare that value with the theoretical density obtained from diffraction analyses where a was determined to be 0.4049 nm.

Procedure Use the method for density determination that was used in Example 3–2.2.

Solution Since the atomic weight is 26.98 amu and aluminum is fcc (Appendix B),

$$\rho_{\text{theor.}} = \frac{4(26.98 \text{ amu})/(0.6022 \times 10^{24} \text{ amu/g})}{(0.4049 \times 10^{-9} \text{ m})^3} = 2.700 \text{ Mg/m}^3$$

$$\frac{\rho}{\rho_{\text{theor.}}} = \frac{2.698}{2.700} = 0.999 \qquad \text{or} \sim 1 \text{ vacancy per 1000 atoms}$$

Additional Information The match is close. Aluminum, like most metals just below their melting temperatures, has about one vacancy in 10^3 atoms. The equilibrium fraction drops to approximately one vacancy in 10^7 atoms at one-half of the absolute melting temperature.

4–2
NONCRYSTALLINE MATERIALS

Long-range order is absent in some materials of major engineering and scientific importance. Included are all liquids, glass, the majority of plastics, and a few metals if the latter are cooled extremely rapidly from their liquid. In principle, we can view this lack of repetitive structure as a volume, or three-dimensional, disorder, and as a continuation of our sequence: point defects, linear defects, and two-dimensional boundaries. We call these materials *amorphous* ("without form") in contrast to crystalline materials.

Liquids

For the most part, liquids are *fluids* (i.e., they *flow* under their own mass). However, just as "molasses in January" is semirigid, various liquids of technical importance can become very viscous and even solid, without crystallizing.

First let us look at the disorder that occurs in a single-component metal as it approaches melting, then transforms into a liquid. We can use lead for our example. As we stated in the preceding section, the greater thermal energy at higher temperatures introduces not only greater thermal vibrations, but also some vacancies. Just short of the melting point, crystalline lead may contain up to 0.1 percent vacancies in its lattice. When the vacancies approach 1 percent in a close-packed structure, "turmoil reigns." The regular 12-fold coordination is destroyed and the long-range order of crystal structure disappears (Fig. 4–2.1).

Energy is required to disrupt the crystalline structure at the time of melting. We call this energy the *heat of fusion* (ΔH_f). This is extra energy beyond the specific heat, or *heat capacity, dE/dT*, that is required for the increase in temperature. The energy for fusion, or melting, differs from material to material, but in general reflects the strength of the atomic bonds. Thus, as we should expect, there is a correlation between the heat of fusion and the melting temperature (Table 4–2.1).

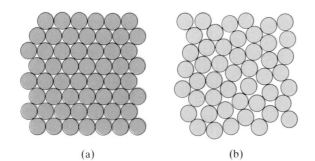

(a)　　　　　　　　(b)

FIG. 4–2.1

Melting (Metal). (a) Crystalline metal with CN = 12. (The 6 in the plane plus 3 above and 3 below.) (b) Liquid metal. Long-range order is lost; CN < 12, and the average interatomic distance increases slightly.

TABLE 4–2.1 Heats of Fusion of Metals

METAL	MELTING TEMPERATURE ° C (° F)	HEAT OF FUSION, joule/mol*
Tungsten, W	3410 (6170)	32,000
Molybdenum, Mo	2610 (4730)	28,000
Chromium, Cr	1875 (3407)	21,000
Titanium, Ti	1668 (3034)	21,000
Iron, Fe	1538 (2800)	15,300
Nickel, Ni	1455 (2651)	17,900
Copper, Cu	1084 (1984)	13,500
Aluminum, Al	660.4 (1221)	10,500
Magnesium, Mg	649 (1200)	9,000
Zinc, Zn	420 (788)	6,600
Lead, Pb	327.4 (621)	5,400
Mercury, Hg	−38.9 (−38)	2,340

* joule/0.6×10^{24} atoms; 4.18 J = 1 cal.

Solidification by crystallization releases the heat of fusion, $-\Delta H_f$, because the disorder of the liquid is eliminated as the atoms reorder themselves into a crystalline array with specific interatomic distances (Fig. 4–2.1).

The disorder that accompanies melting increases the volume of most metals (Fig. 4–2.2) and ionic solids, because those crystals have high packing factors. With the melting, the number of nearest neighbors in an fcc or hcp metal drops from 12 to only 11 or 10; however, the neighbors are not in a repetitive pattern. Consequently, the volume per atom and the average interatomic spacing is increased by a few percent.

The disorder in melting decreases the volume of those materials that have network structures with stereospecific bonds, and therefore low packing factors (Fig. 4–2.3). Materials of this nature include silicon, which has the diamond-cubic structure (Fig. 3–2.6a), and ice, which has a predominance of hydrogen bridges (Fig. 2–3.3). These openly packed crystals collapse into denser liquids when their atoms are thermally excited into the molten state.

Glasses

As we indicated earlier, glasses are sometimes considered to be very viscous liquids, inasmuch as they are noncrystalline. However, only a few liquids can form glasses. Therefore, in an effort to make a distinction, we must look at the structure of glass more critically.

At high temperatures, glasses form true liquids. The atoms are free to move around and to respond to shear stresses. When a commercial glass is supercooled below its melting temperature without crystallizing, thermal contraction is caused by atomic rearrangements that produce more efficient packing of the atoms. This contraction (Fig. 4–2.4) is typical of all liquid phases; however, with more extensive cooling, there is an abrupt change in the thermal expansion coefficient, dV/dT, of glasses. Below a certain temperature, called the *glass-transition temperature,* or more simply the *glass temperature, T_g,* there are no further rearrangements of the atoms and the only volume change is a result of reduced thermal

FIG. 4–2.2

Volume Changes with Temperature (Sodium — bcc, Lead — fcc, Magnesium — hcp). Metals with these structures expand on melting. (From L. H. Van Vlack, *Materials for Engineers: Concepts and Applications,* Addison-Wesley, Reading, Mass., 1982, Fig. 2–4.1, with permission.)

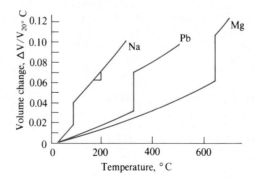

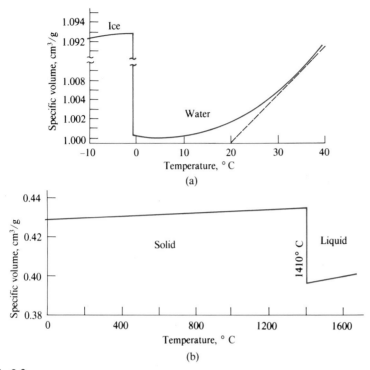

FIG. 4-2.3

Volume Changes with Temperature. (a) Ice (hydrogen-bridge bonding). (b) Silicon (covalently bonded). The coordination number is low for both bondings. Therefore, the packing factors of the solids are low. The structures collapse into smaller volumes as they melt. (From L. H. Van Vlack, *Materials Science for Engineers,* Addison-Wesley, Reading, Mass., 1970, Fig. 4-14, with permission.)

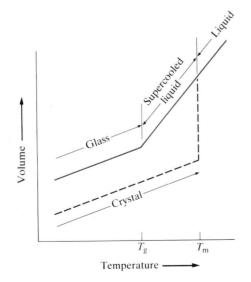

FIG. 4-2.4

Volume Changes in Supercooled Liquids and Glasses. When a liquid is cooled, it contracts rapidly and continuously because, with decreased thermal agitation, the atoms develop more efficient packing arrangements. In the absence of crystallization, the contraction continues below T_m to the glass-transition temperature, T_g, where the material becomes a rigid glass. Below T_g, no further rearrangements occur, and the only further contraction is caused by reduced thermal vibrations of the atoms in their established locations.

vibrations. This lower coefficient is comparable to the thermal expansion coefficient in crystals, where thermal vibrations are the only factor causing volume changes, and no rearrangement occurs.

The term *glass* applies to those materials that have the expansion characteristics of Fig. 4–2.4. Glasses may be either inorganic or organic and are characterized by a *short-range order* (and an absence of long-range order). Figure 4–2.5 presents one of the simplest glasses (B_2O_3), in which each small boron atom fits among three larger oxygen atoms. Since boron has a valence of three and oxygen a valence of two, electrical balance is maintained if each oxygen atom is located between two boron atoms. As a result, a continuous structure of strongly bonded atoms is developed. Below the glass-transition temperature, where the atoms are not readily rearranged, the fluid characteristics are lost and a noncrystalline solid exists. Such a solid has a significant resistance to shear stresses and therefore cannot be considered a true liquid.

The temperature–volume characteristics of Fig. 4–2.4 were originally observed in silicate glasses. However, it soon became apparent that these characteristics have major significance in polymeric materials (Table 4–2.2). Later, we shall look at the glass-transition temperature, T_g, closely. Below the T_g, polymers are hard and brittle and have low dielectric constants. Above the T_g, a plastic becomes flexible and even rubbery, with concurrent changes in its dielectric constant.

Example 4–2.1

From Fig. 4–2.2, calculate the packing factor (a) of solid lead at 326° C; (b) of liquid lead at 328° C. (Assume the hard-ball radius of 0.1750 nm is retained.)

Procedure Lead is fcc with a calculated PF of 0.74 (see Example 3–2.1a); thus, an increase in volume of 3 percent at 326° C and of 7 percent at 328° C decreases the PF accordingly.

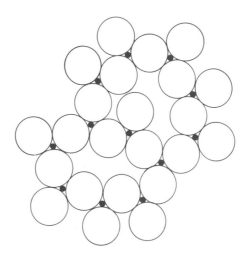

FIG. 4–2.5

Structure of B_2O_3 Glass. Although there is no long-range crystalline order, there is a short-range coordinational order. Each boron atom is among three oxygen atoms. Each oxygen atom is coordinated with two boron atoms.

TABLE 4–2.2 Glass-Transition Temperatures, T_g, of Selected Linear Polymers

POLYMER	T_g, °C (AMORPHOUS)	T_m, °C (IF CRYSTALLIZED)
Polyethylene (HD)	-120	140
Polybutadiene	$-70\pm$	—*
Polypropylene	-15	175
Nylon 6/6	50	265
Polyvinyl chloride	85	210*
Polystyrene	90	240*

* Crystallizes with great difficulty, if at all.

Calculation

(a) $PF_{326°C} = 0.74/1.03 = 0.718$

(b) $PF_{328°C} = 0.74/1.07 = 0.692$

Comment We could suggest that the radius increases by 1.0 l/o from 0.1750 nm at 20° C to 0.1767 nm at 326° C to give the 3 v/o expansion. If so, the packing factor at 326° C would remain at 74 percent, and the packing factor of the liquid would become 0.712. We cannot argue against that suggestion, since one definition of radius is "one-half of the closest interatomic distance." In either event, there is an abrupt discontinuity in packing efficiency at the melting temperature.

4–3
ORDER AND DISORDER IN POLYMERS

Materials such as ice, C_6H_6 ($T_m = 6°$ C), I_2 (Fig. 3–3.3), and dextrose sugar ($C_6H_{12}O_6$) retain their molecular structures when they crystallize. Their crystals may contain the imperfections cited in Section 4–1; however, the basic structure of the molecules is constant. Macromolecules, on the other hand, are subject to variations in their molecular structure. Polyethylene may have 100 carbon atoms along its molecular chain, or it may have 1000. Under stress, the polyethylene molecule may be pulled out nearly straight (except for the bond angles), as presented in Fig. 2–3.2; more commonly, it is kinked and coiled, somewhat akin to a fiber in a cotton boll. These added variables lead to additional types of disorder in polymers. Properties are affected accordingly.

Molecular Weights

Properties vary from different size molecules. For example, butane (C_4H_{10}), octane (C_8H_{18}), and "35-ane" ($C_{35}H_{72}$) have melting temperatures of $-138°$ C,

FIG. 4–3.1

Melting Temperatures Versus Molecule Size in the C_xH_{2x+2} Hydrocarbon Series.

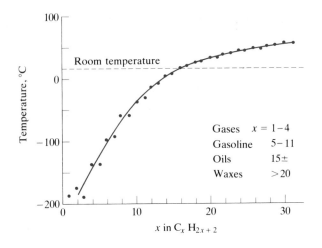

−56° C, and 75° C, respectively (Fig. 4–3.1). The same is true for macromolecules that possess thousands of atoms, so mechanical, chemical, and service stability of polymers also are dependent on molecular size. Figure 4–3.2 shows the tensile strength of polyethylene as a function of the number of *repeat units* (mers) in the average molecule. It will be advantageous for us to quantify molecular size.

Molecular size can be reported in terms of the *degree of polymerization, n.* We do not determine the degree of polymerization by counting the repeat units (mers); rather, we determine the molecular weight, M. By knowing the mer weight, m, of our polymer, we can calculate

$$n = \frac{M}{m} = \frac{\text{molecular weight}}{\text{mer weight}} \qquad \textbf{(4–3.1a)}$$

The dimensions for our calculation are

$$\frac{\text{amu/molecule}}{\text{amu/mer}} = \frac{\text{mers}}{\text{molecule}} \qquad \textbf{(4–3.1b)}$$

FIG. 4–3.2

Strength Versus Molecular Size (Polyethylene). Larger molecules (with more mers per molecule, n) have increased mechanical and thermal properties, and thus are able to withstand more severe service requirements. (Adapted from Wyatt and Dew-Hughes, *Metals, Ceramics and Polymers.*)

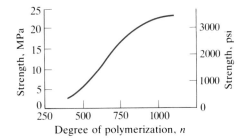

As an example, a polyvinyl chloride molecule may have a molecular weight of 35,000 amu (or 35,000 g/mol). Since the unit weight for C_2H_3Cl is 2(12 amu) + 3(1 amu) + 35.5 amu = 62.5 amu, the degree of polymerization is 35,000 amu/62.5 amu = 560 mers/molecule.

Mass-Average Molecular Weights Molecular weights are obtained from a variety of laboratory tests.* Reported values are, of necessity, average values, since any polymer product contains many molecules, few of which have grown identically. However, we can obtain averages using different calculations, depending on whether we use a weight basis or a number basis. Consider, for example, a syrup made with 18 g of sugar ($C_6H_{12}O_6$ with $M = 180$ amu) and 18 g of water (H_2O with $M = 18$ amu). Based on this 50-50 weight ratio, the *mass-average molecular weight*, $\overline{M}_m$, is

$$0.50(180 \text{ amu}) + 0.50(18) = 99 \text{ amu} \qquad \text{(or 99 g/mol)}$$

Generalizing our calculation, we have

$$\overline{M}_m = \Sigma W_i M_i \tag{4-3.2}$$

where W_i represents the weight fraction, and M_i the molecular weight of each component, i.

Number-Average Molecular Weights You no doubt are aware that 18 g of water will involve 10 times as many molecules as will 18 g of sugar (because the molecular weights are 18 and 180 amu each). A *number-average molecular weight*, $\overline{M}_n$, based on a number fraction, X_i, is

$$\overline{M}_n = \Sigma X_i M_i \tag{4-3.3}$$

Again, M_i is the molecular weight of each component. Since we have 10 times as many water molecules as sugar molecules, the average molecular weight (on a number-fraction basis) is

$$(10/11)(18 \text{ amu}) + (1/11)(180 \text{ amu}) = 32.7 \text{ amu} \qquad \text{(or 32.7 g/mol)}$$

Figure 4-3.3 illustrates this contrast between the two average molecular weights of a polymer where we have a range of various sizes. As indicated in the figure legend, the two presentations for the *same* material differ because there are more small molecules per gram than there are large molecules per gram. By inspecting Fig. 4-3.3 (or by making the calculation of Example 4-3.2), we observe that the number-average molecular weight is less than the mass-average one is. This is a universal situation, whether we consider a sugar-water syrup or any

---

* Principal among these are techniques that (1) involve osmosis, or (2) use light scattering. Molecular weights also are measured indirectly against calibrated standards by viscosity and gel permeation chromatographic procedures.

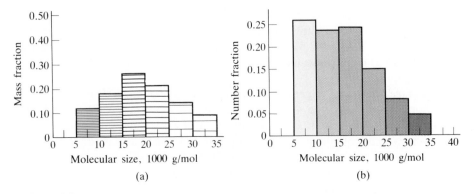

(a)   (b)

FIG. 4-3.3

Polymer Size Distributions (a) Based on mass; (b) Based on numbers. These two presentations are for the same material. In (a), the data indicate that 12 percent of the total *mass* is accounted for by molecules between 5000 amu and 10,000 amu (mid-value = 7500 amu). Other percentages of the total mass are in other size intervals. In (b), we observe that 26 percent of the total *number* of molecules fall in the 5000-amu to 10,000-amu range. These values of 26 percent for (b) versus 12 percent for (a) differ because it takes many of the small molecules to account for 12 percent of the mass. In contrast, 4 percent of the molecules account for 9 percent of the mass that is found in the interval from 30,000 to 35,000. (See Example 4-3.2.)

type of polymer. Only in a *monodisperse* (only one size of molecules) will the two figures be equal. In no case will the number-average be larger.

The ratio of the two average molecular weights is called the *polydispersity index* (PDI); it is a measure of the variation of molecular sizes:

$$PDI = \overline{M}_m / \overline{M}_n \qquad (4-3.4)$$

It ranges from PDI = 1 for laboratory monodisperses to PDI = 4 or 5 for many commercial products. A high value of the PDI indicates a large quantity of small molecules that adversely affect thermal stability in service.

Conformational Disorder

We could calculate from Table 2-2.2 that the average length of a polyethylene molecule with $n = 500$ [i.e., $+C_2H_4+_{500}$] is 154 nm, because each of the 1000 C—C bonds is 0.154 nm long. However, this calculation needs a correction, because the bond angles across the carbon atoms are approximately 120°, (Fig. 2-3.2). Thus, the "sawtooth" length would be (154 nm) sin 60°, or 135 nm. Even this is overly simplified in many polymers, where the single bonds of the carbon chain are free to rotate. This freedom is available at higher temperatures in amorphous (noncrystalline) polymers and in polymers dissolved in a liquid solvent.

With only three C—C bonds in butane (Fig. 4-3.4), the $a-d$ distance is 0.4 nm, whereas the $a-d'$ distance is less than 0.3 nm. The length varies randomly

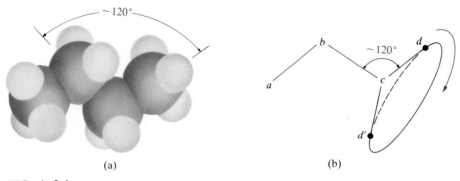

(a)   (b)

FIG. 4-3.4

Bond Rotation (Butane). Although there is a 120° bond angle, the end-to-end distance can vary from a-d to a-d'. Large molecules will have much more variation.

between these limits as the thermal agitation rotates the bond angles. The maximum limit for the end-to-end length of the polyethylene in Fig. 2-3.2 with $n = 500$ is the 135-nm value we calculated; the minimum end-to-end length would be a fraction of a nanometer (in the unlikely event that the molecule were twisted or kinked to bring the two ends into contact). The average, or *root-mean-square* length, $\bar{L}$, in noncrystalline polymers lies between these two extremes (Fig. 4-3.5) and can be calculated on a statistical basis as

$$\bar{L} = l\sqrt{x} \qquad\qquad (4\text{-}3.5)^*$$

where l is the individual bond length, and x is the number of bonds in the chain.

FIG. 4-3.5

Kinked Conformation (Noncrystalline Molecules). Since each C—C bond can rotate (Fig. 4-3.4), a long molecule is normally kinked and has a relatively short mean length, $\bar{L}$ (Eq. 4-3.5).

---

* The mathematics required to derive this equation is beyond the scope of this book. It uses the statistics of a "random-walk" process.

Thus, the doubling of the degree of polymerization of a polyvinyl increases the mean end-to-end length by 40 percent. The twisting and coiling arising from bond rotation is called *conformational* entropy, or disorder. This rotation is limited to single C—C bonds; however, it is important to properties. For example, above the glass-transition temperature, T_g, of Fig. 4–2.4, many polymers may be stretched from their kinked conformation to give high strains without changing the interatomic distances. Rubbers possess this characteristic, so they develop high strains at relatively low stresses. Furthermore, the molecules recoil to their kinked conformation when the stress is removed. Rubber is visibly elastic because it has a very low elastic modulus (Table 1–2.1).

Configurational Variants

Conformational disorder arises from bond rotation. Molecules also may possess more than one *configuration.* The change from one to another involves *bond breaking* and new bond formation. *Isomers,* for example, are two configurations of the same molecular composition. Figure 4–3.6 shows *n*-propyl and *iso*-propyl alcohol, both C_3H_7OH. Since the two configurations differ in structure, they do not have identical properties. For *n*-propyl alcohol, $T_m = -127°$ C, $T_v = 97°$ C, and $\rho = 0.8$ g/cm^3; for *iso*-propyl alcohol, $T_m = -90°$ C, $T_v = 82°$ C, and $\rho = 0.78$ g/cm^3.

Stereoisomers present more complex configurational variants, as shown in Fig. 4–3.7. We use polypropylene, a polyvinyl with $\mathbf{R} = -CH_3$ as our illustration; the figure applies to other polyvinyls as well. The arrangement of the mers show a high degree of regularity along the molecular chain of Fig. 4–3.7(a). Not only is there an addition sequence of $+C_2H_3CH_3+$ units that form the linear polymer, but also there is an identical orientation of each propylene unit. The isomer of Fig. 4–3.7(b) has the same chain of propylene mers, but the geometric (or stereo) arrangement of the $+C_2H_3CH_3+$ units differ. We speak of these two molecules as being *stereoisomers;* that in (a) being *isotactic* (the same pattern), that in (b) being *atactic* (without a pattern.) The properties of these two isomers are different. For example, the isotactic molecules can "mesh together" better and require less volume. We shall see in Section 7–5 that isotactic molecules can crystallize to a greater degree. The consequence is a slightly greater density and a measurably better resistance to adverse thermal and stress conditions in service.

FIG. 4–3.6

Isomers of Propanol. (a) Normal propyl alcohol. (b) Isopropyl alcohol. The molecules have the same composition but different structures. Consequently, the properties are different. Compare with polymorphism of crystalline materials (Section 3–4).

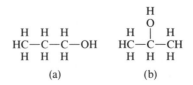

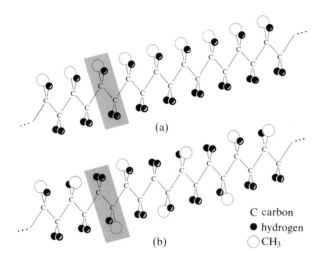

(a)

(b)

C carbon
● hydrogen
○ CH₃

FIG. 4–3.7

Stereoisomers (Polypropy-lene). (a) Isotactic. (b) Atactic. In (a), each $C_2H_3CH_3$ mer has the same spatial (stereo) arrangement. The stereo-isomer of (b) is without a spatial pattern.

A second example of geometric isomerism is found in rubbers that are made of butadiene-type molecules (Table 4–3.1). Natural rubber is polymerized *isoprene* with a mer of

$$(4\text{–}3.6)$$

The two double-bonded carbons have a CH_3 group and a hydrogen on the same side of the chain. This *cis* isomer, with occupancy on the *same* side of the chain, has important consequences in the chain behavior, because it arcs the mer (Eq.

TABLE 4–3.1 Structures of Butadiene-Type Rubbers

$$n\begin{matrix} H \\ C= \\ H \end{matrix}\begin{matrix} R \\ C \end{matrix}\begin{matrix} H \\ -C= \end{matrix}\begin{matrix} H \\ C \\ H \end{matrix} \rightarrow \left[\begin{matrix} H \\ -C- \\ H \end{matrix}\begin{matrix} R \\ C= \end{matrix}\begin{matrix} H \\ C- \end{matrix}\begin{matrix} H \\ C- \\ H \end{matrix}\right]_n \quad (4\text{–}3.3)$$

NAME	R*	EQUATION
Butadiene	—H	
Chloroprene	—Cl	
Isoprene (cis)	—CH₃	(4–3.1)
Gutta percha (trans)	—CH₃	(4–3.2)

* See Table 2–3.1 for vinyl polymers.

4–3.6). Another modification has the CH_3 group and the single hydrogen on *opposite* sides of the chain to give a *trans* isomer:

$$\left(\begin{array}{c} CH_3 \\ H \underset{C}{\overset{C}{\diagdown}} \overset{H}{\underset{C}{\diagup}} \\ C \overset{\diagup}{\diagdown} C \quad H \\ H \qquad H \end{array}\right)_n \qquad \qquad (4\text{–}3.7)$$

Although identical in compositions, these two isomers of $+C_5H_8+$ have different structures and therefore different properties. Natural rubber, with its cis-type structure, has a very highly kinked chain, as a result of the arc within the mer. Thus, it exhibits the very large elastic strain that is typical of rubbers. The polymer of the trans-type called *gutta percha* has a bond-angle pattern that is more typical of linear polymers (Fig. 2–3.2). In effect, the unsaturated positions balance each other across the double bond. As a result, the properties differ markedly.

Branching

Ideally, linear polymer molecules such as those we have studied to date are two-ended chains. There are cases, however, in which polymer chains branch. We can indicate this schematically, as shown in Fig. 4–3.8. Although a branch is uncommon, once formed it is stable, because each carbon atom has its complement of four bonds and each hydrogen atom has one bond. The significance of branching lies in the three-dimensional entanglements that can interfere with plastic deformation. Think of a pile of tree branches compared with a bundle of sticks; it is more difficult to move a branch with respect to its neighbors than to move the individual sticks.

Cross-Linking

Some linear molecules, by virtue of their structure, can be tied together in three dimensions. Consider the molecule of Fig. 4–3.9(a) and its polymerized combination in Fig. 4–3.9(b). Intentional additives (divinyl benzene in this case — but we

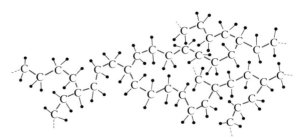

FIG. 4–3.8

Branching (Polyethylene).
The linear molecule of Fig. 2–3.2 can be branched under certain conditions. (The hydrogen atoms are represented by dots.)

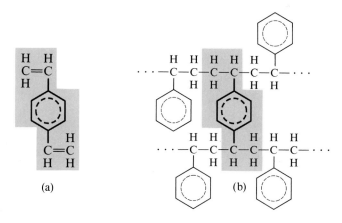

FIG. 4-3.9

Cross-Linking of Polystyrene. The divinyl benzene (a) becomes part of two adjacent chains (b) because it is tetrafunctional; i.e., it has four reaction points.

do not have to remember the name) tie together two chains of polystyrene. This cross-linking causes restrictions with respect to the plastic deformation of polymers.

The *vulcanization* of rubber is a result of cross-linking by sulfur, as shown schematically in Fig. 4-3.10. The effect is pronounced. Without sulfur, rubber is a soft, even sticky, material that, when it is near room temperature, flows by viscous

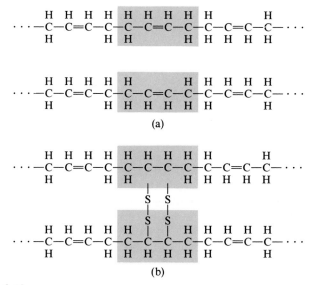

FIG. 4-3.10

Vulcanization (Butadiene-Type Rubbers). Adjacent chains (a) are cross-linked by pairs of sulfur atoms (b). (Other cross-linking configurations are possible; for example, two single sulfur atoms, rather than two pairs of sulfurs, may occur between the carbons of neighboring chains.)

deformation. It could not be used in automobile tires because the service temperature would make it possible for molecules to slide by their neighbors, particularly
at the pressures encountered. However, cross-linking by sulfur at about 10 percent
of the possible sites gives the rubber mechanical stability under the above conditions, but still enables it to retain the flexibility that is obviously required. Hard
rubber has a much larger percentage of sulfur and appreciably more cross-links.
You can appreciate the effect of the addition of greater amounts of sulfur on the
properties of rubber when you examine a hard-rubber product such as a pocket
comb.

Example 4–3.1

A solution contains 15 g of water, 4 g of ethanol (C_2H_5OH), and 1 g of sugar ($C_6H_{12}O_6$). (a)
What is the mass fraction of each molecular component? (b) What is the number fraction of
each molecular component?

Solution

(a) Basis: 20 g

	W_i
H_2O	15 g/20 g = 0.75
C_2H_5OH	4 g/20 g = 0.20
$C_6H_{12}O_6$	1 g/20 g = 0.05

(b) Basis: 20 amu

	molecules	X_i
H_2O	15 amu/18 amu = 0.833	= 0.900
C_2H_5OH	4 amu/46 amu = 0.087	= 0.094
$C_6H_{12}O_6$	1 amu/180 amu = 0.006	= 0.006
	Total = 0.926	

Example 4–3.2

It has been determined that a polyvinyl chloride (PVC) has the molecular weight distribution shown in Figs. 4–3.3(a) and (b)—for mass fraction and number fraction, respectively.
(a) What is the "mass-average" molecular weight? (b) What is the "number-average"
molecular weight? (c) What is the degree of polymerization n based on M_n? (d) What is the
polydispersity index?

Procedure   Consider the six size classes in Fig. 4–3.3 as six molecular types with molecular
weights of 7,500 g/mol; 12,500 g/mol; and so on. Then determine the averages as we did
for the sugar-water syrup to establish Eqs. (4–3.2) and (4–3.3).

Calculation The values we want for (a) and (b) are as follows:

MOLECULAR SIZE INTERVAL, amu	$(M)_i$, MID-VALUE, amu	W_i, MASS FRACTION (FIG. 4–3.3a)	$(W_i)(M_i)$, amu	X_i, NUMBER FRACTION (FIG. 4–3.3b)	$(X_i)(M_i)$, amu
5–10,000	7,500	0.12	900	0.26	1,950
10–15,000	12,500	0.18	2,250	0.23	2,875
15–20,000	17,500	0.26	4,550	0.24	4,200
20–25,000	22,500	0.21	4,725	0.15	3,375
25–30,000	27,500	0.14	3,850	0.08	2,200
30–35,000	32,500	0.09	2,925	0.04	1,300
			$\Sigma = 19,200$ amu/molecule (a) M_m		$\Sigma = 15,900$ amu/molecule (b) M_n

(c) Amu/mer of PVC (Table 2–3.1):

$$(C_2H_3Cl) = 24 + 3 + 35.5 = 62.5 \text{ amu/mer}$$

$$\text{degree of polymerization} = \frac{(15,900 \text{ amu/molecule})}{(62.5 \text{ amu/mer})}$$

$$= 254 \text{ mers/molecule (based on } \overline{M_n})$$

(d) From Eq. (4–3.4), the polydispersity index (PDI) is equal to

$$\overline{M_m}/\overline{M_n} = 19,200 \text{ amu}/15,900 \text{ amu} = 1.2$$

Comments Whenever there is a distribution of sizes, the number-average molecular size is always less than the mass-average value, because of the large number of smaller molecules per gram in the smaller size intervals. The two averages diverge more as the range of the size distribution increases.

Example 4–3.3

A polyethylene $+C_2H_4+_n$ molecule (see Fig. 2–3.2), with the molecular mass of 22,400 amu, is dissolved in a liquid solvent. (a) What is the longest possible end-to-end distance of the polyethylene (without altering the 120° C—C—C bond angle)? (b) What is the shortest? (c) What is the most probable?

Procedure We need to know the number of C—C bonds for both (a) and (c). Since there are two C—C bonds per mer in $(C_2H_4)_n$, we must determine n, the degree of polymerization. From Table 2–2.2, $l = 0.154$ nm.

Calculations

$$\text{Degree of polymerization} = (22{,}400 \text{ amu/molecule})/(28 \text{ amu/mer})$$
$$= 800 \text{ mers/molecule} \qquad \text{(or 1600 bonds)}$$

(a) "Sawtooth length" $= (1600)(0.154 \text{ nm})(\sin 120°/2)$
$$= 210 \text{ nm}$$

(b) Shortest distance $= <1$ nm (with ends in contact)

(c) Eq. (4–3.5): $L = 0.154 \text{ nm } \sqrt{1600} = 6.2$ nm

Comments Since the molecule is under continuous thermal agitation, the probability of its attaining a length of 200 nm is extremely remote. A force would have to be used to stretch it out to that length. Furthermore, the required force would become greater at higher temperatures, because the kinking becomes more persistent with increased thermal agitation.

Example 4–3.4

How many grams of sulfur are needed per 100 g of final rubber product to cross-link completely a polybutadiene $+C_4H_6+_n$ rubber with sulfur according to the pattern started in Fig. 4–3.10?

Procedure In this rubber, the mer is

$$C_4H_6, \text{ or} \begin{array}{ccccc} H & H & H & H \\ | & | & | & | \\ -C & -C & =C & -C- \\ | & & & | \\ H & & & H \end{array}$$

As shown in Fig. 4–3.10, complete cross-linking requires *two* sulfur molecules, S_2, for a *pair* of C_4H_6 mers or an $S/(C_4H_6)$ ratio of 4 to 2, or 2 to 1.

Calculation

$$1 \text{ mol } C_4H_6 = 48 + 6 \ = \ 54 \text{ g}$$
$$2 \text{ atoms S} \qquad\qquad = \ 64 \text{ g}$$
$$\text{product} = 118 \text{ g}$$
$$\text{fraction sulfur} = \frac{64 \text{ g}}{118 \text{ g}} = \frac{x}{100 \text{ g}}$$
$$x = \ 54 \text{ g}$$

Comment Alternate cross-linking structures can develop as mentioned in Fig. 4–3.10.

4-4
SOLID SOLUTIONS

We tend to think of pure, unadulterated substances as ideal; but there are instances where, because of cost, availability, or properties, it is desirable to have impurities present. An example is *sterling silver,* which contains 7.5 percent copper and only 92.5 percent silver (Example 2–1.1). This material, which we rate highly, could be refined to well over 99 percent purity. It would cost more, however, and it would be of inferior quality. Without altering its appearance, the 7.5 percent Cu makes the silver stronger and harder, and therefore more durable—at a lower cost!

Of course, we must consider the properties pertinent to our design. Zinc added to copper produces *brass,* again at a cost lower than that of the pure copper. Brass is harder, stronger, and more ductile than copper (Fig. 4–4.1). On the other hand, brass has lower electrical conductivity than copper does, so we use the more expensive pure copper for electrical wiring and similar applications where conductivity is important.

Zinc atoms join copper in brass as a *solid solution;* that is, the zinc atoms are incorporated into the fcc structure of copper. The zinc no longer possesses its normal hcp structure (Fig. 3–3.2b). This process is comparable to salt (NaCl) dissolving into water to produce a brine. The Na^+ and Cl^- ions do not retain the normal fcc structure that was shown in Fig. 3–1.1. Of course the NaCl brine is a liquid solution, whereas brass is a *solid solution.* Both solutions contain two *components*—NaCl and H_2O in brine, and Cu and Zn in brass.

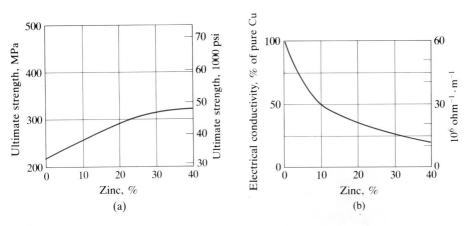

FIG. 4–4.1

Hardness (a) and Conductivity (b) of Cu–Zn Alloys (α-Brasses). Zinc atoms substitute for copper atoms in the fcc structure to form a solid solution. Consequently, the hardness increases and the electrical conductivity decreases.

Solid solutions form most readily when the *solvent* and *solute* atoms have similar sizes and comparable electron structures. For example, the individual metals of brass—copper and zinc—have atomic radii of 0.1278 nm and 0.139 nm, respectively. They both have 28 subvalent electrons and they each form crystal structures of their own with a coordination number of 12. Thus, when zinc is added to copper, its substitutes readily for the copper within the fcc metal structure, until a maximum of more then 35 percent of the copper atoms has been replaced. In this solid solution of copper and zinc, the distribution of zinc is entirely random (see Fig. 4–4.2).

Substitutional Solid Solutions

The solid solution described in the previous paragraph is called a *substitutional* solid solution because the zinc atoms substitute for copper atoms in the crystal structure. This type of solid solution is quite common among various metal systems. The solution of copper and nickel to form cupronickel is another example. Any fraction of the atoms in the original copper structure may be replaced by nickel. Copper–nickel solid solutions may range from no nickel and 100 percent copper, to 100 percent nickel and no copper. There is no *solubility limit.* All copper–nickel alloys are face-centered cubic.

On the other hand, there is a definite limit to the amount of tin that may replace copper to form *bronze,* and retain the fcc structure of the copper. Tin in excess of the *solid solubility* must form another phase. This solubility *limit* will be considered in more detail in Chapter 5.

If there is to be extensive replacement in a substitutional type of solid solution, the atoms must be nearly the same size. Nickel and copper have a complete range of solutions, because both of their individual structures are fcc and their radii are 0.1246 nm and 0.1278 nm, respectively. As the difference in size increases, less substitution can occur. Only 20 percent of copper atoms can be replaced by

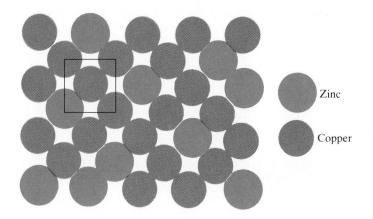

FIG. 4–4.2

Random Substitutional Solid Solution (Zinc in Copper; i.e., Brass). The crystal pattern is not altered.

aluminum, because the latter has a radius of 0.1431 nm as compared with only 0.1278 nm for copper. Extensive solid solubility rarely occurs if there is more than about 15 percent difference in radius between the two kinds of atoms. There is further restriction in solubility when the two components have different crystal structures or valences.

The limiting factor is the *number* of substituted atoms, rather than the *weight* of the atoms that are substituted. However, engineers ordinarily express composition as weight percent. It is therefore necessary that you know how to convert weight percent to atomic percent, and vice versa (see Example 2–1.1).

Interstitial Solid Solutions

In another type of solid solution, illustrated in Fig. 4–4.3, a small atom may be located in the interstices between larger atoms. Carbon in iron is an example. At temperatures below 912° C (1673° F), pure iron occurs as a bcc structure. Above 912° C (1673° F), there is a temperature range in which iron has an fcc structure. In this fcc structure, a relatively large *interstice,* or "hole," exists in the center of the unit cell. Carbon, being an extremely small atom, can move into this hole to produce a solid solution of iron and carbon. At those temperatures where the iron has a bcc structure, the interstices between the iron atoms are much smaller. Consequently the solubility in bcc iron is limited (Section 7–3).

Ordered Solid Solutions

Figure 4–4.2 shows a *random substitution* of one atom for another in a crystal structure. In such a solution, the chance of one element occupying any particular atomic site in the crystal is equal to the atomic percent of that element in the alloy. In that case, there is no *order* in the substitution of the two elements.

It is not unusual, however, to find an *ordering* of the two types of atoms into a specific arrangement. Figure 4–4.4 shows an ordered structure in which most dark "atoms" are surrounded by light "atoms." Such ordering is less common at higher temperatures, since greater thermal agitation tends to destroy the orderly

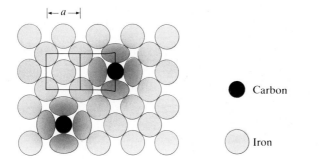

● Carbon

○ Iron

FIG. 4–4.3

Interstitial-Solid Solution (Carbon in fcc Iron).

FIG. 4–4.4

**Ordered Substitutional Solid
Solution.** The majority (but
not all) of the atoms are
coordinated among atoms
unlike themselves. If ordering
is complete, a compound is
formed (Section 4–5).

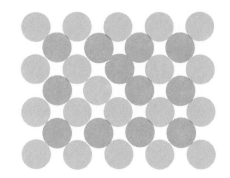

arrangement. For example, the β'-brass of Fig. 3–2.8(a) has a highly ordered
structure at room temperature, in which the copper atoms have zinc neighbors and
vice versa (Fig. 3–2.8b). Above 460° C (860° F), the atoms do not retain this
specific coordination and a random bcc distribution emerges.*

Ordering produces property changes as shown in Fig. 4–4.5. The basis for these
changes will be covered in Sections 8–3 and 11–2.

Example 4–4.1

Bronze is a solid-solution alloy of copper and tin in which 3 percent, more or less, of the
copper atoms are replaced by tin atoms. The fcc unit cell of copper is retained, but is
expanded a bit, because the tin atoms have a radius of approximately 0.151 nm. (a) What is
the weight percent in a 3 a/o tin bronze? (b) Assuming that the lattice constant increases
linearly with the atomic fraction of tin, what is the density of this bronze?

Procedure Select a basis for calculation. One-hundred atoms (25 fcc unit cells) is a natural;
thus, 97 Cu + 3 Sn. (a) Determine the masses. (b) Since $\rho = m/V$, determine $\bar{a}$ for $V = a^3$,
and use the calculated mass.

Calculation

    (a) Mass of copper: 97(63.54 amu) = 6163 amu = 0.945 (or 94.5 w/o)

          Mass of tin: 3(118.69 amu) =   356 amu = 0.054 (or 5.4 w/o)

                      Total mass = 6519 amu

    (b) average radius = 0.97(0.1278 nm) + 0.03(0.151 nm) = 0.1285 nm

$$a = (4)(0.1285)/\sqrt{2} = 0.3634 \text{ nm}$$

$$\rho = \frac{6519 \text{ amu}/(0.602 \times 10^{24} \text{ amu/g})}{25(0.3634 \times 10^{-9} \text{ m})^3} = 9.0 \text{ Mg/m}^3 \; (=9.0 \text{ g/cm}^3)$$

---

* The ordered β'-brass is simple cubic because the unit-cell center does not match the corner (Zn
versus Cu). The random solid solution above 460° C is bcc, because the cell corner and the cell center
are statistically identical. We label this form β-brass (no prime).

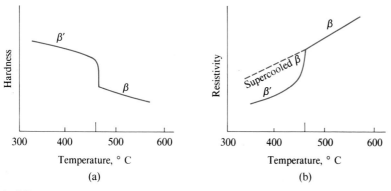

FIG. 4-4.5

Ordering in β-Brass (Schematic). Above 460° C, the 50Cu–50Zn solid solution is random and bcc, β. Below 460° C, the structure orders to β′, which is simple cubic (Fig. 3–2.8). (a) The ordered structure is harder. (b) This nonrandom β′ permits easier electron movements, and therefore has lower resistivity (Chapter 11). The random structure can be preserved by quenching (Chapter 6).

Example 4-4.2

The maximum solubility limit of tin in a Cu–Sn bronze is 15.8 w/o at 586° C. What is the atom percent tin?

Procedure Select a mass basis for calculation—for example, 100,000 amu. Therefore, 15,800 amu Sn + 84,200 amu Cu.

Calculation

Number of Cu atoms: 84,200/63.54 = 1325, which is 90.9 a/o Cu

Number of Sn atoms: 15,800/118.69 = 133, which is 9.1 a/o Sn

Total atoms = 1458

Example 4-4.3

At 1000° C, there can be 1.7 w/o carbon in solid solution with fcc iron (Fig. 4–4.3). How many carbon atoms will there be for every 100 unit cells?

Procedure Since there are 100 unit cells, we have 400 Fe atoms, which account for 98.3 percent of the mass.

total mass = (400 Fe)(55.85 amu/Fe)/0.983 = 22,726 amu

carbon atoms = (22,726 amu)(0.017)/(12.01 amu/C atom) = 32

Comment Note that we did not round off the numbers to two significant figures until we obtained the final answer. Had we done so at 23,000 amu, we would have biased our final answer to the high side (33 carbon atoms).

Additional Information The carbon atom sits as $\frac{1}{2}, \frac{1}{2}, \frac{1}{2}$ locations in about one-third of the unit cells. Since the carbon atom is slightly larger than the hole, it is not possible to have carbon atoms at all equivalent locations, such as at $\frac{1}{2}, 0, 0$.

4-5
SOLID SOLUTIONS IN CERAMIC AND METALLIC COMPOUNDS

Ionic Substitution

Solid solutions can occur in ionic phases as well as in metals. In ionic phases, just as in the case of solid metals, atom or ion size is important. A simple example of an ionic solid solution is shown in Fig. 4–5.1. The structure is that of MgO (cf. NaCl of Fig. 3–1.1) in which the Mg^{2+} ions have been partially replaced by Fe^{2+} ions. Inasmuch as the radii for the two ions are 0.066 nm and 0.074 nm, complete substitution is possible. On the other hand, Ca^{2+} ions cannot be similarly substituted for Mg^{2+} because their radius of 0.099 nm is comparatively large.*

An additional requirement, which is more stringent for solid solutions of ceramic compounds than it is for solid solutions of metallic compounds, is that the valence charges on the replaced ion and the new ion must be identical. For example, it is difficult to replace the Mg^{2+} in MgO with an Li^+, although the two have identical radii, because there would be a net deficiency of charges. Such a substitution could be accomplished only if there were other compensating changes in charge.

Nonstoichiometric Compounds

Many compounds have exact ratios of elements (e.g., H_2O, CH_4, MgO, Al_2O_3, Fe_3C, to name but a few). They have a fixed ratio of atoms; thus they are *stoichiometric.* Bonds form between unlike atoms. As a result, the structure becomes more highly ordered than the pattern we saw in Fig. 4–4.4.

Other compounds deviate from specific integer ratios for the two (or more) elements that are present. Thus, we find that "Cu_2Al" includes 31 a/o to 37 a/o Al (16 to 20 w/o Al), rather than being exactly $33\frac{1}{3}$ a/o Al. Likewise, at 1000° C, "FeO" ranges from 51 to 53 a/o oxygen, rather than being exactly 50 a/o. We call these compounds *nonstoichiometric,* because they do not have a fixed ratio of atoms.

---

* See Appendix B for ionic radii.

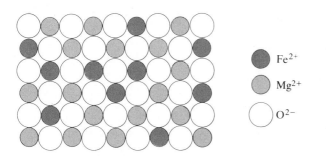

FIG. 4-5.1

Substitutional Solid Solution in a Compound. Fe^{2+} is substituted for Mg^{2+} in the MgO structure.

○ Fe^{2+}

○ Mg^{2+}

○ O^{2-}

Nonstoichiometric compounds always involve some solid solution. In the Cu_2Al example cited, the atoms are nearly enough the same size and are sufficiently comparable electronically that with excess aluminum, some of the copper atoms are replaced by aluminum atoms (up to the maximum of 37 a/o Al). Conversely, in the presence of excess copper, the Cu/Al atom ratio reaches 69/31, by substitution of a few copper atoms into the aluminum sites of Cu_2Al.

Defect Structures

Nonstoichiometry leads to *defect structures* in ionic materials that must maintain a charge balance. We can see this most readily in iron oxide (Fig. 4-5.2). Iron oxide, $Fe_{1-x}O$, is *nonstoichiometric* because some Fe^{3+} ions are invariably present with the predominant Fe^{2+} ions. Since two Fe^{3+} ions have six positive charges, they replace three Fe^{2+} ions; in the process, they leave an ion vacancy, □. The vacancies affect diffusion rates (Chapter 6). More important for electrical considerations, a defect structure permits electrical conductivity by *electron hopping* (Fig. 4-5.3), thus producing semiconduction in this ceramic oxide.

Example 4-5.1

An iron oxide (Fig. 4-5.2) contains 52 a/o oxygen and has a lattice constant of 0.429 nm. (a) What is the Fe^{2+}/Fe^{3+} ion ratio? (b) What is the density? (This structure is like that of NaCl, except for the vacancies, □. See Fig. 3-1.1.)

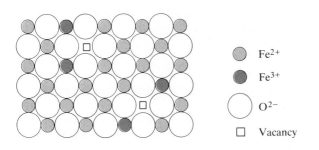

○ Fe^{2+}

● Fe^{3+}

○ O^{2-}

□ Vacancy

FIG. 4-5.2

Defect Structure ($Fe_{1-x}O$). This structure is the same as NaCl (Fig. 3-1.1) except for some iron ion vacancies. Since a fraction of the iron ions are Fe^{3+} rather than Fe^{2+}, the vacancies are necessary to balance the charge. The value of x in $Fe_{1-x}O$ ranges from 0.04 to 0.16, depending on temperature and the amount of available oxygen.

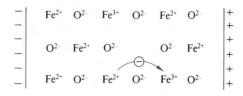

FIG. 4–5.3

Electronic Conduction (Ceramics). Compounds with multiple valence ions can conduct charge by *electron hopping*. In general, such conduction is found among only the transition elements. (Limited conductivity by ionic diffusion also is possible.) (From L. H. Van Vlack, *Materials for Engineers: Concepts and Applications,* Addison-Wesley, Reading, Mass., Fig. 11–5.2, with permission.)

Procedure (a) Choose a basis— 100 atoms (= 52 O^{2-} + 48 iron ions). Make the sum of the charges equal to zero, using y Fe^{3+} and $(48 - y)$ Fe^{2+}. (b) Based on Fig. 3–1.1, the 52 O^{2-} ions require 13 unit cells, but there are only 48 Fe ions present (and 4 □s).

Calculation

(a) Charge balance: $52(2-) + (y)(3+) + (48 - y)(2+) = 0$

$$y = 8 \; Fe^{3+} \qquad 48 - y = 40 \; Fe^{2+}$$

$$Fe^{2+}/Fe^{3+} = 5$$

(b) $\rho = \dfrac{[48(55.8) + 52(16)] \; \text{amu}/13 \; \text{u.c.}}{(0.602 \times 10^{24} \; \text{amu/g})(0.429 \times 10^{-9} \; \text{m})^3/\text{u.c.}} = 5.7 \; \text{Mg/m}^3 \; (= 5.7 \; \text{g/cm}^3)$

Example 4–5.2

A β'-brass is nominally an intermetallic compound, CuZn, with the simple cubic structure shown in Fig. 3–2.8. It also may be called a *partially ordered solid solution,* particularly since it is nonstoichiometric with a range of 46 to 50 a/o zinc at 450° C. Assume 90 percent of the $\frac{1}{2}, \frac{1}{2}, \frac{1}{2}$ sites of Fig. 3–2.8 are occupied by copper atoms in a 46 a/o Zn–54 a/o Cu alloy. What percent of the 0,0,0 sites are occupied by copper atoms?

Solution Basis: 50 unit cells = 50 $\frac{1}{2}, \frac{1}{2}, \frac{1}{2}$ sites (and 50   0,0,0 sites)
                                = 100 atoms = 54 Cu + 46 Zn

$\frac{1}{2}, \frac{1}{2}, \frac{1}{2}$ sites: 45 Cu + 5 Zn

0,0,0 sites: 9 Cu + 41 Zn

Therefore, 9 of the 50, or 18 percent of the 0,0,0 sites, are occupied by copper.

Comments At low temperatures, almost all the neighbors of zinc atoms are copper atoms, and vice versa. However, as the temperature is increased, the atoms start to disorder. In this problem, at 450° C, the 0,0,0 sites are 82 Zn–18 Cu, whereas the $\frac{1}{2}, \frac{1}{2}, \frac{1}{2}$ sites are 10 Zn–90 Cu (all atom percents). Thus, the two sites are not equivalent, and the structure is simple cubic.

Above 470° C, this alloy becomes fully random, with no preference for copper atoms to be surrounded by zinc atoms. Under those conditions, the substitutional solid solution (called β-brass rather than β'-brass) is bcc because the unit-cell centers and corners have equal probability for the same average composition.

4–6
SOLID SOLUTIONS IN POLYMERS (COPOLYMERS)

Many commercial plastics consist of a single phase with two (or more) components, and are therefore solid solutions. These are called *copolymers.* Although the most common copolymers are amorphous nonmetallic solids, we can consider them to be polymeric "alloys."

Polymers may exist as homogeneous solid solutions in two forms (Fig. 4–6.1). The first is a miscible (homogeneous) solution of two distinct molecular species. Except for their molecular size, this solution can be compared with the H_2O and $C_6H_{12}O_6$ (sugar) solution in the discussion for Eq. (4–3.2). The molecules are distinct, but they form a single phase.*

The second type of copolymer contains more than one type of mer within a single polymeric molecule. Each of the vinyl monomers cited in Table 2–3.1 is bifunctional, with two principal carbon atoms. Each polymerizes with the reaction,

$$n\ C_2H_3R \rightarrow +C_2H_3R+_n \qquad (4\text{–}6.1)$$

as indicated for PVC in Fig. 2–3.1. This gives each a linear backbone of carbon atoms. Therefore, it is only natural to ask the question: Can a polymer molecule have more than one type of mer? The answer is definitely "yes." A random arrangement the two (or more) types of mers along the chain is indicated schematically in Figs. 4–6.1(b) and 4–6.2. The properties of these copolymers often are desirable, and therefore are specified in engineering designs. Examples include copolymers of styrene and acrylonitrile (Table 2–3.1), and of vinyl chloride and vinyl acetate (Table 4–6.1).

Random copolymers serve our present purpose of showing solid solution in polymers. Later, in Section 7–5, we shall consider *block* copolymers, where there is a clustering of like mers along the molecular chain.

---

* The common copolymer of this type contains molecules of polystyrene (PS) and polyphenylene oxide (PPO). The latter is a linear polymer, but is not a polyvinyl (Fig. 4–6.1a).

```
                    ···—S—S—S—S—S—S—S—S—···
                ···—S—S—S—S—S—S—S—S—S—S—S—S—S—S—S—S—S—S—S—S—···
       ···—PO—PO—PO—PO—PO—PO—PO—PO—PO—PO—PO—PO—PO—PO—PO—PO—PO—PO—PO—PO—···
        —PO—PO—PO—PO—PO—PO—PO—PO—PO—PO—PO—PO—PO—PO—PO—PO—PO—PO—PO—PO—···
        ···—S—S—S—S—S—S—S—S—S—S—S—S—S—S—S—S—S—S—S—S—···
       ···—PO—PO—PO—PO—PO—PO—PO—PO—PO—PO—PO—PO—PO—PO—PO—PO—PO—PO—PO—···
                         ···—S—S—S—S—S—S—···
```
(a)

```
···—S—A—A—S—A—S—S—A—A—A—S—A—S—S—S—A—A—S—A—···
  ···—S—A—S—S—S—S—A—A—S—A—A—A—S—A—A—S—A—S—A—···
···—S—A—S—S—A—A—A—S—A—S—S—A—A—S—A—A—S—S—A—···
```
(b)

FIG. 4–6.1

Two-Component Polymers (Schematic—the Conformation is Never Straight). (a)
Mutual solubility of two polymers within a single phase. (S—styrene mers; PO—
phenylene oxide mers.) (b) Random copolymer with styrene mers (S) and acrylonitrile
mers (A). [Polystyrene and polyacrylonitrile have vinyl mers (Table 2–3.1). Polyphenylene oxide has mers of (structure).]

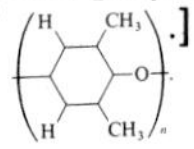

Example 4–6.1

A copolymer contains 10 m/o vinyl acetate and 90 m/o vinyl chloride (m/o = mer percent). (a) What is the w/o vinyl acetate? (b) What is the w/o chlorine?

Solution   From Table 2–3.1:

$$\text{vinyl acetate mer} = \left(\begin{matrix} H & H \\ C - C \\ H & | \\ & Ac \end{matrix}\right) \text{ where Ac is } -OCOCH_3.$$

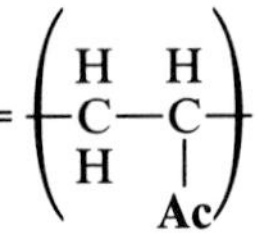

$$\text{vinyl chloride mer} = \left(\begin{matrix} H & H \\ C - C \\ H & | \\ & Cl \end{matrix}\right)$$

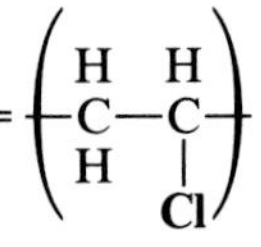

Basis:

$$100 \text{ mers} = 10 \text{ mers VAc} (=40C + 20O + 60H)$$
$$= 90 \text{ mers VC} (=180C + 270H + 90Cl)$$

$$10 \text{ mers VAc} = (12 \text{ amu})(40) + (16 \text{ amu})(20) + (1 \text{ amu})(60)$$
$$= 480 \text{ amu C} + 320 \text{ amu O} + 60 \text{ amu H}$$
$$= 860 \text{ amu}$$

$$90 \text{ mers VC} = (12 \text{ amu})(180) + (1 \text{ amu})(270) + (35.5 \text{ amu})(90)$$
$$= 2160 \text{ amu C} + 270 \text{ amu H} + 3195 \text{ amu Cl}$$
$$= 5625 \text{ amu}$$

(a) w/o vinyl acetate $= 860/(5625 + 860) = 13.3$ w/o

(b) w/o chlorine $= 3195/(5625 + 860) = 49.3$ w/o

Additional Information A copolymer can be viewed as a solid solution of the contributing mers. Just as with the solid solutions in Section 4-4, the overall structural pattern exists with one, two, or several types of components.

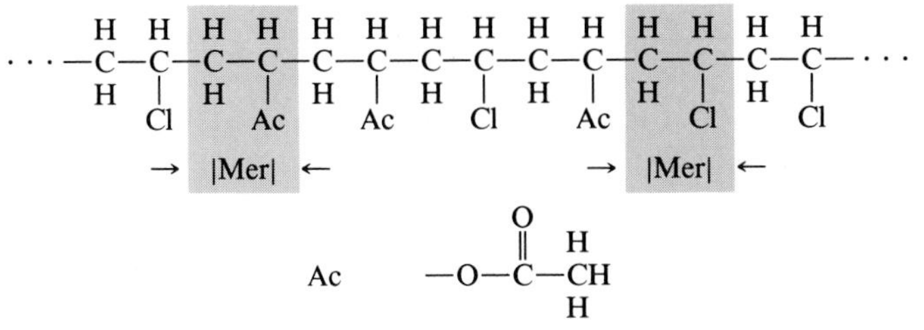

FIG. 4-6.2

Copolymerization of Vinyl Chloride and Vinyl Acetate. When randomly arranged, a copolymer is comparable to a solid solution in metallic and ceramic systems (Figs. 4-4.2 and 4-5.1).

TABLE 4-6.1 Vinyl Chloride-Acetate Copolymers: Correlation between Composition, Molecular Weight, and Applications*

ITEM	W/O OF VINYL CHLORIDE	NO. OF CHLORIDE MERS PER ACETATE MER	RANGE OF AVERAGE MOL. WTS.	TYPICAL APPLICATIONS
Straight polyvinyl acetate	0	0	4,800–15,000	Limited chiefly to adhesives.
Chloride-acetate copolymers	85–87	8–9	8,500– 9,500	Lacquer for lining food cans; sufficiently soluble in ketone solvents for surface-coating purposes.
	85–87	8–9	9,500–10,500	Plastics of good strength and solvent resistance; molded by injection.
	88–90	10–13	16,000–23,000	Synthetic fibers made by dry spinning; excellent solvent and salt resistance.
	95	26	20,000–22,000	Substitute rubber for electrical-wire coating; must be plasticized; extrusion molded.
Straight polyvinyl chloride	100	—	—	Pipes, and similar rigid products.

* Adapted from A. Schmidt and C. A. Marlies, *Principles of High Polymer Theory and Practice.* New York: McGraw-Hill.

Example 4–6.2

A copolymer contains 92 w/o vinyl chloride and eight w/o vinyl acetate. What is the mer fraction of VAc?

Procedure   Since a copolymer (Fig. 4–6.2) is the molecular equivalent to a crystalline solid solution, we will use a procedure comparable to that in Example 4–4.2. Choose a mass basis, and determine the numbers of each component.

Calculation   From Table 2–3.1

$$VC = +C_2H_3Cl+ = 24 + 3 + 35.5 = 62.5 \text{ amu/mer}$$

$$VAc = +C_2H_3OCOCH_3+ = 24 + 3 + 59 = 86 \text{ amu/mer}$$

Basis: 100,000 amu = 92,000 amu VC + 8,000 amu VAc.

VC 92,000/62.5 = 1472 mer = 0.941

VAc 8,000/86 =   93 mers = 0.059

Total = 1565 mers

Comment   This copolymer has ~ 16 PVC mers per acetate (PVAc) mer. Examples of uses are indicated in Table (4–6.1).

S U M M A R Y

In Chapter 3, we discussed the preciseness and regularity of the atomic structure of crystals. Perfect crystals serve as a basis for our consideration of many properties and characteristics of materials, such as density, anisotropy, slip planes, phase stability, piezoelectricity, and semiconducting compounds. At the same time, crystals are not always perfect, and many important properties and behaviors of materials arise from the irregularities. We cited some of these in the preview of the chapter. We must consider the *disorder* in materials as well as their order.

1. *Imperfections* are found in all crystals unless special means are used to reduce them to a low level. It is convenient to categorize them by geometry:

(i)   *Point defects,* which include vacancies and/or interstitials

(ii) *Linear defects,* commonly called dislocations

(iii) *Boundaries,* which may be external surfaces or internal discontinuities between grains or phases; these may be treated as two-dimensional defects

2. Liquids lack the long-range order that characterizes crystals. For some materials, it is possible to avoid crystallization and to retain the amorphous character of a liquid into a solid (i) if the anticipated crystal structure is complex, or (ii) by rapid cooling. We call these amorphous solids *glass.* Although they lack a freezing temperature, amorphous materials possess a glass temperature T_g, which will be important to us when we study the behavior of plastics.

3. The structures of macromolecules are variable. The *degree of polymerization,* and therefore the *molecular weight,* of two individual molecules of a polymer are seldom identical. Bond rotation and bond bending lead to *conformational* variations. *Configurational* variants include *stereoisomers, branching,* and *cross-linking.* Each of these modifications affects the way these large molecules bond into a three-dimensional solid, and therefore influences the properties of a polymer.

4. *Solid solutions* contain atoms of a second component as a solute. These atoms may be present either interstitially among the solvent atoms, or as a substitute atom replacing one of the atoms within the crystal lattice. For extensive solid solution in a crystal, the solute and solvent atoms must be comparable in size and electronic behavior.

5. Solid solutions also exist in compounds. Some *nonstoichiometric* compounds arise from incomplete *ordering.* Others that contain transition elements exist as a result of mixed valances. The latter commonly produce a *defect* structure.

6. Macromolecules may contain more than one component, producing a *copolymer.* As with solid solutions in metals, the properties of copolymers commonly differ from the properties of the basic components. The copolymers provide additional materials and opportunities for the design engineer.

KEY WORDS

Alloy	Grain	Root-mean-square length, $\overline{L}$
Amorphous	Grain boundary	Short-range order
Atactic	Imperfection (crystal)	Slip vector (**b**)
Branching	Interstice	Solid solution
Brass	Isotactic	Solid solution, interstitial
Bronze	Macromolecules	Solid solution, ordered
Butadiene-type compound	Microstructure	Solid solution, substitutional
Cis	Molecular crystal	Solubility limit
Component	Molecular length	Solute
Compound	Molecular weight (mass-average), M_m	Solvent
Copolymer		Stereoisomers
Cross-linking	Molecular weight (number-average), M_n	Sterling silver
Defect structure		Stoichiometric compounds
Degree of polymerization, n	Nonstoichiometric compounds	Surface
Dislocation, edge ($\perp$)	Phase	*Trans* (polymers)
Dislocation, screw (§)	Point defect	Vacancy ($\square$)
Elastomer	Polydispersity index, PDI	Vulcanization
Glass	Polymer	
Glass transition temperature (T_g)		

PRACTICE PROBLEMS

4–P11 The grain boundary area (increases, decreases, remains unchanged) as the grain size increases.

4–P12 Calculate the radius of the largest atom that can fit interstitially into fcc silver without crowding.

4–P13 (a) What is the coordination number of the interstitial site in Problem 4–P12? (b) What structure would result if *every* such site were occupied by a smaller atom or ion?

4–P14 In copper at 1000° C, one out of every 473 lattice sites is vacant. If these vacancies remain in the copper when it is cooled to 20° C, what will be the density of the copper?

4–P15 Why do ion-pair vacancies form as pairs?

4–P16 The area viewed on a 3 × 4 photomicrograph at × 1000 is what percent of the area viewed in a similar photomicrograph at × 100?

4-P17 Cite examples of point defects, linear defects, and two-dimensional defects.

4-P18 (a) Is the slip vector, **b**, parallel or at right angles to a screw dislocation? What is it to an edge dislocation?

4-P21 The density of liquid aluminum at its melting point is 2.37 Mg/m³. Assume a constant atom size; calculate the atomic packing factor.

4-P22 Would you expect the heat of fusion of gold to be nearest to 5000, 10,000, 20,000, or 40,000 J/mol?

4-P23 Account for the change in slope of the upper curve in Fig. 4-2.4.

4-P24 Why is it easier to form a glass in a polymer than it is to do so in a metal?

4-P25 Show that the volume expansion coefficient, α_V, approximates three times the linear expansion coefficient, α_L.

4-P30 Is the elastic modulus of cross-linked butadiene rubber greater than, approximately equal to, or less than that of nonvulcanized butadiene rubber?

4-P31 Two g of dextrose ($C_6H_{12}O_6$) are dissolved in 14 g water. What is the number-fraction of each type of molecule?

4-P32 The following data were obtained from an analysis of a polymeric sample:

Interval midpoint			
Interval midpoint	35,000 amu	1.25 g
" "	25,000	2.65
" "	15,000	2.00
" "	5,000	1.90

Compute the mass-average molecular size.

4-P33 (a) What is the mass-average molecular size of the molecules in Example 4-3.1? (b) What is the number-average molecular size?

4-P34 Polyvinyl chloride ($C_2H_3Cl)_n$ is dissolved in an organic solvent. (a) What is the mean square length of a molecule with a molecular mass of 28,500 g/mol? (b) What would be the molecular mass of a molecule with one-half the mean square length of that in part (a)?

4-P35 Determine the degree of polymerization and the mean square length of the average molecule in Problem 431, if the polymer is polypropylene (number-average).

4-P36 (a) What percent sulfur would be present if it were used as a cross-link at every possible point in polyisoprene? (b) What percent would be present if it were used in polychloroprene?

4-P37 A rubber contains 91 w/o polymerized chloroprene and 9 w/o sulfur. What fraction of the possible cross-links is joined by vulcanization? (Assume that all the sulfur is used for cross-links of the type shown in Fig. 4-3.10.)

4-P38 Sketch three of the four possible isomers of butanol (C_4H_9OH).

4-P39 Sketch the structure of the various possible isomers for octane, C_8H_{18}.

4-P41 Name the components of four common substitutional alloys.

4-P42 An alloy contains 85 w/o copper and 15 w/o tin. Calculate the a/o of each element.

4-P43 There is 5 a/o magnesium in an Al-Mg alloy. Calculate the w/o magnesium.

4-P44 Consider Fig. 4-4.3 to be an interstitial solution of carbon in fcc iron. What is the w/o carbon present?

4-P45 Consider Fig. 4-4.2 to be a substitutional solid solution of copper and gold. (a) What is the w/o Cu present if Cu is the more prevalent atom? (b) What is it if Au is the more prevalent atom?

4-P46 Carbon atoms can be dissolved into the largest interstices of fcc iron. (a) How many of these sites (per unit cell) are available? (b) How many neighboring iron atoms surround these sites?

4-P47 Based on the data in Appendix B, which metallic element among Al, Au, Cd, Ni, and Zn should have the greatest solubility in solid copper?

4-P48 Based on the data in Appendix B, which will have the smallest unit cell dimension, a: brass, bronze, or pure copper?

4–P51 (a) What is the w/o FeO in the solid solution of Fig. 4–5.1? (b) What is the w/o Fe^{2+}? (c) What is the w/o of O^{2-}?

4–P52 If all the iron ions of Fig. 4–5.1 were changed to Co ions, what would be the w/o MgO?

4–P53 A solid solution contains 30 m/o MgO and 70 m/o LiF. (a) What are the w/o Li^+, Mg^{2+}, F^-, and O^{2-}? (b) What is the density?

4–P61 A copolymer has a 5-to-2 mer ratio of styrene and butadiene. What is the weight ratio of these to components?

4–P62 Polyvinyl chloride and polyvinylidene chloride (Table 2–3.1) are copolymerized in a 2-to-1 weight ratio. They form a molecular chain similar to that of Fig. 4–6.1(b). What fraction is of each type?

TEST PROBLEMS

411 Determine the radius of the largest atom that can be located in the interstices of bcc iron without crowding. (*Hint:* The center of the largest hole is located at $\frac{1}{2}, \frac{1}{4}, 0$.)

412 (a) What is the coordination number for the interstitial site in Problem 411? (b) How many of these sites are there per unit cell?

413 The number of vacancies increases at higher temperatures. Between 20 and 1020° C, the lattice constant of a bcc metal increases 0.5 l/o from thermal expansion. In the same temperature range, the density decreases 2.0 percent. Assuming there was one vacancy per 1000 unit cells in this metal at 20° C, estimate how many vacancies there are per 1000 unit cells at 1020° C.

414 Explain the basis for surface energy, and that for grain boundary energy.

421 Based on the slopes of the curves in Fig. 4–2.2, estimate the volume expansion coefficient, α_V, of (a) solid magnesium and of (b) liquid magnesium at the melting point of magnesium.

422 The melting temperature, T_m, in Kelvin of the paraffins ($C_M H_{2M+2}$) of Fig. 4–3.1 is sometimes given by the empirical equation

$$T_m = [0.0024 + 0.017/x]^{-1} \quad (4\text{–}3.8)$$

where x is the number of carbon atoms in the chain. Give the melting point that polyethylene would have with a degree of polymerization, n, (a) of 10, (b) of 100, and (c) of 1000. (*Note:* Mer = C_2H_4.)

431 (a) What is the number-fraction of molecules in each of the four size categories of Problem 4–P32? (b) What is the number-average molecular size?

432 Polyethylene contains equal numbers of molecules with 100 mers, 200 mers, 300 mers, 400 mers, and 500 mers. What is the mass-average molecular size?

433 A polymeric material contains polypropylene molecules that have an average of 700 mers per molecule. What is the theoretical maximum strain this polymer could undergo if every molecule could be unkinked into a straight molecule (except for the ~ 120° bond angles)?

434 Rubber A (200 g) contains 168 g of isoprene $(C_5H_8)_n$ and 32 g sulfur. (a) What fraction of the cross-links are used if all of the sulfur forms those links? Rubber B (217 g) contains 168 g of butadiene $(C_4H_6)_n$ and 49 g of selenium. Although the use of selenium has some major disadvantages, it can cross-link rubber. (It lies immediately below sulfur in the periodic table.) (b) Which rubber, A or B, is more highly cross-linked? (Assume the cross-linking pattern of Fig. 4–3.10.)

435 A rubber contains 54 percent butadiene, 34 percent isoprene, 9 percent sulfur, and 3 percent carbon black. What fraction of the possible cross-links is joined by vulcanization, assuming that all the sulfur is used in cross-linking? (Assume the cross-linking pattern of Fig. 4–3.10.)

436 One kg of divinyl benzene (Fig. 4–3.9) is added to 50 kg of styrene. What is the maximum number of cross-links per gram of product?

437 How much sulfur must be added to 100 g of iso-prene rubber to cross-link 20 percent of the mers? (Assume all of the available sulfur is used, comparable to that in Fig. 4–3.10 for butadiene.)

438 Explain why the number-average molecular weight is always smaller than is the mass-average molecular weight.

439 The elastic strain of rubber is nonlinear with stress, so Eq. (1–2.3) does not apply. Explain this fact on the basis of Fig. 4–3.5.

441 An alloy contains 20 w/o Ni and 80 w/o Cu in substitutional fcc solid solution. Calculate the density of this alloy.

442 The solubility limit for the interstitial solution of carbon in fcc iron is 1.0 w/o at $815°$ C ($1500°$ F). Based on your answer to Problem 4–P46, what fraction of the possible interstitial sites may be occupied at saturation?

443 An alloy with 25 a/o gold and 75 a/o copper forms an fcc solid solution that is random above $380°$ C. Below $380°$ C, it becomes ordered with gold atoms at the corners of the unit cell and with copper atoms at the center of each face. What is the w/o gold?

444 Refer to Problem 443. This alloy is cooled rapidly from 400 to $20°$ C. (a) Estimate its lattice constant. (b) Calculate its density.

445 Fe, Cr, and Ni atoms have radii within $±1$ percent of each other. (a) Which element, Cr or Ni, will have greater solubility in bcc iron at $700°$ C? (b) Which will have greater solubility in fcc iron at $1000°$ C?

446 Silicon that is used as a semiconductor contains 10^{21} boron atoms per m^3. (a) What w/o B is present? (b) How many unit cells of silicon are there per boron atom? (Si is cubic with 8 atoms/unit cell; $a = 0.543$ nm.)

447 Silicon that is used as a semiconductor contains 0.000 001 weight percent phosphorus. How many unit cells are there per phosphorus atom? (Si is cubic with 8 atoms/unit cell; $a = 0.543$ nm.)

451 What is the density of $Fe_{<1}O$ if the Fe^{3+}/Fe^{2+} ratio is 0.14? ($Fe_{<1}O$ has the NaCl structure; ($r_{Fe} + R_O$) averages 0.215 nm.)

452 An intermetallic compound of aluminum and magnesium varies from 49Mg–51Al to 57Mg–43Al (weight bases). What are the atom ratios of these compositions?

453 (a) Which type of vacancy—anion or cation—must be introduced with MgF_2 in order for it to dissolve in LiF? (b) Which type must be introduced with LiF for it to dissolve in MgF_2?

454 A cubic form of ZrO_2 is possible when one Ca^{2+} ion is added in a solid solution for every six Zr^{4+} ions present. Thus, the cations form an fcc structure, and O^{2-} ions are located in the 4-f sites. (a) How many O^{2-} ions are there for every 100 cations? (b) What fraction of the 4-f sites is occupied? (See Fig. 3–4.2.)

455 Is the number of cation vacancies increased, does it remain the same, or is it decreased if the Fe^{3+} content of $Fe_{<1}O$ is increased?

456 One-tenth w/o Fe_2O_3 is in solid solution with NiO. As such, 3 Ni^{2+} are replaced by (2 Fe^{3+} + $\square$) to maintain a charge balance. How many cation vacancies are there per m^3?

457 Is the average cation radius greater, about the same, or less in $Fe_{0.95}O$ compared to $Fe_{0.91}O$?

458 Nonstoichiometric AlMg does not possess a significant number of vacancies; nonstoichiometric FeO does. Explain this difference.

461 The glass-transition temperatures are $90°$ C and $215°$ C for polystyrene (PS) and polyphenylene oxide (PPO), respectively. Assume that amorphous solid solution of Fig. 4–6.1(a) has a T_g that is linear with the mer fraction of the two types of mers. To specify a T_g of $150°$ C, how many lb PPO must the two-component solution contain with each 100 lb of PS?

462 The copolymer of Fig. 4–6.1(b) has a 1-to-1.15 ratio of the two mers, styrene and acrylonitrile. What w/o nitrogen is present?

Chapter 5

PHASE EQUILIBRIA

A *phase diagram* is a collection of curves showing *solubility limits.* On one side of the curve is a single-phase unsaturated solution (liquid or solid). Beyond the curve, the solubility limit is exceeded. Therefore, a second phase must be present and a mixture of two phases exists.

We can learn to use the phase diagrams (1) to predict *what phases* are in equilibrium for selected alloy compositions at desired temperatures; (2) to determine the *chemical composition* of each phase; and (3) to calculate the *quantity* of each phase that is present. Phase diagrams are powerful tools in the hands of scientists and engineers who design materials for specific applications, as well as of other people who must anticipate the stability of specific materials when they design products for service environments.

Throughout this chapter, we shall assume that *equilibrium* is attained; that is, no further reaction is possible between the phases that are present.

5-1
INTRODUCTION

To date, we have considered each phase individually. Many, if not the majority, of technical materials possess *mixtures of phases;* examples are steel, solder, portland cement, grinding wheels, paints, and glass-reinforced plastics. The mixture of two or more phases in one material permits interaction among the phases; therefore, the resulting properties usually are different from the properties of individual phases. It is possible, also, to modify these properties by changing either the shape or the distribution of the phases (see Chapter 7).

Solutions Versus Mixtures

Different *components* can be combined into a single material by means of solutions or of mixtures (A *solution* is a phase with more than one component; a *mixture* is a material with more than one phase.) We discussed solid solutions in Section 4–4, and we are all familiar with liquid solutions. The composition of solutions can vary, because (1) one atom may be substituted for another in the phase structure, or (2) atoms may be placed in the interstices of the structure. The solute does not change the structural pattern of the solvent. A mixture, on the other hand, contains more than one phase (structural pattern). Sand plus added water, rubber with a carbon filler, and tungsten carbide with a cobalt binder are three examples of mixtures. In each of these examples, there are two different phases, each with its own atomic arrangement.

It is possible to have a mixture of two different solutions. For example, in a lead–tin solder, one phase is a solid solution in which tin has replaced some of the lead in the fcc structure, and the other phase has the structure of tin (body-centered tetragonal, or bct). At elevated temperatures, lead atoms may replace a limited number of tin atoms in the bct structure. Thus, just after an ordinary 60–40 solder (60 percent Sn, 40 percent Pb) solidifies, it contains two structures, each a solid solution.

Liquid Solubility Limits

Figure 5-1.1(a) shows the *solubility limit* of ordinary sugar in water; the curve is a *solubility curve.* All compositions shown to the left of the curve will form only one phase, because all the sugar can be dissolved in the liquid phase — *syrup.* With the higher percentages of sugar shown to the right of the curve, however, it is impossible to dissolve the sugar completely, with the result that we have a mixture of two phases, solid sugar and liquid syrup. This example shows the change of solubility with temperature and also demonstrates a simple method for plotting temperature (or any other variable) as a function of composition. From left to right, the abscissa of Fig. 5–1.1(a) indicates the percentage of sugar. We can read the percentage of water directly from right to left, since the total of the component, of course must be equal to 100 percent.

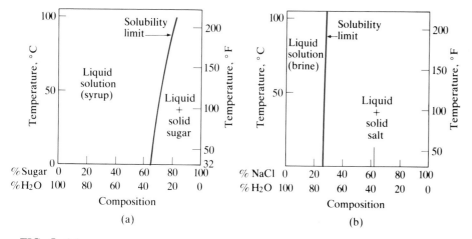

FIG. 5-1.1

Solubility of Sugar in Water. (a) The limit of sugar solubility in water is shown by the solubility curve. Note that the sum of the sugar and water contents at any point on the abscissa is 100 percent. (b) Solubility of NaCl in H_2O.

Figure 5–1.1(b) shows the solubility limit of NaCl in water. As in Fig. 5–1.1(a), there is a region that contains a liquid solution; this particular liquid is called a *brine*. There is a solubility limit that increases with temperature. Beyond the solubility limit, there is a region of liquid brine plus solid salt. Figure 5–1.2 reveals additional features of the H_2O–NaCl *system*. Here the extremes of the abscissa are 100 percent H_2O (0 percent NaCl) and 30 percent NaCl (70 percent H_2O). Note from the figure that (1) the solubility limit of NaCl in a brine solution decreases with decreasing temperature, (2) the solubility limit of H_2O in a brine solution also decreases with decreasing temperature, and (3) intermediate compositions have

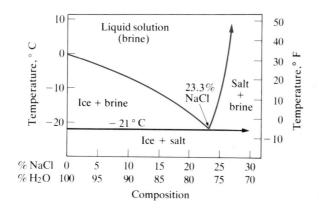

FIG. 5-1.2

Solubility of NaCl Salt in Brine (Right Upward-Sloping Line) and Solubility of Ice in Brine (Left Curve).

melting temperatures lower than those of either pure ice (0° C or 32° F) or of pure salt (800.4° C). Facts (1) and (3) are well known; fact (2), the less familiar limited solubility of ice in the aqueous liquid, can be verified by a simple experiment. A 10–90 salt and water solution can be cooled to less than 0° C (<32° F) and, according to Fig. 5–1.2, it will still be entirely liquid until minus 7° C (19° F) is reached. When such a salty liquid is cooled below −7° C, ice crystals will form and, because the solution cannot contain more than 90 percent H_2O at that temperature, these ice crystals must separate from the liquid. At −20° C (−4° F), the maximum H_2O content possible in a brine solution is 77 percent (23 percent NaCl), as you can verify by making a slush at that temperature and separating the ice from this liquid; the ice will be nearly pure H_2O and the remaining liquid will be saltier (i.e., lower in H_2O) than was the original brine solution.

Solid Solubility Limits

Let us turn our attention to alloys of lead and tin. Face-centered cubic (fcc) lead can dissolve tin into its structure. This substitutional solubility increases from 1 w/o at ambient temperatures to 19 w/o (29 a/o) at 183° C, as is shown in the left curve of Fig. 5–1.3. This lead-rich solid solution is commonly labeled with the Greek letter alpha (α).

Likewise, solid tin can dissolve lead atoms by substitution into its body-centered tetragonal structure, called β. This substitution varies from <1 w/o to 2.5 w/o (1.5 a/o) over the temperature range just cited. Its solubility limit is shown at the right of Fig. 5–1.3, with the abscissa intentionally reversed. (That procedure will let us combine the data of the two solubility curves into one graph in the next section.)

FIG. 5–1.3

Solubility Limits. (a) Tin in the lead-rich, fcc structure, called α. (b) Lead in the tin-rich, bct structure, called β. The abscissa of (b) could also be expressed as 80, 90, and 100 percent Sn, since it is a binary alloy.

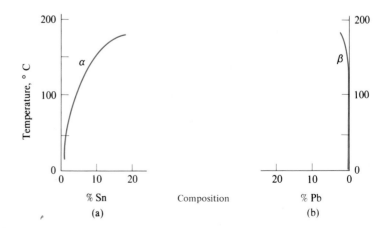

Phases

We have referred to phases already, and we shall do so again numerous times. We are now in a position to define a phase as that part of a material that is *distinct from others in structure or composition.* Consider "ice-water." While of the same composition, the ice is a crystalline solid with a hexagonal lattice; the water is a liquid. The *phase boundary* between the two locates a discontinuity in structure: they are *separate phases.* Or consider silver-plated copper. Both silver and copper are fcc; however, the silver atoms are sufficiently larger than the copper atoms* that there is nearly complete composition discontinuity at room temperature. Thus, they form two separate phases.

Commonly, two phases of a material have distinct differences in both composition and structure—an example is a plastic with a fiberglass reinforcement. In contrast, some phases lose their distinctiveness and dissolve; for example, after dissolving in a cup of coffee, sugar is no longer a separate phase. The same is true of zinc, which, by itself hcp, dissolves in copper (fcc) to produce brass, a single-phase solid solution.

In terms of our discussion here, a *solution* (liquid or solid) is a phase with more than one *component.* A *mixture* is a material with more than one *phase.*

There are many crystalline phases because there are innumerable permutations and combinations of atoms, or groups of atoms. There are relatively few amorphous phases because, lacking long-range order, the atomic arrangements of these phases are less definite and permit a greater range of solution than do crystals.† There is only one gaseous phase. The atoms or molecules are far apart and randomly distributed; as a result, additional vapor components may be introduced into one "structure." No discontinuities are observed in a gas other than at the atomic or molecular level.

Example 5–1.1

According to Fig. 5–1.1(a), a syrup may contain only 67 percent sugar (33 percent water) at 20° C (68° F), but 83 percent sugar at 100° C (212° F). One-hundred g sugar and 25 g water are mixed and boiled until all the sugar is dissolved. During cooling, the solubility limit is exceeded, so (with time) excess sugar separates from the syrup. If equilibrium is attained, what is the weight ratio of syrup to excess sugar at 20° C?

Procedure Basis: 100 g sugar + 25 g H_2O = 80% sugar + 20% water. All of the sugar is dissolved at first, to give a single phase of syrup (Fig. 5–1.1a). When the syrup cools, the *solubility limit* for the 80–20 sugar–water composition is reached at 87° C (189° F). Below this temperature, some of the sugar is rejected (precipitated) from solution.

---

* $R_{Cu} = 0.1278$ nm; $R_{Ag} = 0.1444$ nm.

† The fiberglass and plastic just cited are both amorphous. Although their structures are not sufficiently similar to produce a single phase, each can be a solvent for large quantities of solutes.

Calculation At 20° C, there are x g of syrup and $(125 - x)$ g of excess sugar.

Sugar balance: $0.67x + (1.0)(125 - x) = 100$ g

$x = 75.75$ g syrup

Or H_2O balance: $0.33x + (0)(125 - x) = 25$ g

$x = 75.75$ g syrup

Excess sugar: $125 - x = 49.25$ g

Ratio: syrup/excess $= 75.75$ g/49.25 g $= 1.54$

Comments Typically, some *supercooling* is encountered before the excess phase (in this case, sugar) starts to separate. In fact, in this case, supercooling can proceed to room temperature, so that the start of separation may be delayed considerably. Similar supersaturation commonly occurs in metals and in other materials.

Example 5–1.2

A brine contains 9 percent NaCl (91 percent H_2O by weight). How many g H_2O (per 100 g brine) must be evaporated before the solution becomes saturated at 50° C (122° F)?

Solution Basis: 100 g brine $= 9$ g NaCl $+ 91$ g H_2O. From Fig. 5–1.1(b), solubility of NaCl in brine $= 27$ percent at 50° C.

grams saturated brine $= 9$ g/0.27 $= x/0.73$

$x = 24.3$ g H_2O for saturation

91 g $- 24.3$ g $= 66.7$ g to be evaporated

Alternatively, 9 g of the salt is equal to 27 percent of the brine remaining after evaporation:

$(0.09)(100$ g$) = 0.27(100 - y)$

$y = 66.7$ g to be evaporated

Comment By convention, compositions of liquids and solids are reported in weight percent, unless stated otherwise.

Example 5–1.3

A brine with 40 g H_2O and 10 g NaCl is cooled to $-10°$ C (14° F). (a) This brine is placed in a beaker. How much salt can be dissolved in it? (b) An identical 40–10 brine is placed in a second beaker. How much ice can be added to it without exceeding the solubility limit?

Procedure From Fig. 5–1.2, the brine can range from 13 to 25 percent NaCl (or from 75 to 87 percent H_2O) at $-10°$ C. In (a), calculate the total salt $(= 10$ g $+ x$ g); in (b), calculate the total H_2O $(= 40$ g $+ y$ g).

Calculations

(a)   g NaCl = 0.25(50 + x) = 10 + x

$$x = 3.33 \text{ g NaCl added}$$

(b)   g H_2O = 0.87(50 + y) = 40 + y

$$y = 26.9 \text{ g } H_2O \text{ added}$$

Example 5–1.4

Salt is spread on the street when the temperature is −10° C (14° F). Part of the ice melts, and the brine that forms becomes saturated with H_2O. Ninety g of this brine is splattered over your car. How many grams of NaCl come with it?

Solution   From Fig. 5–1.2, the solubility limit at −10° C is 87 percent H_2O in brine. The balance is NaCl (13 percent).

   0.13(90 g) = 11.7 g NaCl

Comments   A short, transparent millimeter rule will permit you to take a more accurate reading of the solubility limits. Many phase diagrams are drawn to an accuracy of ± 1 percent on a larger scale and then are photoreduced. Thus, your accuracy of calculation is generally limited by the accuracy with which you can read the graphs.

5–2
PHASE DIAGRAMS (QUALITATIVE)

Figure 5–2.1 is a *phase diagram* of the Pb–Sn *system*. It shows which phases are present for all Pb–Sn alloys in the temperature range of 0 to 350° C. Thus, at 100° C, an alloy of 60 w/o Pb (40 w/o Sn) has two phases, called α and β; α and liquid coexist at 200° C; and only liquid is present at 300° C. Likewise, a 20 Pb–80 Sn alloy* also has a mixture of α and β at 100° C; but, it is liquid plus β at 200° C.

A phase diagram is a collection of solubility limit curves. Three pairs of these curves produce the Pb–Sn diagram of Fig. 5–2.1. They are

1. Solubility limit of tin in the fcc phase called α, and the solubility limit of lead in the bct phase called β. These are duplicated from Fig. 5–1.3, and are pertinent at temperatures when *no liquid* is present.

---

* As indicated in the footnote to Example 2–2.1, analyses of condensed phases (solids and liquids) are reported in weight percent, w/o, *unless stated otherwise.* Most phase diagrams of this text will show the corresponding atom percents, a/o, on the upper abscissa, for the reader's convenience. Likewise, the Fahrenheit scale is shown as the right ordinate.

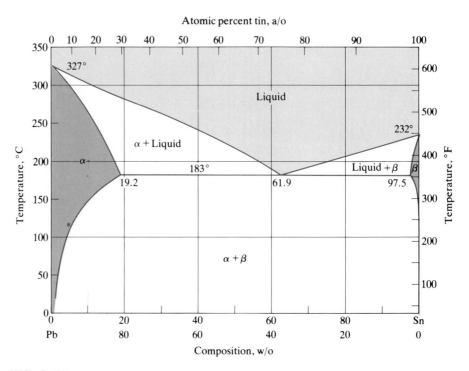

Atomic percent tin, a/o

FIG. 5-2.1

Pb-Sn Diagram. This diagram indicates the phase compositions and permits the calculation of phase quantities for any lead–tin mixture at any temperature. (Adapted from *Metals Handbook,* ASM International.)

2. Solubility limit of tin in the liquid metal (61.9 w/o Sn at 183° C to 100 w/o Sn at 232° C). The solubility limit of lead in the liquid metal (38.1 w/o Pb at 183° C to 100 w/o Pb at 327° C).

3. The solubility limit of tin in α and of lead in β, when liquid is present. The former drops from 19.2 w/o Sn at 183° C to 0 at the melting point of lead (327° C). The latter drops from 2.5 w/o Pb at 183° C to 0 at the melting point of tin (232° C).

Figure 5–2.1 is not only a phase diagram; it is also an *equilibrium diagram.* Its data required the completion of all possible reactions among the phases in this two-component system.

The phase fields in equilibrium diagrams, of course, depend on the particular systems being depicted. When copper and nickel are mixed, the phase diagram is as shown in Fig. 5–2.2. This phase diagram is comparatively simple, since only two phases can be present. In the lower part of the diagram, all alloys form only one solid solution and therefore only one crystal structure, denoted by the Greek letter alpha (α). Both the nickel and the copper have fcc structures. Since the atoms of

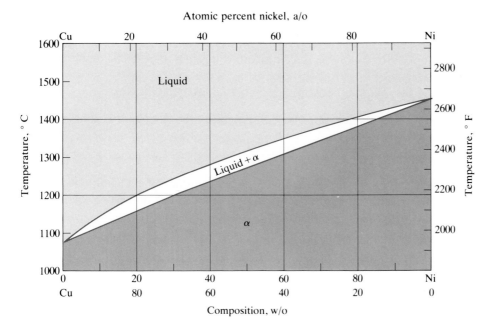

Atomic percent nickel, a/o

Composition, w/o

FIG. 5–2.2

Cu–Ni Diagram. All solid alloys contain only one phase. This phase is fcc. (Adapted from *Metals Handbook,* ASM International.)

each are nearly the same size, it is possible for nickel and copper atoms to replace each other in the crystal structure in any proportion at 1000° C. When an alloy containing 60 percent copper and 40 percent nickel is heated, the solid phase exists until the temperature of about 1235° C (2255° F) is reached. Above this temperature and up to 1280° C (2336° F), the solid and liquid solutions coexist. Above 1280° C, only a liquid phase remains.

Eutectic Temperatures and Compositions

Pure lead melts at 327° C; pure tin melts at 232° C. Alloys of lead and tin melt at lower temperatures—as low as 183° C for the alloy of 61.9 Sn–38.1 Pb. Thus, it is no mere coincidence that a 60–40 *solder* is widely used for electrical connections. An alloy of this composition melts most readily, and thus facilitates production with a minimum of possible damage to adjacent circuit components.

The alloy just described is an *eutectic alloy.* The *eutectic temperature* is 183° C; the *eutectic composition* is 38.1 Pb–61.9 Sn. The eutectic temperature and composition exist at the intersection of the two solubility curves that limit the composition of the liquid. (In addition, the maximum solubility limits in α and β of Figure 5–2.1 are at the eutectic temperature.)

Eutectic Reactions

A solder of eutectic composition (61.9 Sn – 38.1 Pb) changes from a single liquid solution into two solid phases as it is cooled through the eutectic temperature:

$$\text{liquid (61.9 Sn)} \xrightarrow[\text{cooling}]{183^\circ} \alpha \text{ (19.2 Sn)} + \beta \text{ (97.5 Sn)} \tag{5–2.1}$$

Heating reverses the reaction. We can generalize the eutectic reaction:

$$L_2 \underset{\text{heating}}{\overset{\text{cooling}}{\rightleftharpoons}} S_1 + S_2 \tag{5–2.2}$$

where the subscripts imply a sequential change in one of the components. The composition of the liquid lies between the compositions of the two solids.

Freezing Ranges

As shown in the Pb–Sn and Cu–Ni phase diagrams (Figs. 5–2.1 and 5–2.2), the range of temperatures over which freezing occurs varies with the composition of the alloy. This situation influences the plumber, for example, who selects a high-lead alloy as a "wiping" solder because it will not solidify at a specific temperature. The freezing range of an 80–20 Pb–Sn solder is from 270 to 183° C, as compared to only 190 to 183° C for a 40–60 Pb–Sn solder. (With the 80–20 solder, the plumber can wipe a "mushy" solder onto the surface of a pipe without it flowing off. The final freezing occurs after the pipe and the fitting are joined.)

We use the terms *liquidus,* denoting the locus of temperatures above which all compositions are liquid, and *solidus,* denoting the locus of temperatures below which all compositions are solid. Every phase diagram for two or more components must show a liquidus and a solidus, and an intervening freezing range. Whether the components are metals or nonmetals (Fig. 5–2.3), there are certain locations on the phase diagram where the liquidus and solidus meet. For a pure component, a contact point lies at the edge of the diagram. When it is heated, a pure material will remain solid until its melting point is reached, it will then change entirely to liquid before it can be raised to a higher temperature.

The liquidus and solidus also meet at the eutectic temperature and composition. In Fig. 5–2.1, the solder composed of 61.9 percent tin and 38.1 percent lead is entirely solid below the eutectic temperature and entirely liquid above it. At the eutectic temperature, three phases can coexist [$(\alpha + \text{Liq} + \beta)$ in the Pb–Sn system].

OPTIONAL

Isothermal Cuts

A traverse across the phase diagram at a constant temperature (*isotherm*) provides a simple sequence of alternating one- and two-phase fields. Consider the diagram of SiO_2–Al_2O_3 at 1650° C (3000° F) in Fig. 5–2.4; you will see that the sequence is 1–2–1–2–1–2–1. With pure SiO_2, only *one* phase exists (named cristobalite).

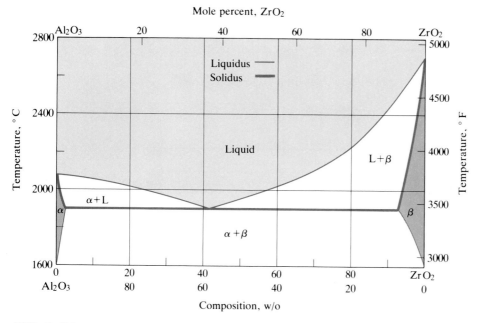

FIG. 5-2.3

The Al₂O₃–ZrO₂ Diagram. As in all phase diagrams, the liquidus delineates the lowest
temperatures for solely liquid. The solidus is the upper limit for completely solid. The
freezing range of solid + liquid lies between the two. The two meet at the eutectic and
where single phases melt. (Adapted from Alper et al., American Ceramics Society)

It holds negligible amounts of Al_2O_3 in solid solution.* Therefore, a second phase
(liquid) appears with the addition of Al_2O_3. The *two-phase* region contains cristobalite and liquid. Between 4 percent Al_2O_3 (96 percent SiO_2) and 8 percent Al_2O_3
(92 percent SiO_2), the liquid can dissolve all the SiO_2 and Al_2O_3 that is present, so
just *one* phase exists. Beyond 8 percent Al_2O_3 (<92 percent SiO_2), the solubility
limit of the liquid for Al_2O_3 is exceeded, and solid mullite precipitates. The *two*
phases liquid and mullite coexist.† The solid-solution range of mullite is from 71
percent Al_2O_3 (29 percent SiO_2) to 75 percent Al_2O_3 (25 percent SiO_2). Only *one*
phase is stable in this range, because it can accommodate both the SiO_2 and Al_2O_3
that are present. A two-phase field, mullite and corundum (Al_2O_3) follows and
extends to within a line's width of the right side of the phase diagram. With only
Al_2O_3, this *one* phase is called corundum.

---

* There is a slight solubility, but it is so low that we are not able to show it on the phase diagram.

† When 8 percent Al_2O_3 is exceeded just slightly, there will be very little mullite. When the right side of
this two-phase field is approached, very little liquid remains.

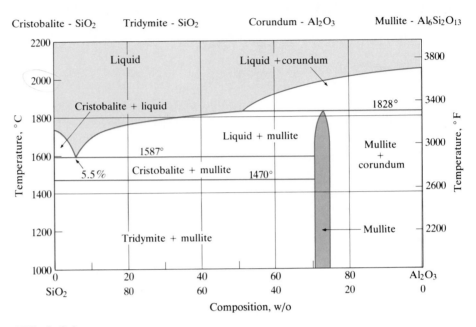

Cristobalite - SiO₂   Tridymite - SiO₂   Corundum - Al₂O₃   Mullite - Al₆Si₂O₁₃

FIG. 5–2.4

SiO₂–Al₂O₃ Diagram. The phase diagrams for nonmetals are used in the same manner as are those for metals. The only difference is the longer time required to establish equilibrium. (Adapted from Aksay and Pask, *Science.*)

Phase Names and Labels

Brass is the phase name of an fcc solid solution of zinc in copper. The *bronze* of historical significance is an fcc alloy of tin in copper. *Sterling silver* contains 92.5 Ag and 7.5 Cu. The fcc structure of silver is maintained, but with copper substitution. We shall encounter other names for phases, such as *ferrite* for a bcc solution where iron is the principal component, and *austenite,* for an fcc phase based on iron. These two iron-based phases have also been labeled by the Greek letters alpha (α) and gamma (γ), respectively.

Greek-letter labels are more common than names, because they are simpler. As we have seen, α and β are the labels for the two solid phases of Pb–Sn alloys (Fig. 5–2.1).* There, α has the fcc structure of lead, but contains tin up to the solubility limit; β has the bct structure of tin, but may contain lead to the limits shown by the curve.

---

* Admittedly, the label α is used for both the bcc phase that is based on iron and the fcc phase that is based on lead. Furthermore, α is used very widely as the first named phase in many other alloys. (See Section 5–6.) However, this is not a problem, since we usually are using specific binary systems. (Remember that x is used repeatedly in algebraic calculations without causing labeling confusion.)

Example 5-2.1

Sterling silver, an alloy containing approximately 92.5 percent silver and 7.5 percent copper (Fig. 5-2.5), is heated slowly from room temperature to 1000° C (1830° F). What phase(s) will be present as heating progresses?

Answer

room temperature to 740° C	$\alpha + \beta$
740° C to 810° C	only α
810° C to 900° C	α + liquid
900° C to 1000° C	only liquid

Example 5-2.2

A combination of 90 percent SiO_2 and 10 percent Al_2O_3 is melted at 1800° C and then cooled extremely slowly to 1400° C. What phase(s) will be present in the cooling process?

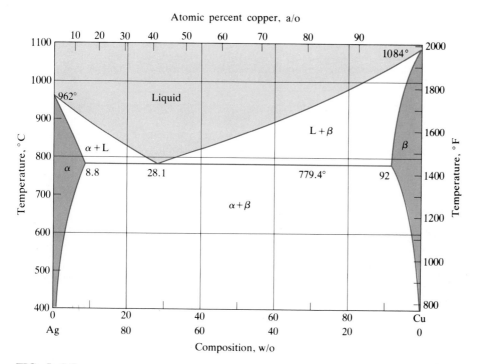

FIG. 5-2.5

Ag-Cu Diagram. (Adapted from *Metals Handbook*, ASM International.)

Answer (See Fig. 5–2.4.)

1800° C to 1700° C	only liquid
1700° C to 1587° C	liquid + mullite ($Al_6Si_2O_{13}$)
1587° C to 1470° C	mullite + cristobalite (SiO_2)
<1470° C	mullite + tridymite (SiO_2)

Comments There must be three phases (liquid + mullite + cristobalite) at 1587° C as the material passes from the (Liq + Mul) field to the (Cri + Mul) field. Likewise, there are three phases at 1470° C.

The cooling will have to be extremely slow, because the process of changing the strong Si–O bonds from one structure to another is very slow.

Pure silica has three common polymorphs: cristobalite and tridymite at high temperatures, and quartz at low temperatures.

Example 5–2.3

Refer to Fig. 5–2.5 for the Ag–Cu system. (a) Locate the liquidus and the solidus. (b) How many phases are present where the two meet?

Answers

(a) Liquidus: 962° at 100 percent Ag, to 779° C at 71.9 percent Ag (28.1 percent Cu), to 1084° at 100 percent Cu. Solidus: 962° at 100 percent Ag, to 779° C at 91.2 percent Ag (8.8 percent Cu), remaining at 779° C to 92 percent Cu (8 percent Ag), to 1084° at 100 percent Cu.

(b) With a single component (only Ag *or* only Cu), two phases are present (solid + liquid) where the liquidus and solidus meet.

   With two components (Ag *and* Cu), three phases (α + liquid + β) are present where the liquidus and solidus meet (at the eutectic).

Comment There are always two phases present in the temperature range between the liquidus and solidus of a two-component, or binary, phase diagram.

Example 5–2.4

Refer to the Al–Mg diagram in Fig. 5–6.4. Apply the 1–2–1–2– . . . rule (a) at 500° C, and (b) at 200° C.

Answers

(a)

	α		$(\alpha + L)$		Liq		$(L + \epsilon)$		ϵ	
% Mg	0		11		30		76		92.5	100
% Al	100		89		70		24		7.5	0

(b) $|\alpha|(\alpha+\beta)|\beta|(\beta+\beta')|\quad\beta'\quad|\quad(\beta'+\gamma)|\gamma|\quad(\gamma+\epsilon)|\epsilon|$
   % Mg 0 3   35 37   ~41 ~42   49 ~56   96 100
         | | | | | | | | | |

Comment The dashed lines on the phase diagram are the best estimates based on present information.

5–3
CHEMICAL COMPOSITIONS OF EQUILIBRATED PHASES

In addition to serving as a "map," a phase diagram shows the chemical *compositions* of the phases that are present under conditions of equilibrium after all reaction has been completed. This information, along with information on the amount of each phase in a two-phase mixture (Section 5–4), constitutes useful data for the scientists and engineers who are involved with materials development, selection, and application in product design.

One-Phase Areas

The determination of the chemical composition of a single phase is automatic: *It has the same composition as the alloy.* This is to be expected, since only liquid is present in a 60 Sn–40 Pb alloy at 225° C (Fig. 5–3.1), the liquid has to have the same 60–40 composition. This determination rule also holds when the location in the phase diagram involves a single-phase solid solution.

Observe that we reported the chemical composition of the individual phase (and of the total alloy) in terms of the *components*—in this case, lead and tin.

Two-Phase Areas

The determination of the chemical compositions of two phases can be handled on a rote basis. We shall do this first; we shall examine the rationale in the next paragraph.

The chemical compositions of the two phases are located at the *two ends* of the *isotherm,* or *tie-line,* across the two-phase area. To illustrate, take an 80 Pb–20 Sn solder at 150° C. As indicated in Fig. 5–3.1, α has a chemical composition of 10 w/o tin (and therefore 90 w/o lead). The composition of the β is nearly 100 w/o tin. Other isotherms on Fig. 5–3.1 permit us to read the chemical compositions of the two phases of any Pb–Sn alloy at any temperature.

The basis for this procedure is simply that the *solubility limit* for tin in α at 150° C is 10 w/o. Our alloy exceeds this limit with its composition of 20 w/o tin.

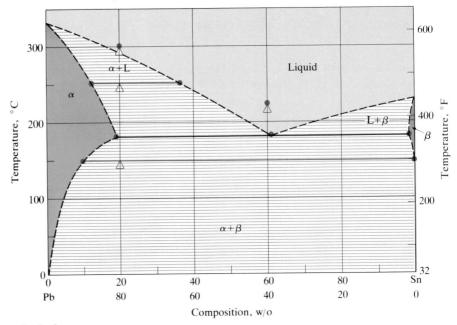

FIG. 5–3.1

Compositions of Phases (Pb–Sn Alloys, Fig. 5–2.1). At 150° C, an 80 Pb–20 Sn alloy contains α and β. The chemical composition of α is dictated by the solubility curve. At this temperature, the solubility limit is 10 percent Sn (and 90 percent Pb) in the fcc α phase. See the text.

Therefore, α is saturated with tin and the excess tin is present in β. Conversely, the solubility limit for lead in β is <1 w/o; therefore, almost all the lead must be in a phase other than β—specifically, it must be in α.

Three-Phase Temperatures and Eutectic Reactions

A liquid that has the analysis of the eutectic composition (38.1 Pb–61.9 Sn, when we consider the Pb–Sn system) separates into two solid phases (α and β) at the eutectic temperature (183° C). Thus, at this temperature *only,* three phases can be in equilibrium. If this alloy is heated, the two solid phases of this solder melt into a one-phase liquid, as we expressed in Eq. (5–2.1). Alternatively, we could have written that equation as

$$\alpha\ (80.8\ \text{Pb}) + \beta\ (2.5\ \text{Pb}) \xrightarrow[\text{heating}]{183° \text{C}} \text{liquid}\ (38.1\ \text{Pb}) \qquad (5\text{–}3.1)$$

Since the three phases of an eutectic reaction can coexist at only a specific temperature, this condition is called *invariant*. There is no choice of a temperature other than 183° C. There is no choice of phases—they must be α, β, and liquid. The compositions are fixed—they are 80.8, 2.5, and 38.1 percent lead, respectively.

Phase Rule

We have just said that conditions are invariant in the two-component system when three phases are present. When only one or two phases are present, we have some freedom with respect to our choices of temperature and composition. For example, in a one-phase field of an equilibrium diagram, both temperature *and* composition can be varied and still retain the same single phase. The *variance* is two.

In a two-phase field, the variance is one, because we have the freedom to make one choice. If we vary temperature, we have no choice of compositions. For a sterling-silver alloy (92.5 Ag–7.5 Cu), for example, we can choose 850° C (1562° F). From Fig. 5–2.5, however, the phase compositions are fixed at 95–5 for α, and 85–15 for liquid. Conversely, if our choice is a specific phase composition within a sterling-silver alloy—say 90 Ag–10 Cu in the liquid—then the temperature must be 890° C (1634° F—from Fig. 5–2.5).

These variances, or *degrees of freedom, F,* are dictated by the *phase rule:*

$$P + F = C + 1 \tag{5–3.2}*$$

where P is the number of phases, and C is the number of components (which is two in a binary system). With three phases present, $F = 0$, thus being invariant.

Example 5–3.1

Consider the Cu–Ni system (Fig. 5–2.2) at 1300° C. (a) What is the solubility limit of copper in solid α? (b) What is the solubility limit of nickel in the liquid? (c) What are the chemical compositions of the phases in a 45 Cu–55 Ni alloy at 1300° C?

Answer (a) 42 w/o Cu (b) 45 w/o Ni (c) α: (42 Cu–58 Ni); L: (55 Cu–45 Ni).

Comment It is the solubility limit that determines the chemical compositions of the phases in a two-phase area.

---

* Equation (5–3.2) is appropriate when temperature is the only external variable. A more general formula is

$$P + F = C + E \tag{5–3.3}$$

where E is the number of external variables, such as temperature, pressure, and, in unusual cases, electric field and magnetic field.

Example 5–3.2

An alloy of 40 Ag–60 Cu (Fig. 5–2.5) is cooled slowly from 1000° C to room temperature. (a) What phase(s) will be present as cooling progresses? (b) Indicate their compositions. (c) Write the eutectic reaction.

Answer

The table shows answers for (a) and (b):

TEMPERATURE, °C	α	LIQUID	β
1000	—	40 Ag–60 Cu	—
800	—	66 Ag–34 Cu	8 Ag–92 Cu
779.4	91.2 Ag–8.8 Cu	71.9 Ag–28.1 Cu	8 Ag–92 Cu
600	96.5 Ag–3.5 Cu	—	2 Ag–98 Cu
400	99 Ag–1 Cu	—	Near 100 Cu
20 (extrapolated)	Near 100 Ag	—	Near 100 Cu

(c) liquid (71.9% Ag) $\xrightleftharpoons{780° C}$ α(91.2% Ag) + β(8% Ag)

Comments The liquid becomes saturated with copper at about 890° C. The first β to separate at this temperature has the composition of 7 Ag–93 Cu.

Example 5–3.3

Figure 5–3.2(a) shows the microstructure of wustite (solid FeO) plus an FeO–SiO$_2$ liquid. The overall composition is 90 FeO–10 SiO$_2$. (a) What is the liquidus temperature for this material? (b) What is the eutectic reaction? (c) What is the composition of the fayalite? (d) What phases in the FeO–SiO$_2$ system undergo a polymorphic change? Indicate their transition temperature.

Procedure Use the FeO–SiO$_2$ diagram of Fig. 5–3.2(b).

Calculation

(a) 1300° C. Above that temperature, only liquid is present for this 90 FeO–10 SiO$_2$ composition.

(b) LIQuid (77 FeO) $\xrightarrow[\text{cooling}]{1175°}$ WUStite (100 FeO) + FAYalite (70 FeO) (5–3.4)

(c) From Fig. 5–3.2(b):

$$Fe_2SiO_4 = 2 \text{ FeO} + SiO_2 \qquad (5–3.5)$$

$$\% \text{ FeO} = \left(\frac{2 \, (55.85 + 16)}{2(55.85) + 28.01 + 2(16)} \right) (100) = 70.5\%$$

(d) TRIdymite (SiO$_2$) $\Longleftarrow$1470° C$\Longrightarrow$ CRIstobalite (SiO$_2$) (5–3.6)

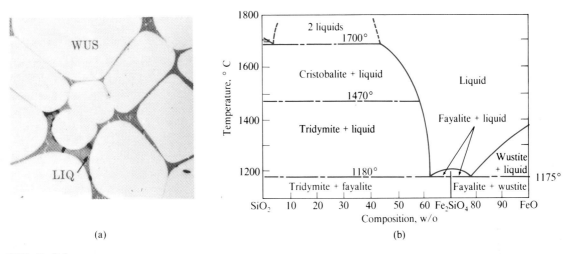

(a)   (b)

FIG. 5–3.2

FeO–SiO₂ (Example 5–3.3). (a) Microstructure of 90 FeO–10 SiO₂ (quenched from 1200° C). (From L. H. Van Vlack, *Physical Ceramics for Engineers,* Addison-Wesley, Reading, Mass., 1964, Fig. 7–4, with permission.) (b) Phase diagram. Iron oxide (FeO) is named WUStite. Iron silicate (Fe_2SiO_4) is named FAYalite. TRIymite and CRIstobalite are both silica (SiO_2). (FeO–SiO₂ phase diagram adapted from Bowen and Schairer, *Amer. Journ. Sci.* **24,** 5th series.)

Comments   This polymorphic reaction proceeds very slowly, since strong Si—O bonds must be broken, and only subtle, second-neighbor changes occur between the two phases. There is a second eutectic composition at 62 FeO–38 SiO₂.

Example 5–3.4

Correlate the microstructures of Fig. 5–3.3(b and d) with the Al–Si phase diagram. The dark phase is β—nearly pure silicon. The white phase is α—the aluminum-rich phase. (*Note:* The photomicrographs of Fig. 5–3.3 were made after the alloys had cooled to room temperature, but they show the structures that were formed during solidification.)

Observations   The 88 Al–12 Si alloy (Fig. 5–3.3b) is a fine mixture of the two phases (α + β). The 50 Al–50 Si alloy (Fig. 5–3.3d) has 45–50 v/o of "massive" β crystals, plus the fine mixture of α + β.

Rationale   (b) The 88 Al–12 Si alloy has the eutectic composition. It was all liquid above 577° C. Below 577° C, this alloy solidified to an intimate mixture of α + β, because the two phases crystallized simultaneously. (d) The 50 Al–50 Si alloy was molten above the liquidus temperature (1050° C). Just above 577° C, this alloy contained a mixture of β and a liquid that had the eutectic composition (88 Al–12 Si). The β had grown in size during the 1050 to 577° C cooling interval. In passing through the eutectic temperature (577° C), the larger β crystals were unchanged; however, the part that was the eutectic liquid crystallized to the same fine mixture of α + β that we see in Fig. 5–3.3(b).

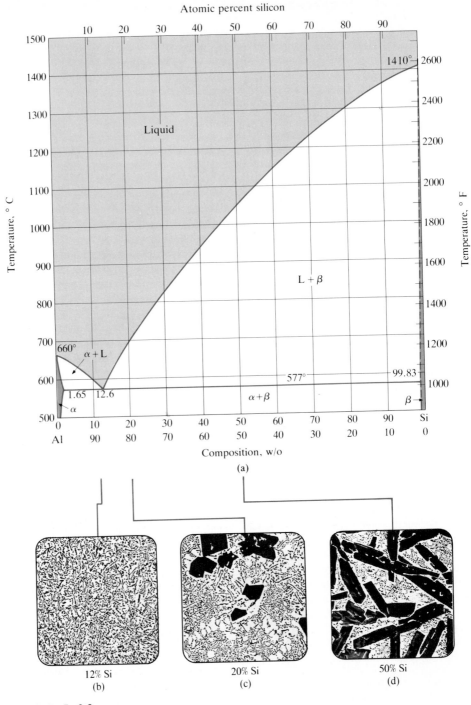

FIG. 5–3.3

Al–Si Alloys (Example 5–3.4). (a) The Al–Si diagram. (Adapted from *Metals Handbook*, ASM International.) (b) 88 Al–12 Si alloy. (c) 80 Al–20 Si alloy. (d) 50 Al–50 Si alloy. (Photomicrographs courtesy of Aluminum Research Laboratories.)

Comments   As the compositions are shifted from the eutectic toward higher silicon contents (b to c to d of Fig. 5–3.3) there is an increase in the amount of the "massive," proeutectic β. Conversely, the amount of final liquid decreases, so there is less of the fine, two-phase $\alpha + \beta$ mixture. (*Proeutectic* means formed before the eutectic reaction occurred.)

5–4
QUANTITIES OF PHASES IN EQUILIBRATED MIXTURES

In addition to identifying the stable, or equilibrium phases (Section 5–2) and to obtaining their chemical compositions from the phase diagrams (Section 5–3), we can determine the *quantity* of each phase that is present in an equilibrated, two-phase mixture. This ability will be useful to us when we consider the properties of multiphase materials later in this text.

One-Phase Areas

Again, as in Section 5–3, we have an automatic situation. With 225 g of an 80 Pb–20 Sn alloy at 300° C, only liquid is present, and its quantity is 225 g. It is equally valid to state that all (or 100 percent) of the alloy is liquid. Thus, we do not have to specify the exact weight of the alloy that is present.

Two-Phase Areas (by Interpolation)

We can use a rote basis to indicate the quantities (or the *quantity fractions*) of the phases that are present in the two-phase areas of phase diagrams. We shall do this first; the rationale will follow. We determine the quantities of the two phases by *interpolating the composition of the alloy along the tie-line between the compositions of the two phases.* To illustrate, we shall again consider an 80 Pb–20 Sn solder at 150° C. As indicated in Fig. 5–4.1, the chemical composition of the alloy (80 Pb–20 Sn) is at a position that is 0.11 of the distance between the chemical composition of α (90 Pb–10 Sn) at this temperature, and the chemical composition of β (<1 Pb and ~100 Sn). Therefore, of the total amount of solder, the quantity fraction of β is 0.11 (and 0.89 of α) at 150° C.* It is equally appropriate to report this as 89 percent α and 11 percent β.

As another example, consider the same alloy at 250° C. On the basis of Fig. 5–4.1, we have α (88.5 Pb–11.5 Sn) and L (63 Pb–37 Sn). The chemical composition of this alloy as a whole (80 Pb–20 Sn) is one-third of the distance between the chemical composition of α and the chemical composition of the liquid. Therefore, of the total amount of solder at 250° C, the quantity fraction of liquid is one-third and that of α is two-thirds.

---

* Thus, if we had 225 grams of solder, there would be ~200 grams of α and ~25 grams of β.

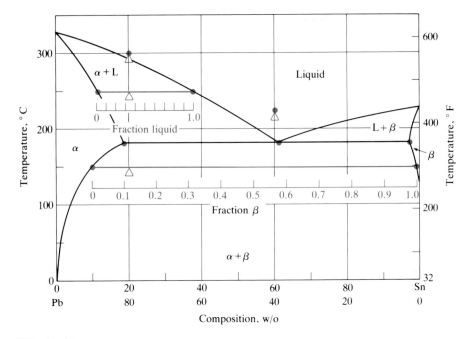

FIG. 5–4.1

Quantities of Phases (Pb–Sn Alloys, Fig. 5–2.1). As observed in Fig. 5–3.1, an 80 Pb–20 Sn alloy contains α and β at 150° C. By interpolation, the fraction of β is 0.11. This same alloy contains 0.33 liquid (and 0.67 α) at 250° C. See text.

Observe that we have reported these quantity figures in terms of *phases* in a mixture — in these cases, α and β, or α and L. This is in contrast to the chemical compositions that were reported in terms of the *components* — Pb and Sn.

To follow this procedure, we would need to have a stretchable scale, because the length of the isotherm between the two solubility limit curves varies continuously. Interpolation by using a millimeter scale offers a simple alternative.*

Lever Rule

If you are alert, you may have noticed an alternative approach that uses values on the abscissa for interpolation. Consider again our 80–20 Pb–Sn alloy at 250° C:

$$\frac{L}{A + L} = \frac{C - C_\alpha}{C_L - C_\alpha} \qquad (5\text{–}4.1)$$

where L and A are the masses of liquid and α, respectively. Compositions, C, of the

---

* Use a small, transparent millimeter scale.

phases are subscripted with L, for liquid and with α for α. The composition of the total alloy carries no subscript. Thus, based on the tin analyses,

$$\% \text{ liquid} = \left(\frac{20 - 11.5}{37 - 11.5}\right) \times 100 = 33\% \qquad \textbf{(5-4.2)}$$

In more general terms,

$$\frac{X}{X + Y} = \frac{C - C_y}{C_x - C_y} \qquad \textbf{(5-4.3)}$$

This is called the *lever rule** because the same equation applies to a lever, where the total composition is the center of gravity at the fulcrum.

Materials Balances

The arithmetic of Eq. (5-4.3) also applies to the calculation of the center of gravity, and we can use it in a *materials balance.* By the term materials balance, we mean that the whole is equal to the sum of the parts. At 250° C, the lead in the total alloy is equal to the lead in the α plus the lead in liquid.

To illustrate, let us use 600 g of our same 80 Pb-20 Sn solder at 250° C. There are 480 g of Pb (and 120 g of Sn). We can consider that there are A grams of α (88.5 Pb-11.5 Sn) and L grams of liquid (63 Pb-37 Sn). Those phase compositions are located in Figs. 5-3.1 and 5-4.1. Of course,

$$A + L = 600 \text{ g}$$

Thus, on the basis of the chemical compositions of the two phases and of the total alloy, the lead balance is

$$0.885A + 0.63L = 0.80(A + L) \qquad \textbf{(5-4.4)}$$

or, in this case, $A = 2L$. This means that we have 200 g of liquid and 400 g of α—the same one-third and two-thirds values we calculated previously.

A generalization of Eq. (5-4.4) is

$$C_x(X) + C_y(Y) = C(X + Y) \qquad \textbf{(5-4.5)}$$

where C_x and C_y are the chemical *compositions* of one of the components of phases x and y.

The Three-Phase Special Case

Lead-tin solders have three phases when they are equilibrated at 183° C, which is the eutectic temperature. In this special case, we *cannot* calculate exactly the quantity fractions of α, β, and the liquid that are present. We can calculate that a 70

---

* It also is sometimes called the *inverse lever rule,* because the amount of each phase is proportional to the length of the opposite end of the lever.

Pb–30 Sn alloy contains 0.86 α at 182° C (and 14 percent β). We also can determine that the same solder contains 0.25 liquid (and 75 percent α) at 184° C. At the eutectic temperature, between 182 and 184° C, the β and some of the α react on heating to give a eutectic liquid. In the process, 1000 grams of solder would have 110 g of α (that is, 860 − 750 g) and 140 g of β consumed to form 250 g of eutectic liquid (61.9 Pb–38.1 Sn). Therefore, with three phases at 183° C, we can indicate only that the quantity of α is between 860 and 750 g; that of β is between 140 and 0 g; and that of liquid is between 0 and 250 g.*

Example 5–4.1

Consider a 90 Cu–10 Sn bronze. (a) What phases are present at 300° C? What are their chemical compositions? What is the fraction of each? (b) Repeat part (a) for 600° C. (c) At what temperature is there one-third liquid? (d) At what temperature is there two-thirds liquid?

Procedure Refer to Fig. 5–6.2, the Cu–Sn diagram. Read the phases and their compositions directly. Interpolate for the fractions, either with an mm scale or by using the abscissa values.

Answers

   (a) α: 93 Cu–7 Sn $(37-10)/(37-7) = 0.9$

      ϵ: 63 Cu–37 Sn $(10-7)/(37-7) = 0.1$

   (b) α: 90 Cu–10 Sn   1.0

   (c) By using an mm scale: $\frac{1}{3}$ L at 900° C

   (d) By using an mm scale: $\frac{2}{3}$ L at 965° C

Example 5–4.2

Using lead, make a materials balance that shows A, which is the quantity of α; and B, which is the quantity of β, in 75 g of a 70 Pb–30 Sn solder at 182° C.

Solution From Eq. (5–4.5), and using the lead analyses of Fig. 5–2.1,

$$0.808A + 0.025B = 0.70(A + B)$$
$$A + B = 75 \text{ g}$$

Solving simultaneously,

$$A = 64.7 \text{ g} \qquad B = 10.3 \text{ g}$$

---

* This special case of three phases for two-component systems has an analog in one-component systems. For example, with H_2O at 0° C (32° F), it is impossible to calculate the fraction that is ice and the fraction that is liquid water.

Comments A materials balance using tin, $0.192A + 0.975B = 0.30(A + B)$, gives the identical answer.

The fractions of the two phases are:

$$A/(A + B) = 0.86 \qquad B/(A + B) = 0.14$$

Example 5–4.3

As a function of temperature, make a histogram of the quantity of the phases in 10.0 g of a 40 Ag–60 Cu alloy (Fig. 5–2.5).

Procedure Plot grams of β as dark bars, grams of α as medium bars, and grams of liquid as light bars. Use intervals of 50° C.

Sample Calculations (Shown on Fig. 5–4.2 as large dots.) We determined the chemical compositions of the phases in Example 5–3.2. At 890° C and above, the alloy is all liquid; therefore, there are 0 g of β. By interpolation,

At 800° C	$0.45(10 \text{ g}) = 4.5 \text{ g}$
780° +	$0.50(10) = 5.0$
780° −	$0.615(10) = 6.15$
600°	$0.60(10) = 6.0$
400°	$0.595(10) = 5.95$
20°	$0.60(10) = 6.0$

Comment Note the discontinuity in quantities at the eutectic temperature (780° C), which is the special case where three phases are present simultaneously.

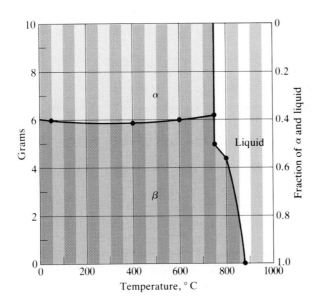

FIG. 5–4.2

Quantity of Phases (10 g of 40 Ag–60 Cu Alloy). The quantity of β was determined by interpolation (Example 5–4.3).

β–**dark bars**
α–medium bars
L–Light bars

The sums of the phases must be equal to 10 g.

5-5
INVARIANT REACTIONS

The eutectic reaction is *invariant.* The temperature is fixed; it is 779.4° C in the Ag–Cu system (Fig. 5–2.5). The phase compositions are fixed—they are α, 91.2 Ag–8.8 Cu; L, 71.9–28.1; and β, 8–92. Also, three is the maximum number of phases that can coexist under equilibrium conditions in a binary system.

Peritectic Reactions

A second invariant reaction, called a *peritectic,* may be observed in the upper part of the Fe–Fe$_3$C diagram (Fig. 5–5.1). On cooling,

$$\delta\,(0.1\%\ \text{C}) + \text{L}\,(0.5\%\ \text{C}) \xrightarrow[\text{cooling}]{1495°\,\text{C}} \gamma\,(0.2\%\ \text{C}) \qquad (5\text{-}5.1)$$

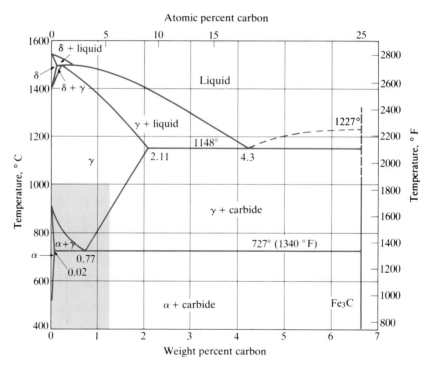

FIG. 5–5.1

Fe–Fe$_3$C Phase Diagram. The lower-left corner receives prime attention in heat-treating of steels (Fig. 7–3.1). (In calculations, 0.77 percent is commonly rounded to 0.8 percent.)

The generic form is

$$S_1 + L_3 \underset{\text{heating}}{\overset{\text{cooling}}{\rightleftharpoons}} S_2 \qquad (5\text{–}5.2)$$

Compare this reaction with the eutectic reaction of Eq. (5–2.2). Each involves three phases of fixed compositions; each is at a specific temperature. However, the two reactions are unlike in that there is a single phase below and two phases above the peritectic temperature. As they did in Eq. (5–2.2), the subscripts imply a sequential change in one of the components.

We can readily identify the peritectic reaction in the Al_2O_3–SiO_2 system (Fig. 5–2.4) as

$$\text{corundum (100\% } Al_2O_3) + \text{liquid (52\% } Al_2O_3) \overset{1828°\,C}{\rightleftharpoons}$$
$$\text{mullite (73\% } Al_2O_3) \quad (5\text{–}5.3)$$

Eutectoid Reactions

A third invariant reaction is the *eutectoid* (eutecticlike) reaction. The similarities between the eutectoid and the eutectic reactions are revealed in the shaded part of Fig. 5–5.1, where a V-notch bounds the one-phase field of γ (an iron-based, fcc solid solution). Below the reaction temperature of 727° C is a two-phase field of α and carbide. The former (α) is nearly pure iron with a bcc structure; the latter is iron carbide, Fe_3C. The eutectoid reaction is

$$\gamma \text{ (0.77\% C)} \overset{727°\,C}{\rightleftharpoons} \alpha(0.02\% \text{ C}) + Fe_3C \text{ (6.7\% C)} \qquad (5\text{–}5.4)$$

In contrast to the eutectic reaction (Eq. 5–2.2), which involves a *liquid* solution, the eutectoid reaction includes a *solid* solution. The generic form of the eutectoid reaction is

$$S_2 \underset{\text{heating}}{\overset{\text{cooling}}{\rightleftharpoons}} S_1 + S_3 \qquad (5\text{–}5.5)$$

Other Invariant Reactions

Monotectic:
$$L_2 \underset{\text{heating}}{\overset{\text{cooling}}{\rightleftharpoons}} S_1 + L_3 \qquad (5\text{–}5.6)$$

Peritectoid:
$$S_1 + S_3 \underset{\text{heating}}{\overset{\text{cooling}}{\rightleftharpoons}} S_2 \qquad (5\text{–}5.7)$$

Syntectic:
$$L_1 + L_3 \underset{\text{heating}}{\overset{\text{cooling}}{\rightleftharpoons}} S_2 \qquad (5\text{–}5.8)$$

Schematic representations are shown in Fig. 5–5.2.

FIG. 5–5.2

Invariant Reactions (Binary Phase Diagrams). Three phases are involved. The temperature and the compositions are fixed. The quantities of the phases are not fixed, as they are in two-phase areas.

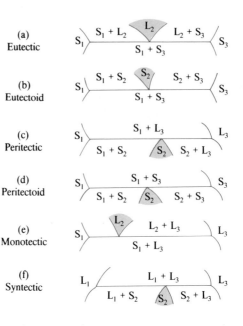

(a) Eutectic

(b) Eutectoid

(c) Peritectic

(d) Peritectoid

(e) Monotectic

(f) Syntectic

5–6

SELECTED PHASE DIAGRAMS

Metallic Systems

You may be relieved to learn that most commercial alloys have compositions that lie in the simpler parts of the phase diagrams. For example, the compositions of most brasses lie in the single-phase α area of Fig. 5–6.1. Likewise, the common bronzes contain less than 10 percent tin. There is little commercial interest in the more complex-appearing central areas* of the Cu–Sn system (Fig. 5–6.2). In Chapter 9, we shall consider alloys such as 95 Al–5 Cu, 90 Al–10 Mg, 90 Mg–10 Al (Figs. 5–6.3 and 5–6.4); each forms one phase at elevated temperatures but crosses a solubility limit curve during cooling. By controlling the rate of separation of the second phase, the engineer can greatly increase the strength of the alloy. The Al–Si system (Fig. 5–3.3) provides the commercial basis for purifying semiconducting and related materials (Example 11–7.1).

---

* Although the central areas are more complex in appearance, all areas are either one-phase or two-phase. We have already considered these. Thus, for any alloy, we can determine (1) *what* the phase(s) are, (2) what their chemical *compositions* are, and (3) what the *quantity* of each phase is.

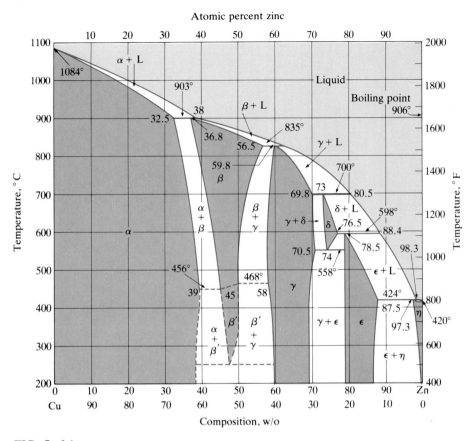

FIG. 5-6.1

Cu-Zn Diagram. (Adapted from *Metals Handbook,* ASM International.)

Phase Diagrams of Ceramics

Ceramists use phase diagrams as much as, if not more than, metallurgists.* Within this text, however, we include only six: Al_2O_3–ZrO_2 (Fig. 5–2.3), Al_2O_3–SiO_2 (Fig. 5–2.4), FeO–SiO_2 (Fig. 5–3.2), Fe–O (Fig. 5–6.5), FeO–MgO (Fig. 5–6.6) and $BaTiO_3$–$CaTiO_3$ (Fig. 5–6.7).† The first, Al_2O_3–ZrO_2, involves very refractory (high-melting-temperature) oxides that have a eutectic temperature just short

---

* See, for example, the six-volume compendium, *Phase Diagrams for Ceramists,* American Ceramic Society.

† Commonly, ceramic products involve three or more components, examples are MgO–Al_2O_3–SiO_2 and K_2O–Al_2O_3–SiO_2 in electrical porcelains. Although they are important, we shall not consider the more complex ternary-phase diagrams.

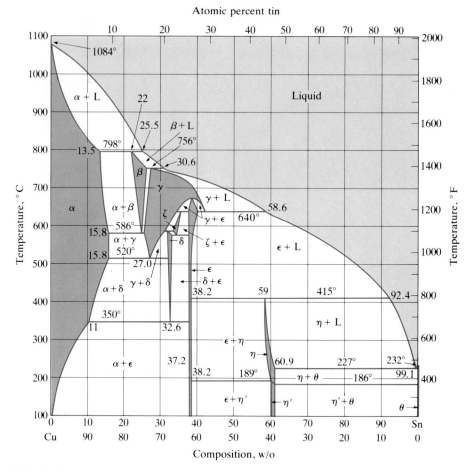

FIG. 5-6.2

Cu-Sn Diagram. (Adapted from *Metals Handbook,* ASM International.)

of 2000° C. The second, $Al_2O_3 - SiO_2$, is pertinent to clay-base ceramics, since the better grade clays are approximately 40 $Al_2O_3 - 60$ SiO_2 after processing. The Fe-O diagram shows the nonstoichiometric range of $Fe_{1-x}O$ (ϵ of Fig. 5-6.5) that was discussed in Section 4-5. The FeO-MgO diagram (Fig. 5-6.6) reveals a complete solid-solution series below solidus temperatures. This compares directly with the Cu-Ni system in Fig. 5-2.2. The $BaTiO_3 - CaTiO_3$ phase diagram (Fig. 5-6.7) shows how the addition of $CaTiO_3$ affects the temperature for the change of tetragonal $BaTiO_3$ (α) to cubic $BaTiO_3$ (β).

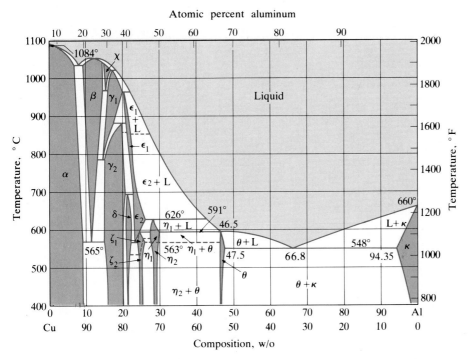

FIG. 5–6.3

Al–Cu Diagram.
(Adapted from *Metals Handbook,* ASM International.)

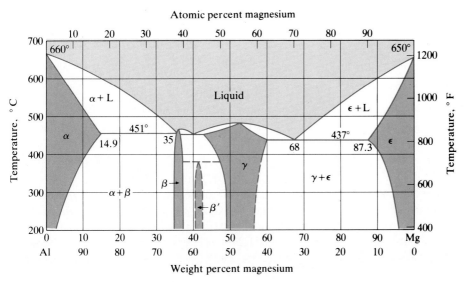

FIG. 5–6.4

Al–Mg Diagram. (Adapted from *Metals Handbook,* ASM International.)

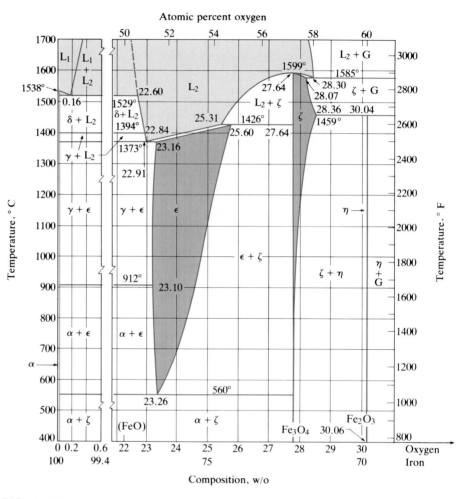

FIG. 5–6.5

Fe–O Diagram. (Adapted from *Metals Handbook,* ASM International.)

Molecular Phase Diagrams

Phase diagrams may involve *molecular* components. Figure 5–1.1(a) showed the solubility of sugar in water, as part of the $C_6H_{12}O_6$–H_2O system. Figures 5–6.8 and 5–6.9 present the two additional systems, phenol–water (C_6H_5OH–H_2O) and phenol–aniline (C_6H_5OH–$C_6H_5NH_2$). The latter contains an intermolecular compound and forms two eutectics. The C_6H_5OH–H_2O system involves two

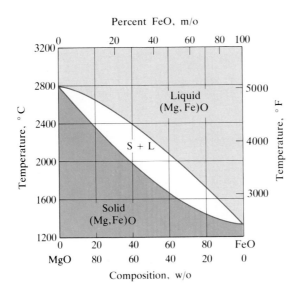

FIG. 5–6.6

MgO–FeO Diagram. As in
Cu–Ni diagram (Fig. 5–2.2),
both the liquid and the solid
phases possess full solubility.
The structure of the FeO–
MgO solid solution is shown
in Fig. 4–5.1.

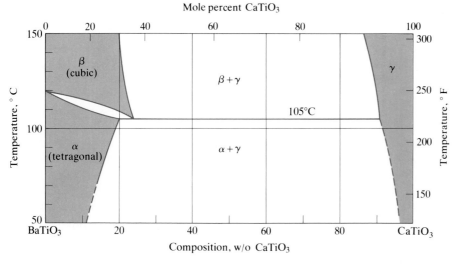

FIG. 5–6.7

BaTiO₃–CaTiO₃ Diagram. (Adapted from DeVries and Roy, American Ceramic Society.)

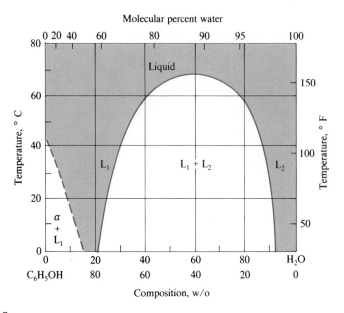

FIG. 5–6.8

Phenol–Water Diagram. Although phenol (C_6H_5OH) solidifies at 43° C, it forms lower-melting-temperature liquids with water. Similar to oil and water, two immiscible liquids exist in the ambient-temperature range. Above 68° C, there is complete miscibility (solubility). (Adapted from W. J. Moore, *Physical Chemistry,* Prentice-Hall.)

immiscible (not mutually soluble) liquids. At 20° C, water can dissolve only 9 percent phenol; conversely, liquid phenol can dissolve only 24 percent water. The mutual solubility increases with temperature, until complete solubility and a single-phase liquid exist above 68° C. Comparable *miscibility gaps* are present in other systems—for example, in oil and water, as well as in less familiar systems such as in Ag–Ni, and Fe–O (Fig. 5–6.5).

Since many polymers are amorphous, we cannot always use x-ray diffraction (Section 3–8) to determine their phase relationships, as we can for solid compositions with crystalline phases. Instead, we resort to other techniques. For example, a solution of polysytrene (PS) and cyclohexane, C_6H_{12}, is a single-phase liquid above 25° C. At slightly lower temperatures (Fig. 5–6.10), the liquid separates into two phases, as evidenced by the material becoming cloudy in appearance. This change is represented schematically by the sketches in Fig. 5–6.10. The small droplets of the emulsion scatter the light when the composition lies within the miscibility gap. Figure 5–6.11 shows the miscibility gap within the polystyrene–polybutadiene (PS–PBD) system. Microstructures containing two such immiscible polymeric phases are called a *polyblends* (Section 7–5); they are widely used in products—ranging from food packages to car fenders—where high impact resistance is required.

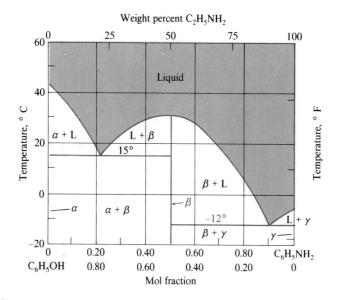

FIG. 5–6.9

Phenol–Aniline Diagram. Phenol (C_6H_5OH, labeled α in this diagram) forms an intermolecular compound (β) with aniline ($C_6H_5NH_2$, labeled γ). This produces two eutectics. (Adapted from W. J. Moore, *Physical Chemistry,* Prentice-Hall.)

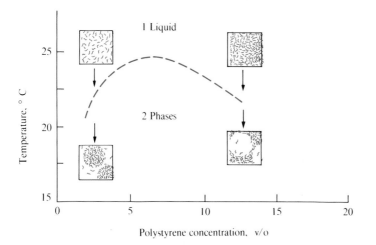

FIG. 5–6.10

Miscibility Gap (Polystyrene–Cyclohexane). Cyclohexane, C_6H_{12}, and polystyrene, PS, are fully miscible (mutually soluble) above 25° C. A miscibility gap exists at lower temperatures to produce two phases. At the left, the resulting emulsion contains PS droplets within a cyclohexane-rich liquid. At the right, the emulsion possesses droplets of the cyclohexane-rich liquid within a PS matrix. The miscibility gap drops and broadens for smaller PS molecular weights. (Adapted from *Polymer Blends and Mixtures,* D. Walsh et al., eds., Martinus Nijhoff Publishers, 1985, p. 430.)

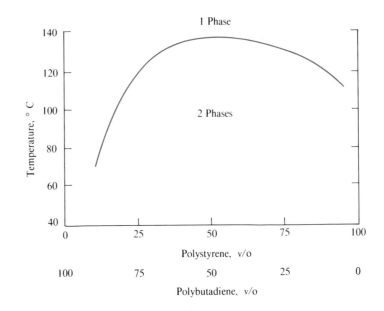

FIG. 5–6.11

Miscibility Gap (Polystyrene–Polybutadiene). Two amorphous phases exist below the miscibility curve, just as they do for phenol and water (Fig. 5–6.8). Because of this phase separation, block copolymers (Section 7–5) of polystyrene (PS) and polybutadiene (PBD) can produce strong, tough products. The PS is a rigid plastic; the PBD is a rubber that absorbs impact energy. (Adapted from *Polymer Blends and Mixtures,* D. Walsh et al, eds., Martinus Nijhoff Publishers, 1985, p. 449.)

S U M M A R Y

Our purpose in this chapter was to understand how to extract necessary data from phase diagrams, so we can select, control, and modify multiphase materials.

1. Phase diagrams contain curves of *solubility limits.* The intervening areas of the two-component diagram are either single-phase *solutions* or *mixtures* of two phases.

2. Two solubility curves cross at a *eutectic* to provide a low-melting liquid. The eutectic reaction is

$$L_2 \underset{\text{heating}}{\overset{\text{cooling}}{\rightleftharpoons}} S_1 + S_3 \qquad (5-2.2)$$

Phase diagrams show *"what phases"* at various composition–temperature combinations.

3. In a single-phase area of a phase diagram, the phase has the same composition as the total material. In a two-phase area, the two compositions are located at the two ends of the isothermal *tie-line*. Chemical compositions are reported in terms of the *component percentages.*

4. When more than one phase is present, the fraction of each phase in the mixture is determined by *interpolation along the isothermal tie-line* between the adjacent solubility limits. (We call this approach the *lever rule.*) The amounts are reported in terms of *phase* percentages (or fractions).

5. Invariant reactions include (1) *eutectic,* (2) *eutectoid,* (3) *peritectic,* (4) *peritectoid,* (5) *monotectic,* and (6) *syntectic* reactions. For each invariant reaction in a binary system, the temperature is fixed, as are the compositions of the three reacting phases.

6. *Commercial* materials are commonly located in the simpler parts of the phase diagrams. We can read any part of a complicated diagram, however, by checking (i) *what phases exist,* (ii) *what the phase compositions are* (when more than one is present) at the two ends of the isothermal tie-line, and (iii) *what the phase amounts are* by interpolation along the isothermal tie-line.

Phase Diagram Index

The *Study Guide* that accompanies this text contains two study sets on the reading of phase diagrams. The approach is more visual, and may be advantageous for some students.

KEY WORDS

Austenite (γ)	Immiscibility	Phase diagram, two-phase area
Component (phases)	Isotherm	Phase diagram, three-phase
Equilibrium	Lever rule (inverse)	temperature
Eutectic composition	Liquidus	Phases, chemical compositions of
Eutectic reaction	Materials balance	Phases, quantities of
Eutectic temperature	Mixture	Solder
Eutectoid composition	Phase diagram	Solidus
Eutectoid reaction	Phase diagram, isothermal cut	Solubility limit
Eutectoid temperature	Phase diagram, one-phase area	Tie-line

PRACTICE PROBLEMS

5–P11 A syrup contains equal quantities of water and sugar. How much more sugar can be dissolved into 100 g of the syrup at 80° C?

5–P12 A molten lead–tin solder has an eutectic composition. Assume 50 g are heated to 200° C. How many g of tin can be dissolved into this solder?

5–P13 One ton of salt (NaCl) is spread on the streets after a winter storm. The temperature is −15° C (5° F). How many tons of ice will be melted by the salt?

5–P21 A 90 Cu–10 Sn bronze (Fig. 5–6.2) is cooled slowly from 1100 to 20° C. What phase(s) will be present as the cooling progresses?

5–P22 A 65 Cu–35 Zn brass (Fig. 5–6.1) is heated from 300 to 1000° C. What phase(s) are present at each 100° C interval?

5–P23 An alloy contains 90 Pb–10 Sn. (a) What phases are present at 100° C, 200° C, and 300° C? (b) Over what temperature range(s) will there be only one phase?

5–P24 Locate the solidus on the Cu–Zn diagram (Fig. 5–6.1).

5–P25 Show the sequential changes in phases when the composition of an alloy is changed from 100 percent Cu to 100 percent Al (a) at 700° C, (b) at 450° C, (c) at 900° C. (See Fig. 5–6.3.)

5–P26 The solidus temperature is an important temperature limit for hot-working processes of metals. Why?

5–P27 (a) Show the sequential changes in phases when the composition of an alloy is changed from 100 percent Al to 100 percent Mg at 300° C. (b) Show them for the changes at 400° C. (See Fig. 5–6.4.)

5–P31 Refer to Fig. 5–6.4. (a) At 500° C, what is the solubility limit of magnesium in solid α? (b) What is it in liquid? (c) What are the chemical compositions of the phase(s) in a 40 Al–60 Mg alloy at 500° C? (d) What are they in a 20 Al–80 Mg alloy?

5–P32 Refer to Fig. 5–2.3. (a) At 2000° C, what is the solubility limit of Al_2O_3 in the liquid? (b) What is it in β? (c) What are the chemical compositions of the phase(s) in a 20 Al_2O_3–80 ZrO_2 ceramic at 1800° C? What are they at 2000° C?

5–P33 What are the chemical compositions of the phase(s) in Problem 5–P22?

5–P34 What are the chemical compositions of the phase(s) in Problem 5–P21?

5–P41 A 90 Al–10 Mg alloy (Fig. 5–6.4) is melted and then is cooled slowly. (a) At what temperature does the first solid appear? (b) At what temperature does it have two-thirds liquid and one-third α? (c) At what temperature does it have one-half liquid and one-half α? (d) At what temperature does it have 99 + percent α with a trace of liquid?

5-P42 Repeat Problem 5-P41, but for a 10 Al-90 Mg alloy (ϵ for α).

5-P43 (a) What composition of Ag and Cu will possess one-quarter α and three-quarters β at 600° C? What composition will possess one-quarter liquid and three-quarters β at 800° C?

5-P44 (a) What composition of Al_2O_3 and ZrO_2 will possess three-quarters α and one-quarter β at 1800° C? (b) What composition will possess three-quarters liquid and one-quarter β at 1950° C?

5-P45 (a) At what temperature will a monel alloy (70 percent nickel, 30 percent copper) contain two-thirds liquid and one-third solid? (b) What will be the composition of the liquid and of the solid?

5-P46 Assuming 1500 g of bronze in Problem 5-P21, what are the masses of solid phase(s) at each 100° C interval?

5-P47 At 175° C, how many g of α are there in 7.1 g of eutectic Pb-Sn solder?

5-P48 With 200 g of 65 Cu-35 Zn, how many grams of α are present at each temperature of Problem 5-P22?

5-P49 At what temperature can an alloy of 40 Ag-60 Cu contain 55 percent β?

5-P51 Write the eutectoid reaction found in the $BaTiO_3$-$CaTiO_3$ system (Fig. 5-6.7).

5-P52 (a) Locate four eutectoids in the Cu-Sn system (Fig. 5-6.2). (b) Write the reactions for two of these.

5-P53 The phenol-aniline system possesses two eutectics. Write an equation for each, using the format of Eq. (5-2.1) with w/o.

5-P54 A "monotectoid" is not listed among the invariant reactions of Section 5-5. Why?

5-P61 An alloy of 50 g Cu and 30 g Zn is melted and cooled slowly. (a) At what temperature will there be 40 g α and 40 g β? (b) At what temperature will there be 50 g α and 30 g β? (c) At what temperature will there be 30 g α and 50 g β?

5-P62 How much mullite will be present in a 60 SiO_2-40 Al_2O_3 brick (10 kg) at the following temperatures under equilibrium conditions? (a) 1400° C. (b) 1580° C. (c) 1600° C.

TEST PROBLEMS

511 (a) Assume 12 g of salt (NaCl) are added to 48 g of ice. Above what minimum temperature will the ice melt completely? (b) Assume 40 g *more* of ice are added to the brine. Above what minimum temperature will this ice melt completely?

512 The solubility limit of zinc in α-brass is 35 percent at 780° C. We want to make 100 kg of 65-35 brass (Cu-Zn) by melting some 80-20 brass and some zinc. How much of each should we melt in the crucible?

513 From Fig. 5-1.2, suggest a method for purifying sea water.

521 One pound (454 g) of a 90 Pb-10 Sn solder is to be completely liquified at 200° C by the addition of more tin. What is the minimum amount of tin that must be added per 100 g of this high-lead solder?

522 A eutectic mixture of Al_2O_3 and SiO_2 is molten at 2000° C. (a) At what temperature(s) will there be only one phase? (b) At what temperature(s) will there be two phases? (c) At what temperature(s) will there be three phases?

523 Locate the liquidus and the solidus as they extend from 1538° C at 100 percent Fe (0 percent oxygen) to 1599° C for Fe_3O_4. (*Note:* There is a two-liquid field where liquid metal separates from liquid oxide.) (See Fig. 5-6.5.)

524 What field in the Cu-Zn diagram (Fig. 5-6.1) contains 78 percent Zn at 575° C?

525 What fields lie to the right of β in the Cu-Al diagram (Fig. 5-6.3)?

531 (a) Write the eutectic reaction for one of the eutectics in the Al-Mg system (Fig. 5-6.4). (b) Repeat for a second eutectic. (Use the format of Eq. 5–2.1.)

532 Write the eutectic reaction for the higher-aluminum eutectic of the Cu–Al system (Fig. 5–6.3). Use the format of Eq. (5–2.1), expressed in copper.

541 The solubility of tin in solid lead at 200° C is 18 percent Sn. The solubility of lead in the molten metal at the same temperature is 43 percent Pb. What is the composition of an alloy containing 60 percent liquid and 40 percent solid α at 200° C?

542 Assume 2 kg of a 90 Pb–10 Sn alloy are cooled slowly from 350 to 20° C. (a) On a graph of % Pb (ordinate) versus ° C (abscissa), plot the composition of α. (b) On a graph of g of α (ordinate) versus ° C (abscissa), plot the amount of α. (The data in Fig. 5–2.1 are accurate to about 1 part/100. You should be correspondingly accurate.)

543 Make a materials balance for P grams of a 60 Ag–40 Cu alloy at 600° C (equilibrium conditions).

544 Based on Figs. 5–2.5 and 5–4.2, make a graph for an alloy containing 80 percent Ag and 20 percent Cu showing (a) the fraction of liquid versus temperatures, (b) the fraction of α versus temperature, and (c) the fraction of β versus temperature.

545 (a) Plot the percent Sn versus temperature in the α phase of a 90 Pb–10 Sn solder. (b) Plot the fraction of α versus temperature in this alloy.

546 We can use interpolation along the horizontal isotherm to determine the quantity fractions of the two phases. We *cannot* obtain a valid answer by interpolating along a vertical composition line. By using a 70 Pb–30 Sn solder, show why the latter approach does not work.

547 Explain to a classmate why, in an 80 Ag–20 Cu alloy, the quantity of α is indeterminate at the eutectic temperature, but that, if you have 50 g of α in 100 g of alloy, the temperature would have to be 779.4° C for equilibrium to exist.

551 Identify three invariant reactions in the Fe–Fe$_3$C system (Fig. 5–5.1).

552 A 1-g ball bearing (99 Fe–1 C) is equilibrated at 750° C. (a) Based on Fig. 7–3.1, how much of each phase is present? (b) How much is present at 700° C?

553 A knife has the composition of the Fe–Fe$_3$C eu-

tectoid. It is slowly cooled from 800 to 20° C. How much of what phases exist at 20° C?

554 An alloy of 60 Cu–40 Sn encounters an invariant reaction that is sufficiently unusual that we have not named it. Write an equation for this unexpected reaction.

561 Plot the amount of α in a 5.8 kg of a 90 Al–10 Mg alloy as a function of temperature.

562 (a) What are the chemical compositions of the phases in a 95 percent magnesium, 5 percent aluminum alloy at 600° C, 400° C, and 200° C? (b) What are the quantities of these phases at each of the temperatures in part (a)? (c) Make a materials balance for the distribution of the magnesium and aluminum in the alloy at 600° C.

563 Make a materials balance for 100 g of a 90 Cu–10 Sn alloy at 200° C (assume equilibrium).

564 An alloy of 70 Al–30 Si is cooled such that the metal contains primary (proeutectic) β and an eutectic mixture of $(\alpha + \beta)$. (a) What fraction of the casting is primary β? (b) What fraction of the eutectic microstructure is β? (See Fig. 5–3.3c.)

565 (a) Determine the compositions of phenol–aniline "alloys" that will contain one-third liquid and two-thirds solid when brought to equilibrium at 20° C. (b) Give the chemical compositions of the liquids.

566 Assume 27 kg (60 lb) of a 92 Cu–8 Sn bronze (to be used to make a bell) is melted and then is cooled very slowly to maintain equilibrium. (a) At what temperature does the first solid appear? (b) What is the composition of that solid? (c) At what temperature is there twice as much solid as liquid? (d) At what temperature does the last liquid disappear? (e) What is the composition of that liquid? (f) What phase(s) exist at 600° C? Give the composition(s) and amount(s) of each phase. (g) What phase(s) exist at 200° C? Give the composition(s) and amount(s) of each phase.

567 Figure 5–6.8 does not include subzero temperatures. Show schematically one of the two possible invariant patterns that could exist.

568 Assume 90 g of a 90 phenol–10 aniline mixture is equilibrated at 16° C, and then is cooled to 10° C after full solidification. (a) How much phenol formed prior to the invariant reaction? (b) What is the α/β ratio in the product of the invariant reaction?

Chapter 6

REACTION RATES

Changes occur in solids, but changes take time. Time is required for the breaking of bonds, for the nucleation of new atomic coordinations, and for the diffusion of atoms to and from the reaction sites.

Equilibrium diagrams are valuable because they tell us *which way* a reaction will proceed, but they give no indication of the *time* requirement. For example, the crystallization of silicate glasses is readily deferred; it is almost infinitely long. In contrast, metallic liquids must be very severely quenched ($> 1000°$/millisecond) to produce a metallic glass.

In this chapter we look at the factors that affect the rates of nucleation and mass transport within solids. Temperature, bonding, and atomic and molecular structures are all pertinent.

6–1

DEFERRED REACTIONS

The phase changes that accompany decreasing temperatures require time for completion. The atoms must relocate. The atomic coordination must be changed in polymorphic reactions, such as $Fe_{fcc} \xrightarrow{\text{cooling}} Fe_{bcc}$ (Section 3–4). In addition, eutectic and eutectoid reactions (Sections 5–2 and 5–5) require the formation of phases with entirely different structures and compositions. The changes dictated as being essential by the phase diagrams may be deferred. Eventually, the reaction should occur; however, in some situations, the reaction time may be almost infinitely long.

Glasses

The glass phases were discussed in Section 4–2. They are amorphous, or noncrystalline, and they retain the structure of a liquid to a sufficiently low temperature that they become rigid solids.

The cooling rate to achieve a glass varies from material to material. A *metallic glass* must be quenched from its liquid state at over 1000° C to ambient temperatures in 1 ms or less. This rate of $> 1,000,000°$ C/s requires special quenching procedures.* This rapid rate is necessary because the ordering process—from that shown in Fig. 4–2.1(b) to that in Fig. 4–2.1(a)—is simple; each atom relocates individually.

The time available to cool a *silicate glass* without crystallizing is much longer, amounting to several hours or more. This long period is required because the strong covalent bonds in the SiO_2 network structure (Fig. 2–4.3) must be broken. Furthermore, there must be a mutual rearrangement of Na^+ and Ca^{2+} ions, and of SiO_4 units, all of which is time-consuming.

Devitrified Glasses

Although the reaction is slow, glasses can be devitrified, or *crystallized.* In fact, products that have trade names such as Pyroceram® or Cervit® take advantage of the deferred crystallization of glass. The viscosity of the supercooled liquid is used to shape the product with an absence of any porosity. These products would remain as a rigid glass at room temperatures. However, a two-step heat-treating process has been established: (1) heating to precipitate nuclei (usually TiO_2 or ZrO_2) that were originally dissolved at high temperatures into the glass melt, then (2) heating at a higher temperature to grow silicate crystals on the nuclei.

---

* *Metallic glasses,* although difficult to make, have some interesting properties that differ from metallic crystals of the same composition, because the glasses have a different structure. For example, some of the iron-based glasses have attractive magnetic permeabilities.

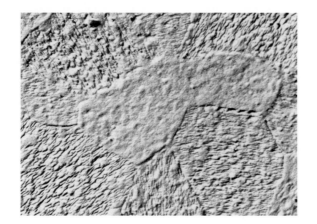

FIG. 6–1.1

Devitrified Glass (Li$_2$O–Al$_2$O$_3$–8 SiO$_2$, Keatite Solid Solution). A silicate glass is shaped into a product as a viscous liquid. Then it receives a two-step heat treatment: (1) nucleation, (2) grain growth. Electron photomicrograph, ×32,000. (Courtesy of H. L. McCollister.)

The final devitrified product possesses a nonporous structure of silicate grains (Fig. 6–1.1), and has better strength characteristics than does the original vitreous material. If the grain size is less than the wavelength of visible light (<0.4 μm), the product remains transparent. These devitrified glass products are commonly called "glass-ceramics."

Supercooled Solid Solutions

According to Fig. 5–1.1, a syrup of 70 percent sugar and 30 percent water becomes saturated when cooled to 45° C (113° F). The syrup can be supercooled to lower temperatures without the separation of sugar. Thus, it is *supersaturated* at the ambient temperature of 20° C. Given time, sugar crystals will form.*

In a like manner, sterling silver can become a supersaturated phase. According to Fig. 5–2.5, this alloy (92.5 Ag–7.5 Cu) is a single-phase solid solution between 750 and 800° C. Rapid cooling to room temperature can trap the copper atoms within the Ag-rich, α phase (Fig. 6–1.2). There is not enough time for them to segregate and form a new phase of β. The α phase becomes supersaturated. This result turns out to be advantageous in several respects when sterling silver is used for coinage or for related noble-metal applications. Although the silver content is diluted with the less expensive copper, the alloy is stronger than is silver alone.

---

* Crystallization can be initiated with less delay if a few particles of fine sugar powder are sprinkled onto the surface of the syrup (Section 6–3).

FIG. 6–1.2

Supercooled Solid Solution (Sterling Silver, 92.5 Ag–7.5 Cu). Processing steps: (1) solution treat at 775° C; (2) cool rapidly so that β does not separate. Properties: (1) harder and stronger than pure silver (Chapter 9); (2) more corrosion resistant than slow-cooled sterling silver (Chapter 14).

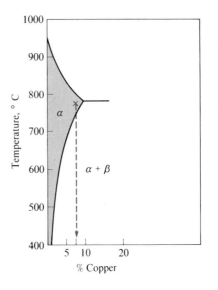

Furthermore, by preventing the formation of β, the alloy retains its corrosion resistance. The same rapid-cooling procedure is widely used as the first step in the precipitation-hardening processes that will be discussed in Section 9–4.

6–2

SEGREGATION DURING SOLIDIFICATION

Under equilibrium conditions, a liquid alloy will start to solidify at the liquidus temperature. The composition of the solidifying phase is different from that of the

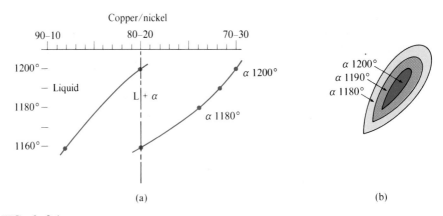

FIG. 6–2.1

Freezing Segregation (80 Cu–20 Ni). (a) Equilibrium (from Fig. 5–2.2). The solid is richer in nickel than is the liquid. (b) Coring (schematic). The initial solid, α, contains 30 percent Ni. It is coated with a more dilute α, so complete equilibrium with the liquid is not achieved.

liquid. Consider, for example, an 80 Cu–20 Ni alloy. The liquidus is at 1200° C according to Fig. 5–2.2, which is partially reproduced in Fig. 6–2.1(a). The initial solid contains 30 percent Ni (and 70 percent Cu). At 1180° C, the separating solid now contains 26 percent Ni. At the solidus (1160° C), as shown in Figure 6–2.1(a), the solid should contain 20 percent Ni, matching the 80–20 ratio with which we started. There is a corresponding change in the compositions of the liquid. Under equilibrium conditions, the final liquid at 1160° C would contain about 12 percent Ni.

Coring

Since the first α in the 80 Cu–20 Ni alloy contained 30 percent Ni, some of the nickel must be removed from that solid to produce the final equilibrated 80–20 phase. Atomic diffusion is required (Section 6–5).

Atoms move much more slowly in a solid than in a liquid, so equilibrium usually is not maintained during the cooling process, and nonhomogeneity is common. Incremental stages of this sequence are shown in Fig. 6–2.1(b). The result, called *coring,* is a graded composition in the solid. Three consequences may follow: (1) segregation remains after solidification is complete; (2) the *average* composition of the solid does not follow the *equilibrium solidus,* but drops below it (Fig. 6–2.2) to give us a *nonequilibrium solidus;* and (3) liquid continues to be present at temperatures below the equilibrium solidus (1160° C). It is not uncommon for liquid to persist during the cooling for another 50 to 100° C, until the average solid, $\bar{\alpha}$, contains 80 percent copper.

Induced Eutectics

Since we are familiar with sterling silver, let us once again consider that alloy. With 7.5 percent Cu, its liquidus is slightly above 900° C (Fig. 5–2.5), and its equilibrium solidus is approximately 800° C. However, solidification segregation will produce a copper-lean α as the nonequilibrium solidus drops below the true solidus (Fig. 6–2.3). Also, as it did in our previous Cu–Ni example, liquid continues to be present below 800° C. In fact, the eutectic temperature is easily

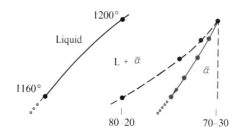

FIG. 6–2.2

Nonequilibrium Solidus (Color). The average solid, $\bar{\alpha}$, contains more nickel than equilibrium prescribes. Liquid is retained to lower temperatures. (See Fig. 6–2.1.)

FIG. 6–2.3

Induced Eutectic (Sterling Silver, 92.5 Ag–7.5 Cu). As in Fig. 6–2.2, the solidified solid does not follow the equilibrium solidus. Therefore, liquid remains below 800° C, and commonly reaches the eutectic composition, which decomposes to $\alpha + \beta$.

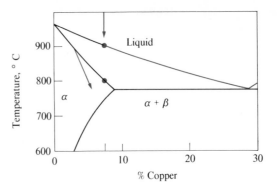

reached and the eutectic reaction is induced. Of course, the phase diagram would not indicate this. Obtaining data from Fig. 6–2.3, the average solid contains only ~5 percent Cu at 780° C; the liquid contains ~ 28 percent Cu. With 7.5 percent Cu overall, interpolation (or the lever rule) predicts 10± percent liquid available for the $(L \rightarrow \alpha + \beta)$ reaction.

Figure 6–2.4 shows a segregated alloy of 96 Al–4 Cu. In this case, the right side of Fig. 5–6.3 is pertinent. The light areas were the copper-lean core. The dark areas were enriched with copper and remained liquid until the eutectic was reached $(L \xrightarrow{\text{548° C}} \theta + \kappa)$.

Homogenization

In general, we wish to avoid segregation of the types mentioned in the previous section*. Otherwise, properties will not be uniform; also, low-melting portions can introduce liquid into an alloy when that liquid cannot be tolerated. Therefore, when we do encounter segregation, we use *homogenization processes.* We reheat the alloy to a temperature just short of the appearance of the liquid. Then we use a *soak,* or an extended *anneal,* so that the atoms can move away from the regions of higher concentration to regions with deficiencies. Equilibrium is approached.

FIG. 6–2.4

Segregation (96A1–4Cu). Uniformity is not attained during the usual solidification process. Therefore, a homogenization step usually is required. (Courtesy of the Aluminum Company of America.)

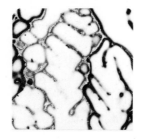

* For an important exception, see zone refining (Section 11–7).

Example 6–2.1

Let 100 g of a 96 Al–4 Cu alloy equilibrate at 620° C (1150° F), forming the κ and liquid, as shown by the phase diagram (Fig. 5–6.3). The alloy is then cooled rapidly to 550° C (1020° F) with no chance for the initial solid to react. A liquid phase will still be present.

(a) What will its composition be?

(b) How many grams will there be of this final liquid?

Procedure Determine the amount and the composition of the 620° C liquid. The solid that is present at 620° C has "no chance to react." It can be ignored in this problem. Therefore, as the liquid is cooled to 550° C, it behaves as a separate alloy and forms a new solid–liquid pair. What is the composition of the liquid in this second-generation pair? How much liquid is there?

Calculations At 620° C, and from Fig. 5–6.3,

Liquid: 88 Al–12 Cu,

κ: 98 Al–2 Cu

$$\text{grams liquid} = 100 \text{ g} \left(\frac{98 - 96}{98 - 88} \right) = 20 \text{ g liquid}$$

At 550° C, and with only the 20 g of alloy reacting,

κ: 94.4 Al–5.6 Cu

(a) Liquid: 67 Al–33 Cu

$$\text{(b) grams liquid} = 20 \text{ g} \left(\frac{94.4 - 88}{94.4 - 67} \right) = 4.6 \text{ g liquid}$$

Comments The final 4.6 g of liquid produces local areas with high copper segregation (33 percent Cu). There also will be nearly 80 g of metal with only 2 percent Cu.

Cooling may be considered to take place in a series of small steps, in which the solute is progressively concentrated in the residual liquid. Conversely, the initial solid is purer than average.

Example 6–2.2

Consider an 80 Cu–20 Ni alloy. Assume that there is no solid diffusion during cooling from the liquidus to 1160° C, with the result that the solid composition follows the nonequilibrium solidus of Fig. 6–2.2. (a) What percent liquid remains at 1160° C? (b) Remove the remaining liquid from the solid, and allow it to cool very slowly, such that equilibrium is realized. At what temperature will the last liquid disappear?

Procedure In (a), the overall composition remains at 80–20, and the average solid is 75 Cu–25 Ni (Fig. 6–2.2). The liquid is 88 Cu–12 Ni (Fig. 6–2.1).

Calculation

(a) At 1160° C: $(80 - 75)/(88 - 75) = 0.38$ (or 38%)

(b) Start again with an 88–12 liquid alloy. Based on the equilibrium diagram of Fig. 5–2.2, *its* solidus is at *1125° C,* which is the temperature at which the final liquid disappears.

Comment In the absence of equilibrium, liquid would continue to be present to still lower temperatures.

Example 6–2.3

An alloy (5.8 kg) contains 6.7 Ag–93.3 Cu. It is melted and then cooled to 950° C. The first solid (β) appears at $\sim$ 1050° C and contains $\sim$ 2 percent Ag. At 950° C, the separating solid contains $\sim$ 6 percent Ag; however, the cooling is sufficiently rapid that the initial solid did not change composition. As a result, the average β composition is $\sim$ 4 percent Ag. (Diffusion is faster in the liquid, so the liquid's composition at 950° C is $\sim$ 26 percent Ag.)

(a) How much liquid is present?

(b) Another step of cooling to 800° C permits solidification to continue in a comparable manner (and without changing the composition of the previous solid). What products are present at 800° C that were not present at 950° C?

(c) Give the approximate amounts of the phases in part (b).

Procedure Use Fig. 5–2.5. (a) The alloy (6.7 Ag–93.3 Cu) contains liquid (26 Ag–74 Cu) and solid β that averages $\sim$ (4 Ag–96 Cu). Therefore, we use the average composition and interpolate. (b) From part (a), we know that the average β separating between 950° C and 800° C will have > 6 percent Ag and < 8 percent Ag, the equilibrium composition. Assume 7 Ag–93 Cu.

Estimates

(a) Liquid at 950° C: $\dfrac{(6.7 - \sim 4)}{(26 - \sim 4)}$ (5.8 kg) $\approx$ 0.7 kg   (and 5.1 kg β)

(b) Liquid ($\sim$ 65 percent Ag); solid (6 percent–8 percent Ag).

(c) The average composition of the second generation of β is $\sim$ 7 percent Ag. It solidified from the 0.7 kg liquid that contained $\sim$ 26 percent Ag.

Liquid at 800° C: $\left(\dfrac{26 - \sim 7}{65 - \sim 7}\right)$ (0.7 kg) $\approx$ 0.2 kg liquid

Therefore,   added $\beta \approx$ 0.5 kg   (or 5.6 kg total β)

Comment This is an example of segregation (see Fig. 6–2.4) Segregation is normal in any solidification process unless opportunity for diffusion is given. Of course, the cooling is continuous, rather than being the two-step sequence of our illustration.

6-3
NUCLEATION

A new grain or phase does not automatically form in a supersaturated solution. It must be nucleated. That is, there must be a "seed" before growth can start. As indicated in the second footnote of Section 6–1, a few particles of sugar powder will initiate the crystallization of sugar from a supersaturated syrup. Likewise, silver iodide (AgI) has been used to "seed" clouds that are supersaturated with moisture. However, nucleation can occur unaided, albeit with delays.

Homogeneous Nucleation

For crystalline sugar to separate from the supersaturated syrup, a boundary must be formed between the two phases; but all grain and phase boundaries require additional energy. That energy does not become available until after the new phase has appeared. The result of this "chicken-or-egg" situation is a delay in phase separation.

We can illustrate the dilemma mathematically by assuming a spherical nucleus emerging from a homogeneous solution. Let γ be the energy (per unit area) that is *required* for the new boundary.* Also let ΔF_V be the energy (per unit volume) that is *released* when the supersaturated phase reacts to give the equilibrium phases.† For a nucleus of radius, r, to form, the net energy for nucleation, ΔF_n is:

$$\Delta F_n = 4\pi r^2 \gamma + \tfrac{4}{3}\pi r^3 (\Delta F_V) \qquad (6\text{-}3.1)$$

Of course, the initial radius r, is atomically small, so $4\pi r^2 \gamma > \tfrac{4}{3}\pi r^3 (\Delta F_V)$. Thus, ΔF_n is positive, and the nucleus does not form spontaneously, since energy is required. Only after a critical radius, r_c, develops does the growth proceed automatically (Fig. 6–3.1).

The value of γ does not vary much with temperature, whereas ΔF_V becomes significantly more negative as the temperature is decreased. Therefore, both (1) the critical radius, r_c, and (2) the energy barrier for nucleation, ΔF_n, decrease with the extent of supercooling (Fig. 6–3.2). At sufficiently low temperatures, nucleation can be triggered by a few atoms statistically clustering as a nucleus, so a small critical radius is exceeded. With added growth, the new phase attains stability. Of course, atom movements are sluggish at low temperatures, so growth is generally slow.

---

* As discussed in Section 4–1, energy is required because the atoms at the interface lack the maximum coordination available to the atoms within the phase.

† Both γ and ΔF_V are the *free energies* of thermodynamics. Of these, γ is positive (required), and ΔF_V is negative (released).

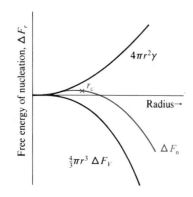

FIG. 6–3.1

Critical Nucleus Radius (Homogeneous Nucleation). Initially, energy is required (ΔF_n of Equation 6–3.1). Thus, growth is not spontaneous until r_c is exceeded. (Adapted from L. H. Van Vlack, *Materials Science for Engineers,* Addison-Wesley, Reading, Mass., Fig. 18–8, with permission.)

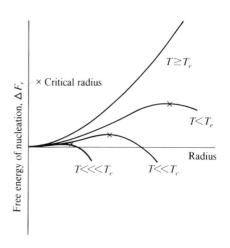

FIG. 6–3.2

Effect of Supercooling on the Critical Radius. The critical radius must be infinite at the equilibrium temperature, T_e. Nucleation is more probable with greater supercooling because the critical radius is reduced. (From L. H. Van Vlack, *Materials Science for Engineers,* Addison-Wesley, Reading, Mass., 1970, Fig. 18–9, with permission.)

Heterogeneous Nucleation

Irregularities in crystal structure, such as point defects and dislocations, possess strain energy (Figs. 4–1.1 and 4–1.3). Nucleation is facilitated by these imperfections if the transformation reduces that strain energy. The released strain energy can reduce the energy requirements for ΔF_n in Eq. (6–3.1). Therefore, nucleation proceeds with a smaller critical radius. Heterogeneous separation of this type is common.

A grain boundary offers a major nucleation site. The disorder among atoms along a grain boundary contains extra energy that can be used in a new phase boundary. Thus, the grain size and the related grain-boundary area are important parameters in analyzing the rates of reactions in solids.

Heterogeneous reactions also are facilitated by *inoculants.* The sugar powder mentioned in the second footnote in Section 6–1 provided an existing surface onto which crystallization could proceed. Silver iodide serves as an inoculant because of the similarity between the unit-cell dimensions of AgI and of ice. The two lattices have near perfect registry and therefore negligible phase-boundary energy. Consequently, ice can crystallize on the silver-iodide surface under conditions where ice would not be able to undergo homogeneous nucleation.

The majority of reactions are initiated by some type of heterogeneous nucleation — on the surface of an inoculant, by the presence of some impurity, at a grain boundary, at crystal imperfections, or simply on the wall of a container.

Example 6–3.1

Consider a phase transformation that occurs at $1025°$ C. The boundary energy, γ, is 0.50 J/m², and the value for ΔF_V for the reaction is -2.1×10^9 J/m³ at $900°$ C. (It is 0 at $1025°$ C.) Determine the critical radius, r_c, for homogeneous nucleation at $900°$ C.

Procedure   The critical radius is at the maximum energy for nucleation, ΔF_n (Fig. 6–3.1). Therefore,

$$d\Delta F_n/dr = 0 \qquad\qquad (6\text{–}3.2)$$

Calculation   From Eqs. (6–3.1 and 6–3.2),

$$d\Delta F_n/dr = 8\pi r\gamma + 4\pi r^2\, \Delta F_V = 0$$

$$r_c = -2\gamma/\Delta F_v \qquad\qquad (6\text{–}3.3)$$
$$= -2(0.50 \text{ J/m}^2)/(-2.1 \times 10^9 \text{ J/m}^3)$$
$$= 0.5 \times 10^{-9} \text{ m} \qquad\qquad \text{(or 0.5 nm)}$$

Comment   This dimension is approximately four times for atomic radius for a metal. Therefore, 60 to 70 atoms would have to cluster as the new structure before it could continue growth spontaneously.

6–4
ATOMIC VIBRATIONS

The atoms in a solid are not static. They are perpetually vibrating about their positions within the molecules or crystals. At ambient temperatures and above, the average kinetic energy of these oscillations increases roughly in proportion to the absolute temperature, T.

We saw with Fig. 2–5.3 that the added energy also increases the mean inter-atomic distances, leading to *thermal expansion.* Further, there is an inverse relationship between the strength of the bonds, as indicated by the melting temperature, T_m, and the thermal expansion coefficient, α. This was shown in Fig. 2–6.2 for three types of materials—metals, glasses, and AX compounds. For these inverse relationships to hold, our comparisons should involve comparable types of materials.

Thermal Energy Distribution

Before discussing the kinetic energies of the vibrating atoms in a solid, let us look first at a gas, such as the air in your room. The total kinetic energy, K.E., of the molecules in a mole of gas is:

$$\text{K.E.} = \tfrac{3}{2} RT \qquad\qquad (6\text{–}4.1)$$

The R of this equation is the same gas constant encountered in introductory chemistry courses, where its value is commonly reported as 1.987 cal/mol·K. If we switch to joules of the SI units and pay attention to individuals instead of to the moles (or 0.602×10^{24}), the value becomes 13.8×10^{-24} J/K.* This is called *Boltzmann's constant* and is identified by the letter k. Thus, the average K.E. of an individual molecule of gas is

$$\text{K.E.} = \tfrac{3}{2}kT \qquad\qquad (6\text{-}4.2)$$

However, this does not imply that all molecules of air in your room have identical energies. Rather, there will be a statistical distribution of energies, as indicated in Fig. 6–4.1. At any particular instant, a very few molecules will have nearly zero energy; many molecules will have energies near to the average energy, and some molecules will have extremely high energies. As the temperature increases, there is (1) an increase in the average energy of the molecules, and (2) an increase in the number of molecules with energies in excess of any specified value.

Our discussion applies to the kinetic-energy distribution of molecules in a gas. A similar principle applies to the distribution of vibrational energy of atoms in a liquid or solid. Specifically, at any particular instant of time, a negligible number of atoms will have near zero energy; many atoms will have energies close to the average energy, $\bar{E}$; and some atoms will have extremely high energies.

We shall be most interested in those atoms that have high energies. Often, we shall want to know the probability of atoms possessing more than a specified amount of energy; for example, we shall want to determine what fraction of the atoms has energy greater than E of Fig. 6–4.2. These are the atoms that can

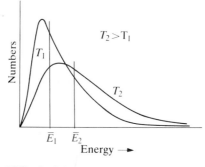

FIG. 6–4.1

Energy Distributions. Both the average energy $\bar{E}$ and the fraction with energies in excess of a specified level are increased as the temperature T is increased.

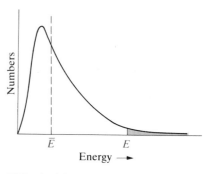

FIG. 6–4.2

Energies. The ratio of the number of high-energy atoms (shaded) to total number of atoms is an exponential function $(-E/kT)$ when $E \gg \bar{E}$ (Eq. 6–4.3b).

---

* $(1.987 \text{ cal/mol} \cdot \text{K})(4.18 \text{ J/cal})/(0.602 \times 10^{24}/\text{mol}) = 13.8 \times 10^{-24}$ J/K.

introduce changes within solids by breaking old bonds and by joining new neighbors. The statistical solution to this problem was worked out by Boltzmann as follows:

$$\frac{n}{N_{tot}} \propto e^{-(E - \bar{E})/kT} \tag{6-4.3a}$$

where k is the previously described Boltzmann's constant. The number n of atoms with an energy greater than E, out of the total number N_{tot} present, is a function of the temperature T. When E is considerably in excess of the average energy $\bar{E}$, the equation reduces to

$$\frac{n}{N_{tot}} = Me^{-E/kT} \tag{6-4.3b}$$

where M is the proportionality constant. The value of E is normally expressed as joules/atom; thus, k is 13.8×10^{-24} J/atom·K. You also can make conversions from other units by means of Appendix A.

Example 6–4.1

At 700° C, the linear expansion coefficients, α, for graphite are as follows:

$$\alpha_{\perp} = 29 \times 10^{-6}/° \text{ C}$$
$$\alpha_{\parallel} = \alpha_{\parallel} = 0.8 \times 10^{-6}/° \text{ C}$$

where $\alpha_{\perp}$ is the expansion coefficient perpendicular to the graphite layers of Fig. 3–3.5, and $\alpha_{\parallel}$ is the coefficient in the two directions parallel to the layers. What is the volume increase in graphite between 600 and 800° C?

Solution Since $V = L^3$, and $V + \Delta V = (L + \Delta L_{\perp})(L + \Delta L_{\parallel})^2$,

$$\Delta V/V = (\Delta L_{\perp}/L) + (2\Delta L_{\parallel}/L) + \cdots$$
$$\alpha_V \Delta T \approx (\alpha_{\perp} + 2\alpha_{\parallel})\Delta T = (29 + 1.6)(10^{-6}/° \text{ C})(200° \text{ C})$$
$$= 0.006 \qquad\qquad (\text{or } 0.6 \text{ v/o})$$

Comments In general,

$$\alpha_V \approx \alpha_x + \alpha_y + \alpha_z \tag{6-4.4}$$

where the three subscripts at the right refer to the linear expansion coefficient of the three coordinate directions. With a cubic crystal,

$$\alpha_V \approx 3\alpha_L \tag{6-4.5}$$

where α_L is the linear expansion coefficient.

Example 6–4.2

At 500° C (773 K), a diffusion experiment indicates that one out of 10^{10} atoms has enough *activation energy* to jump out of its lattice site into an interstitial position. At 600° C

(873 K), this fraction is increased to 10^{-9}. (a) What is the activation energy required for this jump? (b) What fraction of the atoms has enough energy at 700° C (973 K)?

Procedure　The fraction of atoms with energies in excess of a value, E, is an exponential function of temperature. Therefore, we use Eq. (6–4.3b). (a) We have two data pairs to write two equations, each with two unknowns, E and $\ln M$. Solve simultaneously for the activation energy, E. (b) Use the values of E and $\ln M$ to calculate n/N at 700° C.

Calculation

$$\ln(n/N_{tot}) = \ln M - E/kT \qquad\qquad (6\text{–}4.6)$$

(a) $\ln 10^{-10} = -23 = \ln M - (E/(13.8 \times 10^{-24}\ \text{J/atom} \cdot \text{K})(773\ \text{K}))$

and

$\ln 10^{-9} = -20.7 = \ln M - (E/(13.8 \times 10^{-24}\ \text{J/atom} \cdot \text{K})(873\ \text{K}))$

Solving simultaneously,

$\ln M = -2.92$　　　and　　　$E = 0.214 \times 10^{-18}$ J/atom

or, in terms of a mole,

$E = 129{,}000$ J/mol　　　　　　　　　　　　　　　　　(= 30,900 cal/mol)

(b) $\ln(n/N_{tot}) = -2.92 - ((0.214 \times 10^{-18}\ \text{J/atom})/(13.8 \times 10^{-24}\ \text{J/atom} \cdot \text{K})(973\ \text{K}))$,

$n/N_{tot} \approx 6 \times 10^{-9}$

Comments　The relationship shown in Eq. (6–4.6) has a logarithmic value that is linear with reciprocal temperature, $1/T$,

$$y = C - Bx$$

or

$$\ln \frac{n}{N} = C - \frac{E}{k}\left(\frac{1}{T}\right) \qquad\qquad (6\text{–}4.7)$$

where $y = \ln(n/N)$, and $x = (1/T)$. The slope B is E/k. Equation (6–4.7) is called an *Arrhenius* equation. It is widely encountered in any situation where the reaction is thermally activated. Examples include the intrinsic conductivity of semiconductors, the catalytic reaction of emission control, the diffusion processes in metals, the creep in plastics, and the viscosity of fluids. We shall encounter this relationship again.

6–5

ATOMIC DIFFUSION

As the temperature is increased and the atoms in a solid vibrate more energetically, a small fraction of the atoms will relocate themselves in the lattice. Example 6–4.2 related this fraction to temperature. Of course, the fraction depends not only on temperature, but also on how tightly the atoms are bonded in position. The energy requirement for an atom to change position is called the *activation energy.* This energy may be expressed as cal/mol, Q; as J/atom, E; or as eV/atom.

Diffusion Mechanisms

Let us use Fig. 6–5.1 to illustrate activation energy schematically. A carbon atom is small ($r \approx 0.07$ nm) and can sit interstitially among a number of fcc iron atoms. If it has enough energy,* it can squeeze between the iron atoms to the next interstice when it vibrates in that direction. At 20° C, there is only a small probability that it will have that much energy. At higher temperatures, the probability increases (see Example 6–4.2).

Other diffusion mechanisms are sketched in Fig. 6–5.2. When all the atoms are the same size, or nearly so, the vacancy mechanism becomes predominant. The vacancies may be present either as part of a defect structure (Fig. 4–5.2) or because of extensive thermal agitation (see, for example, Example 4–1.1, where aluminum was just 10° below its melting point).

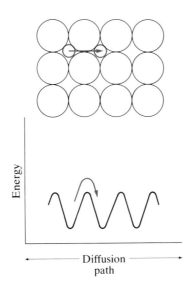

FIG. 6–5.1

Atom Movements. Interstitial mechanism. Additional energy is required because the normal interatomic distances between the large atoms must be enlarged for the interstitial atom to move to the next site.

---

* Approximately 34,000 cal/mol, 0.24×10^{-18} J/atom, or 1.5 eV/atom.

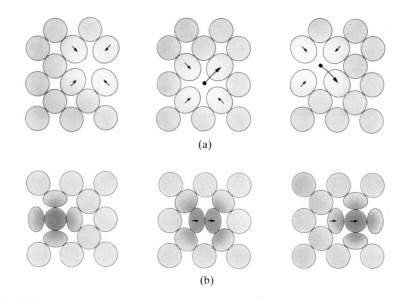

(a)

(b)

FIG. 6–5.2

Diffusion Mechanisms. (a) By vacancies. (b) By interstitialcies. The vacancies move in the opposite direction from the diffusing atoms. (In Chapter 11, we shall observe an analogy for the movements of electron holes.) The movements follow "random-walk" statistics.

Self-Diffusion

Let us consider diffusion in metallic nickel (fcc). In uniform surroundings, each of the 12 surrounding atoms (in the three-dimensional crystal) have equal probability of moving into a vacancy that is present in nickel. Conversely, a vacancy in nickel has equal probability of moving into any of the neighboring lattice points.

Normally, no net diffusion is observed in pure nickel because atom movements are random, and the atoms are all identical. However, using radioactive isotopes, it is possible to determine the diffusion of atoms in their own structure, called, *self-diffusion*. For example, radioactive nickel (Ni[59]) can be plated onto the surface of normal nickel. With time, and as a function of temperature, there is progressive self-diffusion of the tracer isotopes into the adjacent nickel (Fig. 6–5.3). (And there is counter movement of the untagged atoms into the surface layer. The mechanisms of diffusion include those shown in Fig. 6–5.2, as well as ones along grain boundaries where the structure is more open (Fig. 4–1.7).)

The homogenization process shown in Fig. 6–5.3 must be interpreted as follows. Although there is equal probability that an individual atom (or vacancy) will move in any of the coordinate directions, a concentration gradient favors a net movement of tracer atoms to the right in the nickel. At point A in Fig. 6–5.3(b),

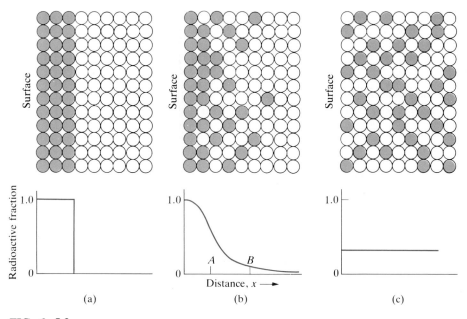

FIG. 6–5.3

Self-Diffusion. Here radioactive nickel (Ni[63]) has been plated onto the surface of nonradioactive nickel. (a) Time, $t = t_0$. (b) Diffusion gradient, $t_0 < t < t_\infty$. (c) Homogenized, $t = t_\infty$.

there are more tagged atoms than there are at point B. Thus, even with the same probability per atom for the individual tagged atoms at A to move right as there is for a tagged atom at B to move left, the difference in numbers produces the net movement as indicated. Greater uniformity develops until eventually (it may be a long time) the tagged atoms become equally dispersed throughout the crystal (as are the untagged atoms).

Diffusivity

The description of self-diffusion and counterdiffusion shown in Fig. 6–5.3 can also be applied to solute atoms in a solid solution. As an example, assume a concentration gradient where there is one carbon atom per 20 unit cells of fcc iron at point 1, and only one carbon atom per 30 unit cells at point 2, which is 1 mm away. Now, since there are random movements of carbon atoms at *each* point, we will find a net flux of carbon atoms from point 1 to point 2, simply because there are half again as many atoms jumping in the vicinity of point 1. (See Example 6–5.4)

The diffusion *flux J* of atoms (expressed in atoms/m² · sec) is proportional to the concentration gradient, $(C_2 - C_1)/(x_2 - x_1)$. In mathematical terms,

$$J = -D\frac{dC}{dx}$$

$(6-5.1a)^*$

The proportionality constant D is called the *diffusivity,* or the diffusion coefficient. The negative sign indicates that the flux is in the downhill-gradient direction. The units are

$$\frac{\text{atoms}}{(\text{m}^2)(\text{sec})} = \left[\frac{\text{m}^2}{\text{sec}}\right]\left[\frac{\text{atoms/m}^3}{\text{m}}\right]$$

$(6-5.1b)$

Diffusivity varies with the nature of the solute atoms, with the nature of the solid structure, and with changes in temperature. Several examples are given in Table 6–5.1. Some reasons for the various values of Table 6–5.1 are the following:

1. Higher temperatures provide higher diffusivities, because the atoms have higher thermal energies and therefore greater probabilities of being activated over the energy barrier between atoms (Fig. 6–5.1).

2. Carbon atoms have a higher diffusivity in iron than do nickel atoms in iron because the carbon atom is a small one (Appendix B).

3. Copper atoms diffuse more readily in aluminum than in copper because the Cu–Cu bonds are stronger than the Al–Al bonds (as evidenced by their melting temperatures).

4. Atoms have higher diffusivity in bcc iron than in fcc iron because the former has a lower atomic packing factor (0.68 versus 0.74). (We shall observe later that the fcc structure has larger interstitial holes; however, the passageways between the holes are smaller in the fcc than in the bcc structure.)

5. The diffusion proceeds more rapidly along the grain boundaries because this is a zone of crystal imperfections and of lower packing (Fig. 4–1.7).

Diffusivities Versus Temperature

The discussion of Section 6–4 related the distributions of thermal energy to temperature. Boltzmann was able to quantify this relation with Eq. (6–4.3), which showed that the number of atoms that have more than a specified amount of

---

* This is called *Fick's first law.* There is also a Fick's second law,

$$\frac{\partial C}{\partial t} = D\left(\frac{\partial^2 C}{\partial x^2}\right)$$

$(6-5.3)$

which shows the rate at which the concentration will change with time. In fact, the values of $\partial C/\partial t$ and $\partial^2 C/\partial x^2$ were determined experimentally in the laboratory to calculate the values of D found in Table 6–5.1

energy increases in proportion to an exponential function that includes that energy and the reciprocal of temperature. With diffusion, the *activation energy* for atom movements corresponds to the energy E of Boltzmann's equation. Thus,

$$D = D_0 e^{-E/kT} \qquad (6\text{–}5.4)$$

where D_0 is the proportionality constant independent of temperature that includes M of Eq. (6–4.3b). The logarithm of the diffusivity is related to the reciprocal of the temperature, $1/T$:

$$\ln D = \ln D_0 - \frac{E}{kT} \qquad (6\text{–}5.5a)$$

The other term k is the same Boltzmann constant as in Eq. (6–4.3); that is, 13.8×10^{-24} J/atom·K.

Table 6–5.2 lists the values of D_0 and the activation energy for a number of diffusion reactions. Since the chemist prefers molar and calorie units, we may rewrite Eq. (6–5.5a) as

$$\ln D = \ln D_0 - \frac{Q}{RT} \qquad (6\text{–}5.5b)$$

Energy is expressed as Q (cal/mol). The gas constant R is 1.987 cal/mol·K.

The 12 sets of data in Tables 6–5.1 and 6–5.2 are plotted in Fig. 6–5.4. These are Arrhenius-type plots.*

Diffusion in Ceramic Oxides

Examples of self-diffusion in Fig. 6–5.4 include Cu in Cu, Ag in Ag, and Fe in Fe. In each case, only one type of atom was involved. All ceramic phases contain at least two atomic species. For example, Mg^{2+} in MgO, and O^{2-} in MgO; or Cr^{3+} in $NiCr_2O_4$, along with Ni^{2+} and O^{2-}.

Figure 6–5.5 shows Arrhenius plots for diffusivities in several oxides. Several comparisons are worthy of note:

1. Co^{2+} ions diffuse more readily in CoO than do O^{2-} ions. Since both carry a double charge and each has CN = 6, the difference in diffusivity must be attributed to ion size. ($r_{Co^{2+}} = 0.072$ nm; $R_{O^{2-}} = 0.140$ nm.)

2. Fe^{2+} ions diffuse more readily in $Fe_{<1}O$ than do Mg^{2+} ions in MgO. Both oxides have divalent ions and the same NaCl-type structure. However, the presence of a few Fe^{3+} ions in $Fe_{<1}O$ provides a defect structure with cation vacancies that facilitate diffusion (Fig. 4–5.2). Thus, even with a somewhat greater size ($r_{Fe^{2+}} = 0.074$ nm vs $r_{Mg^{2+}} = 0.066$ nm), the Fe^{2+} ions have greater diffusivity. Furthermore, the diffusivity of Fe^{2+} in iron oxide is a function of the oxygen pressure. In more-oxidizing environments, iron oxide has more Fe^{3+} and therefore more vacancies to aid cation diffusion.

---

* See the comments after Example 6–4.2.

TABLE 6–5.1 Atomic Diffusivity*

		DIFFUSIVITY, m²/sec	
SOLUTE	SOLVENT (HOST STRUCTURE)	500° C (930° F)	1000° C (1830° F)
1. Carbon	fcc iron	(5×10^{-15})†	3×10^{-11}
2. Carbon	bcc iron	10^{-12}	(2×10^{-9})
3. Iron	fcc iron	(2×10^{-23})	2×10^{-16}
4. Iron	bcc iron	10^{-20}	(3×10^{-14})
5. Nickel	fcc iron	10^{-23}	2×10^{-16}
6. Manganese	fcc iron	(3×10^{-24})	10^{-16}
7. Zinc	Copper	4×10^{-18}	5×10^{-13}
8. Copper	Aluminum	4×10^{-14}	10^{-10} M‡
9. Copper	Copper	10^{-18}	2×10^{-13}
10. Silver	Silver (crystal)	10^{-17}	10^{-12} M
11. Silver	Silver (grain boundary)	10^{-11}	—
12. Carbon	hcp titanium	3×10^{-16}	(2×10^{-11})

* Calculated from data in Table 6–5.2.
† Parentheses indicate that the phase is metastable.
‡ M is calculated, although temperature is above melting point.

3. The high diffusivities of alkali ions in β-Al$_2$O$_3$ is of especial interest, because it provides the basis for batteries with solid electrolytes. The β form of alumina has tunnellike channels through which the relative large alkali ions (e.g., Na$^+$ or K$^+$) can diffuse and carry a positive charge. These open channels make the diffusion process relatively independent of temperature, as we can observe by noting the low slope of the K$^+$ in the β-Al$_2$O$_3$ curve of Fig. 6–5.5. The channels also make β-Al$_2$O$_3$ a viable candidate for a solid electrolyte for future fuel cells.

Example 6–5.1

There is 0.19 a/o copper at the surface of some aluminum and 0.18 a/o copper, 1.2 mm underneath the surface. What will the flux of copper atoms be from the surface inward at 500° C? (The aluminum is fcc, and $a = 0.4049$ nm.)

Procedure We must obtain the concentration gradient of the copper atoms, since the flux is proportional to it. However, concentrations must be expressed in number/unit volume. Therefore, it is necessary first to determine the total number of atoms per m³.

Diffusivity is a function of temperature, according to Eq. (6–5.4) and the data in Table 6–5.2.

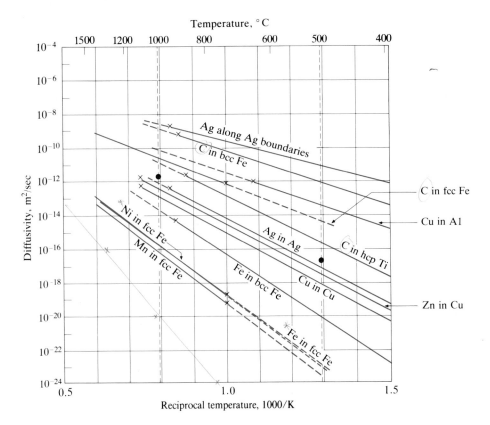

FIG. 6-5.4

Diffusivity Versus Reciprocal Temperature. (See Tables 6-5.1 and 6-5.2.)

Calculation of the concentration gradient

$$\text{atoms/m}^3 = 4/(0.4049 \times 10^{-9} \text{ m})^3$$
$$= 6 \times 10^{28}/\text{m}^3$$

$$(dC/dx)_{\text{Cu}} = \frac{(0.0018 - 0.0019)(6 \times 10^{28}/\text{m}^3)}{0.0012 \text{ m}}$$
$$= -5 \times 10^{27} \text{ Cu/m}^4$$

Calculation of the flux

$$\text{diffusivity} = (0.15 \times 10^{-4} \text{ m}^2/\text{s}) \exp\left[\frac{-0.210 \times 10^{-18} \text{ J}}{(13.8 \times 10^{-24} \text{ J/K})(773 \text{ K})}\right]$$
$$= 4 \times 10^{-14} \text{ m}^2/\text{s}$$

$$\text{flux} = -(4 \times 10^{-14} \text{ m}^2/\text{s})(-5 \times 10^{27} \text{ Cu/m}^4)$$
$$= 2 \times 10^{14} \text{ Cu/m}^2 \cdot \text{s} \qquad \qquad (\text{or } 2 \times 10^8 \text{ Cu/mm}^2 \cdot \text{s})$$

TABLE 6–5.2 Constants for Diffusivity Calculations*
($\ln D = \ln D_0 - Q/RT = \ln D_0 - E/kT$)†

SOLUTE	SOLVENT (HOST STRUCTURE)	D_0, m²/sec	Q, cal/mol	E, J/atom
1. Carbon	fcc iron	0.2×10^{-4}	34,000	0.236×10^{-18}
2. Carbon	bcc iron	2.2×10^{-4}	29,300‡	0.204×10^{-18}
3. Iron	fcc iron	0.22×10^{-4}	64,000	0.445×10^{-18}
4. Iron	bcc iron	2.0×10^{-4}	57,500	0.400×10^{-18}
5. Nickel	fcc iron	0.77×10^{-4}	67,000	0.465×10^{-18}
6. Manganese	fcc iron	0.35×10^{-4}	67,500	0.469×10^{-18}
7. Zinc	Copper	0.34×10^{-4}	45,600	0.317×10^{-18}
8. Copper	Aluminum	0.15×10^{-4}	30,200	0.210×10^{-18}
9. Copper	Copper	0.2×10^{-4}	47,100	0.327×10^{-18}
10. Silver	Silver (crystal)	0.4×10^{-4}	44,100	0.306×10^{-18}
11. Silver	Silver (grain boundary)	0.14×10^{-4}	21,500	0.149×10^{-18}
12. Carbon	hcp titanium	5.1×10^{-4}	43,500	0.302×10^{-18}

* See J. Askill, *Tracer Diffusion Data for Metals, Alloys, and Simple Oxides,* New York: Plenum (1970), for a more complete listing of diffusion data.

† $R = 1.987$ cal/mol·K; $k = 13.8 \times 10^{-24}$ J/atom·K.

‡ Lower below 400° C.

Comments We also can obtain the value of copper diffusivity graphically from Fig. 6–5.4 (and for 500° C, from Table 6–5.1). Since the activation energy enters the exponent of the equation, its three significant figures do not carry over into the value of diffusivity. Usually diffusivity values are significant to only the first or second figure.

Example 6–5.2

A steel contains 8.5 w/o Ni at the center x of a grain of fcc iron, and 8.8 w/o Ni at the edge e of the grain. The two points are separated by 40 μm. What is the flux of atoms between x and e at 1200° C? ($a = 0.365$ nm.)

Procedure Change to atom percent nickel; then calculate the number of Ni atoms/μm³ at the two points. Finally, we need the diffusivity at 1200° C to calculate the flux of Ni atoms.

Calculation

(100 amu)(0.085)/(58.71 amu/Ni) = 0.1448 = 8.1 a/o Ni

(100 amu)(0.915)/(55.85 amu/Fe) = <u>1.638</u> = 91.9 a/o Fe

total atoms = 1.783

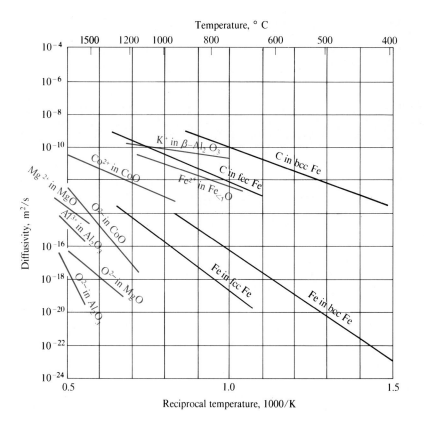

FIG. 6–5.5

Diffusivities in Selected Ceramic Compounds (Versus D in Iron). See the text for comparisons. (Data from various sources.)

A comparable calculation at the edge gives ~ 8.4 a/o Ni.

Per unit cell,

$$C_x = (4 \text{ atoms})(0.081)/(0.365 \times 10^{-3} \ \mu m)^3 = 6.66 \times 10^9/\mu m^3$$

$$C_e = (4 \text{ atoms})(0.084)/(0.365 \times 10^{-3} \ \mu m)^3 = 6.91 \times 10^9/\mu m^3$$

Using Eq. (6–5.5a),

$$\ln D = \ln (0.77 \times 10^{-4} \ m^2/s) - \left(\frac{0.465 \times 10^{-18} \ J}{(13.8 \times 10^{-24} \ J/K)(1473 \ K)} \right)$$

$$D = 9 \times 10^{-15} \ m^2/s = 9 \times 10^{-3} \ \mu m^2/s$$

Using Eq. (6–5.1a),

$$\text{flux} = -(9 \times 10^{-3} \ \mu m^2/s)\left(\frac{(6.66 - 6.91)(10^9/\mu m^3)}{40 \ \mu m} \right)$$

$$= 5.6 \times 10^4 \text{ atoms}/\mu m^2 \cdot s \qquad\qquad (\text{or } 5.6 \times 10^{16} \text{ atoms}/m^2 \cdot s)$$

Example 6–5.3

The diffusivity of aluminum in copper is 2.6×10^{-17} m²/s at 500° C and 1.6×10^{-12} m²/s at 1000° C. (a) Determine the values of D_0, Q, and E for this diffusion couple. (b) What is the diffusivity at 750° C?

Procedure With two unknowns, D_0 and E (or Q), we need at least two data points for a solution. From these we can obtain the answers, either by calculation or graphically. With equal care, the two procedures are equally accurate, because they use the same initial data.

Calculation With Eq. (6–5.5a),

(a) $\ln (2.6 \times 10^{-17}) = \ln D_0 - \dfrac{E}{13.8 \times 10^{-24}(773)}$

$(\ln 1.6 \times 10^{-12}) = \ln D_0 - \dfrac{E}{13.8 \times 10^{-24} (1273)}$

Solving simultaneously, we get

$D_0 = 4 \times 10^{-5}$ m²/s,   and $E = 0.3 \times 10^{-18}$ J/atom

Alternatively, with Eq. 6–5.5b),

$Q = 43{,}000$ cal/mol

(b) From Eq. (6–5.5a),

$\ln D = \ln 4 \times 10^{-5} - \left(\dfrac{0.3 \times 10^{-18} \text{ J}}{(13.8 \times 10^{-24} \text{ J/K})(1023 \text{ K})} \right)$

and therefore,

$D = 2.5 \times 10^{-14}$ m²/s   (or $10^{-13.6}$ m²/s)

Graphically The two data points are located by dots on Fig. 6–5.4. By interpolation at 1/1023 K,

$\log_{10} D = -13.6$ (or $D = 2.5 \times 10^{-14}$ m²/s)

The slope is E/k; the intercept at $1/T = 0$ is log D_0.

Comments Observe that the diffusivity of copper through aluminum is higher than that for aluminum through copper. This is to be expected from our knowledge of the bond strength of the host metals:

$(T_m$ of Cu$) > (T_m$ of Al$)$ therefore, $D_{Al/Cu} < D_{Cu/Al}$

Example 6–5.4

At the surface of a steel bar there is one carbon atom per 20 unit cells of iron. At 1 mm behind the surface, there is one carbon atom per 30 unit cells. The diffusivity at 1000° C is 3×10^{-11} m²/s. The structure is fcc at 1000° C ($a = 0.365$ nm). How many carbon atoms diffuse through each unit cell per minute?

Procedure Determine the number of carbon atoms per unit volume at both locations to obtain the concentration gradient; then calculate the flux in terms of $(0.365 \text{ nm})^2$.

Calculation Carbon concentrations are

$$C_2 = 1/[30(0.365 \times 10^{-9} \text{ m})^3]$$
$$= 0.68 \times 10^{27}/\text{m}^3$$
$$C_1 = 1/[20(0.365 \times 10^{-9} \text{ m})^3]$$
$$= 1.03 \times 10^{27}/\text{m}^3$$

From Eq. (6–5.1a),

$$J = -(3 \times 10^{-11} \text{ m}^2/\text{s}) \left(\frac{(0.68 - 1.03)(10^{27}/\text{m}^3)}{10^{-3} \text{ m}} \right)$$
$$= 1.05 \times 10^{19}/\text{m}^2 \cdot \text{s} \qquad\qquad\qquad (\text{or} \sim 10/\text{nm}^2 \cdot \text{s})$$

Each unit cell has an area of $(0.365 \times 10^{-9} \text{ m})^2$. Therefore,

$$J_{\text{u.c.}} = (10.5/\text{nm}^2 \cdot \text{s})(0.365 \text{ nm})^2 (60 \text{ s/min})$$
$$= 84 \text{ atoms/min}$$

Comments It is apparent that a piece of steel is not a dormant material; rather, numerous changes occur within it.

We use the process described here to *carburize* steel. In Chapter 9, we shall see how this process can be used to advantage to modify the surface hardness of a steel.

S U M M A R Y

Reactions occur within solids. Atoms break bonds and establish new bonds with new neighbors. Thus, the composition and internal structures of materials can be altered, with accompanying changes in properties. The rates at which these reactions occur influence the processing and the stability of materials.

1. Although phase diagrams define the equilibrium phases in a material at specific temperatures and compositions, *time* is required for equilibrium to develop. In fact, the delay may be long enough that the expected reaction does not occur within normal time periods. *Glasses* form by the rapid cooling of a liquid and the avoidance of crystallization. Subsequent controlled crystallization is possible, and is used commercially to produce devitrified glass. *Supercooled solid solutions* result if precipitation is avoided during cooling and supersaturation occurs.

2. *Segregation* may occur when the redistribution of atoms proceeds within a liquid phase but lags within a solid, thus avoiding equilibrium. This appears as *coring*. Segregation can also produce eutectic compositions and reactions in materials that normally would solidify as a single phase. *Soaking* processes are used for *homogenization* to attain equilibrium in a segregated material.

3. The *nucleation* of new phases requires extra energy for the phase boundary. Thus, the initiation of reactions may be retarded until an *imperfection,* an *impurity,* a *grain boundary,* or an *inoculant* can serve as a "seed."

4. At any instant, most atoms and molecules possess near-average energy, but with a range from near zero energy to a few cases of extremely high energies. We are interested in the high end of this statistical distribution, because the atoms at that energy level may be activated to break bonds, and to coordinate with new neighbors within the material. It is only through this mechanism that the internal structure (and hence the properties) of a material can be modified.

5. Diffusion proceeds by thermally activated atoms jumping from one site to another within the material. The diffusion *flux* is proportional to the concentration gradient. The proportionality constant is the *diffusiv-*

ity, D. The Arrhenius relationship holds, in which

$$\ln D = \ln D_0 - E/kT \qquad (6-5.5a)$$

The temperature must be absolute, K.

Diffusion proceeds more rapidly (i) at high temperatures, (ii) when the diffusing atom is small (e.g., C in Fe), (iii) when the packing factor of the host structure is low (e.g., bcc versus fcc), (iv) when the bonds of the host structure are weak (e.g., low-melting-point materials), and (v) when there are imperfections in the material (e.g., vacancies or grain boundaries).

"Atom Movements" is a topic of one of the study sets in the *Study Guide* that accompanies this text. The presentation coordinates sketches with explanations somewhat differently than we have done here. This presentation may be helpful to some readers.

KEY WORDS

Activation energy (E or Q)
Arrhenius equation
Boltzmann's constant (k)
Concentration gradient (dC/dx)
Coring
Devitrification
Diffusion

Diffusion flux (J)
Diffusivity (D)
Energy distribution
Eutectic, induced
Fick's first law
Glass
Heat of fusion (H_f)

Homogenization (soaking)
Nucleation
Nucleation, heterogeneous
Nucleation, homogeneous
Segregation
Supercooled
Thermal expansion coefficient (α)

PRACTICE PROBLEMS

6–P11 The polymerization of ethylene releases energy (Example 2–3.2). This indicates that C_2H_4 is less stable than is $+C_2H_4+_n$. Why is it that ethylene can remain as a less stable phase almost indefinitely?

6–P12 Name a common candy product that is a glassy solid.

6–P13 Sterling silver possesses 7.5 w/o copper. Why is it that amount?

6–P14 Metallic glasses must be quenched from their liquid to ambient temperatures in milliseconds or less. The cooling rate for silica glass can be $\ll 1°$ C/s without crystallization. Why is there a difference?

6–P21 Consider a 40 MgO–60 FeO composition. There is a progressive increase in the FeO content of the liquid as it solidifies. The FeO content of the solid also increases as the temperature drops (Fig. 5–6.6). How can both phases increase in FeO without an overall compositional change?

6–P22 Assume 20 kg of an 8 Al–92 Mg alloy are first melted and then cooled rapidly to 500° C. Since there is no time for diffusion in the solid, the average composition, $\bar{\epsilon}$, of the solid is 5 percent Al. (a) What is the composition of the liquid at 500° C? (Assume rapid diffusion in the liquid.) (b) How much liquid is there at 500° C? (c) At what temperature will the second solid phase appear? (d) What is the composition of the final liquid?

6–P23 A long rod of silicon is "zone-refined" by being moved slowly through a short heater, such that the molten zone moves from one end and toward the other end. The rod solidifies as it leaves the heated zone. Explain how this process could be repeated to remove impurities.

6–P31 Refer to Example 6–3.1. At 975° C, the free-energy value for the reaction, ΔF_V, is -0.8×10^9 J/m³. On a volume basis, how much larger must the nuclei be than they are at 900° C?

6–P32 Nucleation may develop at various micro-structural sites, commonly more readily than by the homogeneous route involved in Example 6–3.1. What are these sites?

6–P41 An aluminum wire is stretched between two rigid supports at 35° C. It cools to 15° C. What additional stress is developed?

6–P42 (a) Estimate the linear expansion coefficient of bcc iron at 900° C from the data in Fig. 3–4.1. (b) Why does this value not match the data of Appendix C?

6–P43 Which will be higher, the *mean energy* or the *median energy* of the gas molecules in your room?

6–P44 Refer to Example 6–4.2. What fraction of the atoms have sufficient energy to jump out of their sites at 1000° C?

6–P45 An activation energy of 2.0 eV (or 0.32×10^{-18} J) is required to form a vacancy in a metal. At 800° C, there is one vacancy for every 10^4 atoms. At what temperature will there be one vacancy for every 1000 atoms?

6–P51 A solid solution of copper in aluminum has 10^{26} atoms of copper per m³ at point X, and 10^{24} copper atoms per m³ at point Y. Points X and Y are 10 μm apart. What will be the diffusion flux of copper atoms from X to Y at 500° C?

6–P52 (a) What is the ratio of diffusivities for carbon in bcc iron to carbon in fcc at 500° C? (b) What is it for carbon in fcc iron to nickel in fcc iron at 1000° C? (c) What is it for carbon in fcc iron at 1000° C to carbon in fcc iron at 500° C? (d) Why are the ratios greater than 1?

6–P53 The inward flux of carbon atoms in fcc iron is 10^{19}/m²·s at 1000° C. What is the concentration gradient?

6–P54 (a) Using the data of Table 6–5.2, calculate the diffusivity of copper in aluminum at 400° C. (b) Check your answer against Fig. 6–5.4. Do the values match?

6–P55 A zinc gradient in copper alloy is 10 times greater than the aluminum gradient in a copper alloy. Compare the flux of solute atoms/m²·s in the two alloys at 500° C. (The data for $D_{Al\ in\ Cu}$ are given in Example 6–5.3.)

6–P56 Aluminum is to be diffused into a silicon single crystal. At what temperature will the diffusion coefficient be 10^{-14} m²/s? ($Q = 73,000$ cal/mol and $D_0 = 1.55 \times 10^{-4}$ m²/s.)

6–P57 Refer to Table 6–5.1. (a) Why are the values higher for couple 2 than for couple 1? (b) Why are they higher for couple 2 than for couple 4? (c) Why are they higher for couple 11 than for couple 10? (d) Why are they higher for couple 8 than for couple 9?

6–P58 Refer to Problem 6–P51. (a) What is the diffusion coefficient of copper in aluminum at 100° C? (b) What will be the diffusion flux of copper atoms from X to Y at 100° C?

6–P59 In Al_2O_3, what is the $D_{Al^{3+}}/D_{O^{2-}}$ ratio at 2000 K? Give two reasons for this significant difference.

TEST PROBLEMS

611 Silicate glasses may be annealed for several hours at $T_m/2$ without devitrifying (crystallizing). Metallic glasses will crystallize in seconds at $T_m/2$ — that is, at one-half of their Kelvin melting temperature. Explain.

621 An aircraft part is cast from a 96 Al–4 Cu liquid into its final shape. In the as-cast condition, it is highly segregated (Fig. 6–2.4). The solidus temperature for this alloy is approximately 1070° F

(575° C); however, the maximum annealing temperature for homogenization is specified as 995° F (535° C). Why is that the temperature? Why is it not higher, so that the process can be faster? The furnace can be controlled to $\pm 15°$ F.

622 A kilogram of a 90–10 bronze contains 900 g copper (100 g Sn). Suggest a procedure for obtaining from this bronze at least 100 g copper that will have an analysis of <2 percent tin. Use melting, partial solidification, and separation steps. (Assume that the liquidus and solidus are straight lines.)

641 Tin (white) is tetragonal. Mean values of its thermal expansion coefficients are $\alpha_a = 15 \times 10^{-6}/°$ C, and $\alpha_c = 10 \times 10^{-6}/°$ C. What is the percent volume contraction between T_m and 20° C?

642 At 900° C, 1 out of 10^{11} atoms, and at 1100° C, 1 out of 10^9 atoms, have sufficient energy for movements within a solid. (a) What is the activation energy in J/atom? (b) What is it in cal/mol?

643 Why must T of Arrhenius-type equations be in absolute temperature, K? We can use ° C for thermal expansion and for thermal conductivity calculations.

644 Refer to Problem 642. What fraction of the atoms have the required energy at 1000° C? Solve graphically.

651 There are 3 a/o carbon at the surface of the iron in Problem 6–P53. What is the a/o carbon at 1500 μm behind the surface? ($a = \sim 0.365$ nm at 1000° C.)

652 Zinc is moving into copper. At point X, there are 2.5 $\times 10^{17}$ Zn/mm^3. What concentration is required at point Y (2 mm from X) to diffuse 60 Zn atoms/mm$^2 \cdot$min at 300° C (Y to X)?

653 What is the weight percent of zinc in the copper at point X of Problem 652?

654 The Ni-in-fcc Fe and the Fe-in-fcc Fe curves of Fig. 6–5.4 cross. (a) By calculation, determine the temperature where they intersect. (b) Suggest why the two curves are nearly coincident.

655 The diffusion of carbon in tungsten has an activation energy of 0.78×10^{-18} J/atom (112,000 cal/mol) and $D_0 = 0.275$ m^2/s. Where does the diffusivity curve lie on Fig. 6–5.4? Suggest why it lies where it does in comparison to the other curves for each carbon diffusion.

656 There are 10^{16} Ni atoms per mm^3 of some iron. (a) What must the concentration gradient be for nickel in fcc iron if a flux of 250 nickel atoms/mm$^2 \cdot$s is to be realized at 800° C? (b) How many times greater will the flux be at 1200° C than at 800° C with the same concentration gradient?

657 Compare the self-diffusivities of iron (fcc), copper, and silver at 60 percent of their melting temperatures—that is, at 0.6 T_m.

658 Based on Fig. 6–5.5, determine the activation energy for the diffusion of either Co^{2+} in CoO, or of O^{2-} in CoO.

659 (a) At what temperature does $D_{Fe^{2+}}$ in $Fe_{<1}O$ become 10,000 times D_{Fe} in bcc iron? (b) At what temperature does it become 10,000 times D_{Fe} in fcc iron? (c) Why is there a difference in the diffusivities?

Chapter 7

MICROSTRUCTURES

A *microstructure* is the geometric arrangement of grains and phases in a material. Variables include the *amount, size, shape,* and *distribution* of these structural features. Typically, the dimensions are such that an optical microscope (up to $\times 2000$)—or even an electron microscope (up to $\times 50,000$)—is necessary for observation.

We can anticipate the amount of each phase on the basis of phase diagrams. The size is dictated by time, temperature, and other kinetic considerations. The shape and distribution are more complex, but can be optimized through appropriate heat treatments.

7–1

SINGLE-PHASE MATERIALS

We use the term *microstructure* to describe the structural level that we normally can visualize only with the aid of a microscope. Optical and electron microscopes extend the lower limits of such structures down to the micrometer and nanometer dimensions, respectively. Although there is no absolute cutoff, we are generally concerned with features that are coarser than the atom-to-atom coordination of crystals. The upper end of the microstructural range is typically of millimeter dimensions. Again, however, there is no sharp cutoff.

A wide variety of materials are *single phase.* Included are window glass and transparent polystyrene drinking cups. These materials are amorphous and therefore lack any conventional microstructure. Silicon boules (for the production of transistor circuits—Fig. 7–1.1) and ruby gems are single crystals, and therefore are single phase. Again, they lack a microstructure in the conventional sense. The majority of single-phase materials however, are *polycrystalline,* and do possess a microstructure. Examples are brass products (Fig. 7–1.2, and the α of Fig. 5–6.1),

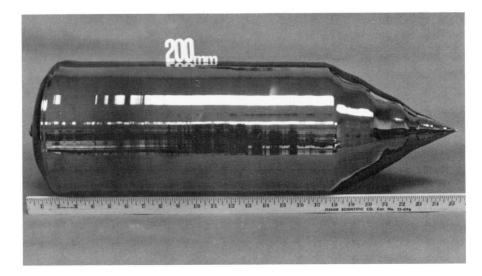

FIG. 7–1.1

Single Crystal (Silicon Boule). This 100-lb (45-kg) single crystal will be cut into 1-mm wafers. Onto each, many hundreds of chips will be produced for computer or similar applications. Being a single crystal, it has no microstructure exceeding the nanometer range. The crystal was grown from a single-crystal seed at the extreme right end (which is the bottom during growth). (Courtesy of D. Golland, Monsanto Chemical Co.)

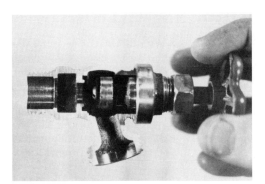

FIG. 7-1.2

Brass (Cutaway Section of a Faucet). This Cu-Zn alloy contains only a single phase, which is fcc. However, it has a microstructure of many grains that formed during solidification.

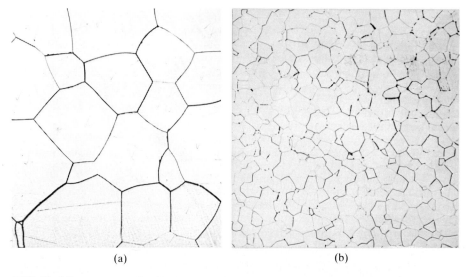

(a)   (b)

FIG. 7-1.3

Grain Boundaries. (a) Molybdenum (×250) (O. K. Riegger). (b) High-density periclase, MgO (×250) (R. E. Gardner and G. W. Robinson, Jr., *J. Amer. Ceram. Soc.*).

and MgO ceramics of Fig. 7-1.3(b). These materials contain many *grains* of the same phase, with a variety of possible microstructures. In this section, we shall consider the microstructures of single-phase, polycrystalline materials.

Grains

The microstructures of single-phase materials can be varied by changes in the *size, shape,* and *orientation* of the grains (Fig. 7-1.4). These aspects are not wholly independent, because the shape and size of grains are both consequences of grain

FIG. 7–1.4

Microstructural Variables of Single–Phase Materials.
(a) versus (b): grain size.
(a) versus (c): grain shape.
(b) versus (d): preferred orientation.

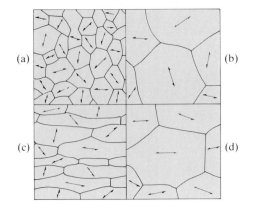

growth. Likewise, grain shape is commonly dependent on the crystalline orientation of grains during growth.

Although it is common to speak of grain size in terms of diameter, few if any grains of a single-phase metal are spherical. Rather, the grains must completely fill space and also maintain a minimum of total boundary area. This is shown in Figs. 7–1.3 and 7–1.4(a), where the grains are described as being *equiaxed* because they have approximately equal dimensions in the three coordinate directions. In addition to equiaxed grains, other commonly encountered *grain shapes* may be plate-like (Fig. 7–1.4c), columnar, or dendritic (i.e., treelike). We shall not attempt to systematize these shapes.

The *orientation* of grains within a metal is typically quite random (Fig. 7–1.4a). There are exceptions, however, that can be important from the standpoint of engineering projects. For example, the [100] directions of iron have a higher magnetic permeability than do the other directions. Therefore, if the grains within a polycrystalline transformer sheet are not random, but are processed to have a *preferred orientation* such that the [100] direction is preferentially aligned with the magnetic field, the transformer will operate with greater efficiency. The metallurgist has learned how to develop this orientation, with the result of billions of dollars worth of savings in electrical-power distribution.

Grain size is important because it varies inversely with the grain-boundary area. It is the latter that affects the behavior of a polycrystalline material—diffusion (Section 6–5), nucleation (Section 6–3), strength (Chapter 9), corrosion (Chapter 14), and so on. The *mean chord length,* $\overline{L}$, is an index of grain size. We can determine it easily by placing a random line of known length across a polished and etched microstructure. The mean chord length is the reciprocal of the number of boundary intersection points per unit length, P_L.

$$\overline{L} = 1/P_L \qquad (7\text{–}1.1)$$

In Fig. 7–1.5, a 50-mm circle has been placed randomly on the photomicrograph of molybdenum taken from Fig. 7–1.3(a). There are 11 intersections. With mag-

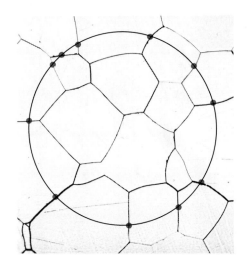

FIG. 7-1.5

Grain-Boundary Area Calculation. Since the magnification is ×250, the length of the circle is $\pi(50$ mm$)/250$, or 0.63 mm (0.025 in.). The circle intersects 11 boundaries in that distance. Therefore, there are 2(11/0.63 mm), or 35 mm^2, of boundary area per mm^3 (see Eq. 7-1.2). (O. K. Riegger.)

nification of ×250, the line on the metal is actually $(50\pi/250) = 0.63$ mm; the value of P_L is $(11/0.63$ mm$) = 17.5$/mm; and $\overline{L}$ is 0.057 mm.

Since the grain-boundary area (and not grain size, *per se*) affects the properties, a measure of the *grain-boundary area* per unit volume, S_V, has a more quantitative meaning. Fortunately, the two measures have a simple relation:

$$S_V = 2\,P_L \qquad\qquad (7-1.2)^*$$

Thus, the grain-boundary area of the molybdenum in Fig. 7-1.3(a) is 35 mm^2/mm^3. You can compare these figures with the mean chord length and the grain-boundary area of the MgO in Fig. 7-1.3(b) as calculated in Example 7-1.1.

ASTM Grain-Size Numbers

Although it is the boundary that affects properties, the *grain size* and the number of grains are more apparent to the viewer. Therefore, a method to determine a grain-size number has been standardized by the American Society for Testing and Materials (ASTM). Although empirical, it is a quantitative and reproducible index. This index uses *2* as a base:

$$N_{(0.01\,\text{in.})^2} = N_{(0.0645\,\text{mm}^2)} = 2^{n-1} \qquad\qquad (7-1.3)$$

---

* At normal temperatures the grain boundaries interfere with slip. Therefore, a *fine-grained* material is stronger than is a *coarse-grained* material. At elevated temperatures, the boundaries can accommodate the dislocations. As a result, the situation is reversed at high temperatures, and creep results (Section 14-5).

FIG. 7–1.6

Grain-Size Numbers (ASTM Comparison Nets). A large grain-size number (G.S.#; Eq. 7–1.3), indicates more grains and more grain-boundary area per unit volume ($\times 100$).

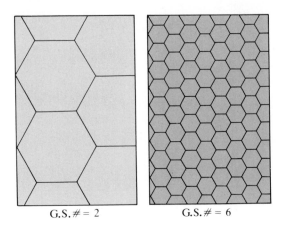

G.S. # = 2   G.S. # = 6

The term N is the number of grains observed in an area of 0.0645 mm^2 (1 in.^{2} at $\times 100$).* The value n is the *grain-size number* (G.S.#). Example 7–1.2 calculates the G.S.# for the molybdenum shown in Fig. 7–1.3(a). Figure 7–1.6 shows two of a series of grain-size nets that can be used for quick visual assignments of a grain-size number to $\times 100$ photomicrographs. These grain-size numbers are pertinent to the heat-treating of steels (Section 9–6), and for the ductility-transition temperatures of steels (Section 8–4).

Grain Growth

The average grain size of a single-phase material increases with time if the temperature produces significant atom movements (Section 6–5). The driving force for *grain growth* is the energy released as an atom moves across the boundary from the grain with the convex surface to the grain with the concave surface. There, the atom is, on the average, coordinated with a larger number of neighbors at equilibrium interatomic spacings (Fig. 7–1.7). As a result, the boundary moves toward the center of curvature. Since small grains tend to have surfaces of sharper convexity than do large grains, they disappear because they feed the larger grains (Fig. 7–1.8). The net effect is grain growth (Fig. 7–1.9).

All crystalline materials—metals and nonmetals—exhibit this characteristic of grain growth. An interesting example of grain growth can be seen in the ice of a snow bank. Snowflakes start out as numerous small ice crystals, lose their identity with time, and are replaced by larger granular ice crystals. A few of the crystals grow at the expense of the many smaller crystals.

The growth rate depends heavily on temperature. An increase in temperature increases the thermal vibrational energy, which in turn *accelerates* the net diffusion of atoms across the boundary from small to large grains (Example 7–1.3). A

---

* The procedure was originally standardized to use a microscope with $\times 100$ lenses, and to count the grains within a 1 in. $\times$ 1 in. area (=0.0001 in^2, or 0.0645 mm^2).

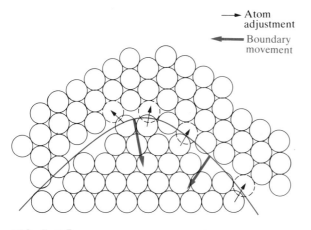

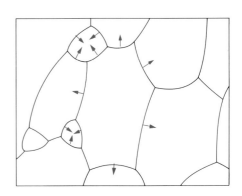

FIG. 7-1.7

Grain-Boundary Movement. The atoms move to the grain
with the concave surface, where they are more stable. As
a result, the boundary is shifted toward the center of
curvature.

FIG. 7-1.8

Grain Growth. The boundaries move toward
the center of curvature (arrows). As a result, the
small grains eventually disappear.

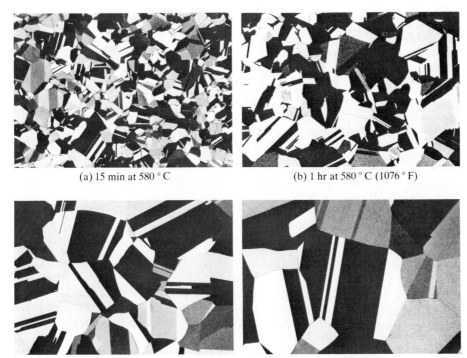

(a) 15 min at 580 ° C

(b) 1 hr at 580 ° C (1076 ° F)

(c) 10 min at 700 ° C

(d) 1 hr at 700 ° C (1292 ° F)

FIG. 7-1.9

**Grain Growth (Brass
at ×40).** (Courtesy of
J. E. Burke, General
Electric Co.)

subsequent decrease in temperature slows down the boundary movement, but *does not reverse it.* The only way to refine (reduce) the grain size in alloys that have only a single phase is to deform the grains plastically and to start new grains by recrystallization (Chapter 9).

Example 7–1.1

Estimate the mean chord length and the grain-boundary area per unit volume in the MgO of Fig. 7–1.3(b).

Solution   Lay a 50-mm straightedge at random across the figure and count the grain boundaries intersected. If, by repeating this procedure five times, you get counts of 13, 17, 12, 14, and 12, you would have 68 counts per 250 mm. However, the magnification is $\times 250$. Therefore, $P_L = 68/\text{mm}$.

Calculation

$$\overline{L} = 1/(68/\text{mm}) = 0.015 \text{ mm}$$

$$S_V = 2(68/\text{mm}) = \sim 140/\text{mm} \qquad\qquad\qquad (\text{or } 140 \text{ mm}^2/\text{mm}^3)$$

Comments   There is about four times as much grain-boundary area per unit volume in the MgO of Fig. 7–1.3 as in the molybdenum. Both of these are *estimates* from samplings. However, with reasonable care, we can be accurate to within ± 10 percent, which is sufficient for most purposes.

Example 7–1.2

Assign an ASTM G.S.# to the molybdenum of Fig. 7–1.3(a).

Procedure   Since the magnification is *not* $\times 100$, we cannot obtain N directly by counting a 1 in.^{2} (or 645 mm^2) sample of the photographed area. However, we can count the grains in the total area and correct for magnification.

Estimation   There are ~ 17 grains in an area of $(59 \text{ mm}/250)^2$. (See comments.)

$$\frac{\sim 17}{0.056 \text{ mm}^2} = \frac{N}{0.0645 \text{ mm}^2}$$

$$N = \sim 20 = 2^{n-1}$$

$$n = 5^+$$

Comments   The photomicrograph in Fig. 7–1.3(a) is, itself, a sample and therefore is subject to statistical variations. As a result, we should not expect utmost precision. Furthermore, it is seldom necessary to be more specific than ± 0.5 in our estimates of the grain-size number.

We can most readily obtain the number of grains in the area that was sampled by (1) counting the grains that lie entirely within the area, (2) adding to the count one-half of the grains at the edges (since these are shared by adjacent areas), and (3) then adding one-fourth of each of the four-corner grains. If we took another sample of the metal in Fig.7–1.3(a), and our counts were as low as 15 or as high as 20, our grain-size estimate would still be $n = 5^+$. The count must double to shift from one G.S.# to the next.

Example 7–1.3

Two identical samples were obtained from the same piece of bronze. The grain size expressed as δ was 50 μm. Sample A was heated to 650° C and required 4.5 hr to double its value of δ to 100 μm. At 700° C, sample B required only 30 minutes for the same change in grain size—50 μm to 100 μm. Estimate the time requirement at 750° C.

Procedure   Each heat-treatment involves atom movements to the same extent, but at different temperatures. Therefore, we expect an Arrhenius relationship for the rates of reaction R versus T. Patterned after diffusion (Eq. 6–5.5a),

$$\ln R = \ln R_0 - E/kT \tag{7–1.4}$$

Since rates and time are reciprocals,

$$\ln t = C + B/T \tag{7–1.5}$$

where C is $-\ln R_0$, and B is E/k.

Calculation

$$\ln 4.5 \text{ hr} = C + B/923 \text{ K} = 1.50$$

$$\ln 0.5 \text{ hr} = C + B/973 \text{ K} = 0.70$$

Solving simultaneously,

$$B = 39,500 \text{ K} \qquad C = -41.3$$

$$\ln t = -41.3 + (39,500 \text{ K}/1023 \text{ K}) = -2.7$$

$$t = 0.07 \text{ hr} \qquad\qquad\qquad\qquad\qquad\qquad \text{(or 4 min)}$$

Comment   The grain size will *not* double when the time is doubled, because the growth rate decreases as the grain size increases.

7–2
PHASE DISTRIBUTION (PRECIPITATES)

Microstructures with two or more phases possess a variety of geometries that are not encountered in single-phase materials, including (1) the relative amounts of the several phases, (2) the distribution of the phases (e.g., whether the minor phase is dispersed, or is present as a grain-boundary network), and (3) the size and shape of the several phases in the mixture. In this section, we shall look at phase distributions that arise by the separation of a minor phase from a supercooled solid solution, called *precipitation*.* In the next section, we shall consider some of the microstructural distributions that develop when two phases form simultaneously through an eutectoid (or eutectic) reaction.

Precipitation Rates

In Section 6–1, we noted that solid solutions such as sterling silver can be cooled sufficiently rapidly that a supersaturated alloy can be retained as a single phase at ambient temperatures. This retention is possible because (1) a new phase does not need to be nucleated (Section 6–3), and (2) atoms are not required to diffuse (Section 6–5) to the growing particle to bring about phase separation. Both of these require time.

On a cursory basis, the rate, R, of separation is the product of nucleation rate, $\dot{N}$, and the growth rate, $\dot{G}$,

$$R = f(\dot{G}\dot{N}) \tag{7–2.1}$$

This relation is shown schematically in Fig. 7–2.1. The faster rates, and therefore the shorter precipitation times, occur at intermediate temperatures. Figure 7–2.2 lets us examine the required times more closely as they involve precipitation from solid solutions. The figure comes from the Pb–Sn phase diagram originally presented in Fig. 5–2.1. At temperatures above 150° C, a 90 Pb–10 Sn alloy may be *solution treated* to produce a single fcc phase, α. Below 150° C, solid precipitation would be expected when this alloy is cooled into the two-phase region of $(\alpha + \beta)$, as shown in the equilibrium diagram (Fig. 7–2.2b). The time requirements vary with temperature, however, as shown in Fig. 7–2.2(a). Immediately below 150° C, nuclei do not form readily, so there is a time delay. Reaction is relatively rapid at 50° C, with a 50 percent completion in a couple of minutes. More time is required at subambient temperatures because diffusion is slow. Fifteen minutes are required for 50 percent completion at 0° C.

This C-type curve is for *isothermal precipitation,* and is rather common in a variety of solid-state reactions. It not only describes the kinetics of phase separation, but also gives us a basis for understanding the origin of certain microstructures.

Intergranular and Intragranular Precipitation

Above the knee of the isothermal precipitation curve (Fig. 7–2.2a), the atoms in this Pb–Sn alloy move readily (120° C $\approx 0.7\ T_m$). Furthermore, the only location for easy nucleation is the grain boundaries, where atoms are already disordered.

---

* Rain is also a precipitate: water droplets were separate from air that is supersaturated with water.

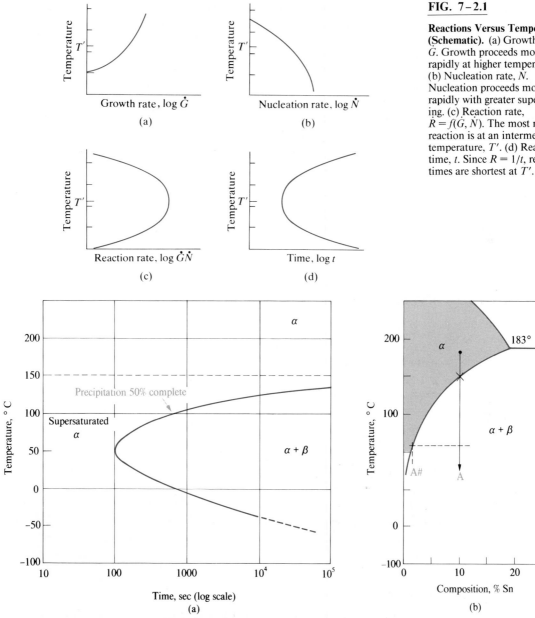

FIG. 7-2.1

Reactions Versus Temperature (Schematic). (a) Growth rate, $\dot{G}$. Growth proceeds more rapidly at higher temperatures. (b) Nucleation rate, $\dot{N}$. Nucleation proceeds more rapidly with greater supercooling. (c) Reaction rate, $R = f(\dot{G}, \dot{N})$. The most rapid reaction is at an intermediate temperature, T'. (d) Reaction time, t. Since $R = 1/t$, reaction times are shortest at T'.

FIG. 7-2.2

Isothermal Precipitation (β from a 90 Pb-10 Sn Solid Solution). (a) The alloy was solution-treated above 150° C to form a single phase (α). (b) The precipitation of β below 150° C requires time, which varies with temperature. Equilibrium develops with sufficient time, so the right end of (a) will match (b). (See Figs. 5-2.1 and 7-2.1.) (Adapted from data by H.K. Hardy and T.J. Heal, *Progress in Metal Physics* 5, Pergamon Press.)

Therefore, the precipitation of β is primarily along the grain boundaries to which the atoms move (Fig. 7–2.3b), and we speak of *intergranular precipitation*.

Below the knee of this curve, the rate of precipitation is limited by diffusion of the atoms. Furthermore, the supersaturation is great, and nucleation will occur wherever possible—at point imperfections such as vacancies or interstitials, along dislocation lines, and adjacent to impurities—all within the interior of the super- saturated grains. Therefore, we see a different microstructure, as depicted by the sketch in Fig. 7–2.3(c) in comparison to that in Fig. 7–2.3(b). This microstructure has a *dispersion* of the minor phase.

Two different samples of the same material, but with these opposing micro- structures, will have significantly different properties (Chapter 9). For example, if the precipitate is hard and brittle and the matrix is ductile, a fracture path could readily propagate through the material Fig. 7–2.3(b) along the grain boundary. Such a material would fail brittlely. Conversely, a dispersion of these hard particles serves to strengthen the ductile matrix in Fig. 7–2.3(c).

Example 7–2.1

Calculate the density of a solder (eutectic) that has been equilibrated at 20° C.

Procedure From Fig. 5–2.1 (Pb–Sn), the solid solder contains β and α that are essentially pure tin and lead, respectively. Therefore, from Appendix B, $\rho_\beta = 7.2$ Mg/m^3 and $\rho_\alpha \approx 11.3$ Mg/m^3 (or 11.3 mg/mm^3).

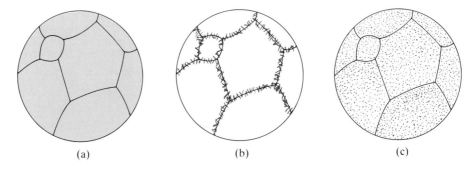

(a)   (b)   (c)

FIG. 7–2.3

Solid-Phase Precipitation (Schematic). (a) Supersaturated solid solution—for example, 90 Pb–10 Sn—rapidly cooled from 180 to 20° C. (b) Grain-boundary precipitation; it requires long-range diffusion to the grain boundaries, where nucleation occurs more readily. (c) Intragranular precipitation. Many minor imperfections throughout the grains nucleate the new phase; therefore, diffusion distances are shorter than in structure (b). Higher precipitation temperatures favor structure (b); lower ones favor structure (c).

Calculation Basis: 100 mg $= 61.9$ mg $\beta + 38.1$ mg α

$\left. \begin{array}{ll} \beta: & 61.9 \text{ mg}/(7.2 \text{ g/mm}^3) = 8.60 \text{ mm}^3 \\ \alpha: & 38.1 \text{ mg}/(11.3 \text{ g/mm}^3) = 3.37 \text{ mm}^3 \end{array} \right\} = 11.97 \text{ mm}^3 \text{ total}$

$\rho = 100 \text{ mg}/(11.97 \text{ mm}^3) = 8.4 \text{ mg/mm}^3$ (or 8.4 Mg/m³)

Alternative From the above calculation, the volume fractions are

$f_\beta = 8.60 \text{ mm}^3/(11.97 \text{ mm}^3) = 0.72$

$f_\alpha = 3.37 \text{ mm}^3/(11.97 \text{ mm}^3) = 0.28$

We can set up a *mixture rule* on a volume basis:

$\begin{aligned} \rho &= \rho_\alpha f_\alpha + \rho_\beta f_\beta \\ &= (0.28)(11.3 \text{ g/cm}^3) + (0.72)(7.2 \text{ g/cm}^3) \\ &= 8.4 \text{ g/cm}^3 \end{aligned}$ (7-2.2)

(or 8.4 Mg/mm³)

Example 7-2.2

An Al–Cu alloy has 2 a/o copper in solid solution κ at 550° C. It is quenched, then reheated to 100° C, where θ precipitates (Cu–Al, Fig. 5–6.3). The θ (CuAl$_2$) develops many *very small* particles throughout the alloy; the average interparticle distance is only 5.0 nm. (a) Approximately how many particles form per mm³? (b) If, by extrapolating from Fig. 5–6.3, we assume that negligible copper remains in fcc κ at 100° C, how many copper atoms are there per θ particle?

Procedure (a) Since particles are 5 nm apart, there is ~ 1 particle/(5 nm)³. (b) We must determine the total number of Cu atoms per unit volume. This number is equal to 2 a/o of the atoms. At 100° C, essentially all the Cu atoms are in the θ particles.

Calculation

(a) $\sim 1 \; \theta/(5 \times 10^{-9} \text{ m})^3 = 8 \times 10^{24} \; \theta/\text{m}^3$ (or $\sim 8 \times 10^{15} \; \theta/\text{mm}^3$)

(b) atoms/m³ $= 4/a^3 = 4/[4(\sim 0.143 \times 10^{-9} \text{ m})/\sqrt{2}]^3$
$= 6 \times 10^{28}/\text{m}^3$

Cu atoms/m³ $= (0.02)(6 \times 10^{28}/\text{m}^3)$
$= 1.2 \times 10^{27}/\text{m}^3$

Cu/particle $= \dfrac{1.2 \times 10^{27} \text{ Cu/m}^3}{\sim 8 \times 10^{24} \; \theta/\text{m}^3} \approx 150 \text{ Cu}/\theta \text{ particle}$

Comment This microstructure is approximately what we shall encounter in precipitation hardening (Section 9–4).

7–3

PHASE DISTRIBUTION (EUTECTOID DECOMPOSITION)

Eutectoid or eutectic reactions appear in the vast majority of phase diagrams. These two invariant reactions have a common feature—two new phases form simultaneously during cooling. We shall use the eutectoid reaction of Eq. (5–5.4) as our prototype. For this, the eutectoid region of the Fe–Fe$_3$C diagram is redrawn on a larger scale in Fig. 7–3.1 (shaded part of Fig. 5–5.1).

Austenite, γ, the fcc polymorph of iron and steels, decomposes on cooling to ferrite, α, plus a carbide (Fe$_3$C, but labeled $\overline{C}$ for convenience):

$$\gamma_{(0.8\% \text{ C})} \rightarrow \alpha_{(0.02\% \text{ C})} + \overline{C}_{(6.7\% \text{ C})} \qquad (7\text{–}3.1)$$

Austenite accepts carbon atoms interstitially. Since the carbon atom is small, the interstices in austenite can accommodate up to 10 a/o carbon (2 + w/o). These small atoms reside in the $\frac{1}{2}, \frac{1}{2}, \frac{1}{2}$ sites within the unit cell (and in their $\pm\frac{1}{2}, \pm\frac{1}{2}, 0$ translations).

Ferrite is the bcc polymorph of iron and steels. Above the γ range (Eq. 3–4.2), it is called δ-ferrite; below the γ range, α-ferrite, or just α. In this lower range, ferrite

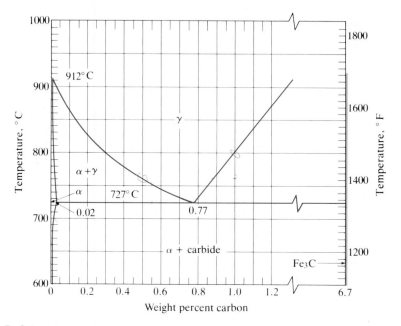

FIG. 7–3.1

The Eutectoid Region of the Fe–Fe$_3$C Phase Diagram. (See Fig. 5–5.1). Steels with ~0.8 percent C are commonly called *eutectoid* steels. *Hypereutectoid* steels are above that value; *hypoeutectoid* steels are below it.

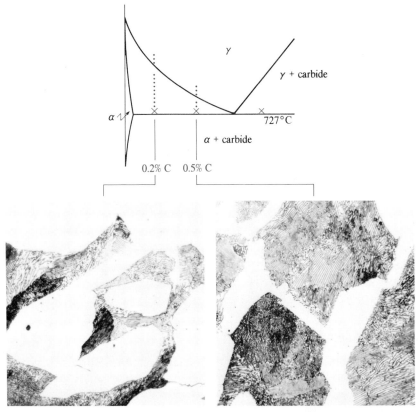

(a) 0.2% C (b) 0.5% C

FIG. 7–3.4

Hypoeutectoid Steels (Annealed) (×500). (a) 0.2 percent C (balance Fe). (b) 0.5 percent C (balance Fe). These steels were first austenized, then cooled slowly. Proeutectoid ferrite (white) formed before the $\gamma \to \alpha + \text{C}$ reaction produced pearlite (lamellar, gray). (Courtesy of U.S. Steel Corp.)

750° C (1380° F) leads to ($\alpha + \gamma$). At 750° C, however, it takes several seconds for the reaction to begin (Fig. 7–3.6). It takes still longer before the decomposition is completed and the α-to-γ equilibrium ratio of 43-to-57 is reached.

The same steel quenched from 850° to 650° C (1560 to 1200° F) and held isothermally leads to ($\alpha + \text{C}$). According to data in Fig. 7–3.6, the first ferrite appears in less than 1 second; carbide appears a couple of seconds later. The last austenite disappears in a little over 10 seconds. More time is required for the decomposition of the austenite at temperatures below 550° C, at which diffusion is slower. The C-type curves of Fig. 7–3.6 are called *isothermal transformation* (**I–T**) curves, or sometimes **TTT** curves for Time–Temperature–Transformation.

FIG. 7–3.5

Hypereutectoid Steel (Annealed ×500). With 1.2 percent carbon (balance iron), Fe₃C separated along the grain boundaries before the austenite decomposed to form pearlite. The amount of pearlite is equal to the amount of austenite that existed at the eutectoid temperature (727° C). (Courtesy of U.S. Steel Corp.)

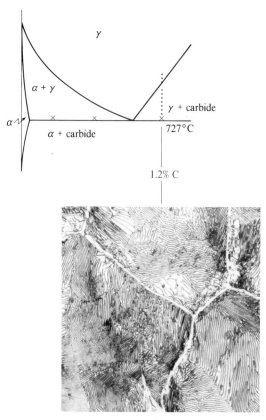

1.2% C

The microstructures of isothermally decomposed austenite vary with the temperature of transformation. In an 0.45 percent carbon steel (balance Fe), proeutectoid ferrite and coarse pearlite form when the reaction is at 700° C (1290° F). The pearlite is finer at 600° C, consistent with our earlier discussion of pearlite formation. Pearlite does not develop below the knee of the I-T curve. Rather, the final product has many fine carbide particles dispersed within a matrix of ferrite. This microconstituent is called *bainite.** Hardness increases with progressively finer microstructures.

---

* At lower temperatures, the carbon diffusion to the grain boundaries is slow. Also, with sufficient supercooling, nucleation occurs at numerous imperfections within the grains. Just as in Fig. 7–2.3(b) and (c), the distribution of the phases is altered.

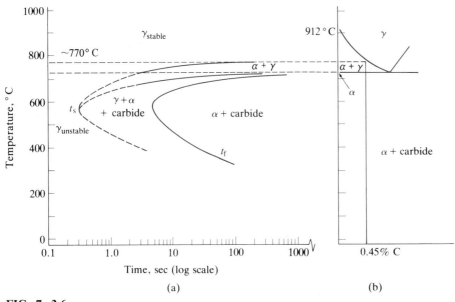

FIG. 7–3.6

Isothermal Austenite Decomposition (0.45 Percent C, Balance Fe). The stable phases of the phase diagram (right) are not obtained immediately, but require the time indicated in (a). (Time for start of reaction t_s; time for finish, t_f.)

Example 7–3.1

Carbon atoms sit in the $\frac{1}{2},\frac{1}{2},0$ positions in bcc iron.

    (a) Using $r_C = 0.077$ nm and $R_\alpha = 0.124$ nm, how much must the nearest iron atoms be displaced to accommodate the carbon?

    (b) A carbon atom sits in the $\frac{1}{2},\frac{1}{2},\frac{1}{2}$ position in fcc iron. Using $R_\gamma = 0.127$ nm (Table 2–5.1) for the radius of γ-iron, how much crowding is present?

Procedure Make a sketch for easier visualization. In (α), the nearest iron atoms are at $\frac{1}{2},\frac{1}{2},\frac{1}{2}$ and $\frac{1}{2},\frac{1}{2},-\frac{1}{2}$. In (γ), the nearest iron atoms are at fcc positions.

Calculation

    (a)
$$a = 4(0.124 \text{ nm})/\sqrt{3} = 0.286 \text{ nm}$$
$$\Delta = (0.124 \text{ nm} + 0.077 \text{ nm}) - (0.286 \text{ nm})/2 = 0.06 \text{ nm}$$

This increases the center-to-center distance of iron by $2(0.06)/0.286 \approx 40\%$

    (b)
$$a = 4(0.127 \text{ nm})/\sqrt{2} = 0.359 \text{ nm}$$
$$\Delta = (0.127 \text{ nm} + 0.077 \text{ nm}) - (0.359 \text{ nm})/2 = 0.025 \text{ nm}$$

This increases the center-to-center distance of iron by $2(0.025)/(0.359) \approx 14\%$

Comments Although bcc iron has a lower packing factor than does fcc, the interstices do not accommodate a carbon atom without crowding. [The low packing factor (0.68) for bcc iron arises from six 6-f sites (related to $\frac{1}{2},\frac{1}{2},0$ and $0,0,\frac{1}{2}$ locations), plus 12 4-f sites (related to $\frac{1}{4},\frac{1}{4},0$). The fcc iron (PF = 0.74) has only four 6-f sites and eight 4-f sites per unit cell (see Fig. 12–3.3).

Example 7–3.2

Determine the amount of pearlite in a 99.5 percent Fe, 0.5 percent C alloy that is cooled slowly from 870° C (1600° F). Basis: 100 g of alloy.

Procedure Since pearlite originates from austenite of eutectoid composition, determine the amount of γ just prior to the eutectoid reaction — that is, at 727° C(+) in Fig. 7–3.1.

From 870 to 780° C: 100 g austenite with 0.5 percent C

From 780 to 727° C(+):   ferrite separates from austenite and the carbon content of the austenite increases to ~0.8 percent C

At 727° C(+):   composition of ferrite = 0.02% C $\left.\right\}$ proeutectoid ferrite
amount of ferrite = 38 g

composition of austenite $\approx$ 0.8% C $\left.\right\}$ γ that transforms to pearlite
amount of austenite = 62 g

Answer

At 727° C(−):   amount of pearlite = 62 g. (It came from, and replaced, the austenite with a eutectoid composition (Fig. 7–3.4b).)

Comments Each of these steps assumes sufficient time for equilibrium to be attained. Ferrite that formed above 727° C (i.e., before the eutectoid reaction) is called *proeutectoid ferrite.* The ferrite that is part of the pearlite, having been formed from austenite with the eutectoid composition, is called *eutectoid ferrite* (see Fig. 7–3.4a and b).

Example 7–3.3

From the results of Example 7–3.2, determine the amount of ferrite and carbide present in that 99.5 Fe–0.5 C alloy (a) at 727° C(−), and (b) at room temperature. Basis: 100 g of alloy. (Some data come from Example 7–3.2).

Procedure There are two choices. (1) All the carbide is in the 62 g of pearlite. Therefore, determine the amount of carbide that it contains. The balance is eutectoid ferrite (plus 38 g of proeutectoid ferrite). (2) By interpolation, determine the g of carbide (or ferrite) in the 100 g of steel.

Calculation

(a) At 727° C(−), and with the eutectoid at ~0.8 percent C:

Amount of carbide: $62\dfrac{0.8-0.02}{6.7-0.02}=7.2\dfrac{\text{g carbide}}{100\text{ g steel}}$

Amount of ferrite: $62-7.2=54.8$ g formed with the pearlite (eutectoid)
$\underline{38\quad}$ g formed before the pearlite
(proeutectoid),
92.8 g total/100 g steel

Alternative Calculation

Amount of carbide: $\dfrac{0.5-0.02}{6.7-0.02}=7.2\dfrac{\text{g carbide}}{100\text{ g steel}}$

Amount of ferrite: $\dfrac{6.7-0.5}{6.7-0.02}=92.8\dfrac{\text{g ferrite}}{100\text{ g steel}}$

(b) At room temperature (the solubility of carbon in ferrite at room temperature may be considered zero for these calculations),

$\dfrac{0.5-0}{6.7-0}=7.5\dfrac{\text{g carbide}}{100\text{ g steel}}$

$\dfrac{6.7-0.5}{6.7-0}=92.5\dfrac{\text{g ferrite}}{100\text{ g steel}}$

Comments The second procedure for part (a) gives the totals more directly but does not reveal the split between the two generations of ferrite, which give different properties to the steel.

As shown, additional carbide is precipitated from the ferrite below the eutectoid point because the solubility of carbon in ferrite decreases to nearly zero. This additional carbide is not part of the pearlite.

Each of these calculations assumes that equilibrium prevails.

Example 7–3.4

A 1045 steel (Fe plus 0.45 percent C) is austenitized at 850° C, then quenched quickly to 400° C, where it was held for isothermal decomposition of the austenite. Determine what phase(s) are present (a) immediately after the quench, (b) after 1 sec, (c) after 10 sec, and (d) after 100 sec.

Procedure The I-T curves of Fig. 7–3.6 are for a 1045 steel.

Answers

In both (a) and (b):   unstable γ.

(c) γ, α, and $\overline{C}$:   the $\gamma \xrightarrow{400°C} \alpha + \overline{C}$ reaction is $\sim \frac{2}{3}$ complete.

(d) α and $\overline{C}$:   the $\gamma \xrightarrow{400°C} \alpha + \overline{C}$ reaction was completed in 25 sec.

7–4

MODIFICATION OF MICROSTRUCTURES

Microstructures can be changed. Engineers have opportunities to tailor their materials to meet specific design needs, but they also must understand how service conditions may alter the internal structure of a material (Chapter 14).

We already examined one example of a change in microstructure; we saw that there was *grain growth* in Fig. 7–1.9. In that case, the driving force was the minimization of grain-boundary energy and the accompanying decrease in grain-boundary area.

Coalescence

Two-phase microstructures can be altered by particle *coalescence,* sometimes called *Ostwald ripening* (Fig. 7–4.1). This process is directly comparable to grain growth in that there is a coarsening of the structure. The driving force is the minimization of *phase-boundary* area.* The mechanism is somewhat more complex than that described for Fig. 7–1.7 and 7–1.8 (grain growth). Atoms from small particles must dissolve into the matrix, then diffuse through that phase to larger particles, where they precipitate. The net effect is fewer, but coarser, particles. The temperature must be high enough to facilitate diffusion, and higher temperatures shorten the coarsening time. As with grain growth, the process is slowed at lower temperatures, but is not reversed.

Spheroidization

A sphere has the minimum surface area per unit volume. Therefore, a driving force exists to spheroidize a minor phase of a two-phase microstructure. Typically, spheroidization proceeds slowly. For example, a change from the pearlite of Fig. 7–4.2(a) to the spheroidite of Fig. 7–4.2(b) takes more than 24 hours at 700° C

---

* The governing principle for coalescence is the same as the one that makes a drop of dew spherical. In each case, the atoms at the phase boundary possess extra energy. A sphere has the least surface per unit volume; therefore, it has less total energy.

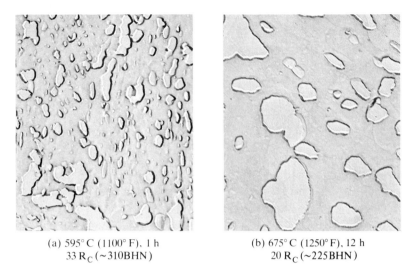

(a) 595° C (1100° F), 1 h
33 R_C (~310BHN)

(b) 675° C (1250° F), 12 h
20 R_C (~225BHN)

FIG. 7-4.1

Grain Growth in a Two-Phase Microstructure (Carbide Particles in a Tempered Steel, ×11,000). Each of these samples came from the same 1080 steel (0.80 percent carbon). The longer time and higher temperature permitted the development of larger (and fewer) carbides in (b). (This steel was previously quenched from austenite.) (*Electron Microstructure of Steel,* American Society for Testing and Materials and General Motors Research Laboratories.)

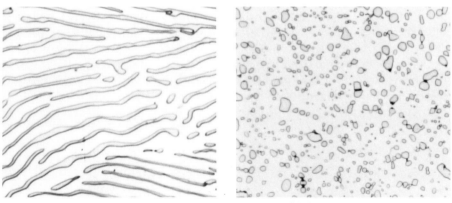

(a) Pearlite, ×2500

(b) Spheroidite, ×1000

FIG. 7-4.2

Phase Distribution in Steels (0.8 Percent C, Balance Fe). (a) Pearlite. (b) Spheroidite. The two samples were taken from the same piece of steel, but were given different heat treatments. Pearlite is stronger than spheroidite; spheroidite is tougher than pearlite. (Courtesy of U.S. Steel Corp.)

(1290° F). Although the two steels of Fig. 7–4.2 have the same phases, the same compositions, and in fact came from the same piece of steel, their properties are markedly different. The pearlite on the left is 50 percent stronger than is the adjacent *spheroidite;* the spheroidite is about twice as tough, meaning that it requires twice as much energy to cause fracturing in it. The implications for design considerations should be obvious.

Martensite

As indicated by Eq. (7–3.1), austenite decomposes during cooling to ferrite plus carbide ($\alpha + \overline{C}$). This assumes that there is time for the carbon to diffuse and to become concentrated in the carbide phase and depleted from the ferrite. If we quench austenite very rapidly, Eq. (7–3.1) can be detoured. We can indicate this as follows:

$$\gamma \text{ (fcc)} \xrightarrow[\text{Cooling}]{\text{Slow}} \alpha \text{ (bcc)} + \text{carbide}$$

$$\downarrow \text{Quench} \quad \nearrow \text{Tempering}$$

$$M \text{ (bct)}$$

(7–4.1)

This alternate route to form (α + carbide) involves a *transition phase of martensite,* M (Fig. 7–4.3). This polymorphic phase of iron is not stable because, given an opportunity, martensite will proceed to form ($\alpha + \overline{C}$). As a result, we do not see martensite on the Fe–Fe$_3$C diagram (Figs. 5–5.1 and 7–3.1). However, martensite is an important phase, as we shall soon see.

Martensite forms above room temperature, but below the eutectoid temperature where the fcc structure of austenite becomes unstable. Austenite changes spontaneously to a body-centered structure in a special way that does not involve diffusion, but rather results from a shearing action. All the atoms shift in concert, and no individual atom departs more than a fractional nanometer from its previous neighbors. Being diffusionless, the change is rapid. All the carbon that was present remains in solid solution. With more than 0.15 w/o carbon, the resulting body-centered structure is tetragonal (bct) rather than cubic.

FIG. 7–4.3

Martensite (×1000). This metastable phase is formed by quenching of austenite. The individual grains are platelike crystals with the same composition as that of the grains in the original austenite. (Courtesy of U.S. Steel Corp.)

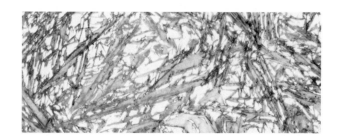

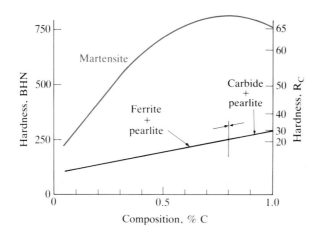

FIG. 7–4.4

Hardness of Annealed Iron–Carbon Alloys (α + Carbide) and Martensite Versus Carbon Content. This difference in hardness is the reason why steel is quenched in processing.

Since martensite of steels commonly has a noncubic structure, and since carbon is trapped in the lattice, slip does not occur readily. Therefore, this martensite is hard, strong, and brittle. Figure 7–4.4 shows a comparison of the hardness of martensite with that of pearlite-containing steels as a function of carbon content. This enhanced hardness is of major engineering importance, since it produces a steel that is extremely resistant to abrasion and deformation. However, martensite is too brittle to use in almost every application. Therefore, we temper it by heating so that ($\alpha + \overline{C}$) forms.

Tempered Martensite

The existence of martensite as a *metastable* phase that contains carbon in solid solution in a bct structure does not alter the iron–carbide phase diagram (Figs. 5–5.1 and 7–3.1). With sufficient time, at temperatures below the eutectoid temperature, the supersaturated solution of carbon in iron continues its progress to the more stable ferrite and carbide (Eq. 7–4.1). This heating process is known commercially as *tempering:*

$$\underset{\text{(martensite)}}{\text{M}} \longrightarrow \underset{\text{(tempered martensite)}}{\alpha + \text{carbide}} \qquad (7\text{–}4.2)$$

The resulting ($\alpha + \overline{C}$) microstructure is not lamellar, as is that of the pearlite that we previously observed, but it contains many dispersed carbide particles (Fig. 7–4.5) because there are numerous nucleation sites within the martensitic steel. This *tempered martensite** is much tougher than the metastable martensite, making it a more suitable product for most applications, although it may be slightly softer.

---

* Note that tempered martensite does not have the crystal structure of martensite. Rather, it is a two-phase microstructure containing *ferrite* and *carbide.* This microstructure originates by the decomposition of martensite.

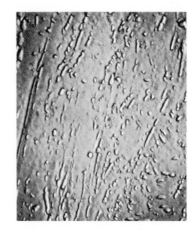

FIG. 7–4.5

Tempered Martensite (Eutectoid Steel, ×15,000). The steel was previously quenched to form martensite, which is a body-centered tetragonal (bct) phase and was >60 R_C. It was then tempered for 1 hour at 425° C (800° F). The tempered martensite is a two-phase microstructure containing carbide particles (light) in a matrix of ferrite (dark). Initially, the martensite was very hard and brittle; however, with the heating, the *tempered martensite* is now only 44 R_C, and is much tougher.
(A. M. Turkalo, General Electric Co.)

The microstructure of tempered martensite becomes coarser with more time or higher tempering temperatures. This trend can be observed in Fig. 7–4.1 where we see the growth (*coalescence*) of the carbides into larger (and fewer) particles.

Example 7–4.1

Compare the interphase-boundary areas (mm²/mm³) in the two tempered steels of Fig. 7–4.1 (×11,000).

Solution Use Eq. (7–1.2) and the perimeter of the photomicrograph. In part (b),

$$P_L = (\sim 20)/(210 \text{ mm}/11,000) \approx 10^3/\text{mm}$$
$$S_V \approx 2000 \text{ mm}^2/\text{mm}^3$$

In part (a),

$$S_V = 2(\sim 50)/(210 \text{ mm}/11,000)$$
$$\approx 5000 \text{ mm}^2/\text{mm}^3$$

Comments The tempered martensite with the greater interphase boundary area is harder. For us to be able to make direct comparisons (such as regarding hardness), other microstructural features (such as grain shape and phase quantities), must be comparable.

Example 7–4.2

The bct unit-cell dimensions are $a = 0.2845$ nm and $c = 0.2945$ nm for martensite that contains 0.8 w/o carbon (3.6 a/o C). The lattice constant of austenite of the same composition and temperature is 0.3605 nm. How much volume change occurs when $\gamma \rightarrow M$?

Calculation   Basis: 4 Fe atoms, or 2 u.c. M, and 1 u.c. γ.

$$\frac{\Delta V}{V_\gamma} = \frac{2(0.2845 \text{ nm})^2(0.2945 \text{ nm}) - (0.3605 \text{ nm})^3}{(0.3605 \text{ nm})^3}$$

$$= +0.018 \hspace{4cm} \text{(or 1.8 v/o)}$$

Comments   This volume change leads to residual stresses within quenched steel. For example, the surface of a steel gear transforms to martensite first (while the center is still a hot, deformable austenite). Shortly thereafter, the center transforms to martensite with an expansion in volume. This expansion stresses the surface, placing the martensite in tension (and compresses the transforming austenite).

Compare and contrast this process with the tempering of glass (Figs. 9–8.3 and 9–8.4).

7–5
MICROSTRUCTURES WITHIN POLYMERS

Common plastics such as polystyrene and polyvinyl chloride are amorphous. In addition, the original method for making polyethylene produced a noncrystalline product. These amorphous materials therefore lack any grains or grain boundaries. Unless they have been modified by a filler or have been stretched to align the molecules (Section 10–2), they have no structural characteristics on a scale larger than that of molecules. As such, they lack a conventional microstructure.

Crystallinity in Polymers

Although the large-tonnage polymeric materials cited in the previous paragraph are amorphous, other polymeric materials may possess significant crystallinity. Even polyethylene may be sufficiently crystalline that the unit cell can be defined by X-ray diffraction (Fig. 3–3.4). With crystallinity, a material has a microstructure that can affect those properties of interest to an engineer.

Originally, scientists viewed a partially crystalline polymer as containing crystalline regions, separated by amorphous borders. (The structure was called "fringed micelles.") Long molecules could be part of both—paralleling other molecules within the crystalline "micelle," much like a string of spaghetti, but extending into the adjacent unordered regions. The latter part of this concept is still considered valid: An individual macromolecule can reside in both crystalline and amorphous regions of a solid. We know now, however, that much of the crystallinity arises, not through the bundling of a large number of parallel molecules, but by the folding of molecular chains on themselves (Fig. 7–5.1). Depending on the processing conditions, folds occur approximately every 10 nm, with 50 to 100 carbon atoms in each reversed segment of the chain. A number of molecules can join the same growing crystal, so the lateral dimension may approach 1 μm or more. Amorphous regions remain, since the folding process is seldom perfect.

The consequences of partial crystallinity are illustrated by the data in Table 7–5.1. Based on Fig. 3–3.4, the calculated density of polyethylene is 1.01 g/cm³ (Example 7–5.1). Polyethylene that is produced by the original process possesses a density of only 0.90 g/cm³; it is called low-density polyethylene (LDPE), because with newer processes it is now possible to produce a polyethylene with a density of 0.96 g/cm³, called high-density polyethylene (HDPE). The reason for the increase in density is the amount of crystallinity that is present — near zero and approximately 50 percent, respectively. Note, however, the effect on properties. There is a 50 percent difference in thermal expansion and conductivity, and well over a 100 percent difference in strength and elastic modulus. Probably equally important is the change in heat resistance; the high-density polyethylene can withstand boiling water for sterilization.

Crystallization is impeded in LDPE, which uses the original, high-pressure processing method that permits some of the mers to enter the chain "incorrectly" (Fig. 7–5.2); whereas the catalyzed reaction that is used for making HDPE leads to the more idealized polymer chain without irregularly spaced —CH₃ side units. In the latter, the molecules of Figs. 2–3.2 and 3–3.4 can "mesh" together more favorably for the long-range order of a crystal, than can the polyethylene with the configuration shown in Fig. 7–5.2. As a result, there is greater density and less free space in the product.

We can generalize the ability for linear polymers to crystallize. A chain with spatial irregularities crystallizes less readily. Thus, atactic polymers (Fig. 4–3.7b) crystallize less completely and possess larger amorphous contents than do isotactic polymers (Fig. 4–3.7a). Chains with large side radicals do not crystallize easily. Polystyrene and polyvinyl chloride are cases in point. The benzene ring of the former, and the large —Cl of the latter, interfere with chain folding in the crystallization process of Fig. 7–5.1; thereby precluding crystallization.

FIG. 7–5.1

Crystallization by Chain Folding. A linear molecule folds back onto itself to produce a crystal. The structure is not perfect, because large molecules may start to fold at separate locations and therefore may become part of two growing crystals. There is also some variation in the lengths of the U-turns, which provides varying amounts of amorphous regions between the crystals.

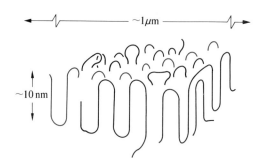

TABLE 7–5.1 Characteristics of Polyethylenes*

PROPERTY	LOW-DENSITY POLYETHYLENE (LDPE)	HIGH-DENSITY POLYETHYLENE (HDPE)
Density, Mg/m³ (=g/cm³)	0.92	0.96
Crystallinity, v/o	Near zero	~50
Thermal expansion, ° C⁻¹	180×10^{-6}	120×10^{-6}
Thermal conductivity (watt/m²)(° C/m)	0.34	0.52
Tensile strength, MPa	5–15	20–40
Young's modulus, MPa	100–250	400–1200
Heat resistance for continuous use, ° C	55–80	80–120
10-min temp. exposure, ° C	80–85	120–125

* At 20° C, except for thermal exposure.

Crystallization is facilitated by the presence of polar groups. For example, crystals form more readily in nylon than in polyethylene. We might not expect this tendency, since much of the nylon chain (Fig. 7–5.3) contains the same (CH_2) segments as does polyethylene. However, an examination of Fig. 3–3.4 reveals that there are negligible opportunities for bonding between adjacent chains of PE. In fact, the hydrogen atoms, which are only exposed protons on the ends of covalent bonds, have a mutual repulsion with the hydrogen atoms of the adjacent chains. In a nylon (Fig. 7–5.3), the single hydrogen with the nitrogen, $\diagdown$N—H, forms a bond with the side oxygen, O=C$\diagup$, in the adjacent molecule:

$$\diagdown\!\!\!N-H \cdots O=C\!\!\!\diagup \qquad\qquad (7\text{–}5.1)$$

The periodic presence of this *hydrogen bridge* helps to align adjacent molecules into the ordered crystalline pattern.

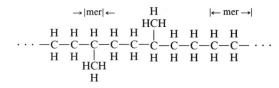

FIG. 7–5.2

Polyethylene Configuration (LDPE). When some of the C_2H_4 mers are not in the linear pattern, the —CH_3 branches interfere with the matching of adjacent chains into the crystalline structure of Fig. 3–3.4.

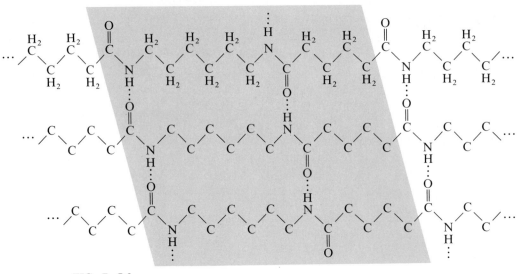

FIG. 7–5.3

Molecular Crystal (Nylon 6/6). The polar C=O groups bond to the next chain through hydrogen bridges. This bonding favors a matching of adjacent molecules and leads to more complete crystallinity than in polyethylene. Within crystals, the molecules are not kinked and coiled, as they are in Fig. 4–3.5 for amorphous polymers. (The nonbridging hydrogens are shown on only the upper molecular chain.)

Polyblends

Plastics with two or more phases are commonly called *polyblends*. They can be formed by mechanical mixing, by phase separation, or from block and graft copolymers. Their development has led to a number of products with enhanced properties.

The simplest blends are made by mechanical mixing. Two-phase mixtures also can be produced by phase separation. For example, a change in temperature, or the removal of a solvent, can introduce immiscibility (mutual insolubility) that physically segregates the two polymeric species from a previously single-phase solution. Such mixtures also can be established through emulsification procedures. The microstructure of Fig. 7–5.4 contains a matrix that is a random copolymer of vinylidene chloride and vinyl chloride (PVDC and PVC in Table 2–3.1). The dispersed phase is a copolymer of styrene and methyl methacrylate (PS and PMMA in Table 2–3.1). Both of the phases are amorphous. Thus, this two-phase system has a miscibility gap such as exists in Fig. 5–6.8 for phenol and water. The polyblend of Fig. 7–5.4 is used in the food-packaging industry, where high transparency and low oxygen transmission is required. Both the amount and the size of the dispersed phase are important for optimum properties.

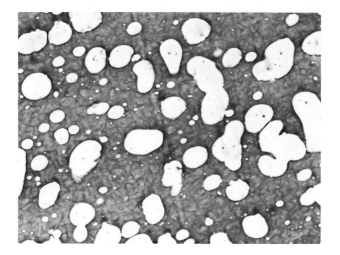

FIG. 7-5.4

Polyblend (Plastic Film for Food Packaging). The dispersed phase is a PS–PMMA copolymer. The matrix is a PVDC–PVC copolymer (Table 2–3.1). Both phases are amorphous and mutually insoluble (immiscible). (Courtesy of Y. C. Sun, The Dow Chemical Company.)

Block copolymers provide an alternative for these two-phase procedures that permits blending of immiscible polymers. Polystyrene (PS) and polybutadiene (PBD) do not form a single phase as do PS and PPO (Fig. 4–6.1a). Rather, they segregate into two distinct phases (Fig. 5–6.11), analogous to oil and water. Furthermore, styrene and butadiene (Tables 2–3.1 and 4–3.1) can be copolymerized into the same molecule. This copolymerization may occur on a random basis (Fig. 4–6.1b), or, with appropriate procedures, it can be grown as a sequence of blocks, as sketched in Fig. 7–5.5(a). The pioneer artificial rubber, Buna-S, is such

··· —S—S—S—S—S—S—BD—BD—BD—BD—BD—BD—S—S—S—S—S—BD—BD— ···

(a)

··· —S—S—S—S—S—S—S—S—BD—BD—BD—BD—S—S—S—S—S—BD—BD— ···
··· —BD—S—S—S—S—S—BD—BD—BD—BD—BD—S—S—S—S—S—S—S—BD— ···
··· —BD—S—S—S—S—S—BD—BD—BD—BD—BD—BD—S—S—S—BD—S—BD—BD— ···
··· —S—S—S—S—S—BD—BD—BD—BD—S—BD—BD—S—S—S—BD—BD—BD—BD— ···
··· —BD—S—S—S—S—BD—BD—BD—S—S—BD—BD—BD—S—S—S—S—BD—BD— ···

(b)

FIG. 7-5.5

Block Copolymers (Schematic). (a) Single chain with blocks of styrene mers (S) and butadiene mers (BD). (b) Domains. Partial phase separation establishes domains that are dominated by one or the other of the components. The polystyrene domains are rigid, since their T_g is above room temperature. The polybutadiene domains are rubbery and introduce toughness to the plastic. Adjacent domains are tied together by the covalent bonds within the molecular chains. Bonds between chains are weaker, secondary bonds.

a copolymer; it contains blocks of styrene mers and blocks of butadiene mers. During processing, similar blocks of each species cluster to form small domains and a two-phase mixture. These domains, which are submicron in size (10 to 100 nm), provide a microstructure (Fig. 7–5.5b) that contains rigid polystyrene and flexible polybutadiene, a combination that leads to greatly increased toughness.

There is a significant difference between the microstructures formed by block copolymerization and those formed by mixing or phase separation. In block copolymers, the same molecular chain extends from one domain (phase) into the next, as shown schematically in Fig. 7–5.5. This structure gives the strong covalent connection of the intramolecular bonds between the two phases. In contrast, the molecules at the phase interface are only weak, secondary intermolecular bonds in the normal polymeric, two-phase mixture.

Graft copolymers provide a variant for the establishment of bonds across phase boundaries. The widely used ABS* plastics are an "alloy" possessing a matrix of a copolymer of stryene and acrylonitrile, as presented in Fig. 4–6.1(b), onto which butadiene mers have been grafted (Fig. 7–5.6). Here again, elastomeric domains form in the product during processing, as they did for the Buna-S rubber in Fig.

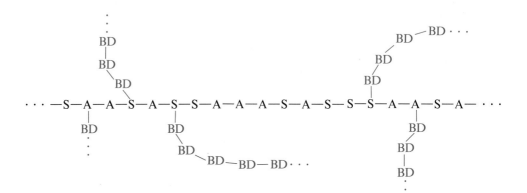

FIG. 7–5.6

Graft Copolymer (Schematic of ABS Polymers). A random copolymer of Styrene and Acrylonitrile possesses Butadiene grafts. Although attached to the main chain, the rubbery butadiene grafts can develop elastic domains as in the block copolymers. These domains provide toughness in the harder S–A plastic.

---

* Acrylonitrile–Butadiene–Styrene. (Tables 2–3.1 and 4–3.1).

7–5.5(b). Likewise, they are tied to the neighboring domains of the styrene–acrylonitrile copolymer by the strong covalent bonds along the molecular backbones.

Example 7–5.1

From Fig. 3–3.4, calculate the density of fully crystalline polyethylene.

Solution A $(C_2H_4)_n$ mer is parallel to the two ends of the rectangular cell, for an equivalent of one mer per unit cell. Likewise, a mer is parallel to the two sides. Together, they comprise a total of two mers per unit cell.

$$\rho = \frac{2(24 + 4 \text{ amu})/(0.602 \times 10^{24} \text{ amu/g})}{(0.253 \times 0.740 \times 0.493)(10^{-27} \text{ m}^3)} = 1.01 \times 10^6 \text{ g/m}^3 \qquad (\text{or } 1.01 \text{ g/cm}^3)$$

Comments Densities normally lie in the 0.92 to 0.96 g/cm³ range, depending on the degree of crystallinity (Table 7–5.1). As a result, additional space is present. The chemist calls this *free space.*

Example 7–5.2

(a) Determine how much more free space exists in the LDPE of Table 7–5.1 than there is in the HDPE. (b) How does the free space relate to the properties given in Table 7–5.1?

Procedure The "free space" is the volume in excess of the true volume of the crystal. Make use of the density data in Table 7–5.1 and in Example 7–5.1.

Calculation

(a) Basis: 1 g PE.

   HDPE volume: $(1 \text{ g})/(0.96 \text{ g/cm}^3) = 1.042 \text{ cm}^3$

   LDPE volume: $(1 \text{ g})/(0.92 \text{ g/cm}^3) = 1.087 \text{ cm}^3$

   $\Delta V/V = (1.087 - 1.042 \text{ cm}^3)/1.042 \text{ cm}^3 = 4.3 \text{ v/o}$

(b) The free space decreases the levels of stress or thermal energy required for deformation. Therefore, the LDPE has the lower strength, elastic modulus, and thermal resistance. Since ambient temperatures are above T_g, molecular rearrangements supplement thermal vibrations to expand the volume of the amorphous portions of the polyethylene; therefore, $\alpha_{LDPE} > \alpha_{HDPE}$.

Comment The thermal conductivity of HDPE is greater than that of LDPE because heat is conducted through nonmetallic solids by elastic waves. These waves travel more readily in the ordered structure of crystalline materials.

OPTIONAL

S U M M A R Y

Microstructures involve the geometry of *grains* and *phases* that are present in a material. The majority of single-phase microstructures are *polycrystalline.* Their dimensions range from nanometers and micrometers —in which case they can be visualized only with electron and optical microscopes—to millimeters.

1. Microstructural variables within single-phase materials include grain *size,* grain *shape,* and crystal *orientation.* Materials with a small grain size have large *grain-boundary areas.* The grain boundary plays an important role in diffusion, nucleation, strength, corrosion, and so on.

2. Precipitation from a solid solution is a function of the *nucleation* rate and of *growth* rate. Since isothermal nucleation is slowest just below the solubility limit, and the isothermal growth rate by diffusion is slowest at low temperatures, the most rapid precipitation rates (shortest times) are at intermediate temperatures, where both nucleation and diffusion are moderately fast. This presents a C-type *isothermal-precipitation* curve that is representative of a number of reactions. Along the upper arm, nucleation and growth occur primarily along grain boundaries to produce *intergranular* precipitation. The precipitates are dispersed *intragranularly* with reactions along the lower arm of the C-curve.

3. Eutectoid reactions, such as found in the $Fe-Fe_3C$ diagram (Fig. 7–3.1), produce two new phases simultaneously. *Pearlite* is a *lamellar* mixture of *ferrite* and *carbide* formed by the decomposition of *austenite of eutectoid composition. Hypoeutectoid* steels transform to *proeutectoid* ferrite and pearlite. C-type, *isothermal-transformation* (I-T) curves reveal that the most rapid decomposition of austenite is in the middle temperature range. *Coarse pearlite* forms if enough time and higher temperatures are available for the growth of thick lamallae. More rapid cooling produces *fine pearlite* lamellae.

4. Microstructures can be modified during processing steps (Chapter 9) and during service (Chapter 14). Dispersions of minor phases will *coalesce* into coarser structures. Also, elongated grains and lamellae of minor phases will *spheroidize.* These changes affect the strength and toughness of a material. *Martensite* is a transition phase that forms when not enough time is available to decompose to austenite to $(\alpha + \overline{C})$. It is hard, but extremely brittle. Therefore, for practical applications, it is tempered. *Tempered martensite* has a microstructure with a dispersion of fine, hard carbide particles in a tough ferrite matrix. Being both strong and tough, it has a microstructure appropriate for many uses.

5. Many polymers lack a microstructure of grains or phases because they are amorphous. Partial crystallization is realized in other polymers, commonly with useful changes in properties. Crystalline and amorphous regions may share individual macromolecules. We can control a two-phase polymer with a microstructure of glassy and elastomeric components to produce a material that is both rigid and tough.

K E Y   W O R D S

ASTM Grain Size (G.S.#)	Hypoeutectoid	Phase
Austenite (γ)	Intergranular precipitation	Phase, transition
Austenite decomposition	Intragranular precipitation	Phase boundary
Block copolymer	Isothermal precipitation	Polyblends
Carbide ($\overline{C}$)	Isothermal transformation	Polycrystalline
Coalescence	I-T curve	Precipitation
Ferrite (α)	Martensite (M)	Preferred orientation
Graft copolymer	Mean-chord length, ($\overline{L}$)	Proeutectoid
Grain	Microstructure	Spheroidite
Grain-boundary area (S_V)	Mixture	Tempered martensite
Hypereutectoid	Pearlite (P)	T-T-T curve

PRACTICE PROBLEMS

7–P11 Estimate the grain-boundary area per unit volume for the iron in Fig. 4–1.8. The magnification is ×500.

7–P12 (a) Assume that the ASTM G.S. #6 of Fig. 7–1.6 represents a two-dimensional cut through a polycrystalline solid. Estimate the corresponding grain-boundary area. (b) Repeat part (a) for G.S. #2.

7–P13 With ASTM G.S. #6, how many times as much grain boundary area is there as with ASTM G.S. #3?

7–P14 The grain size increases noticeably at higher temperatures. Describe what happens with respect to grain size at lower temperatures.

7–P21 Cite the geometric variables that exist (a) in single-phase microstructures, and (b) in multiphase microstructures.

7–P22 An alloy of 95 Al–5 Cu is solution-treated at 550° C, then cooled rapidly to 400° C where it is held for 24 hours. During that time, it produces a microstructure with 10^6 particles of θ per mm^3. The θ particles ($CuAl_2$) are nearly spherical and have approximately 60 percent greater density than the κ matrix. (a) Approximately how far apart are these particles? (b) What is the average particle dimension?

7–P23 A microstructure has nearly spherical particles of β with an average dimension $\bar{d}$ that is 10 percent of the average distance $\bar{D}$ between the centers of the adjacent particles. (a) What is the volume percent of β? (b) What is the ratio $\bar{d}/\bar{D}$ with 0.5 v/o β?

7–P24 An Al–Cu alloy contains 2.5 v/o θ ($\rho = 4.4$ Mg/m^3) in a matrix of κ, which is essentially pure Al. What is the density of the alloy?

7–P25 Why do curves for isothermal reactions involving supercooled phases commonly have a C shape, such as that in Fig. 7–2.2?

7–P31 Write the eutectoid reaction found in the $BaTiO_3$–$CaTiO_3$ system (Fig. 5–6.7).

7–P32 (a) Locate three eutectoids in the Cu–Al system that contain less than 20 percent Al (Fig. 5–6.3). (b) Write the reactions for two of these eutectoids.

7–P33 (a) What phases are present in a 99.8 Fe–0.2 C steel at 800° C? (b) Give the compositions of these phases. (c) What is the fraction of each?

7–P34 Refer to Fig. 5–4.2. Make a similar presentation of phase fractions for a 79 $BaTiO_3$–21 $CaTiO_3$ ceramic (Fig. 5–6.7).

7–P35 The maximum solubility of carbon in ferrite (α) is 0.02 w/o. Show by calculation how many unit cells there are per carbon atom.

7–P36 Iron carbide has a density of 7.6 Mg/m^3 (=7.6 g/cm^3). Its unit cell is orthorhombic (Section 3–1), and contains 12 iron atoms plus four carbon atoms. What is its volume?

7–P37 Describe the phase changes that occur on heating a 0.45 percent carbon steel from room temperature to 1200° C. (Ferrite, α; austenite, γ; carbide, $\bar{C}$.)

7–P38 Without referring to an Fe–Fe_3C phase diagram, indicate the compositions of the phases in Problem 7–P37 at selected temperatures. (Knowing these, you will find the Test Problems easier to solve.)

7–P39 Why do eutectoid reactions generally require longer times than do polymorphic reactions. (For example, compare Eqs. 5–5.4 and 3–4.2).

TEST PROBLEMS

711 For the grain-boundary area per unit to be doubled, (a) by how much must the mean chord length, $\bar{L}$, be changed? (b) By how much must the ASTM grain-size number be increased, and (c) by what factor must the number of grains that can be viewed at ×100 increase?

712 Joe Moe concluded that the ASTM grain size of a metal was #3. However, he had neglected to observe that the magnification was ×200, not ×100. (a) What was the correct grain-size number? (b) Refer to Fig. 4–1.8. What is the ASTM G.S.#?

713 Joe Moe plans to check Example 7–1.3 by heating the bronze in boiling water to double the grain size. What are his chances for success?

714 During grain growth in a single-phase material, the grain boundaries move toward their center of curvature. Explain.

721 The density of a Pb–Sn alloy is 9.0 Mg/m³ (=9.0 g/cm³) after equilibration at 20° C. (a) What is the volume fraction β? (b) What is the percent Sn?

722 Describe the factors that lead to intergranular precipitation, and those that lead to intragranular precipitation.

723 Sterling silver (92.5 Ag–7.5 Cu) is quenched after solution treatment, then reheated to 400° C until equilibrium is attained. At equilibrium, it contains a β precipitate (approximately spherical in shape) in an α matrix. (a) If the representative dimension d of the β particle is 0.12 μm, how many particles will there be per mm³? (b) How many will there be if $d = 0.04$ μm? (c) What is the average distance between particles?

724 An 86 Pb–14 Sn solder is held at 185° C until equilibrium is established. The average volume of the grains is 10^6 μm³. The alloy is then cooled *rapidly* to room temperature, where β precipitates within the initial α grains. The resulting matrix in those former α grains is 99 Pb–1 Sn ($\rho = 13.3$ Mg/m³), and the β particles ($\sim$ 100 percent Sn) are separated by an average distance of 0.1 μm. (a) Estimate the number of particles there are per original grain. (b) What is the volume fraction of β? (c) Approximately how many tin atoms are there per β particle?

731 At 860° C, the maximum solubility of carbon in austenite, γ, is 1.15 w/o. How many unit cells per carbon atom?

732 (a) Determine the phases present, the composition of each of these phases, and the relative amounts of each phase for 1.0 percent carbon steel at 810, 760, and 700°C. (Assume equilibrium.) (b) How much pearlite is present at each of these temperatures?

733 Calculate the percent ferrite, carbide, and pearlite, at room temperature, in iron–carbon alloys containing (a) 0.4 percent carbon, (b) 0.8 percent carbon, and (c) 1.2 percent carbon.

734 Refer to Fig. 5–4.2. Make a similar presentation of phase fractions for a 99.7 Fe–0.3 C steel between 20° C and the peritectic at 1500° C.

735 Repeat Problem 734 for a 98.9 Fe–1.1 C steel.

736 Assume equilibrium. For a steel of 99.4 Fe–0.6 C, determine (a) the lowest temperature of 100 percent γ; (b) the fraction that is γ at 730° C, and its composition; (c) the fraction that is pearlite at 720° C, and its total composition; (d) the fraction that is proeutectoid ferrite at 730° C; and (e) the fraction that is proeutectoid ferrite at 720° C after cooling from 730° C.

737 The eutectic microstructure of θ and κ of the Cu–Al system is lamellar, and is similar in appearance to pearlite of the Fe–Fe₃C system. The phase, θ, is an intermetallic compound, CuAl₂, with 4 Cu (8 Al) per unit cell, which is tetragonal with $a = 0.604$ nm, and $c = 0.486$ nm. The phase, κ, has a density of 2.73 g/cm³ when saturated with copper. Which will be thicker within the lamellar microstructure, the θ layers or the κ layers?

738 (a) Determine the amount of pearlite in a 99.45 Fe–0.55 C alloy that is cooled slowly from 870° C. Basis: 8.3 kg alloy. (b) From your results in part (a), determine the amount of ferrite and carbide in the pearlite at 720° C.

739 An oxide containing 75 w/o iron and 25 w/o oxygen is solution-treated at 1300° C (2370° F). It is then cooled slowly to room temperature. Compare and contrast the reactions and possible microstructures with those found after slow cooling of a 1060 steel from 900° C (1650° F).

741 On the basis of this chapter, would you choose a high- or low-carbon steel for an automobile fender? Give your reasons.

742 Martensite contains two iron atoms per unit cell. Its bct structure has $c = 0.290$ nm, and $a = 0.285$ nm when 0.5 w/o C is present. Make a sketch of a unit cell, and point out the locations within the unit cell where the small carbon atoms can reside with least strain.

743 Refer to Problem 742. Calculate the density of martensite that contains 0.5 w/o carbon.

751 A polyethylene with no evidence of crystallinity has a density of 0.90 Mg/m³. Commercial grades of low-density polyethylene (LDPE) have 0.92 Mg/m³, whereas HDPE has a density of 0.96 Mg/m³. Estimate the volume fraction of crystallinity in each material.

752 From the densities in Problem 751, calculate the amount of "free space" in LDPE and in HDPE.

...MATION AND
...URE

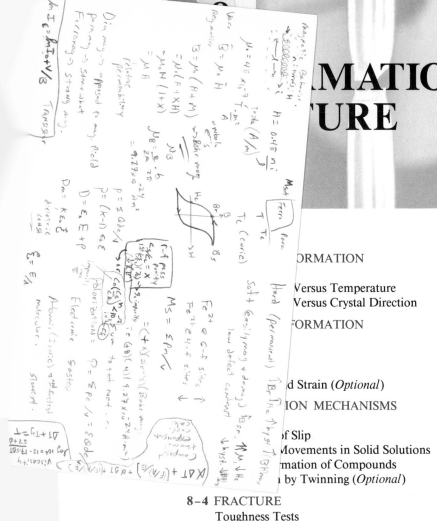

...ORMATION

...Versus Temperature
...Versus Crystal Direction

...FORMATION

...d Strain (*Optional*)

...ION MECHANISMS

...f Slip
...Movements in Solid Solutions
...rmation of Compounds
... by Twinning (*Optional*)

8-4 FRACTURE
Toughness Tests
Ductility-Transition Temperature
Fracture Toughness K_{Ic}
Design Considerations

Most materials are subjected to stresses and the accompanying deformation during processing and use. Plastic deformation is required in processing most metals. Furthermore, the engineer must provide for elastic deformation of structural parts in designs. Even electrical materials, such as a wire, undergo deformation during manufacture and use. Normally, the originally intended role of any material is terminated if there is fracture, whether it is from an overload, from a sudden impact, or from sharp thermal gradients.

In this chapter, we shall consider the nature and mechanism of deformation—both elastic and plastic (recoverable and permanent). Then, we shall be in a better position to anticipate the rigidity and deformation strength of materials for processing and design. When we design to avoid fracture, we must consider stress concentrations and toughness.

8–1
ELASTIC DEFORMATION

Elastic deformation is a reversible strain. If a stress is applied in tension, the material becomes slightly longer; removal of the load permits the material to return to its original dimension. Conversely, when it is under compression, the material becomes slightly shorter. The dimensions of the unit cells change when the material undergoes elastic strain (Fig. 8–1.1).

Elastic Moduli

When only elastic deformation exists, the strain is proportional to the applied stress (Fig. 8–1.2). The ratio of stress to strain is the *modulus of elasticity* (Young's modulus), and is a property of the material (Eq. 1–2.3). The greater the forces of attraction between atoms in a material, the higher the modulus of elasticity (Fig. 2–6.1).

Any lengthening or compression of the crystal structure in one direction, due to a uniaxial force, produces an adjustment in the dimensions at right angles to the force. In Fig. 8–1.1(a), for example, a small contraction is indicated at right angles to the tensile force. The negative ratio between the lateral strain e_y and the direct tensile strain e_z is called *Poisson's ratio, v*:

$$v = -\frac{e_y}{e_z} \qquad (8–1.1)$$

Engineering materials may be loaded in shear as well as in tension (and compression). In shear loading, the two forces are parallel but are not aligned (Fig. 8–1.3b). As a result, the *shear stress, τ*, is the shear force, F_s divided by the sheared area, A_s:

$$\tau = F_s/A_s \qquad (8–1.2)$$

FIG. 8–1.1

Elastic Normal Strain (Greatly Exaggerated). Atoms are not permanently displaced from their original neighbors. (a) Tension (+). (b) No strain. (c) Compression (−).

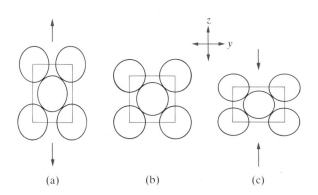

(a)   (b)   (c)

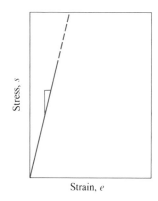

FIG. 8–1.2

Stress Versus Strain (Elastic). With only elastic strain, the two are proportional, and Eq. (1–2.3) applies. The slope of the curve is the modulus of elasticity (Young's modulus), $E = s/e$.

A shear stress produces an angular displacement, α. We define *shear strain, γ,* as the tangent of that angle; that is, as x/y in Fig. 8–1.3(b). The recoverable or elastic shear strain is proportional to the shear stress:

$$G = \tau/\gamma \qquad\qquad (8-1.3)$$

where G is the *shear modulus.* Also called the modulus of rigidity, the shear modulus is different from the modulus of elasticity, E; however, the two are related at small strains by

$$E = 2G(1 + v) \qquad\qquad (8-1.4)$$

Since Poisson's ratio v is normally between 0.25 and 0.5, the value of G is approximately 35 percent of E.

A third elastic modulus is the *bulk modulus, K.* It is the reciprocal of the compressibility β of the material and is equal to the hydrostatic pressure P_h per unit

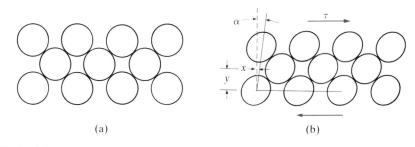

(a)   (b)

FIG. 8–1.3

Elastic Shear Strain. Shear couples produce a relative displacement of one plane of atoms past the next. This strain is elastic as long as atoms keep their original neighbors. (a) No strain. (b) Shear strain.

of volume compression, $\Delta V/V$:

$$K = \frac{P_h V}{\Delta V} = \frac{1}{\beta} \tag{8-1.5}$$

The bulk modulus is related to the modulus of elasticity as follows:

$$K = \frac{E}{3(1 - 2\nu)} \tag{8-1.6}$$

You will derive this equation in Problem 817.

Elastic Moduli Versus Temperature

Elastic moduli decrease as temperature increases, as shown in Fig. 8–1.4 for four common metals. In terms of Fig. 2–5.2(a), a thermal expansion reduces the value of dF/da, and therefore decreases the modulus of elasticity. The discontinuity in the curve for iron in Fig. 8–1.4 is due to the change from bcc to fcc at 912° C (1673° F). Not surprisingly, the more densely packed fcc polymorph requires greater stresses for a given strain; that is, the elastic modulus is greater for fcc. Also note from Fig. 8–1.4 that higher-melting-temperature metals have greater elastic moduli.

Elastic Moduli Versus Crystal Direction

Elastic moduli are *anisotropic* within materials; that is, they vary with crystallographic direction. As an example, iron has an average modulus of elasticity of about 205 GPa (30,000,000 psi); however, the actual modulus of a crystal of iron varies from 280 GPa (41,000,000 psi) in the [111] direction to only 125 GPa (18,000,000 psi) in the [100] direction (Table 8–1.1). The consequence of any such anisotropy becomes significant in polycrystalline materials. Assume, for example, that Fig. 8–1.5(a) represents the cross-section of a steel wire in which the average stress is 205 MPa (30,000 psi). If the grains are randomly oriented, the

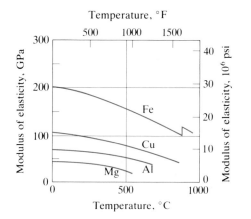

FIG. 8–1.4

Modulus of Elasticity Versus Temperature. (Adapted from A. G. Guy and J. J. Hren, *Elements of Physical Metallurgy,* Addison-Wesley.)

TABLE 8–1.1 Moduli of Elasticity (Young's Modulus)*

METAL	MAXIMUM		MINIMUM		RANDOM	
	GPa	10^6 psi	GPa	10^6 psi	GPa	10^6 psi
Aluminum	75	11	60	9	70	10
Gold	110	16	40	6	80	12
Copper	195	28	70	10	110	16
Iron (bcc)	280	41	125	18	205	30
Tungsten	345	50	345	50	345	50

* Adapted from E. Schmid and W. Boas, *Plasticity in Crystals.* English translation, London: Chapman Hall.

elastic strain is 0.001, because the average modulus of elasticity is 205 GPa (30,000,000 psi). However, in reality, the stress varies from 125 MPa (18,000 psi) to 280 MPa (41,000 psi) as shown in Fig. 8–1.5(b), because grains have different orientations, but each is strained equally (0.001). Of course, this means that some grains will exceed their yield strength before other grains reach their yield strength.

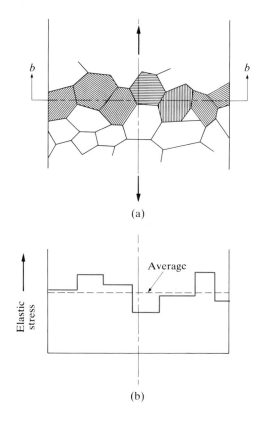

(a)

(b)

FIG. 8–1.5

Stress Heterogeneities (Schematic). Elastic stresses vary with grain orientation, because the moduli of elasticity are not isotropic.

Example 8–1.1

A plate of steel has a 100.0-cm × 100.0-cm square scribed on its surface. It is loaded in one direction (perpendicular to opposite sides of the square) with a 200-MPa (29,000-psi) stress. (a) What are the dimensions of the scribed area? (Poisson's ratio of steel = 0.29).

Without the initial stress being removed, a second tension stress of 410 MPa (60,000 psi) is applied at right angles to the first — that is, perpendicular to the other edges of the square. (b) What are the new dimensions of the scribed area?

Procedure Since no preferred orientation is indicated, we shall assume the random grain value of the modulus of elasticity (Table 8–1.1). The strains are additive when the two stresses are applied.

Calculation

(a) From Eq. (1–2.3): $e_z = 200$ MPa/205,000 MPa $= 0.000975$

From Eq. (8–1.1): $e_y = -0.29(0.000975) = -0.00028$

1000 mm $(1 + 0.000975) \times$ 1000 mm $(1 - 0.00028) = 1001.0$ mm $\times$ 999.7 mm

(b) $e_y = -0.00028 + 410$ MPa/205,000 MPa $= 0.00172$

$e_z = 0.000975 - 0.29(410/205,000) = 0.00040$

1000 mm $(1 + 0.0004) \times$ 1000 mm $(1 + 0.00172) = 1000.4$ mm $\times$ 1001.7 mm

Additional information We can write a general equation for elastic deformation in three dimensions from Eqs. (1–2.3 and 8–1.1):

$$e_x = \frac{s_x}{E} - \frac{v s_y}{E} - \frac{v s_z}{E} \qquad\qquad (8\text{–}1.7)$$

Example 8–1.2

What is the percentage volume change in iron if it is hydrostatically compressed with 1400 MPa (200,000 psi)? (Poisson's ratio = 0.29.)

Procedure We need the bulk modulus $(= P_h/(\Delta V/V))$ which is obtainable from E and v (Eq. 8–1.6). An alternative solution makes use of Eq. (8–1.7). Since $e_x = e_y = e_z (= \Delta L/L)$, and with $\Delta V/V \approx 3(\Delta L/L)$, we get the same answer.

Solution

$K = (205,000 \text{ MPa})/3(1 - 0.58))$
$\quad = 162,700$ MPa.

$\Delta V/V = -1400$ MPa/162,700 MPa $= -0.86$ v/o

Alternative solution

$e_x = (1 - 2v)(-1400 \text{ MPa})/(205,000 \text{ MPa}) = -0.00287$

$\Delta V/V \approx 3e = -0.86$ v/o

Comment The approximation $\Delta V/V = 3(\Delta L/L)$ originates from $1 + \Delta V/V = (1 + \Delta L/L)^3$, and it is valid when the Δs are small.

8–2
PLASTIC DEFORMATION

Figure 8–1.2 shows only elastic strain. This situation is typical of brittle (nonductile) materials such as cast iron, glass, and phenol–formaldehyde polymers. *Ductile* materials undergo some *plastic* (permanent) strain before fracture. For example, if a steel beam is loaded, it will first deflect elasticly. The deflection disappears when the load is removed. An overload will permanently bend the beam in the locations where the stresses exceed the yield strength of the steel. In this case, the bent beam has failed, but it has not fractured. In contrast, on a production line, the yield strength of a sheet of steel may be intentionally exceeded to bend the sheet into the shape of a car fender. At this stage, the metal has yielded, but it has not failed because production requires considerable plastic strain. It is necessary in both production and service to know (1) the critical stress requirements to initiate permanent deformation, and (2) the amount of plastic strain available before eventual fracture of a ductile material.

Strengths

The laboratory test bar of Fig. 8–2.1(a) was used to produce the accompanying *stress–strain diagram*. The *stress, s,* is the force per unit area expressed in N/m^2 (or $lb_f/in.^2$), but we more commonly expressed it in MPa (or psi). The *strain, e,* is $\Delta L/L$, and therefore is dimensionless. The increasing stress produces only elastic strain in the linear part of the test. Beyond the *proportional limit,* plastic strain accompanies and generally exceeds the elastic strain:

$$e = e_{el} + e_{pl} \qquad (8–2.1)$$

The *yield strength, S_y,* is the critical stress required to introduce yielding, or plastic strain. The point of deviation from the elastic slope is gradual in some materials, such as aluminum and zinc. It also may be abrupt, as it is in structural steels. In either case, continued loading produces *both* elastic and plastic strain.*

Inasmuch as the first increment of plastic strain that is detected is dependent on the sensitivity of the testing equipment, an unstandardized test procedure could

---

* In many steels, where initial yielding is pronounced, the engineer sometimes speaks of a *yield point,* or an elastic limit, rather than of the yield strength. Although used, the latter term is not fully appropriate, because continued loading produces additional elastic strain.

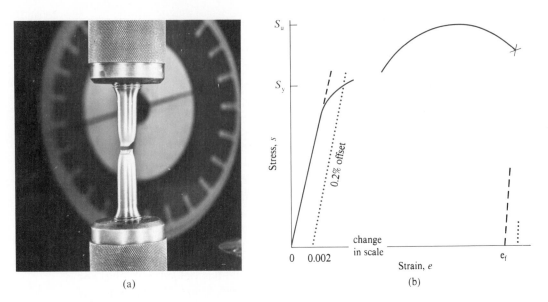

(a)   (b)

FIG. 8–2.1

Tensile Test. (a) Standardized test bar after ductile fracture. (Courtesy of U.S. Steel Corp.) (b) Stress–strain diagram (ductile material). The yield strength, S_y, is the stress that initiates plastic deformation (commonly defined by 0.2 percent permanent offset). The ultimate strength, S_u, is the maximum stress based on the original, or nominal, area.

lead to considerable data variability from test bar to test bar, and from laboratory to laboratory. Pragmatically, the engineer standardizes the testing by defining the yield strength as the stress required to produce a tolerable amount of strain— commonly 0.2 percent. This strength is obtained by plotting the 0.2 percent *offset* from the linear portion of the stress–strain curve, then reading the intercept (Fig. 8–2.1b).

Almost all ductile materials become stronger when they are deformed plasticly. We call this process *strain hardening.* It is used advantageously in the design of engineering materials (Section 9–3). This added strength appears in Fig. 8–2.1(b) as the concurrent increase in stress and strain beyond the yield strength.

As presented in the Fig. 8–2.1(b), the stress–strain curve reaches a maximum and then falls off. The maximum is called the ultimate tensile strength, or more simply the *ultimate strength, S_u.* The ultimate strength is the limiting stress used by the engineer. Beyond this point, the test bar shows a reduction in cross-sectional area, called *necking.* As a result, the load-carrying capability of the bar is reduced until the final fracture occurs. A *breaking strength, S_B,* can be calculated, but has little engineering significance because the nominal test-bar area is not pertinent at this stage.

Hardness

As we indicated in Section 1-2, *hardness* is the resistance to penetration. Various procedures are used to measure hardness. These depend on the material, its thickness, the indentor used, and the load applied. More common hardness indices are the Brinell Hardness Number (BHN), and the Rockwell hardnesses (R). The latter has several scales—R_c, R_b, R_f, and so on—which have appropriate loads and indentors for harder steels, softer brasses, and thin sheet metal. Although hardness is not a basic property of a material, we shall consider hardness data as useful indices of strength. For example, with steels, a rule of thumb is that the ultimate strength in psi is 500 times the BHN value. Since the hardness can be measured *in situ,* and does not require the machining of a test bar, its index is useful in quality control and service checking.

Ductility

We use the term *ductility* to define the permanent strain that is realized before the test bar fractures. As she does with strength, the engineer has more than one way to define this strain. *Elongation* is the linear plastic strain accompanying fracture:

$$El_{\text{gage length}} = (L_f - L_o)/L_o \qquad (8-2.2)$$

It is imperative that for the engineer to identify the gage length, because the plastic strain is almost invariably localized (Fig. 8-2.2).

A second measure of ductility is the *reduction of area,* R of A, at the point of fracture:

$$R \text{ of } A = (A_o - A_f)/A_o \qquad (8-2.3)$$

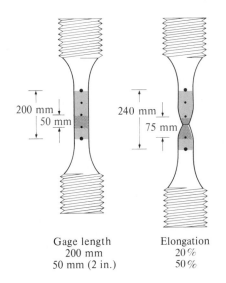

Gage length	Elongation
200 mm	20%
50 mm (2 in.)	50%

FIG. 8-2.2

Elongation Versus Gage Length. Since final deformation is localized, an elongation value is meaningless unless the gage length is indicated. For routine testing, a 50-mm (2-in.) gage length is common.

Both measures of ductility normally are expressed in percent. Values for these two measures of ductility parallel each other: both are high in ductile materials, and are nil in a brittle material. However, there is no established mathematical relationship between the two, because the final plastic deformation is highly localized.

True Stress and Strain

The stress–strain curve as commonly presented (Fig. 8–2.1b) is based on force per nominal, or *original, area*. As previously noted, however, necking occurs before a ductile material fractures. After necking starts, the *true stress, σ*, is higher than is the *nominal stress, s*. Likewise, the *true strain, ε*, differs from the *nominal strain, e*. We must modify the nominal stress–strain curve of Fig. 8–2.1(b) if we want a *true stress–strain* curve (Fig. 8–2.3).

The design engineer almost always uses the nominal *s/e* data, rather than the true *σ/ε* data. The nominal data are used partly because design calculations for engineering products are based on original dimensions. More important, it would be impractical to reduce the load as the material plasticly deforms during the last moments before complete failure.*

Although products are never designed on the basis of the true fracture strength, σ_f, (Fig. 8–2.3), knowledge of this stress is useful in the design of deformation processes. We shall observe in the next chapter that it is possible to make materials in which the ultimate strength, S_u, is raised toward that limit.

FIG. 8–2.3

True Stress and Strain. The true stress, σ, is based on the actual area rather than on the nominal (original) area. Therefore, the true fracture stress, σ_f, exceeds the nominal breaking strength, S_b. Also, because the strain is localized, the true strain at the point of fracture, ϵ_f, exceeds the nominal strain for fracture, e_f. (See Example 8–2.2.)

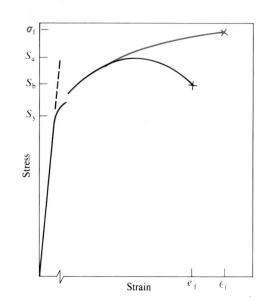

---

* In a facetious but illustrative vein, you as the design engineer will probably not volunteer to crawl out on the bridge to measure the diameter of a tie bar just before it breaks and the bridge falls into the river so that the *true* stress and strain are known. Sometimes, it does not pay to know the truth!

Example 8–2.1

A copper test bar has a 2.00-in. (51-mm) gage length. After testing, the gage marks are 2.82 in. (71.6 mm) apart. What is the ductility?

Calculation

$$\text{elongation}_{2.00 = \text{in.}} = (2.82 \text{ in.} - 2.00 \text{ in.})/2.00 \text{ in.}$$
$$= 0.41 \qquad\qquad\qquad (\text{or } 41\%)$$

Comments The reduction of area is expected to be high, but it cannot be calculated from the elongation data. A calculation in SI units will give the same ductility.

Example 8–2.2

A copper wire has a nominal breaking strength of 300 MPa (43,000 psi). Its ductility is 77 percent reduction of area. Calculate the true stress σ_f for fracture.

Solution Based on the original area, A_o,

$$\frac{F}{A_o} = 300 \text{ MPa} \qquad F = (300 \times 10^6 \text{ N/m}^2)A_o$$

$$\sigma_f = \frac{F}{A_{tr}} = \frac{F}{(1 - 0.77)A_o} = \frac{(300 \times 10^6 \text{ N/m}^2)A_o}{0.23A_o} = 1300 \text{ MPa}$$

or

$$\frac{F}{A_{tr}} = \frac{(43,000 \text{ psi})A_o}{0.23A_o} = 187,000 \text{ psi}$$

Comment We can determine the *true strain* ϵ from the cross-sectional dimensions. If we define the true strain ϵ as

$$\epsilon \equiv \int_{l_o}^{l} \frac{dl}{l} = \ln\left(\frac{l}{l_o}\right)$$

and assume constant volume, $Al = A_o l_o$, then

$$\epsilon = \ln\left(\frac{l}{l_o}\right) = \ln\left(\frac{A_o}{A}\right) \qquad\qquad (8\text{–}3.4)$$

This equation gives a definition of true strain that holds for all strains and is independent of gage length.

Example 8–2.3

A 212-cm copper wire is 0.76 mm (0.03 in.) in diameter. Plastic deformation started when the load was 8.7 kg. (a) What was the force provided in N and lb$_f$? (b) When loaded to 15.2 kg (33.9 lb$_f$), the total strain was 0.011. The wire was then unloaded. What is the length of the wire after unloading? (c) What is the yield strength of the copper?

Procedure The yield strength is the critical stress to initiate plastic strain. Therefore, we must calculate the stress with the 8.7-kg load for (c). After yielding is initiated, any additional stress produces *both* elastic and plastic deformation. Thus, both are included in the 0.011 strain; we can calculate the elastic strain from the modulus of elasticity and the 15.2-kg load.

Calculation

(a) $(8.7 \text{ kg})(9.8 \text{ m/s}^2) = 85.25 \text{ N}$ $(8.7 \text{ kg})(2.2 \text{ lb}_f/\text{kg}) = 19.1 \text{ lb}_f$

$$\text{area} = \pi(0.76 \times 10^{-3} \text{ m})^2/4 = 0.45 \times 10^{-6} \text{ m}^2$$

(b) Stress at 15.2 kg: $(15.2 \text{ kg})(9.8 \text{ m/s}^2)/(0.45 \times 10^{-6} \text{ m}^2) = 331 \text{ MPa}$

Elastic strain at 15.2 kg: $(331 \text{ MPa})/(110{,}000 \text{ MPa}) = 0.003$

Plastic strain at 15.2 kg: $0.011 - 0.003 = 0.008$ permanent

Wire length after unloading: $1.008(212 \text{ cm}) = 213.7 \text{ cm}$

(c) Yield strength: $S_y = (85.2 \text{ N})/(0.45 \times 10^{-6} \text{ m}^2) = 190 \text{ MPa}$

Comment Elastic strain continues to increase after plastic deformation starts, because the bonds of all atoms experience increased forces, even if they do not establish new neighbors by plastic deformation.

8-3
DEFORMATION MECHANISMS

Cubic metals and their nonordered alloys deform predominantly by *plastic shear,* or *slip,* in which one plane of atoms slides over the next adjacent plane. Plastic shear also is one of the methods of deformation in hexagonal metals, and in a few ceramic materials. Shear deformation even occurs when compression or tension forces are applied, because the stresses may be resolved into shear stresses.

Slip Systems

Slip occurs more readily along certain crystal directions and planes than along others. This is illustrated in Fig. 8–3.1, where a single crystal of an hcp metal was deformed plastically. The shear stress required to produce slip on a crystal plane is called the *critical shear stress, τ_c.*

The predominant sets of *slip systems* in several familiar metals are summarized in Table 8–3.1. A slip system includes the *slip plane (hkl),** and a *slip direction [uvw].* There are a number of slip systems, because of the multiple planes in a family of planes and the multiple directions in a family of directions (Sections 3–6 and 3–7). Two facts stand out in Table 8–3.1.

---
* See Section 3–7 for (*hkil*) indices of hexagonal crystals.

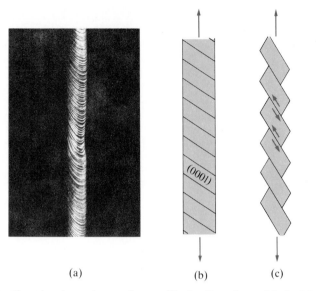

FIG. 8-3.1

Slip in a Single Crystal (hcp). Slip paralleled the (0001) plane, which contains the shortest slip vector. (See Section 3-7 for (*hkil*) indices of hexagonal crystals.) (Constance Elam, *Distortion of Metal Crystals*, Oxford: Clarendon Press.)

(a)   (b)   (c)

1. The slip direction in each metal crystal is the direction with the highest linear density of lattice points, or the shortest distance between lattice points

2. The slip planes are planes that have wide interplanar spacings and, therefore, high planar densities (Section 3-8)

Mechanism of Slip

Figure 8-3.2 shows a simplified mechanism of slip. If we were to calculate the strength of metals on this basis, the result would indicate that the strength of metals should be approximately $G/6$, where G is the shear modulus. Since metals have only a fraction of that strength, a different slip mechanism must be operative. All experimental evidence supports a mechanism involving dislocation movements.

TABLE 8-3.1 Predominant Slip Systems in Metals

STRUCTURE	EXAMPLES	SLIP DIRECTIONS	SLIP PLANES	NUMBER OF INDEPENDENT SLIP SYSTEMS
bcc	α-Fe, Mo, Na, W	$\langle\bar{1}11\rangle$	$\{101\}$	12
bcc	α-Fe, Mo, Na, W	$\langle\bar{1}11\rangle$	$\{211\}$	12
fcc	Ag, Al, Cu, γ-Fe, Ni, Pb	$\langle\bar{1}10\rangle$	$\{111\}$	12**
hcp	Cd, Mg, α-Ti, Zn	$\langle11\bar{2}0\rangle$	$\{0001\}$*	3
hcp	α-Ti	$\langle11\bar{2}0\rangle$	$\{10\bar{1}0\}$*	3

* See Section 3-7 for (*hkil*) indices of hexagonal crystals. ** See Example 8-3.1 and Fig. 8-3.10.

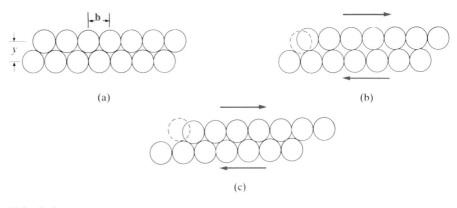

(a)   (b)

(c)

FIG. 8–3.2

An Assumed Mechanism of Slip (Simplified). Metals actually deform with less shear stress than this mechanism would require.

If we use Fig. 8–3.3 as a model of a dislocation and place a shear stress along the horizontal direction, the dislocation can be moved (Fig. 8–3.4) with a shearing displacement within the crystal. (See also Fig. 4–1.5.) The shear stress required for this type of deformation is a small fraction of the previously cited value of $G/6$, and it matches observed shear strengths.

The mechanism of slip requires the growth and movement of a dislocation line; therefore, energy is required. The energy, E, of a dislocation line is proportional to the length of the dislocation line, l, the product of the shear modulus, G, and the square of the unit slip vector $\mathbf{b}$ (Section 4–1):

$$E \propto lG\mathbf{b}^2 \tag{8–3.1}$$

Thus, the easiest dislocations to generate and expand for plastic deformation are those with the shortest unit slip vector, $\mathbf{b}$ (Fig. 4–1.2), particularly since the $\mathbf{b}$ term

FIG. 8–3.3

Edge Dislocation. "Bubble-raft" model of an imperfection in a crystal structure. Note the extra row of atoms. (Bragg and Nye, *Proc. Roy. Soc. (London)*.)

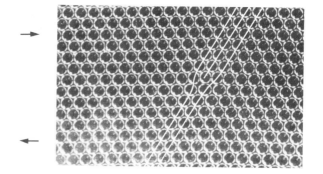

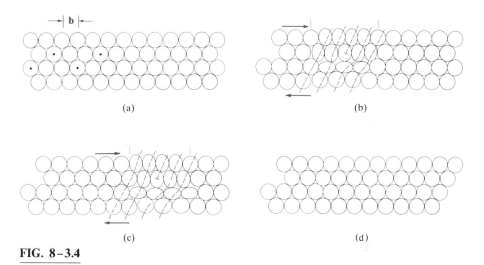

(a)

(b)

(c)

(d)

FIG. 8-3.4

Slip by Dislocations. In this model, only a few atoms at a time are moved from their low-energy positions. Less stress is therefore required to produce slip than would be needed if all the atoms moved at once as proposed in Fig. 8-3.2.

is squared. The directions in a metal with the shortest slip vector will be the directions with the greatest linear density of atoms. The lowest value for the shear modulus, G, accompanies the planes that are farthest apart and hence have the greatest planar density of atoms. Thus, we can develop the rule of thumb that predicts that the lowest critical shear stresses will occur *on the most densely packed planes and in the most densely packed directions.*

Dislocation Movements in Solid Solutions

The energy associated with an edge dislocation (Fig. 4–1.3) is the same whether the dislocation is located at (b) or at (c) in Fig. 8–3.4. Therefore, no net energy is required for the movement between the two.* Such is not the case when solute atoms are present. As shown in Fig. 8–3.5, when an impurity atom is present, the energy associated with a dislocation is less than it is in a pure metal. Thus, when a dislocation encounters foreign atoms, its movement is restrained because energy must be supplied to release it for further slip. As a result, solid-solution alloys always have higher strengths than do pure metals (Fig. 8–3.6). We call this process *solution hardening* (Section 9–2).

---

* This statement does not apply if (1) the movement includes an increase in the length of the dislocation, or (2) there is a pile-up of dislocations.

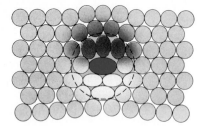 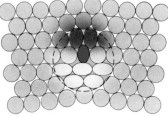 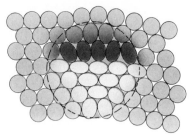

(a) Larger impurity atom   (b) Smaller impurity atom   (c) Same size atom

FIG. 8–3.5

Solid Solution and Dislocations. An odd-sized atom decreases the stress around a dislocation. As a result, energy must be supplied and additional stress applied to detach the dislocation from the solute atom. This process accounts for solution hardening.

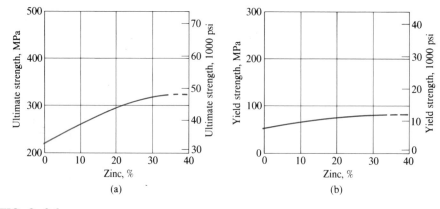

(a)   (b)

FIG. 8–3.6

Solution Hardening (Annealed α-Brass). The addition of zinc to copper increases both the yield and the ultimate strengths. (Thirty-five percent zinc is the practical limit for commercial brasses. See Fig. 5–6.1.)

Plastic Deformation of Compounds

An atom in a single-component metal always has like atoms for neighbors, even during deformation. A compound, however, has two or more types of atoms with a preference for unlike neighbors (see, for example, Fig. 3–1.1). On many potential slip planes, deformation brings together like atoms and separates a fraction of the unlike atoms (Fig. 8–3.7). Within a compound, this means higher energy, which shows up as a resistance to shear. Consequently, the critical shear stresses on some planes are sufficiently high that slip is essentially impossible. In effect, the number

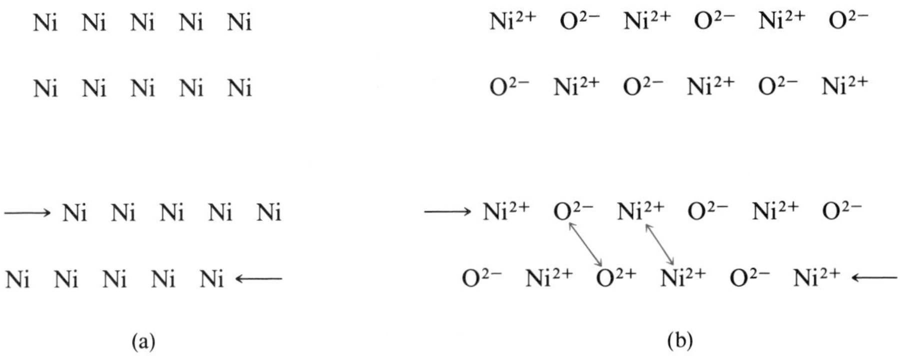

Ni Ni Ni Ni Ni Ni^{2+} O^{2-} Ni^{2+} O^{2-} Ni^{2+} O^{2-}

Ni Ni Ni Ni Ni O^{2-} Ni^{2+} O^{2-} Ni^{2+} O^{2-} Ni^{2+}

$\longrightarrow$ Ni Ni Ni Ni Ni $\longrightarrow$ Ni^{2+} O^{2-} Ni^{2+} O^{2-} Ni^{2+} O^{2-}

Ni Ni Ni Ni Ni $\longleftarrow$ O^{2-} Ni^{2+} O^{2+} Ni^{2+} O^{2-} Ni^{2+} $\longleftarrow$

(a)   (b)

FIG. 8-3.7

Comparison of Slip Processes (Metallic Nickel and Nickel Oxide). More force is required to displace the ions in NiO than is needed for the atoms in nickel. The strong repulsive forces between like ions become siginificant. Nickel also has more slip systems than does nickel oxide.

of possible slip systems is reduced, and ductility decreases. This result is revealed in Table 8-3.2, which cites the slip systems for several relatively simple compounds. For example, only six sets are operative in MgO. In more complex compounds— such as in $Ni_8Fe_{16}O_{24}$, used in magnet ceramics, and in $PbZrO_3$, used in piezoelectric transducers—the possibility of slip is negligible at normal temperatures; hence, the materials behave in a brittle manner.

As revealed in Fig. 3-1.1, the shortest repeating distance (i.e., the slip vector) in an NaCl-type crystal is along the several $\langle 110 \rangle$ directions. Quite expectedly, therefore, the $\langle 110 \rangle$ directions have been found by experiment to be the slip directions. The most common slip plane for NaCl-type crystals is one of those in the $\{110\}$ form. This is particularly true in those compounds, such as LiF and MgO, that have small, "nondeformable" ions. Other NaCl-type compounds with large ions, such as PbS and MnSe, possess other slip planes, but still have the $\langle 110 \rangle$ slip directions (Table 8-3.2).

TABLE 8-3.2 Slip Systems in Simple Compounds

STRUCTURE	EXAMPLES	SLIP DIRECTIONS	SLIP PLANES	NUMBER OF COMBINATIONS
NaCl	LiF, MgO, MnS, TiC	$\langle 110 \rangle$	$\{1\bar{1}0\}$	6
NaCl	PbS	$\langle 110 \rangle$	$\{001\}$	6
NaCl	MnSe	$\langle 110 \rangle$	$\{\bar{1}11\}$	12
CsCl	CsCl	$\langle 100 \rangle$	$\{001\}$	6
Al_2O_3	Al_2O_3	$\langle 11\bar{2}0 \rangle$	$\{0001\}$	3

We conclude that both intermetallic and ceramic compounds are inherently less deformable than are pure metals and their solid solution phases, because critical shear stresses are high and the number of slip planes is small. Consequently, fracture stresses are commonly exceeded before plastic deformation is initiated (Section 8–4).

Deformation by Twinning

A crystal is *twinned* if it possesses a mirror boundary. In Fig. 8–3.8, the (110) plane of a bcc metal is sketched in two dimensions in the plane of the paper. Vertical to this plane is a twin boundary, across which the atoms are arranged in the same, but reflected, pattern.

Twinning can be induced by shear stresses, as indicated in Fig. 8–3.9. This twinning produces a permanent displacement, but is limited to relatively low strains, because, once repositioned, the atoms cannot move farther by twinning. For this reason, twinning deformation is overshadowed in most materials by

FIG. 8–3.8

Twin (Bcc). The (110) plane is shown. The boundary is a ($\bar{1}$12) plane emerging perpendicularly through the plane of the sketch. The two parts are mirror images of one another, and the ($\bar{1}$12) plane is common to both parts of the twin. The (110) and ($\bar{1}$12) planes intersect along the [1$\bar{1}$1] direction. (From L. H. Van Vlack, *Materials Science for Engineers,* Addison-Wesley, Reading, Mass., Fig. 6–22, with permission.)

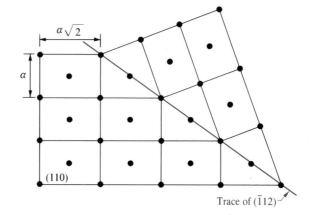

FIG. 8–3.9

Mechanical Twinning [Bcc—(110) Plane]. A shear stress in the [1$\bar{1}$1] direction produces strain by twinning. Deformation by twinning is significant in hcp metals, because there are fewer hexagonal slip systems (Table 8–3.1).

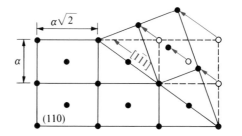

plastic slip. However, hexagonal materials have only a few slip systems (Table 8-3.1). In magnesium, zirconium, zinc, and a few other hcp metals, twinning strain is the predominant mechanism for processing to final shapes and dimensions.

Example 8-3.1

With a sketch, show the 12 slip systems that are included in the $\langle\bar{1}10\rangle\{111\}$ slip systems of fcc metals.

Answers See Fig. 8-3.10. Each of the four planes of the {111} family has three slip directions. Thus, we have

$[\bar{1}10](111)$	$[1\bar{1}0](11\bar{1})$	$[110](1\bar{1}1)$	$[110](\bar{1}11)$
$[10\bar{1}](111)$	$[101](11\bar{1})$	$[10\bar{1}](1\bar{1}1)$	$[101](\bar{1}11)$
$[0\bar{1}1](111)$	$[011](11\bar{1})$	$[011](1\bar{1}1)$	$[01\bar{1}](\bar{1}11)$

Comment The rear planes are parallel to the front planes and therefore involve the same slip systems; for example, $[110](\bar{1}11)$ is the same slip system as $[110](1\bar{1}\bar{1})$. Likewise, $[\bar{1}10](111)$ and $[1\bar{1}0](111)$ are the same slip system, but of opposite sense.

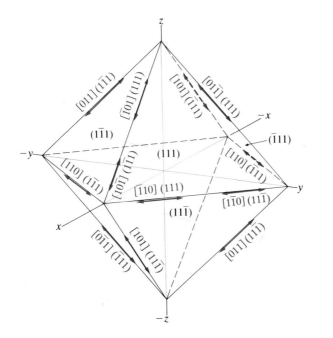

FIG. 8-3.10

The 12 Slip Systems of $\langle 1\bar{1}0\rangle\{111\}$. (See Example 8-3.1).

FIG. 8-3.11

Slip Vectors on (111).
See Example 8-3.2.

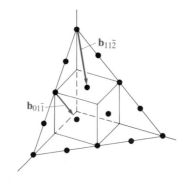

Example 8-3.2

Both $[01\bar{1}]$ and $[11\bar{2}]$ lie in the (111) plane of fcc aluminum. Therefore, both $[01\bar{1}](111)$ and $[11\bar{2}](111)$ slip are conceivable. (a) Make a sketch of the (111) plane and show the $[01\bar{1}]$ and $[11\bar{2}]$ unit slip vectors. (b) Compare the energies of the dislocation lines that have these two displacement vectors.

Solution

(a) See sketch (Fig. 8-3.11).

(b) Refer to Eq. (8-3.1):

$$\frac{E_{01\bar{1}}}{E_{11\bar{2}}} = \frac{lG\mathbf{b}_{01\bar{1}}^2}{lG\mathbf{b}_{11\bar{2}}^2}$$

Both involve the same slip plane (111); therefore, the same shear modulus applies. Based on unit length l,

$$\frac{E_{01\bar{1}}}{E_{11\bar{2}}} = \left(\frac{\mathbf{b}_{01\bar{1}}}{\mathbf{b}_{11\bar{2}}}\right)^2 = \left(\frac{a/\sqrt{2}}{a\sqrt{6}/2}\right)^2 = \frac{1}{3}$$
$$E_{01\bar{1}} = \tfrac{1}{3}E_{11\bar{2}}$$

Comment With this $\frac{1}{3}$ ratio, slip occurs appreciably more readily with the first of the two potential slip systems for fcc metals.

8-4

FRACTURE

The ultimate mechanical failure is fracture. We commonly categorize fracture as being either ductile or brittle (i.e., nonductile). Little energy is required to break brittle materials, such as glass, polystyrene, and some of the cast irons. Conversely, tough materials, such as rubber and many steels, absorb considerable energy in the fracture process. The contrast is important, since the service limit in many engineering products is not the yield or ultimate strength; rather it may be the energy associated with fracture propagation.

Brittle fracture requires energy to separate atoms and to expose new surface along the fracture path. *Ductile fracture* requires not only the energy just cited, but much additional energy to deform plastically the material adjacent to the fracture path.

One measure of toughness is the area under the stress–strain curve (Fig. 8–4.1). With no plastic deformation, that area is $se/2$, or $s^2/2E$, and the energy is solely elastic. The calculation is not as simple for the ductile material in Fig. 8–4.1(b), because the plastic strain of the deformable material far exceeds the limited elastic strain. The result is a much higher energy consumption before fracture. The units of energy as just described are the product of stress and strain:

$$(\text{N/m}^2)(\text{m/m}) = \text{joule/m}^3 \qquad (\text{or ft} \cdot \text{lbs/in}^3)$$

that is, joules per unit volume. In reality, the energy consumption is very nonuniform within the fractured test piece; negligible energy is absorbed in undeformed regions, whereas a major quantity is absorbed in the vicinity of the fracture. Furthermore, this distribution is markedly affected by the size and shape of the test specimen. Notches are especially critical in determining the energy requirements for fracture.

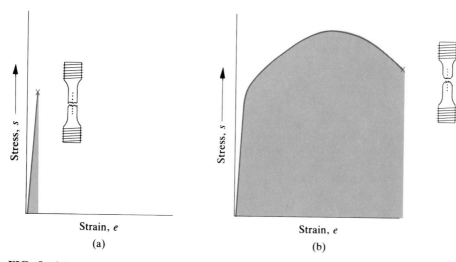

Stress, s Strain, e
(a)

Stress, s Strain, e
(b)

FIG. 8–4.1

Fracture. (a) Brittle fracture involves little or no plastic deformation. (b) Ductile fracture requires energy for plastic deformation. Toughness, the energy requirement, is equal to the area under the s–e curve.

Toughness Tests

Figure 8–4.2 shows an impact tester that has been widely used to obtain toughness data. The energy absorbed in the fracture is calculated from the height of the followthrough swing of the heavy pendulum (Example 8–4.1). A Charpy V-notch test specimen (Fig. 8–4.2b) is commonly used for comparative purposes. Note, however, that these values are a function of size and shape, as well as of the materials being compared.

Ductility-Transition Temperature

Many materials exhibit an abrupt drop in ductility and toughness as the temperature is lowered. In glass and other amorphous materials, this change corresponds to the glass-transition temperature. Of course, metals are crystalline and do not

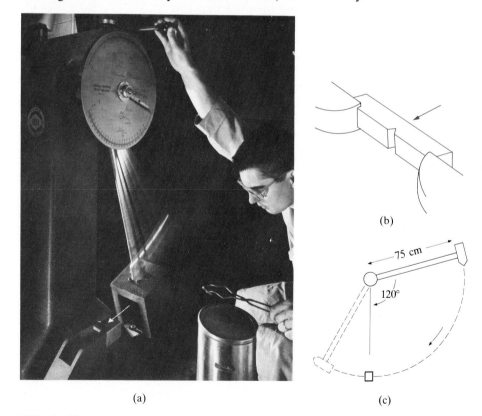

(a)   (c)

FIG. 8–4.2

Toughness Test. The notched test specimen—arrow in (a) and sketched in (b)—is broken by the impact of the swinging pendulum (c). The amount of energy absorbed is calculated from the arc of the followthrough swing (Example 8–4.1). (Courtesy of U.S. Steel Corp.)

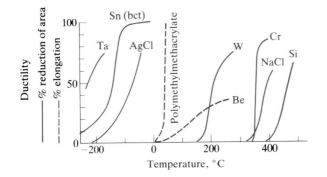

FIG. 8-4.3

Ductility Versus Temperature (Tensile Tests). Except for fcc metals, most materials lose ductility abruptly at decreasing temperatures. For a given material, the transition temperature is higher for higher strain rates—for example, in impact loading. (After data by A. H. Cottrell, *The Mechanical Properties of Matter,* Wiley.)

have a glass-transition temperature. However, they may have a *ductility-transition temperature, T_{dt}* (Fig. 8-4.3), that divides a lower temperature regime where the fracture is said to be nonductile from a higher temperature range where considerable plastic deformation accompanies failure.

Figure 8-4.4 shows the energy absorbed during impact fracture by two different steels as a function of temperature. As with the glass-transition temperature (Section 4-2), the ductility-transition temperature is not fixed, but rather depends on size and shape, the rate of loading, presence of impurities, and so on. However, the engineer will be quick to choose steel C over steel B of Fig. 8-4.4, if it is to be used in a welded ship in the North Atlantic winter waters. A crack, once started in the steel with a high ductility transition temperature, could continue to propagate with a low energy fracture until the ship was broken apart. Several unfortunate naval catastrophies occurred before design engineers learned how to minimize stress concentrations, and metallurgists found that fine-grained steels have lower

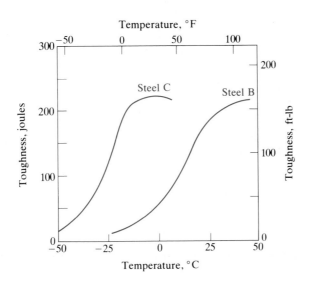

FIG. 8-4.4

Toughness Transitions. For each steel, there is a marked decrease in toughness at lower temperatures. The transition temperature is significantly lower for steel C (fine-grained) than for steel B (rimmed). (Adapted from Leslie, Rickett, and Lafferty, *Trans. AIME.*)

ductility-transition temperatures than do coarse-grained steels. Fortunately, fcc metals do not display an abrupt change in ductility as the temperature is lowered. Thus, aluminum, copper, and austenitic stainless steels can be used in cryogenic applications. Unfortunately, other properties and costs preclude the widespread use of these materials as replacements for large structures, pipelines, and so on.

Fracture Toughness, K_{Ic}

The impact tests that we described in the previous section have been available for several decades. They give qualitative data that may be used advantageously for comparative purposes; however, they do not give property data that can be used for design purposes. To obtain these values, we must consider *fracture mechanics.* Fracture always starts at some point of *stress concentration,* which may be adjacent to a rivet hole, along a keyway of a shaft, at some reentrant angle in the product or structure, along a scored line on a piece of window glass, or at a flaw in the material itself. In each of these cases, the stress is concentrated because the load cannot be uniformly distributed across the full area. The load must be redistributed around the end of the missing cross-section. This situation is illustrated schematically in Fig. 8–4.5(a) for an elliptical hole. As shown for the crack in Fig. 8–4.5(b), the

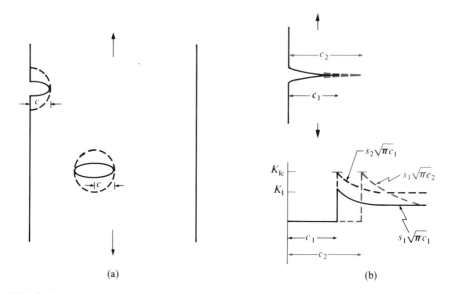

(a)   (b)

FIG. 8–4.5

Stress Intensification. (a) Elliptical hole. The stress is intensified adjacent to the hole because the load must be redistributed around the end of the hole. (b) Stress-intensity factor, K_I. Fracture occurs when the stress intensity attains the critical value, K_{Ic}, for the material. As indicated in Eq. (8–4.1), that value may be reached by either a greater stress or a deeper crack.

stress intensity is greatest at the tip. The intensity increases in proportion to the nominal stress, s (after correction for true area), and the square root of the crack's depth, c. More specifically,

$$K_I = Ys\sqrt{\pi c} \qquad\qquad (8\text{-}4.1)$$

where K_I is the *stress-intensity factor* that is independent of the nature of the material. (The term Y provides a correction for the thickness-to-width ratio of the material. For our purposes, it is slightly more than unity for metal plate products.) With the $s\sqrt{c}$ relationship of Eq. (8-4.1), the units of the stress-intensity factor are $\text{MPa}\cdot\sqrt{m}$ (or $\text{Psi}\cdot\sqrt{in.}$). With more intense stresses or with deeper cracks, the stress intensity becomes sufficient for the fracture to progress spontaneously. This threshold stress intensity is a property of the material. It is called the *critical stress-intensity factor, K_{Ic},* or the *fracture toughness* of the material.

The loading in Figs. 8-4.5(a) and 8-4.6 is called mode I; hence the subscript to the symbol, K in Eq. (8-4.1). Other modes include II, shear forces parallel to the crack surface and perpendicular to the crack front, and III, shear forces parallel to the crack surface as well as to the crack front (tearing).

Design Considerations

There are several design implications from the relationships of the last section. Cracks can be tolerable, providing $Ys\sqrt{\pi c}$; that is, K_I, does not exceed the critical stress-intensity factor, K_{Ic}, for the material.* Conversely, the material will fracture if $Ys\sqrt{\pi c}$ exceeds K_{Ic}, even though S_y or S_u are not exceeded.

The engineer must consider both strength and fracture toughness because materials with higher strengths typically have low fracture toughness, and vice versa.

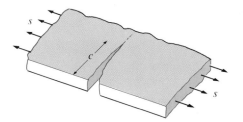

FIG. 8-4.6

Fracture Stresses (Mode I). Stresses that open an edge crack in tension introduce an intensity factor, K_I, that increases with $\sqrt{\pi c}$ as indicated in Eq. (8-4.1). If K_I exceeds a critical value, K_{Ic}, the crack will advance catastrophically.

---

* We can make a comparison with our previous observations on stress, which is a function of force and area ($s = F/a$), and is independent of the material. When the stress reaches a critical value for a given material, however, plastic deformation occurs. We call that critical stress for the material the yield strength, S_y. Here the critical value of *stress intensity, K_{Ic},* is the fracture toughness of the material.

Thus, conditions for fracture may be encountered before the yield stress is reached, so full advantage cannot be taken of the high strength. Such considerations lead us to place a high value on careful design, as well as on materials that have no surface or internal imperfections.

Example 8–4.1

An impact pendulum on a testing machine weighs 10 kg and has a center of mass 75 cm from the fulcrum. It is raised 120° and released. After the test specimen is broken, the followthrough swing is 90° on the opposite side. How much energy did the test material absorb?

Calculation Refer to Fig. 8–4.2.

$$\Delta E = 10 \text{ kg}(9.8 \text{ m/s}^2)(0.75 \text{ m})[\cos(-120°) - \cos 90°]$$
$$= -36.8 \text{ J} \qquad\qquad\qquad (lost \text{ by the pendulum})$$
$$= +36.8 \text{ J} \qquad\qquad\qquad (absorbed \text{ by the sample})$$

Comments In general, materials have more toughness at high temperatures than they do at low ones. In fact, there is an abrupt decrease in the toughness of many steels as they are cooled below ambient temperatures. The temperature at which this discontinuous decrease occurs is called the *ductility–transition temperature* (Fig. 8–4.4).

Example 8–4.2

A steel has a yield strength, S_y, of 1100 MPa (160,000 psi), an ultimate strength, S_u, of 1200 MPa (174,000 psi), and a fracture toughness, K_{Ic}, of 90 MPa·$\sqrt{m}$ (82,000 psi·$\sqrt{in.}$). (a) A plate has a 2-mm (0.08-in.) edge crack. Will it fail by fracturing before it fails by yielding? (b) What are the deepest tolerable cracks that may be present without fracturing before yielding? [Assume that the geometric factor, Y, is equal to 1.1 for steel plates loaded in tension as in Fig. 8–4.6.]

Procedure Determine the stress *s* required to raise the stress intensity factor K_I to the critical value of 90 MPa·$\sqrt{m}$. In (b) use 90 MPa·$\sqrt{m}$ and 1100 MPa (K_{Ic} and S_y, respectively) to calculate the critical limiting crack length.

Calculation Use Eq. (8–4.1).

(a) $s = (90 \text{ MPa} \cdot \sqrt{m})/(1.1\sqrt{0.002\pi \text{ m}})$
$= 1030 \text{ MPa}$

This stress of 1030 MPa (150,000 psi) to cause fracture is less than S_y; so this steel will fracture before it yields.

(b) $c\pi = [(90 \text{ MPa} \cdot \sqrt{m})/(1100 \text{ MPa})(1.1)]^2$

$c = 0.0017 \text{ m}$ \qquad\qquad\qquad\qquad\qquad\qquad (or 1.7 mm)

Comment Why should the answer be rounded down rather than up?

SUMMARY

1. Elastic deformation is a *reversible* strain, and is proportional to the applied stress. There are three elastic moduli: *Young's modulus, E,* is the axial stress per unit strain; the *shear modulus, G,* is the shear stress per unit strain, and the *bulk modulus, K,* is the hydrostatic stress (or pressure) per unit volume change. We can calculate one from another by using *Poisson's ratio, v.* The elastic moduli are generally anisotropic, and usually decrease with increasing temperatures.

2. *Plastic strain* is not reversible; it is permanent. The critical stress to initiate plastic strain is called the *yield strength, S_y.* Further stressing produces both elastic and plastic strains, with the latter predominant. The *ultimate strength, S_u,* is the maximum stress per unit of *original area. Ductility* is the plastic strain prior to fracture. True stress and true strain are calculated on the basis of the *true area.* They exceed the nominal stress and strain in ductile materials, because the plastic deformation reduces the cross-sectional area.

3. Plastic strain is not reversible. Plastic deformation occurs in crystalline materials predominantly by *dislocation movements* along slip planes. A minor amount of plastic deformation may arise from *twinning.* Those *slip systems* with the lowest *critical shear stresses* are the most densely packed directions on the most densely packed planes. Compounds have inherently high shear stengths. *Solution hardening* results from dislocations becoming anchored on solute atoms.

4. Fracture may be ductile or brittle; that is, it may occur with or without plastic deformation. The former is tougher. Toughness tests measure the energy required for fracture of standardized test specimens. Steel and a number of other materials exhibit a *ductility-transition temperature, T_{dt},* below which fracture occurs with little energy absorption.

The *stress-intensity factor (K_I* of Eq. 8–4.1) relates the nominal stress and the depth of a crack to the stress concentration at the tip of the crack. *Fracture toughness, K_{Ic},* is the critical value of K_I that permits crack propagation. It is a property of a material, and it commonly dictates the stress limits on high-strength materials that have low ductility.

KEY WORDS

Anisotropic
Annealing point
Bulk modulus (K)
Critical shear stress
Deformation, elastic
Deformation, plastic
Ductility
Ductility-transition temperature (T_{dt})
Elastic strain (e_{el})
Elastomer
Elongation (El.)
Fracture, brittle
Fracture, ductile

Fracture toughness
Hardness
Plastic strain (e_{pl})
Poisson's ratio (v)
Proportional limit
Reduction of area (R of A)
Shear modulus (G)
Shear strain (γ)
Shear stress (τ)
Slip direction
Slip plane
Slip system
Strain (e)

Strength (S)
Stress (s)
Stress-intensity factor (K_I)
Stress-intensity factor, critical (K_{Ic})
Stress–strain diagram
Toughness
True stress (σ)
True strain (ϵ)
Twin (crystal)
Ultimate strength (S_u)
Yield strength (S_y)
Young's modulus (E)

PRACTICE PROBLEMS

8–P11 A 7.6-mm (0.30-in.) diameter 1040 steel bar, which was initially 2.27 m (89.4 in.) long, supports a weight of 3963 N (890 lb$_f$). What is the difference in length if the steel bar is replaced by a 70–30 brass bar?

8–P12 What is the volume change of a rod of brass when it is loaded axially by a force of 233 MPa (33,800 psi)? (Poisson's ratio is 0.37.)

8–P13 What is the bulk modulus of the brass in Problem 8–P12? What is the shear modulus?

8–P14 If iron has an axial stress of 208 MPa (30,200 psi), what will be the highest local stress within a polycrystalline copper bar?

8–P15 A precisely machined steel rod has a specified diameter of 18.6 mm (0.732 in.). It is to be elastically loaded longitudinally with a force of 670,000 N (150,000 lb$_f$). (a) By what percentage will its diameter change? (b) By what percentage will the cross-sectional area change? (Poisson's ratio is 0.29.)

8–P16 The elastic modulus of copper drops from 110 GPa (16 × 10^6 psi) at 20° C (68° F) to 107 GPa (15.5 × 10^6 psi) at 50° C (122° F). What is the change in the total length of a 1575-mm (62-in.) bar if the stress is held constant at 165 MPa while the temperature rises those 30° C?

8–P17 Based on Problems 8–P12 and 8–P13, what shear angle, α, does a shear stress of 262 MPa (38,000 psi) produce in brass?

8–P18 Distinguish among the three elastic moduli.

8–P21 "The yield strength is the stress where strain switches from elastic to plastic." Comment on the validity of this statement.

8–P22 A wire of a magnesium alloy is 1.05 mm (0.04 in.) in diameter. Plastic deformation starts with a load of 10.5 kg (which is _____ N), or 23 lb$_f$. The total strain is 0.0081 after loading to 12.1 kg (26.6 lb$_f$). (a) How much permanent strain has occurred with a load of 12.1 kg (26.6 lb$_f$)? (b) Rework this problem with English units.

8–P23 A brass test specimen has a reduction of area of 35 percent. (a) What is the true strain, ϵ? (b) What is the

ratio of the true stress, σ, to the nominal stress, s, for fracture, S_b?

8–P31 Identify the 12 $\langle \bar{1}11 \rangle \{101\}$ slip systems for a bcc metal.

8–P32 (a) Sketch the atomic arrangement on the (110) plane of MnS, which has the structure of NaCl (Fig. 3–2.5). (b) Identify the six $\langle 1\bar{1}0 \rangle \{110\}$ slip systems for this compound. (c) How long is the slip vector, **b**, in these systems?

8–P33 Differentiate between the terms *resolved shear stress* and *critical shear stress.*

8–P34 The engineer commonly defines the stress for yield strength after a 0.2 percent offset strain has occured (Fig. 8–2.1b). Based on Fig. 8–1.5, why is this offset necessary?

8–P35 Alloys of zinc and of magnesium are more widely used as casting alloys than as wrought (plastically deformed) alloys. Suggest a valid reason why this is so.

8–P36 Refer to Table 8–3.2. Why are there $\langle 100 \rangle \{001\}$ slip systems in CsCl structures, and not $\langle 111 \rangle \{\bar{1}10\}$ slip systems? There are more atoms per mm in the $\langle 111 \rangle$ directions than in the $\langle 100 \rangle$ directions.

8–P41 What would the followthrough angle have been in Example 8–4.1 if, before testing the sample, the initial angle of the pendulum had been set at 105°?

8–P42 The value of K_{Ic} for a steel is 186 MPa·m$^{1/2}$ (169,000 psi·in$^{1/2}$). What is the maximum tolerable crack when the steel carries a nominal stress of 800 MPa (116,000 psi)? (The geometric factor, Y, is 1.1.)

8–P43 Convert K_{Ic} = 135,000 psi·in$^{1/2}$ to MPa·m$^{1/2}$.

8–P44 The testing machine of Fig. 8–4.2 is described in Example 8–4.1. What is its maximum capacity expressed in the amount of energy absorbed by a specimen?

8–P45 A small hole drilled through a steel plate ahead of a crack will stop the crack's progress until repairs can be made. Explain how the hole accomplishes this feat.

TEST PROBLEMS

811 Provide a ball-park figure for Young's elastic modulus (a) of chromium and (b) of gold.

812 An aluminum rod is stressed 210 MPa (30,000 psi) in tension. Its initial diameter was 24.02 mm (0.946 in.). What is the change in diameter as a result of the load? Its bulk modulus is 73 GPa.

813 Assume a material deforms elastically in tension with no change in volume; that is, $(1 + e_x)(1 + e_y)(1 + e_z) = 1$. Calculate Poisson's ratio.

814 Aluminum that has an elastic modulus of 70 GPa (10^7 psi) and a Poisson's ratio 0.34 is under hydrostatic pressure of 83 MPa (12,000 psi). What are the dimensions of the unit cell?

815 Which will have the greater compressibility, β, iron with a Poisson's ratio of 0.29, or brass with a ratio of 0.37?

816 A test bar 12.83 mm (0.5051 in.) in dia. with a 50-mm gage length is loaded axially with 200 kN (45,000 lb$_f$) and is stretched 0.458 mm (0.018 in.). Its diameter is 12.79 mm (0.5035 in.) under load. (a) What is the bulk modulus of the bar? (b) What is the shear modulus?

817 A precisely ground cube (100.00 mm)³ of an aluminum alloy is compressed in the x direction 70 MPa. (a) What are the new dimensions if Poisson's ratio is 0.3? (b) A second compression of 70 MPa is concurrently applied in the y direction. Now, what are the dimensions? (c) Derive Eq. (8–1.7). [*Hint:* Consider that the cube is under a stress of $s_x = s_y = s_z = P_h$ and recall that $(1 + \Delta V/V) = (1 + \Delta L/L)^3 = 1 + 3e + \cdots$.]

818 A vulcanized rubber ball decreases 1.2 percent in diameter when it is under 1000 psi hydrostatic pressure. A test bar of the same type of rubber stretches 2.1 percent when it is stressed 175 psi in tension. What is Poisson's ratio of the rubber?

821 A hardened steel fractures with a stress of 1515 MPa (217,000 psi) and a total strain of 0.027. What is the ductility? (Elastic strain is recovered.)

822 A 70 Cu–30 Zn brass has an elastic modulus of 110 GPa (16,000,000 psi) and a yield strength of 140 MPa (20,000 psi). (a) How much load can a 2.74-mm (0.108-in.) wire of this alloy support without yielding? (b) If a load of 55 kg (121 lb$_f$) is

supported by a 30.5-m (100-ft) wire of this size, what is the total extension?

823 A 0.50-in. (12.7-mm) diameter test bar of the brass identical to that in Problem 822, had a 0.29-in. (7.4-mm) diameter at its fracture point. The testing machine recorded 1030 lb$_f$ (468 kg) when the test bar broke. What were the (a) true stress, and the (b) true strain when the bar fractured?

824 A testing machine recorded a load of 53,000 lb$_f$ when the deformation (2-in.) was 24 percent. The elongation (2-in.) at fracture was 47 percent, but the load was only 41,000 lb$_f$. Explain.

825 A 1040 steel wire has a diameter of 0.89 mm (0.035 in.). Its yield strength is 875 MPa (127,000 psi) and its ultimate strength is 1070 MPa (155,000 psi). An aluminum alloy also is available with a yield strength of 255 MPa (37,000 psi), and an ultimate strength of 389 MPa (56,400 psi). (a) How much (what percentage) heavier, or lighter, will an aluminum wire be than a steel wire to support a 35-kg (77-lb) load with the same elastic deformation as the steel wire? (b) How much (what percentage) heavier, or lighter, must an aluminum wire be to support the same maximum load without permanent deformation? (c) How much heavier or lighter must it be to support the load without breaking? (*Hint:* Establish a ratio of masses without calculating the actual areas.)

826 The percent reduction of area cannot be calculated from the percent elongation (nor vice versa). Why is this true, given that both are measures of ductility?

827 Change the experimental data for the standard test bar (diameter = 0.5051 in., or 12.7 mm) of Fig. 8–2.2 from 240 mm and 75 mm to 219 mm and 59 mm, respectively. The final diameter at the break is 9.3 mm. Calculate *four* ductility values.

831 Identify eight of the 12 $\langle \bar{1}11 \rangle \{211\}$ slip systems for bcc metals.

832 Select a (112) plane of a bcc metal. Sketch it such that you can illustrate a $\langle 111 \rangle$ slip direction that lies in that plane.

833 Under certain conditions, the $\langle 011 \rangle \{100\}$ slip systems can operate in NaCl-type structures to supplement the $\langle 1\bar{1}0 \rangle \{110\}$ systems. (a) How many slip systems belong to this $\langle 011 \rangle \{100\}$ set? List them. (b) What is the slip vector in PbS, which has the NaCl-type structure and these two slip systems?

834 Impurity atoms inhibit dislocation movements. Explain.

835 A 95 Cu–5 Sn bronze has the same elastic modulus as does pure copper (Appendix C). This bronze, however, has a yield strength that is three times as great as that of pure copper. Discuss reasons for this difference.

836 Your company is about to make a bid on a contract for 1500 feet of metal rods that can support 2000 lb$_f$ loads without permanent deformation. Copper or a 70 Cu–30 Zn brass could be used. Assume that the purchase price per pound of copper is twice that of zinc. (a) What will the cost ratio be for purchasing the raw materials for the two metals? (Processing costs are to be calculated separately.) (b) Recalculate for $\$_{Cu} = 1.2\ \$_{Zn}$.

841 A test specimen is expected to absorb two-thirds as much energy during fracture as did the specimen in Example 8–4.1—that is, 24.5 J. What angle of followthrough should be expected?

842 The steel of Problem 8–P42, with a 9-mm (0.35-in.) crack, can support what maximum nominal stress without the crack advancing?

843 A steel with a K_{Ic} value of 190 MPa·m$^{1/2}$ also has a yield strength of 900 MPa. How deep will a crack be before the steel is subject to fracture as a mode of failure? (The geometric factor, Y, is 1.05.)

844 A steel has a yield strength of 690 MPa (100,000 psi), and a K_{Ic} value of 70 MPa·m$^{1/2}$ (63,700 psi·in.$^{1/2}$). What will be the limiting design stress if the minimum tolerable crack is 2.5 mm (0.10 in.) and no plastic deformation is permitted? (The plate dimensions require a geometric factor of 1.1.)

845 Engineers who previously designed riveted ships paid little attention to transition temperatures, unlike those who currently design welded ships. Why?

846 Explain why rubber is brittle at liquid-nitrogen temperatures (77 K).

Chapter 9

SHAPING, STRENGTHENING, AND TOUGHENING PROCESSES

An early step of materials processing often is one of producing the desired shape. Commonly, shaping is achieved by solidification of a liquid in a mold, by plastic deformation, or by sintering powders.

Creating stronger and tougher materials is a perpetual goal of design and materials engineers. Several approaches may be used, either singly or jointly. They include *solid-solution* hardening, *strain* hardening, *precipitation* hardening, and strengthening by *rigid particles*. The latter three are closely associated with the microstructure. Thus, they are (1) controllable by heat treatments, and (2) subject to alteration in service (Chapter 14).

Brasses and steel will be our principal prototypes, not only because they are widely studied, but also because they are amenable to a number of heat-treating procedures. Optional topics explore the *heat treatment* of steels, and their *hardenability*. In the final section of the chapter, we introduce an area of active development—that of making normally brittle ceramics both stronger and tougher.

———————

///OPTIONAL/// ## 9-1

SHAPING PROCESSES

Except in special cases, engineering materials must be shaped before they are assembled into a technical device or product. The procedures are too numerous for us to give consideration to each. However, there are several general categories that warrant attention. We shall examine three—casting, plastic forming, and sintering.

Casting

This procedure starts with liquids or semifluid solids, which solidify in a mold. Commonly, the starting liquid is a molten metal that crystallizes during solidification; however, it may be a thermoplastic resin, or it may be a slurry or suspension, from which water or other liquid is extracted. This latter process has a long history for clay-based ceramics. Polymers also may start out as a fluid, in which a chemical reaction between two components produces a solid plastic.

Molds must be used in all casting processes. They provide a cavity with the negative shape of the product. Molten brass was poured into a sand mold that was bonded with a phenolic resin to make the faucet of Fig. 7–1.2. In the injection molding of polymers (Section 10–2) and in the comparable die-casting process— used for many aluminum and zinc alloys—molten material is forced into a die cavity. Since these materials have a low melting temperature, the die (a permanent mold) can be reused. In contrast sand molds are used only once for higher-melting-temperature alloys. Molds for clay-based ceramics are commonly made out of plaster; the porous plaster absorbs liquid to dewater the adjacent slurry.

Almost invariably, volume changes accompany the solidification of a casting. Ceramic powders are compacted as the water is extracted. Likewise, most metals shrink into a more closely packed structure as bcc, fcc, or hcp solids are formed. Not only is there a 4 to 6 percent solidification shrinkage, but also there is a comparable volume contraction between the freezing temperature and ambient temperatures (Fig. 4–2.2). A few materials expand on freezing. Water is the notable example. This freezing expansion of water is of little consequence in material processing; however, it does have engineering importance for concrete structures and for various other low-temperature applications. As we discussed in

Section 4–2, the materials that expand during crystallization typically have stereospecific bonds in a network structure. This structure can account for the expansion of covalently bonded silicon during freezing (Fig. 4–2.3).

When shrinkage accompanies freezing, the casting process must provide means of feeding more liquid into the casting during solidification. The liquid is added through *risers* in the mold design (Fig. 9–1.1). Likewise, the engineer must ensure that the *sprues* (feeders) do not solidify before the mold cavity is fed and solidified.

Plastic Forming

Ductile materials, most commonly single-phase metals, can be shaped by plastic deformation. The mechanism is through slip along crystal planes, involving dislocation movements, as described in Section 8–3.

A typical procedure is to cast an *ingot,* which is simply a solidified block of metal that is deformed by *mechanical working* or *forming* to produce a rod, wire, tube, plate, forging, or other shape. During the mechanical processing, when the shape is changed, stresses must be applied that are above the yield strength described in the previous chapter. Consequently, the mechanical processing on the initial large ingot is commonly performed at high temperatures at which the

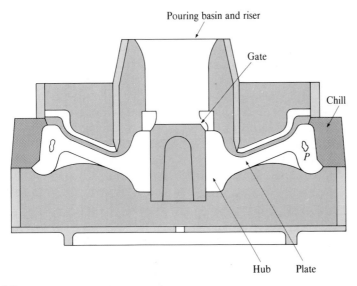

Pouring basin and riser

Gate

Chill

P

Hub   Plate

FIG. 9–1.1

Mold for Metal Casting (Railroad-Car Wheel). The shrinkage of several volume percent during solidification is accommodated in two ways. (1) An expendable *riser* (a reservoir) feeds the casting during solidification. (2) The mold is sand, except for the *chill* around the rim. The low thermal conductivity of the sand delays the solidification in the wheel's plate, while the chill accelerates the solidification of the rim. In that manner, liquid metal continues to feed the mold. The porosity that is portrayed by *P* in the thicker rim section is avoided. (Adapted from Flinn, *Fundamentals of Metal Casting,* Addison-Wesley, Reading, Mass.)

material is softer and more ductile. At those temperatures, less energy is required for deformation and there is less chance of fracturing during processing. *Rolling, forging,* and *extrusion* (Fig. 9–1.2) are among the hot deformation processes.

After the primary deformation steps, the processing often is continued at ambient temperatures. This is possible because the dimensions are now smaller and the required forces and energy are reduced accordingly. Also, furnace and fuel costs are avoided. Further, most metals oxidize rapidly at high temperatures, whereas oxidation generally can be avoided at low temperatures. Finally, we shall see in this chapter that the strength of the final product generally can be increased if the deformation is performed while the metal is at normal temperatures. Wire *drawing, spinning* (Fig. 9–1.3), and *stamping* are among the common secondary deformation processes performed at ambient temperatures.

Sintering of Ceramics and Metals

Sintering is the process of bonding particles by heat. It is the main agglomeration process for almost all ceramics (except glass), for making powder metal products (Fig. 9–1.4), and for bonding certain polymeric materials (e.g., Teflon). The MgO of Fig. 7–1.3(b) was produced by sintering of MgO powder. Sintering involves not only the bonding of powder particles, but also the elimination of the initial porosity to give a more dense product. (See Example 9–1.1.)

The principle involved in sintering solid particles in the absence of any liquid is the same as that of grain growth, which is the reduction of surface and boundary energies, and therefore the minimization of boundary areas. As shown in Fig. 9–1.5(a), there are two surfaces between any two particles before sintering. After sintering, there is a single grain boundary. The two surfaces are high-energy boundaries; the grain boundary has much less energy. Thus, this sintering reaction occurs naturally if the temperature is high enough for a significant number of atoms to diffuse.

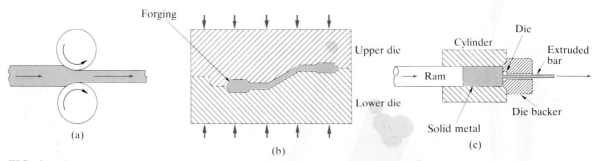

FIG. 9–1.2

Mechanical Working (Primary). (a) Rolling. (b) Forging. (c) Extrusion. These processes can produce severe plastic deformations at high temperatures. (L. H. Van Vlack, *Textbook of Materials Technology,* Addison-Wesley, Reading, Mass., with permission.)

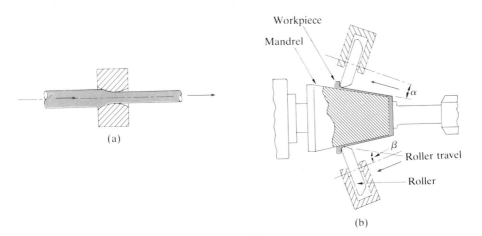

(a)

Workpiece

Mandrel

α

β

Roller travel

Roller

(b)

FIG. 9-1.3

Mechanical Working (Secondary). (a) Wire drawing. (b) Spinning. Most metals are strengthened as they are deformed at ambient temperatures. Therefore, these processes are generally limited to products with smaller cross-sections than those for which we use the primary processes of Fig. 9-1.2. (L. H. Van Vlack, *Textbook of Materials Technology*, Addison-Wesley, Reading, Mass., with permission.)

The actual mechanism of sintering is shown in Fig. 9-1.6 in photographs taken through a scanning electron microscope. The points of contact between particles grow into areas of contact by the diffusion of atoms. The particles therefore are pulled closer together, producing an accompanying shrinkage and a reduction of porosity. Figure 9-1.7 shows quantitative data for these volume changes. As anticipated, changes proceed more rapidly at higher temperatures.

FIG. 9-1.4

Powdered Metal Gear. The metal particles are sintered into a coherent structure.

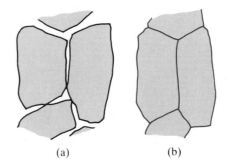

(a)   (b)

FIG. 9-1.5

Solid Sintering. (a) Particles before sintering have two adjacent surfaces. (b) Grains after sintering have one boundary. The driving force for the sintering is the reduction of surface area (and therefore of surface energy).

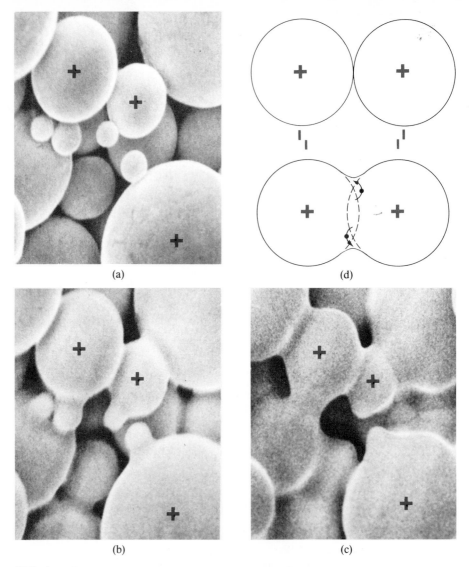

(a)

(d)

(b)

(c)

FIG. 9-1.6

Sintering (Nickel Powder). The initial points of contact in (a) become areas of contact in (b) and (c) while the material is heated to 1100° C. (d) The atoms diffuse from the contact points to enlarge the contact area. (Vacancies diffuse in the opposite direction.) The particles of powder move closer together, and the amount of particle surface is reduced. (R. M. Fulrath, dec., University of California, Berkeley.)

Example 9-1.1

A ceramic magnet has a porosity of 28 v/o before sintering and a density of 5.03 g/cm³ after sintering. The true density is 5.14 g/cm³ (= 5.14 Mg/m³). (a) What is the porosity after firing (sintering)? (b) If the final dimension should be 16.3 mm, what should be the die dimension?

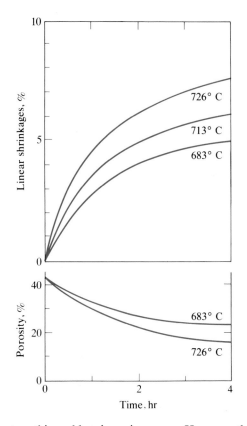

FIG. 9–1.7

Sintering Shrinkage (NaF). Powdered
NaF (− 330 mesh) loses porosity as it
shrinks during sintering. (Adapted from
Allison and Murray, *Acta Metallurgica.*)

Procedure We can set up this problem in various ways. However, the volumes of 1 g are
convenient because they let us compare volume changes directly.

Calculation

(a) true volume of 1 g = $1/5.14 = 0.1946$ cm³

   fired volume of 1 g = $1/5.03 = 0.1988$ cm³

   final porosity = $(0.1988 − 0.1946)/0.1988 = 2\%$

(b)   true volume = $0.72 V_0 = 0.98 V_f$

$$\frac{V_0}{V_f} = \frac{L_0^3}{L_f^3} = \frac{0.98}{0.72}$$

$$L_0 = L_f \sqrt[3]{0.98/0.72}$$
$$= (16.3 \text{ mm})(1.108) = 18.1 \text{ mm}$$

Comment Note that, if the die is set up for a 10 percent shrinkage (18.1 mm to 16.3 mm),
then processing procedures must be consistent, so that there is *always* a 28 percent porosity
in the pressed (presintered) stage. Otherwise, the dimensional specifications will be missed.
Processing variables must be controlled closely.

Example 9–1.2

A ceramic magnet has a true density of 5.41 Mg/m³ (=5.41 g/cm³). A poorly sintered sample weighs 3.79 g dry, and 3.84 g when saturated with water. The saturated sample weighs 3.08 g when suspended in water. (a) What is its *true volume?* (b) What is its *bulk volume* (total volume)? (c) What is its *apparent (open) porosity?* (d) What is its *total porosity?*

Background There are three measurements of volumes:

$$\text{true volume} = \text{total (bulk) volume} - \text{total pore volume}$$

$$\text{apparent volume} = \text{total volume} - \text{open pore volume}$$
$$= \text{true volume} + \text{closed pore volume}$$

$$\text{bulk (total) volume} = \text{true volume} + \text{total pore volume}$$

Since $\rho = m/V$, there also are three densities that relate to the three volumes.

We can use Archimedes' principle: "An object is buoyed up by the weight of the displaced fluid."

Calculation

(a) true volume $= m/\rho_{tr} = 3.79$ g/(5.41 g/cm³) $= 0.70$ cm³

(b) From Archimedes' principle,

water displaced by bulk sample $=$ buoyancy $= 3.84$ g $- 3.08$ g $= 0.76$ g

$\therefore$ total volume $= 0.76$ g/(1 g/cm³) $= 0.76$ cm³

This includes material plus all pore space.

(c) apparent porosity $=$ (open pore volume)/(total volume)
$= [(3.84 - 3.79$ g)/(1 g/cm³)]/0.76 cm³
$= 0.066$ (or 6.6 v/o, bulk basis)

(d)   total porosity $= (V_{total} - V_{true})/V_{total}$
$= (0.76 - 0.70$ cm³)/0.76 cm³
$= 0.079$ (or 7.9 v/o bulk basis)

Comment The *closed porosity* is 1.3 v/o. The *bulk density* is 3.79 g/0.76 cm³ $= 5.0$ g/cm³. We also can speak of an *apparent density:*

mass/(bulk volume $-$ open pore volume)

This density is (3.79 g)/(0.76 cm³)(1 $-$ 0.066) $= 5.34$ g/cm³   (or 5.34 Mg/m³)

9–2

SOLUTION HARDENING

Figure 8–3.6 showed that zinc additions to copper increase the alloy's strength. Thus, brasses are stronger than pure copper. Using this hardening, the engineer can design not only a stronger alloy, but also, in the case of brass, a more ductile one. This combination facilitates processing and leads to greater fracture toughness during service. Furthermore, because zinc is less expensive than is copper, we gain these advantages with no increase in cost.*

Other alloys show similar hardening and strengthening behaviors (Fig. 9–2.1). Unlike zinc, nickel can be added to copper in unlimited quantities, since there is no solubility limit (Fig. 5–2.2). Just as nickel increases the strength of copper, copper increases the strength of nickel. The maximum strength of cupronickel alloys is realized at 70 Ni–30 Cu. Although expensive, this alloy has extensive use in marine applications, where corrosion resistance is required.

Of course, solution hardening is predicated on there being phase relationships that permit solid solubility. We can obtain guidance from phase diagrams. Thus, we could anticipate the solution-hardening range available for the Cu–Zn and

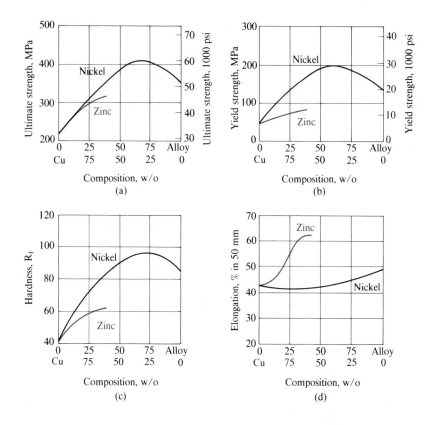

FIG. 9–2.1

Solution Hardening (Annealed Copper Alloys). Impurity atoms serve as anchor points for dislocations. Therefore, a greater shear stress is required, strengthening the alloy.

---

* Not all properties are enhanced by alloying. Electrical and thermal conductivities invariably are decreased (Section 11–2). Thus, if a copper wire were replaced by a brass wire of the same size, there would be a greater loss of power.

Cu–Ni alloys (Fig. 9–2.1) from Figs. 5–6.1 and 5–2.2, respectively. The processing of these alloys normally involves the melting of specified compositions, followed by a homogenization treatment (Section 6–2) to ensure uniformity.

An Ag–Cu equilibrium diagram (Fig. 5–2.5) might not predict solution hardening as an option for strengthening sterling silver for use at ambient temperatures. As we discussed in Section 6–1, however, it is possible to solution-treat this 92.5 Ag–7.5 Cu alloy at 775° C to form a single phase, α, to be followed by quenching. This α solid solution is retained because diffusion is too slow for β precipitation. Both increased strengthening and decreased cost are results. Note, however, that a similar procedure cannot be used to strengthen lead with tin at ambient temperatures. The weakly bonded lead permits tin diffusion in relative short periods, as shown in Fig. 7–2.2. Precipitation of β occurs readily, so the resulting microstructure is a mixture of two soft phases — nearly pure lead, α, and essentially pure tin, β.

Figure 9–2.2 shows the effect of solute concentration on the hardnesses of ferrite (α-iron), and yield strengths of copper. Note that the abscissa values are in atom percent, because the strengthening is proportional to the *number* of solute atoms present (and not to their masses). The effectiveness of each atom on solution hardening is partly a function of the size mismatch, whether larger *or* smaller (Fig. 8–3.5). Thus, the strengthening effects of tin ($R = 0.151$ nm) and beryllium ($R = 0.11$ nm) are significantly greater in copper ($R = 0.1278$ nm) than are the effects of nickel ($R = 0.124$ nm) and zinc ($R = 0.139$ nm).*

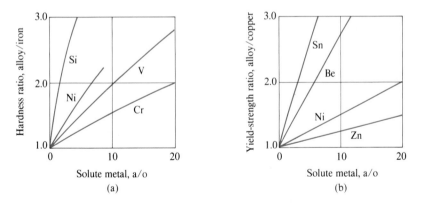

FIG. 9–2.2

Solution Hardening. (a) Iron. (b) Copper. The hardness and strength ratios for the alloy versus the unalloyed metal depend on the size mismatch of solute versus solvent atoms and the stress field which results. (R. S. French and W. R. Hibbard, *Trans. A. I. M. E.,* Vol. 188, L. H. Van Vlack, *Materials Science for Engineers,* Addison-Wesley, Reading, Mass., with permission.)

---

* This mismatch also introduces a limit to the solubility; thus, although tin is an effective hardening agent in copper, only a few atomic percent of these large tin atoms can be contained in solid solution (Fig. 5–6.2).

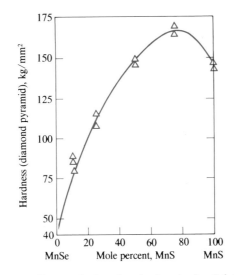

FIG. 9–2.3

Solution Hardening in Binary Compounds (MnSe and MnS). Each compound has the same fcc_{NaCl} structure. Substitution either of sulfur for selenium or of selenium for sulfur increases the hardness to give a maximum at 25 MnSe–75 MnS. (Adapted from P. G. Riewald.)

A second factor that affects solution hardening is the "rigidity" of the atoms. Nickel, which by itself has a large shear modulus (83 GPa), strengthens copper more on an atom-for-atom basis than does zinc (48 GPa). We can view the zinc atom as being more deformable in the strain field of a moving dislocation.

Solution hardening also occurs in binary compounds. This process is illustrated in Fig. 9–2.3 for solutions of MnS and MnSe. These compounds have the NaCl structure that we saw in Fig. 3–1.1. All the cations are Mn^{++} ions. There is a full range of substitution between S^{2-} and Se^{2-} ions in the anion locations. Each end member is hardened by anion substitution, with a maximum similar to that found in the Cu–Ni system (Fig. 9–2.1b).

Example 9–2.1

Specifications for a product that is produced by your company call for a cupronickel alloy with the ultimate strength of at least 370 MPa (54,000 psi) and a yield strength of at least 140 MPa (20,000 psi). The annual production involves about 6 tons of this alloy per year. Which Cu–Ni alloy would you buy from your supplier, who has compositions available in 5 percent composition increments?

Procedure Determine the composition "window" that will meet the requirements. Then make the best selection within that range.

Solution From Fig. 9–2.1, we determine that

For $S_u > 370$ MPa: 45% Ni ↔ 93% Ni
For $S_y > 140$ MPa: 30% Ni ↔ 100% Ni

Use the 50 percent nickel (50 percent Cu) alloy, which meets both specifications. It is cheaper than alloys with a higher percentage of nickel. The 45 Ni–55 Cu alloy is on the edge of the window and does not allow any margin.

Comment If you are uncertain about the relative cost of copper and nickel, think of the change that you have in your pocket.

Example 9–2.2

How much vanadium (per 100 lb of alloy) is necessary to double the hardness of iron?

Procedure As indicated in Fig. 9–2.2, hardness and strength of an alloy are a function of the number of atoms.

Calculation From Fig. 9–2.2(a), the hardness is doubled at 10 a/o. Basis: 100 atoms alloy = 90 atoms Fe and 10 atoms V.

$$(90 \text{ atoms})(55.85 \text{ amu/Fe}) = 5026 \text{ amu} = 0.908$$

$$(10 \text{ atoms})(50.94 \text{ amu/V}) = \underline{\ 509 \text{ amu}} = \underline{0.092} \qquad \text{(or 9.2 lb/100 lb)}$$

$$\text{sum} = 5535 \text{ amu} = 1.000$$

9–3
STRAIN HARDENING AND ANNEALING

We said in Section 8–2 that almost all ductile materials become stronger when they are deformed plasticly. We observe this added strength on a stress–strain diagram (Fig. 8–2.1b) as the concurrent increase in stress and strain beyond the yield strength. This increase in strength is called *strain hardening*. It provides opportunities for the engineer to modify and control the properties of ductile materials, particularly metals.

Cold Work

It is convenient to refer to the percent of *cold work* as an index of plastic deformation. Cold work is the amount of plastic strain introduced during processing, expressed by the percent decrease in cross-sectional area from deformation; that is,

$$\% \text{ CW} = \left[\frac{A_o - A_f}{A_o} \right] 100 \qquad (9\text{–}3.1)$$

where A_o and A_f are the original and final areas respectively, as illustrated in Fig. 9–1.3(a).

The copper of Fig. 9–3.1 has been plastically deformed. The deformation shows up as traces of slip planes on a previously polished surface. The effect of plastic deformation also may be revealed by an electron microscope. Figure 9–3.2 shows dislocations in a stainless steel that had been severely cold-worked. The entanglement of lines are dislocations, the numbers and lengths of which increase greatly with additional cold work. As seen in Fig. 9–3.3, the total length of

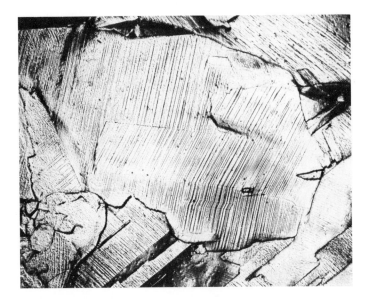

FIG. 9–3.1

Plastically Deformed Polycrystalline Copper (×25). The traces of the slip planes are revealed at the polished surface of the metal. (National Bureau of Standards. Reproduced by permission from B. Rogers, *The Nature of Metals,* 2nd ed., American Society for Metals, and Iowa State University Press, Chapter 13.)

dislocation lines markedly affects the shear stress. We may conclude that *although dislocations account for plastic deformation* (Fig. 8–3.4), *they interfere with the movements of other dislocations.* The dislocation entanglements, or "traffic jams," increase the critical shear stress, τ_C, and therefore the strength of the material.

Data of the type shown in Fig. 9–3.3 are given in a more practical engineering format in Figs. 9–3.4, 9–3.5, and 9–3.6.

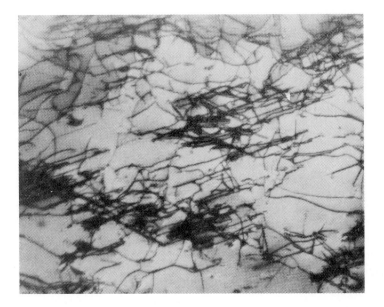

FIG. 9–3.2

Dislocations in Plastically Deformed Metal (Stainless Steel, ×30,000). Additional deformation introduces more dislocations. Also, additional dislocations interfere with further deformation. *Strain hardening* is the result. (Courtesy of M. J. Whelan.)

FIG. 9–3.3

Dislocation Density Versus Critical Shear Stress. An increased number of dislocations provides interference to dislocation movements. (Adapted from Wiedersich, *Journ. of Metals.*)

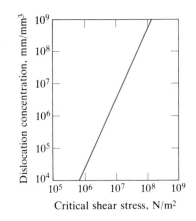

Although strain hardening increases both the yield strength, S_y, and the ultimate strength S_u, it reduces the ductility. Part of the deformation occurred during cold working of the material, before the test bar was made and the gage marks established (Fig. 8–2.2). Thus, less ductility is observed during testing. Cold working increases the yield strength more than it does the ultimate strength (Fig. 9–3.6).

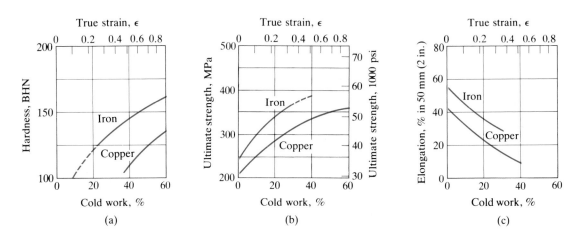

FIG. 9–3.4

Cold Work Versus Mechanical Properties (Iron and Copper). Cold work is the amount of plastic strain, expressed as the decrease in cross-sectional area (Eq. 9–3.1).

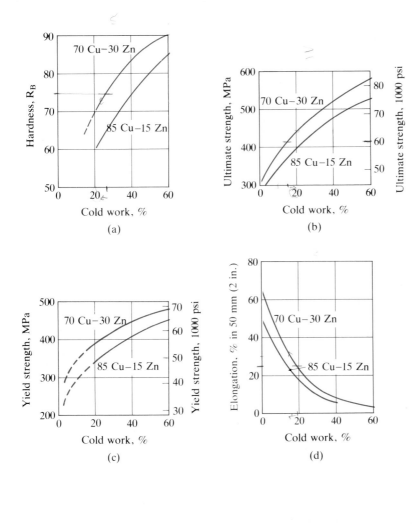

FIG. 9–3.5

Cold Work Versus Mechanical Properties (Brasses).

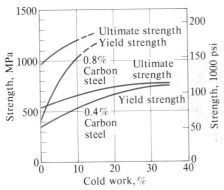

FIG. 9–3.6

Cold Work Versus Strength of Plain-Carbon Steels.

Recrystallization

Recrystallization is the process of growing new crystals from previously deformed crystals. Materials that have been plastically deformed, like those shown in Figs. 9 – 3.1 and 9 – 3.2, have more energy than do unstrained materials because they are loaded with dislocations and point imperfections. Given a chance, the atoms will move to form a more perfect, unstrained array. Such an opportunity arises when the metal is heated, through the process called *annealing*. The greater thermal vibrations of the lattice at higher temperatures permit a reordering of the atoms into less distorted, softer grains. Figure 9 – 3.7 shows the progress of this recrystallization.

Brass was cold worked 33 percent and therefore strain hardened (Fig. 9 – 3.7a). Several samples were heated to 580° C (1080° F) for a few seconds. New grains (i.e., new fcc crystals of brass) are detected in the sample that was heated for only 3

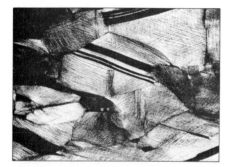

(a) Cold-worked 33%

(b) 3 sec at 580 °C (~ 1080°F)

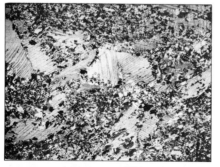

(c) 4 sec at 580°C

(d) 8 sec at 580°C

FIG. 9 – 3.7

Recrystallization of Strain-Hardened Brass (×40). (Courtesy of J. E. Burke, General Electric Co.)

sec (Fig. 9 – 3.7b). The initial recrystallization started along traces of the earlier slip planes. In 4 sec, the brass has been nearly half recrystallized (Fig. 9 – 3.7c), and it is completely recrystallized in 8 sec (Fig. 9 – 3.7d). Were we to examine the new crystals of Fig. 9 – 3.7(d) by an electron microscope, we would observe a greatly reduced number of dislocations. Also, the hardness has dropped significantly from 165 BHN initially to less than 100 BHN.

Recrystallization Temperatures

The recrystallization process requires movements and rearrangements of atoms. These rearrangements for recrystallization occur more readily at high temperatures. In fact, we observe in Fig. 9 – 3.8 that a marked decrease in strength has occurred in a sample held for 1 hr at 300° C (570° F). Initially cold worked 75 percent, this sample was almost completely recrystallized in that period of time. In contrast, samples held for 1 hr at temperatures below 200° C ($\sim$ 400° F) retained almost all their higher strength, which was obtained during the 75 percent cold work. We thus speak of a *recrystallization temperature, T_R*, in this case approximately 270° C, where the microstructure and strength change drastically.*

The temperature for recrystallization is dictated by several factors. A ball-park figure is that T_R lies between three-tenths and six-tenths of the absolute, K, melting

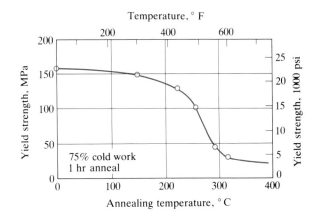

FIG. 9–3.8

Yield Strength Versus Recrystallization (Aluminum). Initially cold-worked 75 percent, the metal was reheated to the indicated temperatures for 1 hr. This was enough time to recrystallize the metal at 300° C and above. The yield strength decreases and the strain hardening disappears with the development of the new grains. (See Fig. 9–3.7.) (Adapted from *Aluminum*, Vol. 1, American Society for Metals.)

---

* There is a slight loss of strength and a major recovery of electrical conductivity at temperatures slightly below the recrystallization temperature. Called *recovery*, these lower-temperature changes come about because the point imperfections (vacancies, interstitials, etc., that were introduced by cold work) move to dislocation edges within the strained crystals. The point imperfections do not affect deformation significantly; therefore, only slight softening occurs during this recovery stage. The mean free paths of the electrons are significantly lengthened by the disappearance of the point defects; therefore, the resistivity drops with their removal (*Chapter 11*).

temperature; that is, 0.3 T_m to 0.6 T_m. The rationale for this generality is that the diffusivity D for self-diffusion is directly related to the melting temperature for the metal.* However, *time* is a second factor that affects recrystallization temperatures. For example, the recrystallization of a commercially pure aluminum alloy that had been cold worked 75 percent may be completed in 1 min at 350° C (623 K and 662° F), but requires 60 min at 300° C and 40 days at 230° C. We expect this difference, because the flux of atoms is proportional to the diffusivity (Eq. 6–5.1a), which in turn is a function of temperature (Eq. 6–5.4).

A third factor that affects the recrystallization temperature is the amount of *strain hardening.* As shown by the hardness data in Fig. 9–3.9, the recrystallization temperature drops from above 320° C (> 600° F) for a 65–35 brass with 20 percent cold work to approximately 280° C (~535° F) for the same brass with 60

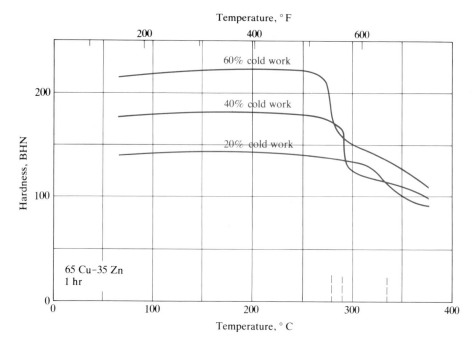

FIG. 9–3.9

Softening During Annealing (65 Cu–35 Zn Brass). The hardnesses were determined at 20° C after heating to the indicated temperatures for 1 hr. The more highly strain-hardened brass softens at a lower temperature and with less thermal energy. (ASM data.)

---

* A solution of Problem 657 gives diffusivity values at 0.6 T_m for iron, copper, and silver of $10^{-17.5}$, $10^{-17.3}$, and $10^{-17.4}$ m²/s, respectively. (Each of these calculations was for fcc structures; therefore, they can be compared directly.)

percent cold work (each with 1 hr of annealing time). An explanation takes cognizance of the fact that a highly strain-hardened metal has more stored energy in the form of vacancies and dislocations (Fig. 4–1.3) than does one with little cold work. With this energy available, it takes less thermal energy for atoms to rearrange themselves into an annealed grain; that is, recrystallization can occur at lower temperatures.

Finally, *pure* materials recrystallize at temperatures lower than those for solid solutions. Thus, highly pure electrical copper wire anneals much more readily than does a comparably deformed brass wire, even though the melting temperature would predict a small opposite relationship. We shall not present the rationale, which involves dislocation movements.

Recrystallization Rates

The kinetics of recrystallization typically follows an S-shaped, or *sigmoidal,* curve. For example, the 3.25 percent Si electrical steel of Fig. 9–3.10 shows undetectable recrystallization in the first 100 sec at 600° C, but is 50 percent recrystallized in 20 min (1200 sec). More than 2 hr ($\sim 10^4$ sec) are required for completion at this temperature.

Recrystallization proceeds similarly at the other temperatures of Fig. 9–3.10, but with different time frames. Observe that the 50 percent recrystallization time is relatively easy to identify, because the reaction is most rapid at that point. When we relate the time of that reaction point to the recrystallization temperature, we observe an Arrhenius behavior (Fig. 9–3.10b). This is expected, because atom movements govern the reaction, and they, in turn, are dependent on thermal activation.

The mathematical relationship in Fig. 9–3.10(b) is

$$\ln t = C + B/T \qquad (9\text{–}3.2)$$

where C and B are constants. We can relate this equation to Eq. 6–5.5 if we recognize that a fast reaction rate R (a short time, since $R = 1/t$) corresponds to rapid diffusion. Thus,

$$\ln R = \ln R_0 - E/kT = -\ln t \qquad (9\text{–}3.3)$$

where C and B of Eq. (9–3.2) are $-\ln R_0$ and E/k, respectively.

Processing Strain-Hardenable Materials

By definition, strain hardening must involve a change in shape. It also involves an input of mechanical energy — work. Hence, we refer to cold work when the shaping process is performed at ambient temperatures. Metals also can be hot worked, in which plastic deformation is performed at elevated temperatures. However, the distinction between cold working and hot working rests not on temperature alone, but rather on the relationship of the processing temperature to the recrystallization

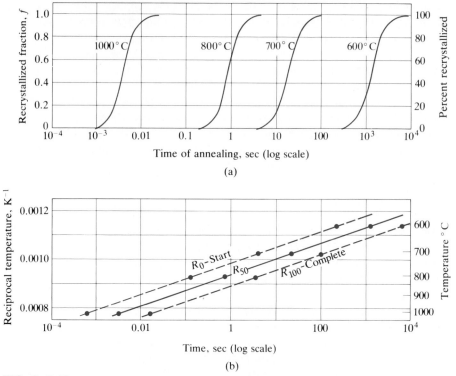

FIG. 9–3.10

Recrystallization Time (3.25 Percent Si, Electrical Steel). (a) The rate increases (time shortens) at higher temperatures. (Adapted from data by G. R. Speich and R. M. Fisher, U.S. Steel Corp.) (b) Arrhenius plot. R_0: start of recrystallization; R_{50}: 50 percent recrystallized; R_{100}: recrystallization completed. (R_{50} usually is the more precise value, since R_0 and R_{100} are asymptotes.)

temperature. *Hot working* includes shaping processes that are performed above the recrystallization temperature; *cold working* includes shaping processes that are performed below that temperature. Thus, the temperature for cold working of copper may be higher than that for hot working of lead.

The choice of the recrystallization temperature as the point for distinguishing between hot and cold working is quite logical from the production point of view. Below the recrystallization temperature, the metal becomes harder and less ductile, with additional deformation during processing. More power is required for deformation and there is a greater chance for cracking during the process. Above the recrystallization temperature, the metal will anneal itself during, or immediately after, the mechanical working process. Thus, the metal remains soft and relatively ductile during processing.

Strain hardening by cold work is of prime importance to the design engineer. It permits the use of smaller parts with greater strength. Of course, the product must not be used at temperatures that will anneal the metals.

Cold work reduces the amount of plastic deformation that a metal can undergo subsequently during a shaping operation. The hardened, less ductile, cold-worked metal requires more power for further working and is subject to cracking. Therefore, *cold work–anneal cycles* are used to assist production. Example 9–3.3 describes such a process.

The loss of ductility during cold working has a useful side effect in machining. With less ductility, the chips break more readily (Fig. 9–3.11), thus facilitating the cutting operation. As a result, the engineer may commonly specify cold working for "screw-stock" metal; that is, metal that requires cutting in automatic screw machines.

Example 9–3.1

An annealed iron rod was originally 12.3 mm (0.48 in.) in diameter. It is cold worked by drawing it through a die with a hole that is 10.4 mm (0.41 in.) in diameter. (a) What was the ductility of the rod before drawing? (b) What is it after drawing?

Procedure An annealed metal has 0 percent cold work. Calculate the cold work during drawing from the change in diameter. In both cases, obtain the ductility from Fig. 9–3.4(c).

Calculations

(a) 55% elongation (50 mm)

(b) cold work $= \dfrac{(\pi/4)(12.3)^2 - (\pi/4)(10.4)^2}{(\pi/4)(12.3)^2} = 0.285$ (or 28.5%)

From Fig. 9–3.4(c),

28% elongation (50 mm)

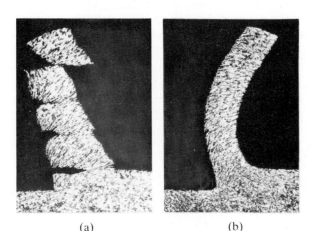

(a)   (b)

FIG. 9–3.11

The Cutting of Metal Turnings by a Machine Tool. The strain-hardened metal in (a) formed the more desirable chips, whereas the annealed metal in (b) formed continuous turnings. (Courtesy of Hans Ernst, Cincinnati Milacron, Inc.)

Comment A limited amount of extrapolation generally is permissible. Also, we can inter-polate between curves to obtain estimates of data for other alloys—for example, 80 Cu – 20 Zn in Fig. 9 – 3.5, or 90 Cu – 10 Zn, if Figs. 9 – 3.4 and 9 – 3.5 are used together.

Example 9 – 3.2

A cold-worked copper or a brass may be used in an application with the specifications of $S_u \geq 345$ MPa ($\geq 50{,}000$ psi), and a ductility of greater than 20 percent elongation in 50 mm (2 in.). Choose a metal.

Answers The cold-work (CW) data provide the following specification ranges:

	COPPER	BRASS (85 – 15)	BRASS (70 – 30)
Figure	9 – 3.4	9 – 3.5	9 – 3.5
Ultimate strength	$\geq 45\%$ CW	$\geq 9\%$ CW	$\geq 4\%$ CW
Ductility (% Elong.)	$\leq 22\%$ CW	$\leq 16\%$ CW	$\leq 22\%$ CW
Range	—	9 – 16% CW	4 – 22% CW

Comments The design engineer has a choice of brasses. Other factors being equal, he would choose the higher-zinc brass (70 Cu – 30 Zn), because zinc is less expensive than is copper. Furthermore, the specification range of 4 to 22 percent cold work gives more flexibility in processing the metal.

Example 9 – 3.3

A 70 – 30 brass rod is required to have a diameter of 5.0 mm (0.197 in.), an ultimate strength of more than 420 MPa (61,000 psi), and a 50-mm elongation of more than 18 percent. The rod is to be drawn from 9.0-mm (0.355-in.) diameter stock that had been previously annealed. Specify the final processing steps for making the 5.0-mm rod.

Procedure To "meet the spec," there must be at least 16 percent cold work for S_u, and no more than 24 percent cold work for ductility (Figs. 9 – 3.5b and d). We can use 20 percent. The *last* step should use an annealed bar, diameter $= d$, that has its area reduced 20 percent to diameter $= 5.0$ mm.

Calculation

$$CW = 0.20 = \frac{d^2\pi/4 - (5.0)^2\pi/4}{d^2\pi/4}$$

$$d = 5.6 \text{ mm} \hspace{4cm} (=0.22 \text{ in.})$$

Comments Hot work from 9.0 mm to 5.6 mm (or cold work and anneal in one or more cycles). The rod should be in the annealed condition at 5.6-mm diameter. Cold draw 20 percent to 5.0-mm diameter. [*Note:* The elongation data in the figures of this chapter are for the standard 50-mm test bar of Fig. 8–2.2, in which the diameter is 0.505 in. (12.8 mm). Since the necking before breaking (Fig. 8–2.2) is a function of the diameter, ductility values for wires normally require a correction.]

Example 9–3.4

The time–temperature relationship for the completed recrystallization of the aluminum in Fig. 9–3.12 follows an Arrhenius-type relationship. Establish an appropriate empirical equation.

Procedure The Arrhenius relationship is

$$\ln t = C + B/T \tag{9-3.4}$$

We need to determine the constants C and B. Pick t and T for points near the two ends of the R_f curve in Fig. 9–3.12. Solve simultaneously for C and B.

Calculations

$$T = 250°\,C \qquad t = 200\ hr \qquad \ln 200 = C + B/523$$
$$T = 327°\,C \qquad t = 0.14\ hr \qquad \ln 0.14 = C + B/600$$
$$C = -52 \qquad B = 30{,}000\ K$$
$$\ln t = -52 + 30{,}000\ K/T$$

Comment Just as the various diffusion couples of Table 6–5.2 had their own values of D_0 and E, these values of C and B apply to only *this* commercially pure aluminum that had been cold worked 75 percent.

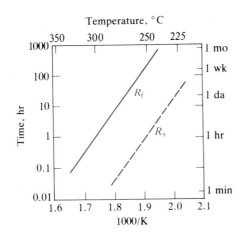

FIG. 9–3.12

Recrystallization Time Versus Temperature (Aluminum, 75 Percent Cold Worked). Dashed line: start of recrystallization, R_s. Solid line: recrystallization finished (Example 9–3.4). An Arrhenius relationship ($\ln t$ versus $1/T$) is followed because recrystallization requires atom movements. (Adapted from *Aluminum,* Vol. 1, American Society for Metals.)

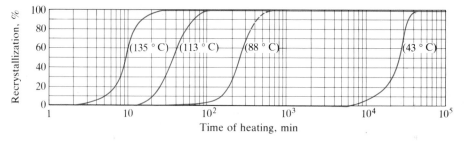

FIG. 9–3.13

Isothermal Recrystallization (99.999 Percent Pure Copper, Cold-Rolled 98 Percent). Refer to Problem 939. (Adapted from A. G. Guy and J. J. Hren, *Elements of Physical Metallurgy,* Addison-Wesley, after Decker and Harker.)

9–4
PRECIPITATION HARDENING

A significant increase in hardness may develop during the initial stages of precipitation from a supersaturated solid solution. This strengthening is controllable, and therefore has engineering significance.

Recall from our discussion in Section 7–2 that low-temperature precipitation occurs throughout the interior of the parent grain (Fig. 7–2.3c). Also recall that these microstructural changes are delayed and may require extended periods. Thus, this hardening that accompanies this *start* of precipitation is commonly called *age hardening.* We use the terms *precipitation hardening* and *age hardening* interchangeably.

Age Hardening

The prime requirement for an alloy that is to be age hardened is that solubility decrease with decreasing temperature, so that a supersaturated solid solution can be obtained, as shown in Figs. 7–2.2(b) and 9–4.1(a). Numerous metal alloys have this characteristic.

The process of age hardening involves a *solution treatment* followed by a *quench* to supersaturate the solid solution. Usually, the quenching is carried to a temperature where the precipitation rate is exceedingly slow. After the quench, the alloy is reheated to an intermediate temperature* at which precipitation is initiated in a reasonable length of time. These are the two steps, *XA* and *AB,* in Fig. 9–4.1 and in Table 9–4.1.

---

* But below the "knee" of the C-curve (Fig. 7–2.2a), to produce *intra*grain precipitation (Figs. 7–2.3(c) and 9–4.1"B").

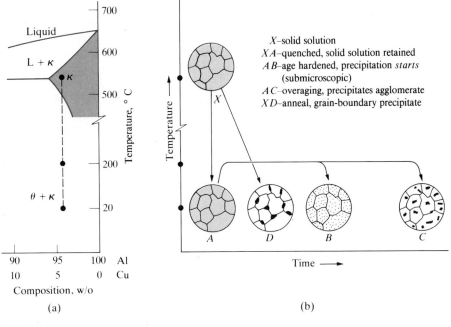

FIG. 9–4.1

Age-Hardening Process (96 Percent Al–4 Percent Cu Alloy). See Table 9–4.1. The precipitates are still submicroscopic at the time of maximum hardness.

Observe the enhanced properties for the age-hardened alloy *(XAB)* in Table 9–4.1, as compared to annealing of the same alloy *(XD)*. The former has several times the yield strength of the annealed material and, at the same time, possesses greater ductility. As a result, the toughness is increased markedly by age hardening. Greatest ductility is obtained when only one phase is present—that is, after solution treatment (plus quenching to preserve the single phase).

TABLE 9–4.1 Properties of an Age-Hardenable Alloy (96 Al–4 Cu)

	TREATMENT (SEE FIG. 9–4.1)	ULTIMATE STRENGTH		YIELD STRENGTH		DUCTILITY, % IN 5 cm (2 in.)
		MPa	(psi)	MPa	(psi)	
XA	Solution treated (540° C) and quenched (20° C)	240	(35,000)	105	(15,000)	40
XAB	Age hardened (200° C, 1 hr)	415	(60,000)	310	(45,000)	20
XAC	Overaged	~170	(25,000)	~70	(10,000)	~20
XD	Annealed (540° C)	170	(25,000)	70	(10,000)	15

An interesting example of the utility of the age-hardening process is the way it is used in airplane construction. Aluminum rivets are easier to drive and fit more tightly if they are soft and ductile, but in this condition they lack the desired strength. Therefore, the manufacturer selects an aluminum alloy that can be quenched as a supersaturated solution, but that will age harden at room temperature. The rivets are inserted while they are still relatively soft and ductile, and they harden further after they have been riveted in place. Since hardening sets in fairly rapidly at room temperature, there arises the practical problem of delaying the hardening process if the rivets are not to be used almost immediately after the solution treatment. Here, the engineer can take advantage of the known effects of temperature on the reaction rate. After the solution treatment, the rivets are stored in a refrigerator, where the lower temperature will delay hardening for reasonable lengths of time.

Detailed studies have produced the following interpretation of the age-hardening phenomenon. The supersaturated atoms (Cu atoms in Example 7–2.2 and in Fig. 9–4.1"B") tend to accumulate along specific crystal planes in the manner indicated in Fig. 9–4.2(b). The concentration of the copper (solute) atoms in these positions lowers the concentrations in other locations, producing less supersaturation and therefore a more stable crystal structure. At this stage, the copper atoms have not formed a phase that is wholly distinct; a *coherency* of atom spacing exists across the boundary of the two structures. Dislocation movements proceed with difficulty across these distorted regions; consequently, the metal becomes harder and more resistant to deformation under high stresses.

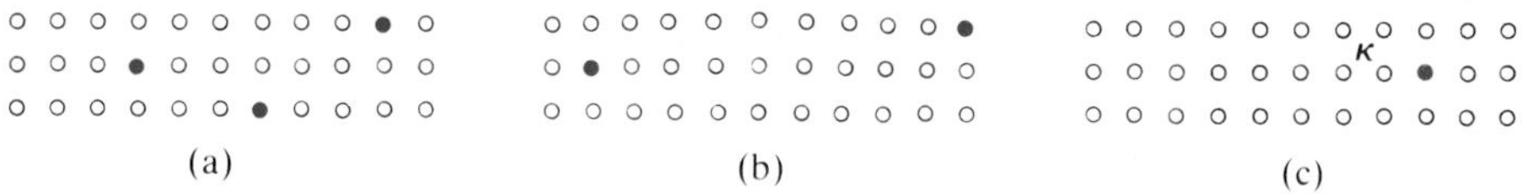

FIG. 9–4.2

Age-Hardening Mechanism. (a) κ solid solution. (b) Age-hardened; the θ precipitation has been initiated. Since the two structures are coherent at this stage, there is a stress field around the precipitate. (c) Overaged. There are two distinct and noncoherent phases, κ and θ. With limited numbers of solute atoms, maximum interference to dislocation movements occurs in part (b). (A. G. Guy and J. J. Hren, *Elements of Physical Metallurgy,* Addison-Wesley.)

Overaging

A continuation of the local segregation process over long periods leads to true precipitation and to *overaging,* or softening. For example, the development of a truly stable structure in an alloy of 96 percent aluminum and 4 percent copper involves an almost complete separation of the copper from the fcc aluminum at room temperature. According to Fig. 5–6.3, nearly all the copper forms $CuAl_2$ (θ in Fig. 9–4.2c). Because the growth of the second phase provides larger areas that have practically no means of slip resistance, a marked softening occurs.

Figure 9–4.3 shows data for the aging and overaging of a commercial aluminum alloy (2014). The initial hardening is followed by softening as the resulting precipitate is agglomerated. Two effects of the aging temperature may be observed: (1) precipitation, and therefore hardening, starts very quickly at higher temperatures; (2) overaging, and therefore softening, occurs more rapidly at higher temperatures. These two phenomena overlap to affect the maximum hardness that is attained. Lower temperatures permit greater increases in hardness, but longer times are required.

Combined Hardening

Occasionally, it is desirable to combine two methods of hardening. The cold working of an alloy that has previously been age hardened increases the hardness still further. However, practical difficulties are encountered in this process. Age hardening increases resistance to slip and therefore increases the energy required for cold working; it also decreases ductility, so rupture occurs more readily during

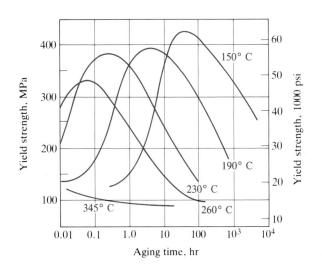

FIG. 9–4.3

Aging and Overaging (2014 Aluminum). Softening occurs as the precipitated particles grow. This proceeds more rapidly at elevated temperatures. (*Aluminum,* American Society for Metals.)

cold working. A possible alternative is to cold work the metal prior to the precipitation-hardening treatment. The metal is cold worked more readily, and the age-
hardening reaction occurs at a lower temperature, because the dislocations serve as
nuclei for the precipitation. However, the temperature of the aging process that
follows cold working may relieve some of the strain hardening and cause a slight
loss in hardness. Although this process does not produce hardnesses as great as
those obtained from one of the reverse order, the final hardness is greater than that
developed by using either method alone (Table 9–4.2).

Example 9–4.1

Use information in Fig. 9–4.3 to estimate the temperature required to reach the maximum
hardness for that aluminum alloy in 10,000 hr ($\sim$14 months).

Procedure As they are in recrystallization, atom movements are involved. Therefore,
assume that the Arrhenius relationship applies (Eq. 9–3.2). Use the peaks of the 150° C
curve (30 hr) and the 260° C curve (3 min or 0.05 hr). (Interpolation must be on a log basis.)

Calculation At 150° C,

$$\ln t = \quad \ln 30 = C + B/423 = 3.4$$

At 260° C,

$$\ln t = \ln (0.05) = C + B/533 = -3.0$$

Solving simultaneously,

$$C = -27.6 \quad B = 13,100 \text{ K}$$
$$\ln 10^4 \text{ hr} = 9.21 = -27.6 + 13,100/T$$
$$T = 356 \text{ K} = 83° \text{ C}$$

TABLE 9–4.2 Ultimate Strengths of a Strain- and Age-Hardened Alloy
(98 Cu–2 Be)

Annealed, 870° C	240 MPa	35,000 psi
Solution treated, 870° C and cooled rapidly	500	72,000
Age hardened only	1200	175,000
Cold worked only (37%)	740	107,000
Age hardened, then cold worked*	1380	200,000
Cold worked, then age hardened	1340	195,000

* Cracked

Comments We can check our Arrhenius assumption at 230° C (503 K) and 190° C (463 K):

$$\ln t_{230} = -27.6 + 13{,}100/503 = -1.56, \ t = 0.2 \text{ hr}$$
$$\ln t_{190} = -27.6 + 13{,}100/463 = 0.69, \ t = 2 \text{ hr}$$

The agreement with the experimental data in Fig. 9–4.3 is not exact, but it is reasonable; however, an extrapolation to still longer times or lower temperatures is an approximation.

9–5
SECOND-PHASE STRENGTHENING

It is easy for us to appreciate that the presence of a second, minor phase can increase the strength of a material, since we are well aware that sand will stiffen the asphalt on a roadway. Each of us can think of other such examples, such as adding fillers to plastics. However, we should not generalize from those situations, because we also are aware that the addition of small amounts of water will soften and plasticize clay.* We must look at the stress-bearing interactions of two phases more closely. In this section, we shall direct our attention to those situations where a second phase can strengthen a material. Using this knowledge, we can prescribe processing that will lead to the enhancement of mechanical properties.

Carbide-Containing Steels

Annealed plain-carbon steels provide an example. They contain a soft, ductile ferrite (bcc iron) and a hard, brittle carbide, Fe_3C. Figure 9–5.1 shows the increase in hardness and strengths with an increase in carbon contents of the steels. (Since the carbon is present as Fe_3C, and this carbide exists within the pearlite, alternative scales are presented on the abscissa.) While the carbide increases the hardness and strength (Fig. 9–5.1a), ductility and toughness decrease (Fig. 9–5.1c).

Phase shape and *phase distribution* take many forms and have a major influence on mechanical properties. We shall use only one example, again looking at steels. Parts of the two microstructures from Fig. 7–4.2 are repeated in Fig. 9–5.2. The two samples came from the same original source, but received different heat treatments. The resulting property differences are pronounced, as we can see in Fig. 9–5.2(c and d). The rigid lamellar carbides reinforce the soft, ductile ferrite more than do the dispersed carbides of spheroidite, which accounts for the greater strength of the pearlite-containing steels. The carbide lamellae also decrease the ductility and lead to a loss in toughness. A crack can propagate readily along brittle carbide layers of the pearlite by making short jumps through any intervening

---

* See Section 10–2 for a discussion of *fillers* and *plasticizers* in polymeric materials.

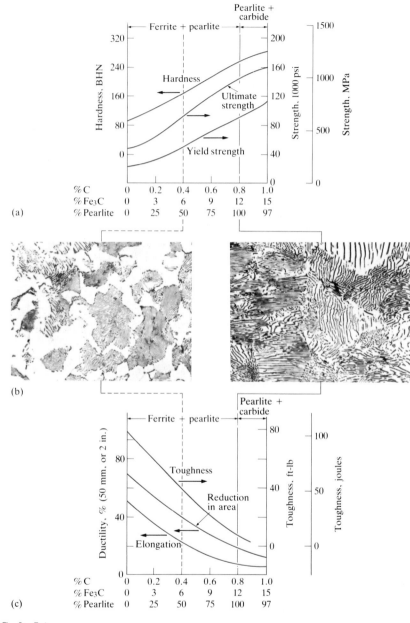

FIG. 9-5.1

Properties Versus Microstructure (Annealed, Plain-Carbon Steels). (a) Hardness and strength versus the amount of carbon, Fe₃C, and pearlite. (b) Microstructure of 0.40 percent C (left) and 0.80 percent C (right) annealed steels (×500). (Courtesy of U.S. Steel Corp.) (c) Ductility and toughness versus the amount of carbon, Fe₃C, and pearlite.

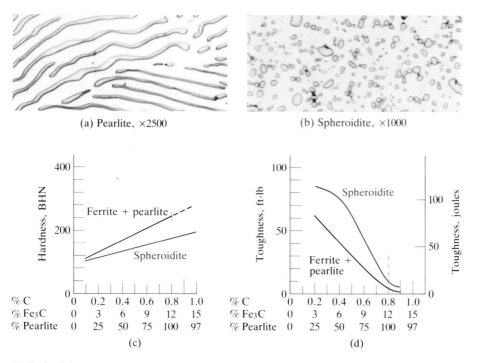

(a) Pearlite, ×2500

(b) Spheroidite, ×1000

(c)

(d)

FIG. 9–5.2

Properties Versus Phase Distribution (Plain-Carbon Steels). (a) Pearlite (0.8 percent C). (b) Spheroidite (0.8 percent C). (c) Hardness. (d) Toughness. (Photomicrographs courtesy of U.S. Steel Corp.)

tough ferrite. In the spheroidite, any fracture must progress through the tough ferrite matrix for a significant percentage of its path. Thus, more energy is required for fracture and greater toughness is realized.

The *dimensions* of the second phase also influence the hardnesses and strengths of a material. Just as a fine sand stiffens an asphalt more than does an equal volume fraction of coarse sand, a steel with a fine pearlite structure is harder than is steel with a coarse pearlite (Fig. 9 – 5.3). (As we discussed in Section 7 – 3, we can control the thickness of the pearlite lamellae by adjusting the rate of cooling, as well as by adjusting the temperature of the austenite decomposition.)

Plastic Constraint

For strengthening to occur, the second phase must be more rigid than the major phase. As such, the rigid phase is able to prevent deformation in the softer material. Let us illustrate by a simple experiment. Identically prepared test bars of a steel and of a lead-tin solder are prepared. The materials have yield strengths of 300 and 30

FIG. 9–5.3

Effect of Microstructural Dimensions on Hardness of Steel. The harder, finer pearlite was formed by faster cooling.

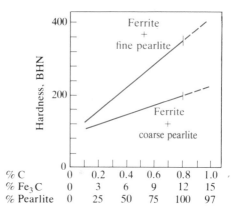

% C	0	0.2	0.4	0.6	0.8	1.0
% Fe$_3$C	0	3	6	9	12	15
% Pearlite	0	25	50	75	100	97

MPa, respectively. Their corresponding ultimate strengths are 400 and 40 MPa. A second identical test bar of steel is cut in half at its midpoint, and then is rejoined by a layer of the solder. This reconstituted test bar is stronger than was the original solder—easily 70 MPa (10,000 psi). If the solder joint is very thin, such a bar could maintain a stress of as much as 140 MPa (20,000 psi). It would appear that, in some cases, a chain is *stronger* than its weakest link!

We can explain this strengthening phenomenon on the basis of *plastic constraint.* The more rigid steel prevents the adjacent ductile solder from deforming. The solder is constrained at the interface between the two materials. Admittedly, away from the interface, some plastic deformation can occur. For that reason, a thick solder joint is not as strong as is a thin one.*

The amount of plastic constraint that is available for strengthening is directly proportional to the interface area provided by the rigid phase. Thus, an increase in the carbide content of the steels in Fig. 9–5.1 increases the strengths and hardnesses. Pearlite (Fig. 9–5.2a), with its lamellar carbides, is stronger and harder than is spheroidite, in which the interfacial area has been minimized (Section 7–4). The fine pearlite of Fig. 9–5.3 has many layers of carbide, and therefore has a large carbide-ferrite interfacial area. This increases the hardnesses as shown in Fig. 9–5.4. Softening accompanies the coalescence of carbides or other hard phases into fewer, but larger, particles within the microstructure, as shown by the data in Fig. 7–4.1.

Control of these microstructures is the means by which we modify steels to attain the desired properties.

---

* This same principle holds for adhesive bonding. Assuming that there is complete coverage and that other factors are equal, a thin adhesive layer will provide a stronger bond than will a thick one.

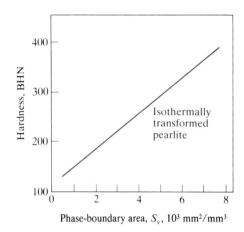

FIG. 9–5.4

Strength Versus Microstructure (1080 Pearlitic Steel). The steel is harder with more ferrite–carbide boundary area (thinner lamellae). (Courtesy of F. Rhines, *Metal Progress*.)

Phase-boundary area, S_v, 10^3 mm^2/mm^3

9–6
HEAT TREATMENTS OF STEELS

///*OPTIONAL*///

Since austenite can decompose in several different ways, the engineer has a choice of different microstructures, and therefore a variety of properties, for steel products. These are summarized in Table 9–6.1. The resulting properties vary significantly.

TABLE 9–6.1 Transformation Processes for Steels*

PROCESS	PURPOSE	PROCEDURE	PHASE(S)
Annealing	To soften	Slow cool from γ-stable range	α + carbide
Quenching	To harden	Quench to miss I–T curves	Martensite†
Interrupted quenching	To harden without cracking	Quench, followed by slow cool from M_s to M_f	Martensite†
Austempering	To harden without forming brittle martensite	Quench, followed by isothermal transformation above the M_s	α + carbide
Tempering	To toughen (usually with minimal softening)	Reheating of martensite	α + carbide

* See Fig. 9–6.8.

† Steels containing martensite must be toughened by the tempering process.

Annealing Processes

The term *annealing* originally was used by craftspeople who discovered the benefits of heating some materials to elevated temperatures, then cooling the materials slowly (as opposed to quenching them).

The structural changes that occur during annealing are not the same for all materials. When glass is annealed, the glass-transition temperature is exceeded slightly, so residual strains are relieved (Fig. 4–2.4 and Section 9–8) and the possibility of fracture is lessened. Only slight structural changes occur; the hardness of the product is not changed. We anneal cold-worked brass to soften it and to reestablish its ductility. We now know that the brass is recrystallized during the annealing process (Section 9–3). The same results are achieved when cold-drawn steel wire or cold-rolled steel sheet is *process annealed*. The *full anneal* of a steel for the machining of a gear blank, however, has different results. In this case, the ductility may be reduced, with the consequence that chips form more readily, facilitating machining.

Process Anneal Products formed by cold working are inherently small in cross-section.* As a practical matter, this annealing process is limited to subeutectoid temperatures of steels. If higher annealing temperatures were used, austenite would form. Because of the very large surface-to-volume ratio of wire and sheet products, cooling could be rapid enough to form the brittle martensite discussed in Section 7–4 (Eq. 7–4.1). Of course, the desired ductility would not be realized. The temperature limitation for process annealing is not a problem, since the annealing times below 727° C (1340° F) are not excessively long.

Full Anneal The *full anneal* process normally is used for products that are to be machined subsequently, such as transmission gear blanks. Thus, the steels typically have medium to high carbon contents (0.35 to 0.65 percent carbon). The steel is first austenitized; then furnace cooled. The temperature of austenization varies with the carbon content, as shown in Fig. 9–6.1. The 25 to 30° C (50° F) margin above the so-called A_{C3} line is used to ensure complete austenization in the center, as well as at the surface, of the product. *Furnace cooling* is slow cooling; it produces very coarse pearlite that is as soft as possible, with a minimum of ductility. Typically, the product receives additional heat treatments after machining to restore hardness and strength.

Normalizing A third annealing process used for steels is *normalizing*. Its purpose is the homogenization of the alloy steels (see Section 6–2.) Austenization is performed at approximately 50 to 60° C (100° F) above the A_{C3} line (Fig. 9–6.1). The purpose is to accelerate the diffusion required for the homogenization of the

---

* It would require impossibly heavy equipment and energy expenditure to cold roll a structural I-beam!

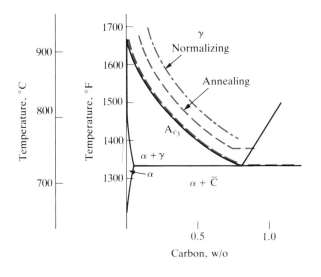

FIG. 9-6.1

Annealing and Normalizing (Plain-Carbon Steels). The heat-treating temperature varies with the carbon content. For annealing, the temperature is sufficiently high to ensure complete disappearance of ferrite. The steel is then cooled slowly to form a coarse pearlite and a relatively soft product. For normalizing, the steel is heated to a somewhat higher temperature to promote more rapid atom diffusion and micro-structural uniformity (only austenite). Excessive heating, however, would permit undesirable grain growth. Following austenization, the normalized steel is air-cooled to produce uniformly fine pearlite. Low hardness values are not a primary objective. (See Fig. 9-6.2.)

substitutional atoms (Ni, Cr, Mo, V, etc., for Fe), while avoiding the excessive grain growth that would occur at still higher temperatures. The heating is followed by air cooling.

Stress Relief Quenched steels, and steels that have portions of their surfaces machined, are subject to distortion and possible cracking from the presence of residual stresses. A *stress relief* is necessary to remove these stresses. In contrast to recrystallization and austenization, it is not necessary to relocate large numbers of atoms. Therefore, the necessary temperature is lower, as shown in Fig. 9-6.2, which summarizes the annealing process for steels.

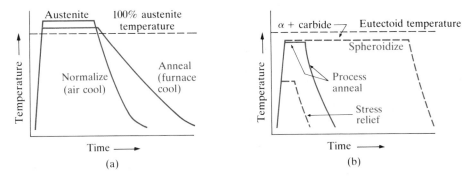

FIG. 9-6.2

Steel Heat-Treatment Processes (Schematic). (a) Austenization processes. (b) Subeutectoid processes.

Quenching and Tempering Processes

In addition to using annealing treatments, we can modify the structure and properties of steels by quenching, followed by tempering. *Quenching* produces martensite, which we described in Section 7–4 as a very hard, but brittle, phase. The subsequent *tempering* of martensite produces a two-phase microstructure of ferrite and finely dispersed carbides ($\alpha + \overline{C}$). This microstructure is called *tempered martensite,* and can be both hard and tough—a favorable combination.

The control of quenching and tempering processes is based on the type of information presented in Fig. 9–6.3. The figure shows the isothermal-transformation (I-T) curve of Fig. 7–3.6, to which M_s and M_f lines have been added. These are the temperatures at which the ($\gamma \rightarrow M$) reaction starts and finishes.*

We know from Eq. (7–4.1) that quenching avoids the ($\gamma \rightarrow \alpha + \overline{C}$) reaction that would be predicted by the phase diagram. From Figs. 9–6.3 and 9–6.4, we can see that quenching must be extremely rapid past the knee of the I-T curve if martensite is to form. In practice, anything larger than "knife blades and screw drivers" cannot be processed out of plain-carbon steels to produce the desired tempered martensite. At best, larger products could develop only a martensitic surface during quenching. (See Section 9–7.)

Low-alloy steels provide a solution to the problem of quenching rates. When alloying elements (Cr, Ni, Mo, Si, Mn, and others) are added to a steel, the rate for the ($\gamma \rightarrow \alpha + \overline{C}$) decomposition is reduced markedly. (Time is increased.) For

FIG. 9–6.3

Isothermal-Transformation Diagram for SAE 1045 Steel. The stable phases of the phase diagram (right) are not achieved immediately. In this steel, however, the austenite decomposition reaction is faster than it is in a eutectoid steel (Fig. 9–6.4).

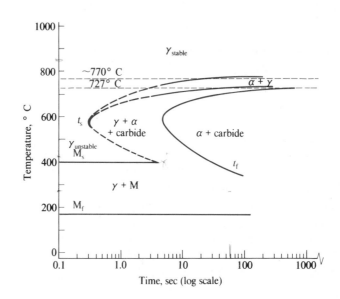

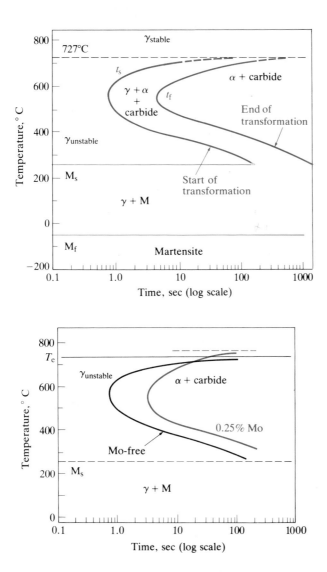

FIG. 9-6.4

Isothermal-Transformation Curves for Austenite Decomposition (SAE 1080). (Adapted from U.S. Steel Corp. data.)

FIG. 9-6.5

Transformation Retardation. Molybdenum, like other alloying elements, retards the *start* of transformation of austenite. (*Note:* The addition of Mo also raises the eutectoid temperature, T_e.)

example, only 0.25 percent molybdenum extends the available time for the $(\gamma \rightarrow M)$ reaction by a factor of four (Fig. 9-6.5), and it is possible to increase the alloy content to higher concentrations as required (Table 9-6.2).* Thus, the engineer can specify steels that contain tempered martensite for large gears (Fig. 9-6.6), die blocks, aircraft landing struts, and so on.

---

* Of course, alloys are expensive additions to steels. In most cases, it is more effective to add small amounts of several alloying elements, rather than large amounts of one. Most low-alloy steels receive quench-and-tempering processing.

TABLE 9–6.2 Nomenclature for AISI and SAE Steels

AISI OR SAE NUMBER*	COMPOSITION	UNS**
10xx	Plain-carbon steels†	G10xx0
11xx	Plain-carbon steels (resulfurized for machinability)	G11xx0
15xx	Manganese (1.0–2.0%)	G15xx0
40xx	Molybdenum (0.20–0.30%)	G40xx0
41xx	Chromium (0.40–1.20%), molybdenum (0.08–0.25%)	G41xx0
43xx	Nickel (1.65–2.00%), chromium (0.40–0.90%) molybdenum (0.20–0.30%)	G43xx0
44xx	Molybdenum (0.5%)	G44xx0
46xx	Nickel (1.40–2.00%), molybdenum (0.15–0.30%)	G46xx0
48xx	Nickel (3.25–3.75%), molybdenum (0.20–0.30%)	G48xx0
51xx	Chromium (0.70–1.20%)	G51xx0
61xx	Chromium (0.70–1.10%), vanadium (0.10%)	G61xx0
81xx	Nickel (0.20–0.40%), chromium (0.30–0.55%), molybdenum (0.08–0.15%)	G81xx0
86xx	Nickel (0.30–0.70%), chromium (0.40–0.85%), molybdenum (0.08–0.25%)	G86xx0
87xx	Nickel (0.40–0.70%), chromium (0.40–0.60%), molybdenum (0.20–0.30%)	G87xx0
92xx	Silicon (1.80–2.20%)	G92xx0

* xx: carbon content, 0.xx percent.

** The Unified Numbering System (UNS) is more extensive than are the AISI–SAE numbers, since it encompasses all present *commercial* alloys (currently approaching 10,000). However, the two systems are compatible. Among the alloys of the UNS, plain-carbon and low-alloy steels carry the prefix G, plus a fifth digit for future variants. Thus, AISI–SAE 4017 becomes G40170.

† All plain-carbon steels contain 0.50 percent ± manganese, and residual amounts (<0.05 w/o) of other elements.

Alloying elements slow down the rate of the ($\gamma \rightarrow$ pearlite) reaction because the alloying atoms, like the carbon atoms, must be relocated during the austenite decomposition (see Fig. 7–3.3). In austenite, they are uniformly distributed as substitutional atoms in the fcc structure of iron. In pearlite, these alloying atoms must "choose" between the ferrite or the carbide. Nickel and silicon segregate into the ferrite within the pearlite; chromium, molybdenum, and vanadium segregate into the carbide. Unlike carbon, all these elements have large atoms that diffuse comparatively slowly and defer pearlite formation. (Compare the diffusivities of Mn and Ni with that of C in fcc iron in Fig. 6–5.4. The latter is seven orders of magnitude greater at 600° C.)

FIG. 9–6.6

Quenched and Tempered Gear (Low-Alloy Steel). The $(\gamma \rightarrow \alpha + \overline{C})$ reaction is too rapid in a plain-carbon steel to form martensite in this gear by quenching. Small alloy additions (up to 1 to 2 percent—Table 9–6.2) retard the austenite decomposition and permit martensite formation (Eq. 7–4.1). This is followed by tempering to toughen the steel. (See Fig. 9–6.7.) (Courtesy of Amax Corp.)

Tempering as a process can be scheduled separately in production; once formed, the metastable martensite can be retained indefinitely at ambient temperatures. However, the martensite can originate from only quenched austenite. Therefore, quenching must precede tempering.

The initial carbide particles that form during tempering must be viewed with an electron microscope, since they are in the submicron size range. Particles of that size possess large interphase boundary areas, and therefore accentuate strengthening by plastic constraint. Hardness indices as high as 65 on the R_C scale are realized, if the carbon content is greater than 0.5 percent (Fig. 7–4.4). The carbide particles coalesce slowly at 200° C, as revealed by the slow drop in hardness of the AISI–SAE 1080 (eutectoid) steel in Fig. 9–6.7. Higher temperatures lead to more rapid carbide growth and to accelerated softening. *Overtempering* can result, and usually is not desired.

Interrupted Quenches In Section 3–4 we observed that there is a 1.4 percent volume change as iron transforms from fcc austenite to bcc ferrite (or vice versa). There is a similar change in the transformation from austenite to martensite, which introduces complications in the quenching process. The outside of a steel product expands with the $(\gamma \rightarrow M)$ reaction while the inside is still hot and ductile. Therefore, no stresses are induced. When the austenite in the more slowly cooled center transforms, however, expansion must occur within a rigid, brittle surface shell of martensite. *Quenching cracks* commonly result from the induced stresses.

FIG. 9–6.7

Hardness of Tempered Martensite (SAE 1080 Steel Quenched to 65 R$_C$). Softening occurs as the carbide particles grow (and decrease in number), giving greater intervening ferrite distances (Fig. 7–4.1). This change in microstructure also reduces the interphase-boundary area, and therefore decreases the amount of plastic constraint.

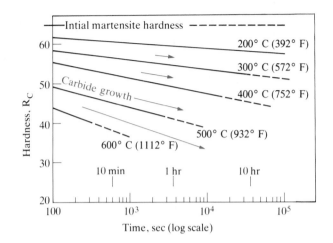

By examining the isothermal-transformation diagram, we can see an alternative to the direct quench. That alternative is an interrupted quench (also called *martempering* or *marquenching*).

In this process, the steel is quenched rapidly past the knee of the transformation curve to avoid $(\alpha + \overline{C})$ formation, but cooling is interrupted just above the M_s temperature. Cooling is then continued at a slow rate through the martensite range to ambient temperatures, so that the surface and the center of the steel can transform more or less simultaneously, thus avoiding quenching cracks. Slower cooling is possible at these lower temperatures because the $(\alpha + \overline{C})$ transformation is delayed, whereas the martensite forms directly with the drop in temperature.

This process is more complicated than is direct quenching from the production viewpoint, because the cooling rate must be shifted from a quench to a "hold," and then to a slow cooling rate. As it is of the earlier direct quench, martensite is the product, and it must be tempered to secure toughness.

Properties

Figure 9–6.8 summarizes the different steel-treating processes. The resulting properties vary significantly, as shown in Table 9–6.3 for an AISI–SAE 4140 steel (0.40 percent C plus 0.4 to 1.2 percent Cr and 0.08 to 0.25 percent Mo). This steel is relatively tough, but only moderately strong, in both the annealed and the spheroidized conditions. Martensite, although very hard and basically strong, lacks toughness; therefore, it may crack readily. Tempered martensite, however, can have high strength *and* good toughness if it is heat treated appropriately. The costs of the heat treatments are justified in many applications.

Of course, there are many other steels—plain-carbon, low-alloy (Table 9–6.2), stainless, and so on. Therefore, an engineer not only has a large number of choices, but also profits from the versatility of each.

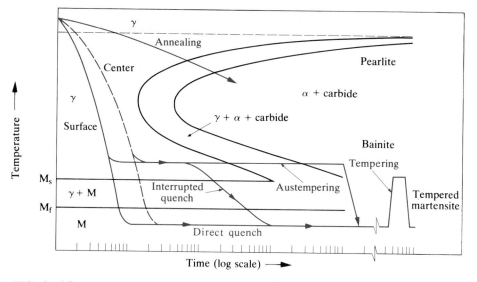

FIG. 9-6.8

Transformation Processes. *Annealing:* The normal $(\gamma \rightarrow \alpha + \overline{C})$ transformation occurs. *Direct quench:* Martensite forms, first in the surface, then in the center. Severe stresses result. *Interrupted quench:* Time is available for the surface and center to transform nearly simultaneously, thus avoiding the quenching cracks that result from direct quenching. *Tempering:* Both the direct and the interrupted quench must be followed by a tempering process to complete the transformation. *Austempering:* Quenching avoids pearlite formation, but the $(\gamma \rightarrow \alpha + \overline{C})$ transformation may still occur above the M_s. The resulting microstructure is bainite (Section 7-3).

TABLE 9-6.3 Effects of Heat Treatments (SAE 4140)

	TENSILE STRENGTH		TOUGHNESS	
MICROSTRUCTURE	MPa	(psi)	J	(ft-lb)
Annealed, $(\alpha + \overline{C})$ Lamellar carbides	655	(95,000)	55	(40)
Spheroidite, $(\alpha + \overline{C})$ Large "spherical" carbides in ferrite matrix	480	(70,000)	110	(80)
Martensite, M	~1400	~(200,000)	<3	<(2)
Tempered martensite, $(\alpha + \overline{C})$ Dispersed carbides in ferrite matrix:				
500° C (930° F), 1 hr	1275	(185,000)	55	(40)
600° C (1110° F), 1 hr	1035	(150,000)	110	(80)

Example 9-6.1

Three AISI–SAE 1045 steel (Table 9–6.2) wires underwent the following thermal steps in the indicated sequences. Give the phases after each step and their approximate chemical analyses. (The symbol # means that the wire was held at that temperature until equilibrium was reached.)

		Time Held
Wire (a)	1) Heated to 820° C (1510° F)	#
	2) Quenched to 560° C (1040° F)	0
	3) Held at 560° C	1 min
	4) Reheated to 820° C	#
Wire (b)	1 and 2) Same as wire (a), (1 and 2)	#
	3) Held at 560° C	1 sec
	4) Quenched to 430° C (805° F)	0
Wire (c)	1) Heated to 730° C (1345° F)	#
	2) Quenched to 430° C (805° F)	0
	3) Quenched to 330° C (625° F)	10 sec
	4) Held longer at 330° C	#

Procedure For equilibrium, use the phase diagrams. For isothermal transformation, use the I-T diagram. A quench does not allow for diffusion; however, γ can change to martensite below the M_s.

Answers

Wire (a) 1) γ 0.45 C–99.55 Fe
2) Same as (1), but austenite is metastable (Fig. 9–6.3)
3) α negligible carbon $\overline{C}$ 6.7 percent carbon
4) γ same as (1)

Wire (b) 3) $\gamma(0.45\%\ C) + [\alpha(\text{negligible carbon}) + \overline{C}(6.7\%\ C)]$
4) Same as (3); $(\alpha + \overline{C})$ will *not* revert to austenite below the eutectoid temperature

Wire (c) 1) $\alpha(0.02\%\ C) + \gamma(0.8\%\ C)$; see the phase diagram.
2) Note that nothing is expected to happen to the ferrite as it cools. The austenite is of eutectoid composition; therefore, we should turn to Fig. 9–6.4 for its transformation. At zero time, unstable γ is present.
3) Still, $\alpha(0.02\%\ C)$ + unstable $\gamma(0.8\%\ C)$
4) $\alpha + \overline{C}(6.7\%\ C)$

Example 9-6.2

A thin (<0.5 mm), 1-g sample of 1045 steel (Fe + 0.45 percent carbon) undergoes the following steps during a heat-treating process. (The symbol # means equilibrium was reached.)

Time Held

	Time Held
1) Heated to 730° C	#
2) Quenched to 550° C	10 sec
3) Quenched to 100° C	0 sec

What phases are present after each step? What are their carbon contents? Approximately how much of each phase is present?

Procedure Follow the same procedure as in Example 9–6.1.

Solution and Comments

1) 0.44 g α(0.02% C)   0.56 g γ(0.8% C)

We now have a mixture of α and γ. On quenching, nothing happens to the ferrite (except that it gets colder). The austenite is 1080 (Fe + 0.80 percent carbon) steel on a microscopic scale. Therefore, we must use the I-T diagram for an SAE 1080 (Fig. 9–6.4); we cannot use the 1045 diagram (Fig. 9–6.3).

2) 0.44 g α(0.02% C)
   0.56 g $\gamma \rightarrow$ 0.5 g α(0.02%) + 0.06 g $\overline{C}$(6.7% C)
   total $\alpha = 0.94$ g α

We are now through with Fig. 9–6.4, since the γ-decomposition is complete.

3) 0.94 g α(0.02% C)   0.06 g $\overline{C}$(6.7% C)

The quench to 100° C does not permit the last, small amount (0.02 percent) of carbon to separate from the ferrite.

Example 9–6.3

Two drill rods (4-mm dia.) of 1080 steel (Table 9–6.2) receive the following sequences of treatments:

	(1)	(2)
Austenitize	775° C	775° C
Quench to	275° C	275° C
Held at 275° C for	45 sec	45 min
Cooled to 30° C in	30 sec	30 sec

(a) Indicate the phase(s) that will be present after each step.
(b) What microstructure does each final product have?
(c) What process name does each have?

Procedure Same as for Example 9–6.1.

Answer

(a) Phases

Step	Rod (1)	Rod (2)
Austenitize	γ	γ
Quench	(γ) unstable	(γ) unstable
Held	(γ) unstable	$\alpha + \overline{C}$
Cooled	$M + (\gamma)$	$\alpha + \overline{C}$
(b) Microstructure	M + retained γ	Bainite
(c) Process	Interrupted quench (or marquenching)	Austempering

Comments   The hardness of the martensite for rod (1) is 65 R_C; however, it must be tempered to become toughened. The final product loses some of that hardness (Fig. 9–6.7). When bainite is formed at 275° C (525° F), its hardness is ~50R_C. It does not need to be tempered, because it already contains a fine dispersion of carbides within the ferrite matrix.

Example 9–6.4

There is still 5 v/o austenite present after quenching an AISI–SAE 1080 steel. This retained austenite is present as small residual grains ($\delta \approx 1~\mu$m) within a martensite matrix. What pressure must be overcome within the metal for one of these small grains of this retained austenite to complete its transformation to martensite?

Procedure   This last austenite must expand as it changes to martensite. In Example 7–4.2, we calculated a 1.8 v/o expansion. For calculation purposes, assume that these final γ grains transform to martensite and then are compressed (-1.8 v/o) to the confines of the surrounding martensite. The hydrostatic pressure is proportional to compression—$P_h = K(\Delta V/V)$, as presented in Eq. (8–1.5). In Example 8–1.2, the value of K for steel was calculated to be 162,700 MPa (or 23,600,000 psi).

Calculation

$$P_h = K(\Delta V/V)$$
$$= (162{,}700~\text{MPa})(-0.018)$$
$$= 2930~\text{MPa compression} \qquad\qquad\qquad \text{(or 425,000 psi)}$$

Comments   This calculation assumes that the adjacent metal is absolutely rigid. Actually, it relaxes some and reduces the pressure slightly. Even so, the compressive stresses are sufficient to stop the ($\gamma \rightarrow M$) reaction short of completion.

OPTIONAL   ## 9–7

HARDENABILITY OF STEELS

It is important to distinguish between *hardness* and *hardenability: Hardness* is a measure of resistance to plastic deformation by indentation. *Hardenability* is the "ease" with which hardness may be attained.

Figure 9-7.1 shows the maximum possible *hardnesses* for increasing amounts of carbon in steels; these maximum hardnesses are obtained only when 100 percent martensite is formed. A steel that transforms rapidly from austenite to ferrite plus carbide has low *hardenability* because $(\alpha + \overline{C})$ is formed at the expense of the martensite. Conversely, a steel that transforms very slowly from austenite to ferrite plus carbide has greater hardenability. Hardnesses nearer the maximum can be developed with less severe quenching in a steel of high hardenability, and greater hardnesses can be developed at the center of a piece of steel even though the cooling rate is slower there.

Hardenability Curves

For any given steel, there is a direct and consistent relationship between hardness and cooling rate. The relationship, however, is highly nonlinear. Furthermore, the theoretical bases for quantitative analyses are complex.* Fortunately, it is possible to use a standardized test that lets the engineer make necessary predictions of hardnesses for many applications in a minute or two, and hardness comparisons between steels at a glance. This is the *Jominy end-quench test.* In this test, a round bar with a standard size is heated to form austenite and is then end quenched with a water stream of specified flow rate and pressure, as indicated in Fig. 9-7.2(a). Hardness values along the cooling-rate gradient are determined on a Rockwell hardness tester, and a *hardenability curve* is plotted (Fig. 9-7.2b).

The quenched end is cooled very fast and therefore has the maximum possible hardness for the particular carbon content of the steel that is being tested. The

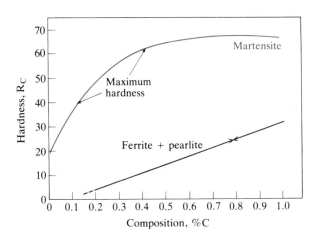

FIG. 9-7.1

Maximum Hardness Versus Carbon Content of Plain-Carbon Steels, Showing Maximum Hardnesses Arising from Martensite Compared with Hardness Developed by Pearlitic Microstructures. To produce maximum hardness, the reaction $(\gamma \rightarrow \alpha + \overline{C})$ must be avoided during quenching.

---

* Variables include each and every alloying element and/or impurity, grain size, and austenitizing temperature. Also, the cooling rates are measured at 700° C. This rate decreases at lower temperatures and approaches zero before cooling is complete.

FIG. 9–7.2

End-Quench (Jominy) Test. (After A. G. Guy and J. J. Hren, *Elements of Physical Metallurgy,* Addison-Wesley.)

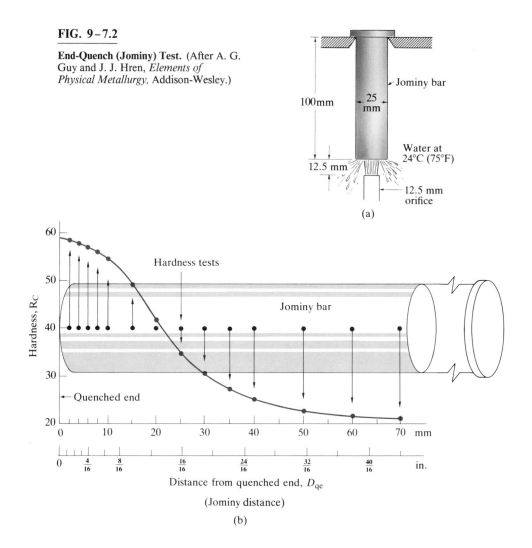

cooling rates at points behind the quenched end are slower (Fig. 9–7.3), and consequently the hardness values are lower (Fig. 9–7.2b). The cooling-rate data of Fig. 9–7.3 generally are valid for all types of plain-carbon and low-alloy steels, since they have comparable values for density, heat capacity, and thermal conductivity — the three properties that affect thermal diffusivity.*

Figure 9–7.4 shows hardenability curves for several common grades of steels. They are plots of hardness versus cooling rates. The rates are shown in ° C/sec on the upper abscissa. In general, however, it is more convenient to use the distance

---

* Stainless-type steels do not follow the pattern shown in Fig. 9–7.3, since their high-alloy contents reduce their thermal conductivities significantly without a comparable effect on density or heat capacity. However, these steels are seldom quenched for hardness requirements.

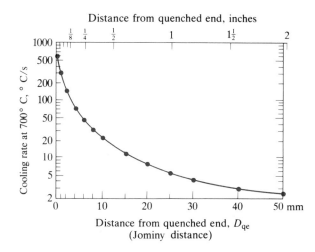

Distance from quenched end, inches

Distance from quenched end, D_{qe}
(Jominy distance)

FIG. 9-7.3

Cooling Rates (at 700° C) Versus the Distance, D_{qe}, from the Quenched End of a Jominy Bar. [Since the cooling rate decreases continuously as the temperature drops, 700° C is selected as a reference temperature for comparisons. It is approximately the eutectoid temperature for most low-alloy steels and therefore is critical for the $(\gamma \rightarrow \alpha + \overline{C})$ reaction.]

from the quenched-end, or D_{qe} (called the *Jominy distance*), because it can be plotted directly from the laboratory data. We will use this simplified procedure.*

We can observe several things from Fig. 9-7.4. The low-alloy steels (SAE 4140 and 4340) have greater hardenability than do the plain-carbon steels (10xx); that is, for a *given cooling rate,* their hardnesses are nearer the maximum possible. Specifically, for a 0.40 percent C steel, the maximum hardness is ~60R$_C$ as indicated in Fig. 9-7.1. At $D_{qe} = 10$ mm (where CR = 25° C/sec), the hardnesses of 4340 and 4140 are 55R$_C$ and 48R$_C$, respectively; the hardness of 1040 steel is only 26R$_C$. Expectedly, higher-carbon steels are harder (1060 versus 1040 versus 1020); this is true with rapid cooling rates ($D_{qe} = 0$ mm) as well as with slow cooling rates ($D_{qe} = 30$ mm).

Use of Hardenability Curves

End-quench hardenability curves are of great practical value because (1) if we know the cooling rate of a steel in any quench, we can read the hardness directly from the hardenability curve for that steel, and (2) if we can measure the hardness at any point, we can obtain the cooling rate at that point from the hardenability curve for that steel.

Figure 9-7.4 presents the end-quench hardenability curve for an AISI–SAE 1040 steel with the grain size and composition indicated.† The quenched end has nearly maximum hardness for 0.40 percent carbon steel because the cooling was

---

* Thus, identical distances from the quenched end on two different hardenability curves have the same specific cooling rates.

† These data apply to this SAE 1040 composition (and grain size). A slight variation is possible in the chemical specifications of any steel (e.g., in an SAE 1040 steel, C = 0.37-0.44, Mn = 0.60-0.90, S = 0.05, P = 0.04, and Si = 0.15-0.25). As a result, two different SAE 1040 steels may have slightly different hardenability curves.

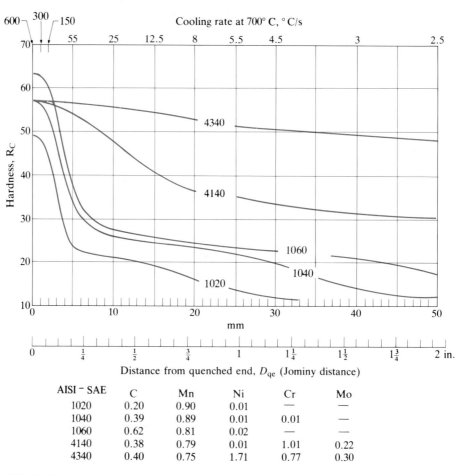

FIG. 9–7.4

Hardenability Curves for Five Steels with the Indicated Compositions (ASTM G.S. #8). The steels were end-quenched as shown in Fig. 9–7.2. In commercial practice, the hardenability curve of each type of steel varies because of small variations in composition. As a result, hardenability tests are commonly made for each heat of steel that is produced for quench-and-temper applications. (Adapted from U.S. Steel Corp. data.)

very rapid and only martensite was formed. However, close behind the quenched end, the cooling rate was not rapid enough to avoid some ferrite and carbide formation, and so maximum hardness was not attained at that point.

In the laboratory, it is also possible to determine the cooling rates within bars of steel. Table 9–7.1, for example, shows the cooling rates at eutectoid temperatures for the surfaces, mid-radii, and centers of 75-mm (~3-in.) rounds quenched in mildly agitated water and oil. These cooling rates were determined by thermocouples embedded in the bars during the quenching operation. Similar data may be obtained for bars of other diameters. These data are summarized in Fig. 9–7.5.

TABLE 9–7.1 Cooling Rates in a Steel Bar 75 mm (~3 in.) in Diameter (at 700° C)

POSITION	AGITATED WATER QUENCH		AGITATED OIL QUENCH	
	° C/s	D_{qe}*	° C/s	D_{qe}*
Surface	~100	3	~20	11
$\frac{3}{4}$-radius	27	9	9.5	18
Mid-radius	14	14	7.5	21
Center	11	17†	5.5	25

* Distance from the quenched end of a Jominy bar that has the same cooling rate at 700° C (Jominy distance).

† Observe that during a water quench the center of a 75-mm diameter bar, which is 37 mm from the surface, cools at the same rate as when $D_{qe} = 17$ mm, and therefore much faster than the steel that is 37 mm from the quenched end of a Jominy bar (11° C/sec versus 3.5° C/sec). Of course, heat is removed radially from the 75-mm bar, but primarily from one end of the Jominy bar.

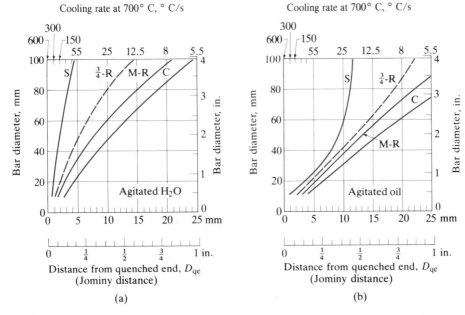

FIG. 9–7.5

Cooling Rates in Round Steel Bars (a) Quenched in Agitated Water, and (b) Agitated Oil. Top abscissa, cooling rates at 700° C; bottom abscissa, equivalent positions on an end-quench test bar. (C, center; M-R, mid-radius; S, surface; Dashed line, approximate curve for $\frac{3}{4}$ radius.) The high heat of vaporization of water produces a severe quench in that quenching medium.

By using the data of Fig. 9–7.5 and a hardenability curve, we can predict the *hardness traverse* that will exist in a steel after quenching. For example, the center of the 75-mm (~ 3-in.) round bar quenched in oil has a cooling rate of 5.5° C/sec. Since the center of this large round bar has the same cooling rate as does a Jominy test bar at a point 25 mm (~ 1 in.) from the quenched end, the hardnesses at the two positions will be the same. Thus, if the bar is AISI–SAE 1040 steel (Fig. 9–7.4), the center hardness will be 23R_C. Figure 9–7.4 also shows that the following center hardnesses may be expected for 75-mm bars of other oil-quenched steels (cooled at 5.5° C/sec):

AISI–SAE:	1040	4140	1020	4340	1060
R_C:	23	36	14	52	24

Several determinations of quenched hardnesses are given in the examples.

Tempered Hardness

The results of Examples 9–7.3 and 9–7.4 are hardnesses of quenched steel. As indicated in Fig. 9–6.7, the hardness decreases with continued tempering because the carbide particles coalesce. The data of that figure are for a plain-carbon eutectoid steel (SAE 1080). Alloy steels temper more slowly.* Data for tempering rates are available in metallurgical books written for use by the engineer who must specify heat-treating processes for specific steels.

Example 9–7.1

Determine the cooling rate for the center and midradius of a 20-mm (0.79-in.) diameter round steel bar when quenched in agitated water. (The bar is long enough that there is no end effect.)

Procedure Use Fig. 9–7.5, which has plots of cooling rates versus bar diameters and radial positions for water quenching (part a) and for oil quenching (part b). Since the cooling rate (CR) decreases with temperature (and therefore time), we commonly specify the rate as the temperature passes 700° C (upper abscissa). Alternatively, and more conveniently, since most CR plots are highly nonlinear, we index the cooling rates to positions on the standardized Jominy end-quench test (bottom abscissa).

---

* For example, a "high-speed" tool steel contains elements, such as V, Cr, W, and Mo, that make very stable carbides. These carbide particles grow and coalesce appreciably more slowly than Fe_3C does in a plain-carbon steel. Thus, tool steels can be used at higher temperatures (higher-speed operation) before overtempering, softening, and consequent destruction of the cutting edge becomes critical.

Example 9–7.4

Figure 9–7.7 shows the points in the cross-section of a V-bar of AISI–SAE 1060 steel in which the following hardness readings were obtained after oil quenching: A—31 R_C, B—30 R_C, C—28 R_C, D—27 R_C, E—26 R_C, F—63 R_C. What hardness values would be expected for an identically shaped bar of AISI–SAE 4068 steel?

Procedure The hardenability curve for the 1060 steel of Fig. 9–7.4 permits us to know the cooling rates at each point. The 4068 steels will have the same cooling rates for the same locations, since the thermal diffusivity for the two steels are comparable, as pointed out in the text. Obtain the 4068 hardnesses for each D_{qe} from the 4068 hardenability curve in Fig. 9–7.8.

Answers

POINT	FROM FIG. 9–7.4		FROM FIG. 9–7.8	
	AISI–SAE 1060	EQUIVALENT COOLING RATE D_{qe}	EQUIVALENT COOLING RATE D_{qe}	AISI–SAE 4068
A	31 R_C	7 mm	7 mm	62 R_C
B	30 R_C	7.5 mm	7.5 mm	61 R_C
C	28 R_C	9 mm	9 mm	60 R_C
D	27 R_C	12 mm	12 mm	57 R_C
E	26 R_C	15 mm	15 mm	50 R_C
F	63 R_C	1 mm	1 mm	64 R_C

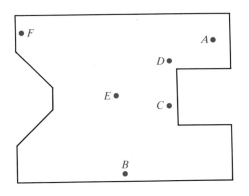

FIG. 9–7.7

V-Bar Cross-Section. See Example 9–7.4.

FIG. 9–7.8

Hardenability Curves for AISI–SAE 40xx Steels. Except for carbon content, the composition is the same for each. Additional carbon gives harder martensite and harder $\alpha + \overline{C}$, in keeping with Fig. 9–7.1.

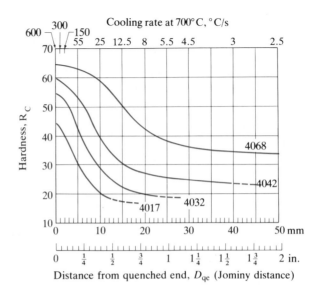

Distance from quenched end, D_{qe} (Jominy distance)

Example 9–7.5

An AISI–SAE 1020 steel rod (diameter 32 mm, or $1\frac{1}{4}$ in.) is *carburized* to 0.62 percent C at the surface, to 0.35 percent C at 2 mm below the surface, and is unaltered beyond the depth of 4 mm.

(a) Determine the hardness profile of the steel after quenching in water.

(b) What would the profile have been without carburizing?

Procedure We use the same procedure as in Example 9–7.3, except that we replace the 1020 hardenability curve (Fig. 9–7.4) by 1062 and 1035 data in the carburized zone. Interpolate where necessary.

Answers

	D_{qe}	(a)		(b)	
Surface	1.5 mm	0.62% C	63 R_C	0.20% C	47 R_C
2 mm	~2.5 mm	0.35% C	~50 R_C	0.20% C	~40 R_C
4 mm	3 mm	0.20% C	34 R_C	0.20% C	34 R_C
MR	4 mm	0.20% C	26 R_C	0.20% C	26 R_C
Center	6.5 mm	0.20% C	22 R_C	0.20% C	22 R_C

Comment The harder, carburized surface zone (case) provides better wear resistance. The bulk of the steel remains tough.

9–8
STRONG AND TOUGH CERAMICS

Ceramic materials have high shear strengths and low ductility for the reasons we discussed in Section 8–3. Therefore, they are inherently strong in compression, but apparently weak in tension. This weakness arises because any geometric irregularity leads to a stress concentration. Without ample ductility, there is no energy consumed by plastic deformation. A crack, once started, may grow spontaneously when the critical stress intensity is exceeded (Eq. 8–4.1). If no surface scratches or other flaws are present, the tensile strength will be as high as the compressive strength. This is realized with glass fibers, which, if protected from surface abrasion, attain strengths greater than steel.

This relationship of high compressive strength and low tensile strength is important to the design engineer. Concrete, brick, and other ceramics are used primarily in compressive locations (Fig. 9–8.1). When it is necessary to subject materials such as glass to bending (and therefore to tensile forces), it may be necessary to increase dimensions. For example, the viewing glass of a television picture tube may be as much as 15 mm thick.

Induced Compression

An alternate procedure is to induce compressive stresses into the material. *Prestressed* concrete is an example. In prestressing, reinforcing rods place a product under initial compression before the service load is applied. The bending moment does not build up tensile stresses until the introduced compressive stresses are exceeded.

Surface compression may be induced into *laminated composites* by thermal treatments. For example, consider a laminate of three layers of glass, A-B-A. The two surface layers, A, are of a glass with a lower thermal-expansion–contraction coefficient than the center layer, B. These three layers are heated sufficiently as a "sandwich" to weld together the layers. During subsequent cooling, the B layer should contract more than the A layers (Fig. 9–8.2). However, the three are bonded together, so they are forced to have the *same* $\Delta L/L$. The A layers develop compressive stresses; the B layer develops compensating tensile stresses. The interior B layer is not subject to surface damage so the stress concentrations do not

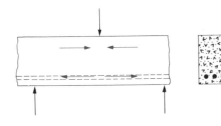

FIG. 9–8.1

Reinforced Concrete Beam. This beam uses the nonductile material in the compressive positions.

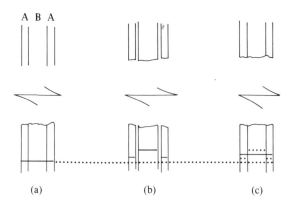

FIG. 9–8.2

Laminated Glass (Cross-Section). (a) The glass is heated to weld the layers into a sandwich, and to eliminate any residual stresses. (b) The B layer would have contracted more than the two A layers, *if* the two glasses had *not* been joined. (c) Since they are bonded together, the B layer is strained into tension, and the outer layers are compressed. In bending, the compressive forces of the A layers must be overcome before tension is encountered, thus strengthening the product.

occur. As with prestressed concrete, the bending moment does not build up tensile stresses at the outer surface until the induced compressive stresses are exceeded. One of the popular dinnerware brands uses this type of composite.

Surface compression also is induced in tempered glass. This engineered glass is used for glass doors, rear windows of cars, ophthalmological lenses, and other high-strength applications.* To produce *tempered glass,* the glass plate is heated to a temperature high enough to permit adjustments to stresses among the atoms, then is quickly cooled by an air blast or oil quench (Fig. 9–8.3). The surface contracts because of the drop in temperature and becomes rigid, while the center is still hot and can adjust its dimensions to the surface contractions. When the center cools and contracts slightly later, compressive stresses are produced at the surface (and tensile stresses are produced in the center). The stresses that remain in the cross-section of the glass are diagrammed in Fig. 9–8.4. A considerable deflection must be applied to the glass before tensile stresses can be developed in the surface of the glass where cracks start.† In effect, since the compressive stresses must be overcome first, the overall strength of the glass is greatly enhanced.

---

* The laminated windshields of U.S. cars are not designed for increased strength or toughness, but include a polymeric layer that prevents the shards of fractured glass from becoming bayonets.

† If a crack penetrates through the compressive skin (e.g., by scratching) into the tension zone shown in Fig. 9–8.4, the crack may become rapidly self-propagating. The aftermath of this effect can be observed in a broken rear window of a car, where the crack pattern is a mosaic, rather than spearlike shards.

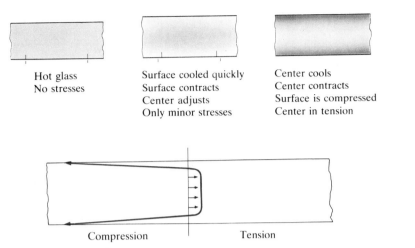

FIG. 9–8.3

Dimensional Changes in "Tempered" Glass.

Hot glass
No stresses

Surface cooled quickly
Surface contracts
Center adjusts
Only minor stresses

Center cools
Center contracts
Surface is compressed
Center in tension

FIG. 9–8.4

Surface Compression of "Tempered" Glass. These compressive stresses must be overcome before the surface can be broken in tension.

Compression   Tension

"High-Tech" Ceramics

Many ceramics possess an excellent stability at high temperatures because of their strong internal bonding. Thus, they have been used for many years as *refractories* in furnaces that melt and refine metals (Fig. 14–7.1), and that generate power from steam. Historically, these and related applications have involved static loading. Many modern designs, from space shuttles to automotive pistons to gas turbines, require dynamic loading at high temperatures for efficient operation — often above the temperature capabilities of commonly available metals. For this reason, we have turned our attention to ceramics. The challenge has been to make these materials stronger and tougher. Significant success has been achieved with "high-tech" ceramics. Design engineers, however, always will request additional progress to meet ever higher temperature specifications.

Stronger ceramics are achieved primarily through more sophisticated process control. Fewer flaws of even microscopic size result in fewer stress concentrations that lead to fracturing. This approach has more than doubled the strength of a number of quality ceramics. Strength also can be increased through the control of induced stresses. Strengths of glasses are increased by as much as 300 percent by this procedure.

The development of *tougher* ceramics, which withstand dynamic energy, is more complex, but is being achieved by several procedures. First, note that the energy required to fracture a nonductile material is the area under the elastic portion of the stress–strain curve, as sketched in Fig. 9–8.5(a). This is simply $se/2$, or $s^2/2E$, as we indicated in the early part of Section 8–4. Thus, a doubling of the strength of a ceramic quadruples the material's toughness. Even so, the energy requirements are low, because most ceramics have high elastic moduli, E.

Current approaches to toughening increase the apparent strain before fracture is completed. This process can involve one or more of the following mechanisms: (1) *matrix microcracking,* (2) *fiber debonding,* and (3) *phase-transformation*

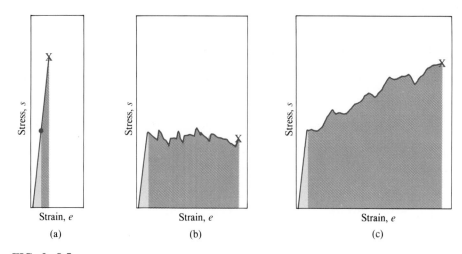

FIG. 9–8.5

Toughening of Ceramics. The energy for fracture is equal to the area under the stress–strain curve. (a) A doubling of the strength by imperfection control quadruples the energy requirements. (b) Microcracking and fiber debonding introduce *apparent strain* that adds to the stress–strain energy product of energy consumption before complete fracture at X. (c) Phase-transformation toughening (Al_2O_3–ZrO_2 composites). Expansion accompanies the tetragonal-to-monoclinic transformation of ZrO_2 (Eq. 3–4.3) to introduce compression strengthening at the crack tip, thereby providing significantly greater toughness.

toughening. We can summarize the first two by noting that the integrated fractured surfaces are increased markedly, so more surface energy has to be provided. Furthermore, during the process, microdisplacements appear as *apparent strains*. The net effect is to increase the area under the stress–strain curve and the resulting energy consumption (Fig. 9–8.5b).

Phase-transformation toughening almost always involves the use of ZrO_2 particles in a ceramic matrix (e.g., Al_2O_3). Zirconia has several polymorphic transformations as it cools—cubic to tetragonal and later tetragonal to monoclinic. The latter step (Eq. 3–4.3) is a martensitic-type reaction (Section 7–4) that involves a volume expansion. This step is easily avoided during the cooling process after sintering. As with martensite, this phase transition occurs by a shear displacement. Therefore, the application of large shear stresses to the composite during the fracture process nucleates the tetragonal-to-monoclinic phase change. The accompanying volume expansion induces compressive stresses that strengthen the material ahead of the crack tip. Furthermore, the shear displacements add to the area under the stress–strain curve and therefore increase energy consumption (Fig. 9–8.5c).

The critical stress-intensity factor, K_{Ic}, for a composite of Al_2O_3 and ZrO_2 can be as high as 15 to 20 MPa·$\sqrt{m}$ as compared to 1 MPa·$\sqrt{m}$ for commercial window glass and over 50 MPa·$\sqrt{m}$ for many steels.

Example 9–8.1

A 2.5-mm (0.10-in.) iron sheet that is to be used as a lining for a household oven is coated on *both* sides with a glassy enamel. The final processing occurs above the glass-transition temperature at 500° C (930° F), to give a 0.5-mm (0.02-in.) coating. The glass has a modulus of elasticity of 70,000 MPa (10^7 psi) and a thermal expansion of $8.0 \times 10^{-6}/°$ C.

(a) What are the stresses in the glass at 20° C?

(b) What are the stresses at 200° C? (Assume no plastic strain.)

Procedure Since $\Delta L/L =$ thermal expansion + elastic strain, and in this case $(\Delta L/L)_{gl} = (\Delta L/L)_{Fe}$, we can write

$$\alpha_{gl}\Delta T + s_{gl}/E_{gl} = \alpha_{Fe}\Delta T + s_{Fe}/E_{Fe}$$

Calculation

(a) Using data from the Procedure and from Appendix C,

$$s_{Fe}/205{,}000 \text{ MPa} - s_{gl}/70{,}000 \text{ MPa} = (8.00 - 11.75)(10^{-6}/° \text{ C})(-480° \text{ C})$$

But $A_{Fe} = 2.5A_{gl}$ and $F_{Fe} = -F_{gl}$, so $s_{gl} = -2.5s_{Fe}$. Thus,

$$s_{Fe}\left[\frac{1}{205{,}000} + \frac{2.5}{70{,}000}\right] = 0.0018$$

Solving, we obtain

$$s_{Fe} = +44 \text{ MPa} \qquad \text{(or 6400 psi)} \qquad (+: \text{tension})$$
$$s_{gl} = -110 \text{ MPa} \qquad \text{(or } -16{,}000 \text{ psi)} \qquad (-: \text{compression})$$

(b) By similar calculations for $\Delta T = (500 - 200)°$ C,

$$s_{gl} = -69 \text{ MPa (or } -10{,}000 \text{ psi)} \qquad \text{and} \qquad s_{Fe} = +27.5 \text{ MPa (or 4000 psi)}$$

Comments We assumed unidirectional strain. In reality, plane (i.e., two-dimensional) strain occurs. The necessary correction gives a higher stress by the factor of $(1 - v)^{-1}$, where v is Poisson's ratio. These residual stresses are desirable. With the glass in compression, the composite is stronger. (See Fig. 9–8.4.)

S U M M A R Y

1. We can *shape* a material in numerous ways. Among the more common processes are *casting, plastic forming,* and *sintering.* Casting starts with either a liquid or a semifluid suspension. Volume changes during solidification must be considered. Plastic forming involves crystalline slip. Temperature is a common process variable. Sintering requires sufficient atom movements to coalesce adjacent particles.

2. *Solution hardening* is widely used for strengthening materials. It arises when dislocations anchor to solute atoms, whether these are larger *or* smaller.

3. *Cold work* hardens almost all materials. Although this hardening increases the strength of materials, it generally decreases the ductility.

A *strain-hardened* (cold-worked) material recrystallizes during annealing. The *recrystallization temperature*, T_R, depends on (i) the available time, (ii) the amount of strain hardening, and (iii) the purity—each of which, when increased, facilitates recrystallization. In a *hot-working* process, the temperature is high enough for recrystallization to occur essentially simultaneously with deformation. *Cold-work–anneal* cycles are used in production to accomplish extensive reshaping and to avoid cracking from loss of ductility.

4. *Precipitation hardening,* also called age hardening, is achieved by *solution treating,* followed by rapid cooling to produce a *supersaturated* solid solution. As a second phase *starts* to separate (precipitate), the material gains significant hardness. The required time is a function of temperature.

Overaging and softening occur as the precipitating particles coalesce into a coarser microstructure. Thus, age-hardened materials have limitations on their service temperatures.

5. A minor amount of a rigid phase can strengthen a ductile material. *Plastic constraint* produces this *second-phase strengthening.* The amount of strengthening increases with an increase of interfacial area between the reinforcing phase and the matrix.

6. Heat treatments of steels include several types of reheating processes. *Process annealing, spheroidizing,* and *stress relief* are all performed below the eutectoid temperature. The first removes the strain hardening of cold work. The second spheroidizes the carbides to toughen high-carbon steels. The third removes residual stresses to avoid distortion and possible cracking.

Both *full annealing* and *normalizing* treatments *austenitize* the steel as the first step. In the former, the material is cooled slowly to provide as soft a product as possible. Normalizing is a homogenization process.

The *quenching* of steels also is preceded by austenitization. In plain-carbon steels, the quenching must be completed within approximately 1 sec if $(\alpha + \overline{C})$ is to be avoided in favor of martensite (which is subsequently *tempered* for toughening). Alloying additions slow the $(\gamma \rightarrow \alpha + \overline{C})$ reaction, so that larger parts may be quenched and tempered. Most *low-alloy* steels receive quench-and-temper processing. Extended heating coalesces the carbide particles in tempered martensite and softens the steel.

7. *Hardness* is a measure of resistance to plastic deformation by penetration; *hardenability* is the ease with which maximum hardness is attained. *Hardenability curves* are based on the fact that a given steel always develops the same microstructure (and therefore hardness) with a given, standardized cooling rate.

8. We can strengthen and toughen ceramics (i) by minimized stress risers, (ii) by induced surface compression, (iii) by controlled microcracking, and (iv) by controlled phase-transformation alloying.

The instructor who wishes to put greater emphasis on ferrous metallurgy can assign the following two study sets included in the *Study Guide: Austenite Decomposition,* and *Microstructures of Steels.*

KEY WORDS

Age hardening	End-quench test (Jominy bar)	Microcracking
Annealing (strain-hardened metal)	Extrusion	Normalizing
	Hardenability	Overaging
Annealing (steels)	Hardenability curve	Plastic constraint
Austenization	Hardness traverse	Precipitation hardening
Carburize	Hot-working	Quench
Casting	Induced compression	Rate $(R = t^{-1})$
Cold work (percent)	Interrupted quench	Recovery
Cold working	Isothermal transformation (I-T)	Recrystallization
Drawing	Jominy distance (D_{qe})	Recrystallization temperature (T_R)

Rolling	Spheroidization	Stress relief
Sintering	Steel, low-alloy	Tempered glass
Solution hardening	Steel, plain-carbon	Tempering
Solution treatment	Strain hardening	

PRACTICE PROBLEMS

9–P11 Based on the data of Section 4–2, what is the maximum percent porosity that could exist in a magnesium casting, if no riser were used?

9–P12 A 1.00-in. × 0.25-in. aluminum bar (commonly sold in a hardware store) was extruded from a 100-lb, 6-in.–diameter aluminum billet. Assuming 1 percent end scrap, how many feet of product were obtained? (1 g/cm³ = 62.4 lb/ft³ = 0.0361 lb/in.³)

9–P13 A ceramic insulator will have a 1 v/o porosity after sintering and should have a length of 13.7 mm. During manufacturing, the raw powders can be compressed to contain 24 v/o porosity. What should the die dimension be?

9–P14 A powdered metal part has a porosity of 23 percent after compacting of the powders and before sintering. What linear shrinkage allowance should be made if the total porosity after sintering is expected to be 2 percent?

9–P15 A ceramic wall tile, 5 mm × 200 mm × 200 mm, absorbs 2.5 g of water. What is the apparent porosity of the tile?

9–P16 A brick weighs 3.3 kg when dry, 3.45 kg when saturated with water, and 1.9 kg when suspended in water. (a) What is the apparent porosity? (b) What is the bulk density? (c) What is the apparent density?

9–P21 Copper increases the strength of nickel, even though nickel is stronger than copper. Why?

9–P22 Which is stronger, a copper alloy with 2 w/o tin, or one with 2 w/o beryllium?

9–P23 You are designing a seat brace for a motorboat. Iron is excluded because it rusts. Select the most appropriate alloy from Fig. 9–2.1. The requirements include an ultimate strength of at least 310 KPa (45,000 psi), a ductility of at least 45 percent elongation (in 50 mm), and a low cost. (*Note:* Zinc is less expensive than copper, which in turn is less expensive than nickel.)

9–P31 Two identical 20-m (65-ft) aluminum rods are each 14.0 mm (0.551 in.) in diameter. One of the two is drawn through a 12.7-mm (0.50-in.) die. (a) What are the new dimensions of that rod? (b) Assume test samples of identical size are machined from each rod (deformed and undeformed) and are marked with 50-mm gage lengths. Which one, if either, will have the greater ductility? (c) Which one will have the greater yield strength?

9–P32 (a) How much cold work was performed on the aluminum rod in Problem 9–P31? (b) Change the metal in Problem 9–P31 to copper. What are the ultimate strength and the ductility for the deformed metal?

9–P33 A copper wire 2.5 mm (0.10 in.) in diameter was annealed before being cold drawn through a 2.0-mm (0.08-in.) die. What ultimate strength does the wire have after cold drawing?

9–P34 Use the data of Figs. 8–3.6, 9–3.4, and 9–3.6. Sketch a plausible curve of yield strength versus cold work for copper.

9–P35 A copper wire must have a diameter of 0.7 mm and an ultimate strength of >345 MPa (50,000 psi). It is to be processed from a 10-mm rod. What should the diameter be for annealing before the final cold draw?

9–P36 A 70–30 brass wire is to be made by cold drawing with an ultimate strength of more than 415 MPa (60,000 psi), a hardness of less than 75 R_b, and an elongation of greater than 25 percent on a standard test bar. The diameter of the wire as received is 2.5 mm (0.10 in.). The diameter of the final product is to be 1.0 mm (0.04 in.). Prescribe a procedure for obtaining these specifications. (*Note:* The elongation data in the figures of this chapter are for the standard 50-mm test bar of Fig. 8–2.2, in which the diameter is 0.505 in. Since the necking is a function of the diameter, a value from a wire would require a subsequent correction.)

9–P37 A rolled 66 Cu–34 Zn brass plate 12.7 mm (0.5 in.) thick has a ductility of 2 percent elongation by the standard test bar (Fig. 8–2.2) when it is received from the supplier. This plate is to be rolled to a sheet with a final thickness of 3.2 mm (0.125 in.). In this final form, it is to have an ultimate strength of at least 483 MPa (70,000 psi) and a ductility of at least 7 percent elongation. Assume that the rolling process does not change the width. Specify *all steps* (including temperatures, times, thickness, etc.) that are required.

9–P38 Aluminum has been shaped into a cake pan by spinning (Fig. 9–1.3b). Assume that the data from Fig. 9–3.12 apply to this cold-worked aluminum. Will the metal start to recrystallize while in a 180° C (350° F) oven? Solve both (a) graphically, and (b) mathematically.

9–P39 Assume that annealing should be completed in 1 sec to permit the hot working of the aluminum in Fig. 9–3.12. What temperature is required?

9–P41 Explain why a 92 percent copper, 8 percent nickel alloy can (or cannot) be age hardened.

9–P42 Maximum hardness is obtained in a metal when an aging time is 10 sec at 380° C, or 100 sec at 315° C. Neither of these constraints is satisfactory for production, because we cannot be certain the parts will be uniformly heated. Recommend a temperature for the maximum hardness in 1000 sec (16 to 17 min), a time compatible with production requirements.

9–P43 How long should it take the metal in Problem 9–P42 to reach maximum hardness at 100° C?

9–P44 An aircraft manufacturer receives a shipment of aluminum-alloy rivets that are already age hardened. Can they be salvaged? Explain your answer.

9–P51 More rapid cooling produces a pearlite with twice as many layers as the one shown in Fig. 9–5.2(a). How much (what percentage) change will there be in the amount of interphase boundary that is available for plastic constraint?

9–P52 Extended annealing produces a spheroidite (Fig. 9–5.2b) with carbide particles that are twice as large in diameter (see L of Eq. 7–1.1.) How much (what percentage) change will there be in the amount of interphase boundary that is available for plastic constraint?

9–P61 Determine the temperature that should be used to normalize (a) a 1030 steel, (b) a 1080 steel, and (c) a 1-percent carbon steel?

9–P62 A small piece of 1080 steel is heated to 800° C, quenched to −60° C, reheated immediately to 300° C, and held 10 sec. What phases are present at the end of this time?

9–P63 A piece of 1045 steel is quickly quenched from 850° C to 400° C and held (a) for 1 sec, (b) for 10 sec, and (c) for 100 sec. What phase(s) will be present at each time point?

9–P64 Sketch a plausible isothermal-transformation diagram for a steel of 1.0 w/o C and 99.0 w/o Fe.

9–P65 A small piece of 1080 steel has its quench interrupted for 20 sec at 300° C (570° F) before final cooling to 20° C. (a) What phase(s) are present after the 20 sec? (b) What phase(s) are present at 20° C?

9–P66 Why did we *not* use Fig. 9–6.3 for wire (c), step 2, of Example 9–6.1?

9–P67 A 1020 steel is equilibrated at 770° C (1420° F), then is quenched rapidly to 400° C (750° F). How long must the steel be held at that temperature to reach the midpoint of the austenite decomposition?

9–P68 Compare and contrast the following terms: (a) *martensite* and *tempered martensite,* (b) *tempered martensite* and *spheroidite,* (c) *tempered martensite* and *pearlite.*

9–P69 Why does austenite in a 0.25% Mo steel (plus 0.7% C) transform to $(\alpha + C)$ less rapidly than it does in an 0.7% C steel?

9–P71 Explain the difference between *hardness* and *hardenability.*

9–P72 (a) What is the cooling rate (reported in ° C/sec) at the midradius of a 50-mm (1.97-in.) round steel bar, which was quenched in agitated oil? (b) What is it when reported as the distance from the quenched end of a Jominy bar? (c) Repeat parts (a and b), but for the $\frac{3}{4}$-radius and a water quench.

9–P73 Why should the steel (or the water) be agitated when a gear is water quenched?

9–P74 (a) What is the quenched hardness at the midradius of a 50-mm (1.97-in.) round steel bar of an SAE 1040 steel quenched in agitated oil? (b) What is it when the steel is quenched in agitated water?

9–P75 (a) What hardness would you expect at the center of a 50-mm (2-in.) round bar of an SAE 1040 steel if that bar were quenched in agitated oil? (b) What hardness if it were quenched in agitated water?

9–P76 A 40-mm (1.6-in.) bar of an SAE 1040 steel (i.e., with diameter 40 mm, and length ≫ 40 mm) is quenched in agitated water. (a) What is the cooling rate through 700° C at the surface? (b) What is it at the center? (c) Plot a hardness traverse.

9–P77 Explain why low-alloy steels, rather than plain-carbon steels, are commonly specified for components that are to be heat treated by quenching and tempering.

9–P78 An 80-mm (3.15-in.) round SAE 4340 steel bar is quenched in agitated oil. Plot the hardness traverse.

9–P79 Repeat Example 9–7.5, but with oil quenching.

9–P81 Glass-coated (porcelain-enameled) steel is used as an oven liner. Why is this combination chosen? How should the properties of the glass and of the steel match? How should they be unlike?

9–P82 A 1-mm-diameter fiber with the composition of plate glass (Appendix C) is coated with 0.1 mm of borosilicate glass, so the fiber is now 1.2 mm in diameter. Assuming no initial stresses at 200° C, what longitudinal stresses are developed when the composite fiber is cooled to 20° C?

9–P83 Repeat Problem 9–P82, but interchange the two locations of the glasses. (a) Which will have the higher stress, plate or glass? (b) Comment on the strength of this composite glass rod.

T E S T P R O B L E M S

911 An electrical porcelain product has only 2 v/o porosity as sold. It had 33.1 v/o porosity (total) after pressing and drying. How much linear shrinkage occurred during processing?

912 A magnetic ferrite for an electric-motor component is to have a final dimension of 43 mm (1.69 in.). Its volume shrinkage during sintering is 27 percent (unsintered basis). What initial dimension should the powdered compact have?

913 Refer to Fig. 9–1.6. (a) How much linear shrinkage occurred during the sintering of this powder — from part (a) to part (c)? (b) How much volume shrinkage occurred?

914 An insulating brick weighs 1.77 kg (3.9 lb) dry, 2.25 kg (4.95 lb) when saturated with water, and 1.05 kg (2.3 lb) when suspended in water. (a) What is the apparent porosity? (b) What is the bulk density? (c) What is the apparent density?

921 Predict the yield strength of a 95 Cu–5 Sn bronze on the basis of the figures in Section 9–2.

922 Specifications for a product call for a minimum ultimate strength of 280 MPa (41,000 psi) and a maximum hardness of 60 R_f. Which of the Cu–Ni or Cu–Zn alloys in Fig. 9–2.1 is the most suitable?

923 The hardness of an iron–silicon alloy is to match the hardness of an Fe plus 2 w/o Ni alloy. Estimate how much silicon must be added per 100 g of iron.

931 (a) An iron rod is to have a Brinell Hardness Number (BHN) of at least 125 and an elongation of at least 28 percent (50 mm or 2 in.). How much cold work should the iron receive? (b) A pure iron sheet 7.5 mm (0.30 in.) thick is annealed before cold rolling. It is rolled to 6.0 mm (0.24 in.), with negligible change in width. What will be the ultimate strength of the iron after the cold rolling?

932 A 2.55-mm (0.10-in.) annealed copper rod is drawn through a 2.29-mm (0.090-in.) circular die. How much (what percentage) does the ultimate strength increase?

933 (a) Change the metal in Problems 9–P31 and 9–P32 to 75 Cu–25 Zn. What are the ductilities of the two rods? (b) This metal is to be used in an application where an ultimate strength of at least 430 MPa (63,400 psi), and a ductility of at least 10 percent (50 mm), are required. Specify the amount of cold work.

934 Specify a brass for an application that requires at least 10 percent elongation (50 mm), and at least a 350-MPa (50,000-psi) yield strength.

935 (a) How much zinc should there be in a cold-worked brass to give a ductility of at least 20 percent elongation *and* a hardness of at least 70 R_b (=97 R_f)? (b) How much should it be cold worked? (Your answer should not exceed 36 percent Zn because of processing complications.)

936 A round bar of an 85–15 brass, with a 5-cm diameter, is to be reduced to a rod 1.25 cm in diameter. Suggest a procedure to be followed if a final ultimate strength of 400 MPa (or greater) is to be achieved, as well as a final ductility of at least 10 percent (in a 50-mm gage length).

937 A company is making a 70–30 brass plate (thickness of 1 cm, or 0.4 in.) by rolling a 1.5-cm (0.6-in.) plate to meet the following specifications: $S_u \geq 415$ MPa (>60,000 psi); hardness $\leq R_b 75$; and elongation ≥ 25 percent (2 in., or 50 mm). Prescribe a procedure for obtaining these specifications.

938 A round bar of brass (85 percent Cu, 15 percent Zn), 50 mm (2.0 in.) in diameter, is to be processed to a diameter of 10 mm (0.4 in.). Specify a procedure so the product has a hardness of 72 R_b, an ultimate strength greater than 415 MPa (60,000 psi), and a ductility of greater than 10 percent elongation in the standard gage length.

939 Assume that annealing should be completed in 1 sec or less for the hot working of the high-purity copper of Fig. 9–3.13. What temperature is required? Solve by equation and check your answer with a graphical solution.

941 Explain why the following alloys can (or cannot) be considered for age hardening: (a) 97 percent aluminum, 3 percent copper. (b) 97 percent copper, 3 percent zinc. (c) 97 percent nickel, 3 percent copper. (d) 97 percent copper, 3 percent nickel. (e) 97 percent aluminum, 3 percent magnesium. (f) 97 percent magnesium, 3 percent aluminum.

942 (Refer to Fig. 9–4.3.) Select a plotting procedure that lets you estimate (graphically) the time that should be required for that material to reach peak hardness at 100° C. Use your procedure to determine that time for this 2014 aluminum.

943 (a) Estimate graphically (as in Problem 942) the time that would be required to attain 200 MPa (29,000 psi) at 100° C. (b) Establish an equation for the required time.

944 A slight amount of age hardening is realized when a steel (99.7 w/o Fe, 0.3 w/o C) is quenched from 700° C (1300° F) and reheated for 3 hr at 100° C. Account for the hardening.

945 Repeat Problem 943, but for 300 MPa (43,000 psi) at 125° C.

946 The solubility of tin in lead (α) matches that required for precipitation hardening (Fig. 5–2.1). Yet tin does not age harden lead satisfactorily. Suggest a reason why it does not.

947 As a practical matter, why is it simpler to first quench a 5-lb, solution-treated, 94 Al–4 Cu alloy aircraft part to room temperature, then reheat it to 125° C for precipitation, rather than to quench it directly from 525 to 125° C, then cool it to R.T.?

951 Refer to Fig. 7–4.1. Estimate the ratio of interphase-boundary areas in the two microstructures.

952 According to the Ag–Cu phase diagram, sterling silver can be treated to produce approximately 7 v/o of β particles in a matrix of α. However, very little hardening occurs, especially when compared with 7 v/o of Fe_3C in a matrix of ferrite (e.g., spheroidite). Why is there a difference?

961 Point out the differences between normalizing a steel and giving it a full anneal. What are the technical bases for these differences?

962 Some AISI–SAE 1045 steel (Table 9–6.2) is quickly quenched from 850 to 425° C and held 5 seconds before quenching again to 20° C. (a) What phase(s) will be present just before the second quench? Give the composition of each. (b) What phase(s) will be present immediately after the second quench?

963 Answer Problem 962, but for 275° C instead of 20° C.

964 (a) Why does the "start" curve in Fig. 9–6.3 have two branches on its upper arm? (b) Sketch an isothermal-transformation diagram for an SAE 1020 steel.

965 A small wire of AISI–SAE 1045 steel is subjected to the following treatments as *successive* steps:

    1. Heated to 875° C, held there for 1 hr
    2. Quenched to 250° C, held there 2 sec
    3. Quenched to 20° C, held there 100 sec
    4. Reheated to 550° C, held there 1 hr
    5. Quenched to 20° C and held

Describe the phases or structures present *after each step* of the heat-treatment sequence.

966 (a) Repeat Problem 965 with steps (1), (2), (5). (b) Repeat Problem 965 with steps (1), (3), (4), (5). (c) Repeat Problem 965 with steps (1), (2), (4), (5).

967 Refer to Fig. 9–6.5. A steel without molybdenum is equilibrated at 740° C, then quenched to 400° C and held for ~ 5 sec for γ decomposition to start. What sequence would be necessary to achieve the same results if 0.25 percent Mo were present? Explain your answer.

968 An SAE 1080 steel is to be quenched and tempered to a hardness of 50 R_C. (a) What should the austenitizing temperature be? (b) How long should the steel be tempered at 400° C (750° F)?

969 Draw temperature (ordinate) and time (abscissa) plots for the following heat treatments. Indicate the important temperatures, relative times, and reasons for drawing the curves as you do. (a) Normalizing an SAE 1095 steel, contrasted with annealing the same steel. (b) Solution-treating a 95 Al–5 Cu alloy, contrasted with aging the same alloy. (c) Spheroidizing an SAE 1080 steel, contrasted with spheroidizing an SAE 10 · 105 steel.

971 A bar of SAE 1040 steel has a surface hardness of $48R_C$, and a center hardness of $24R_C$. How fast were the two locations cooled through 700° C?

972 The center hardness of six bars of the same 0.40 percent C steel are indicated in the following table. From these data, plot the hardenability curve for the steel. (*Hint:* There should be only *one* curve.)

DIAMETER	WATER QUENCH	OIL QUENCH
25 mm (1.0 in.)	50 R_C	38 R_C
50 mm (2.0 in.)	32 R_C	30 R_C
100 mm (3.9 in.)	29 R_C	28 R_C

973 (a) How would the hardness traverse of the 1040 steel shown in Fig. 9–7.6 vary if the steel were quenched in still oil? (b) How would it vary if the steel were quenched in still water? (c) How would it vary if the steel had a coarser austenite grain size? Explain your answers.

974 Plot a hardness traverse for an SAE 1060 steel bar (diameter of 38 mm, or 1.5 in.) that has been quenched in (a) agitated water, and (b) agitated oil.

975 (a) An SAE 40xx steel is to have a hardness of 58 R_C after it is quenched at the rate of 30° C/sec. What is the required carbon content? (b) What will the hardness be if the steel is quenched twice as fast?

976 A 60-mm (2.4-in.)–diameter steel rod of an SAE 4022 steel has been carburized to 0.65 percent C at the surface and to 0.38 percent C at 3 mm below the surface, and remains at 0.22 percent C at 7.5 mm below the surface. Determine the hardness profile of the steel bar after oil quenching.

977 Why does the center of a 50-mm (2-in.)–diameter steel bar water quench to a greater hardness than the steel that is 25 mm behind the end of a Jominy end-quench test bar? Which will quench more rapidly?

978 A spline gear had a hardness of 50 R_C at its center when it was made of an SAE 4068 steel shown in Fig. 9–7.8. What hardness would you expect the same gear to have if it were made out of an SAE 4140 steel?

979 A 50-mm–diameter bar of an SAE 1080 steel is quenched and forms a martensite "rim" that is 2.5 mm thick. While the austenitic center is hot, it can still adapt to the ($\gamma \rightarrow$ M) volume change (Example 7–4.2). The more slowly cooled center changes directly to (α + C). What hoop stress is placed on the martensite rim? [For SAE 1080, $\rho_\gamma = 7.99$ Mg/m^3 (=7.99 g/cm^3); $\rho_M = 7.85$ g/cm^3; and $\rho_{(\alpha+C)} = 7.84$ g/cm^3.]

981 A glaze for dinnerware is chosen such that it has a slightly lower thermal expansion coefficient than the underlying porcelain. Explain why this glaze makes the product stronger.

982 A glass "sandwich" is made with three layers of glasses A and B fused together; $\alpha_A = 9 \times 10^{-6}/$° C; $\alpha_B = 7 \times 10^{-6}/$° C. (a) Which glass, A or B, should be the "bread," and which should be the "filling"? (b) The sandwich is made with the filling four times as thick as each layer of bread. Consider one dimension only (parallel to the flat surface). What is the maximum compressive stress that can be induced into the surface by thermal treatment? [For both glasses, $T_g = 520$° C; $E = 10^7$ psi (70 GPa).]

Chapter 10

POLYMERS AND COMPOSITES

All solid materials respond to the imposition of external forces and energy. Typically, the initial response is elastic; beyond a critical stress, there may be permanent displacements; fracture is the ultimate response. Although this three-step pattern is general, the mechanistic details vary with the nature of the bonding within the material. This variance is particularly striking with polymers, because these materials possess a combination of strong covalent bonds and weak secondary bonds. The two types of bonds may exist side by side, and even connect different neighbors of the same atoms. In addition, many polymers either are amorphous or are only partially crystalline.

In this chapter, we shall first consider the mechanism of strain and flow, then turn our attention to the processing of polymeric materials, since much of the processing is a function of shaping and reshaping procedures. Polymers have greater strength and toughness when they are part of a reinforced composite. Wood —the most widely used material—is a natural composite and has the merit of being replenishable.

10–1

DEFORMATION AND FLOW OF AMORPHOUS MATERIALS

Polymers are sufficiently different structurally that their mechanical behaviors do not always duplicate those of metals and of nonsilicate ceramics. First, polymeric solids are commonly amorphous. This not only precludes plastic deformation by slip as discussed in Section 8–3, but also commonly introduces considerable "free space" into the material (Example 7–5.2). Second, many of our polymers are linear, with strong covalent, *intra*molecular bonds and weak, secondary *inter*molecular bonds (Section 2–2). Third, a single bond can be rotated: $C - \overset{\curvearrowright}{\underset{\smile}{C}} C$ (Fig. 4–3.4).

Elastic Strain

As occurs in all solids, stresses introduce elastic strains. Elastic strain develops when the stress is applied, remains constant at constant stress, and disappears when the stress is removed. Figure 10–1.1 uses a spring as an example.

Typically, the initial elastic modulus for most polymers is low when compared with those of other materials (Appendix C). In part, this low value arises from the free space that is commonly present in amorphous materials. Dimensional changes are made without straining the atom-to-atom bonds. More important for linear molecules, however, is the straightening of kinked and coiled molecules (Fig. 10–1.2) and the unfolding of molecular crystals (Fig. 7–5.1). This extension can occur with little stretching of the atom-to-atom bonds.

With additional strain, the covalent bonds come into play, and the s/e ratio therefore increases; the elastic modulus increases. Thus, the $s\text{–}e$ relationship becomes significantly nonlinear. This increase is very marked for rubbers *(elastomers)*, since their conformation is particularly tortuous (Eqs. 4–3.6 and 4–3.7). The nonlinearity is less for cross-linked polymers, and for those with three-dimensional, network structures (Fig. 10–1.3).

Typically, the elastic modulus decreases with increasing temperature. This is true for metals, ceramics, and polymers other than elastomers (Fig. 8–1.4). In elastomers the elastic modulus increases as the temperature rises because the

FIG. 10–1.1

Elastic Strain. (a) Strain develops when the stress is applied, remains constant at constant stress, and disappears when the stress is removed. (b) Spring model (L. H. Van Vlack, *Materials Science for Engineers,* Addison-Wesley, Reading, Mass., with permission.)

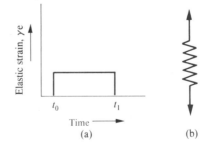

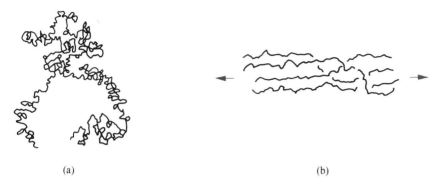

(a)   (b)

FIG. 10–1.2

Straightening of Kinked and/or Folded Molcules (Linear Polymers). (a) Without
tension. (b) Pulled.

greater thermal agitation favors a return to the more stable, kinked conformation,
in opposition to the directional pull on the stretched chains (Fig. 10–1.4).

Viscous Flow

Fluids and amorphous solids are subject to *viscous flow* independent of any crystal
structure. In simple Newtonian fluids, the velocity gradient of flow, dv/dy, is
proportional to the applied *shear stress, τ*:

$$dv/dy = f\tau \qquad \textbf{(10–1.1)}$$

where f is the *fluidity* of the material. More commonly, we speak of *viscosity, η*, the
reciprocal of fluidity. Therefore,

$$\eta = \tau/(dv/dy) \qquad \textbf{(10–1.2a)}$$

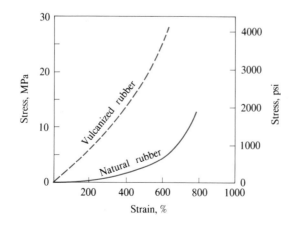

FIG. 10–1.3

**Stress–Strain Curves for Rubber
(Isoprene).** The elastic modulus,
ds/de, is low in natural rubber
(unvulcanized) until the
molecules become aligned with
the direction of stress. The
vulcanized rubber is cross-linked
with less opportunity for
extending the molecules. Therefore, it has a higher modulus. (L.
H. Van Vlack, *Materials for
Engineers: Concepts and
Applications,* Addison-Wesley,
Reading, Mass., with permission.)

FIG. 10–1.4

Young's Modulus Versus Temperature (Stretched Rubber). The retractive forces of kinking are greater with more thermal agitation. Therefore, the elastic modulus increases with temperature. (See Fig. 8–1.4.)

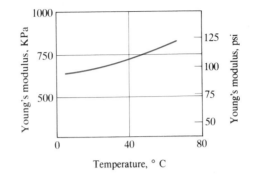

as sketched in Fig. 10–1.5. The SI units for viscosity are

$$\overline{N}/m^2/((m/s)/m) = Pa \cdot s \tag{10–1.2b}$$

Handbook data commonly are given in poises; 10 poises = 1 Pa·s.

Fluidity and viscosity values are temperature sensitive, since they require the movement (by shear) of atoms and molecules with respect to their neighbors.* In fact, we can modify the diffusion relationship of Eq. (6–5.4) to

$$f = f_0 \, e^{-E/kT} \tag{10–1.3a}$$

for the strict Newtonian fluid. Or, inversely,

$$\eta = \eta_0 \, e^{E/kT} \tag{10–1.3b}$$

where E is an *activation energy* for the flow of atoms or molecules. More commonly, we rearrange Eq. (10–1.3b) to the Arrhenius form:

$$\ln \eta = \ln \eta_0 + E/kT \tag{10–1.3c}$$

As with diffusion, the temperature must be in Kelvin, K, and k is Boltzmann's constant (13.8×10^{-24} J/K).

FIG. 10–1.5

Viscosity, η, is the Ratio of Shear Stress, τ, to Velocity Gradient, v/y. Since it is the reciprocal of fluidity, f, it decreases with increased temperature (Eq. 10–1.3). (L. H. Van Vlack, *Materials for Engineers, Concepts and Applications,* Addison-Wesley, Reading, Mass., with permission.)

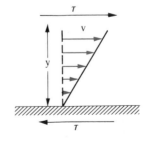

---

* This sensitivity is revealed by the units of Eq. (10–1.2b), which could be restated for fluidity as m^2/s (from the diffusion relationship, Eq. (6–5.1b) per unit force: $m^2/s \cdot N$.

TABLE 10–1.1 Viscosities of
Noncrystalline Materials (20° C)

MATERIAL	VISCOSITY, η Pa·s*
Air	0.000018
Pentane, C_5H_{12}	0.00025
Water	0.001
Phenol, C_6H_5OH	0.01
Syrup (60% sugar)	0.055
Oil, machine	0.1 to 0.6
Glycerin	0.9
Sulfur (120° C)	10^2
Window glass (515° C)	10^{12}
Window glass (800° C)	10^4
Polymers, T_g†	$\sim 10^{12}$
Polymers, $T_g + 15°$ C	$\sim 10^8$
Polymers, $T_g + 35°$ C	$\sim 10^5$

* 1 Pa·s = 10 poises.
† T_g = glass-transition temperature.

Table 10–1.1 lists the viscosities of a variety of liquids and supercooled liquids to help us reference our thinking. In comparison to 0.5 Pa·s for heavy motor oil, materials with $> 10^4$ Pa·s are semisolids that can maintain self-support for limited periods.

Flow starts when the shear stress is applied, continues at a constant rate under a constant stress, and has no recovery after the stress is removed. Figure 10–1.6 uses a dashpot (loose-fitting piston) as an example.

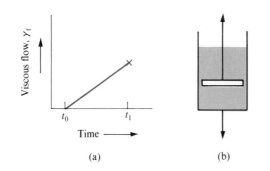

Viscous flow, γ_f

t_0 t_1

Time ⟶

(a)   (b)

FIG. 10–1.6

Ideal (Newtonian) Fluid. (a) Displacement. (b) Dashpot model. The flow is proportional to time. The flow is not reversible when the shear stress is removed at t. (L. H. Van Vlack, *Materials Science for Engineers,* Addison-Wesley, Reading, Mass., with permission.)

Polymer Viscosity

General laboratory experience has shown that the glass-transition temperature, T_g, coincides with a viscosity of approximately 10^{12} Pa·s, as indicated in Table 10-1.1. Molecular rearrangements are precluded at lower temperatures, as we discussed in Section 4-2. Since polymers do not display strict Newtonian behavior, we can use the following empirical relationship that holds for a number of linear polymers:

$$\log_{10} \eta = 12 - (17.5 \times \Delta T)/(52 + \Delta T) \qquad \text{(10-1.4a)*}$$

Here, ΔT is the added temperature above T_g, that is, $(T - T_g)$. As the temperature is raised, the value of $\log_{10} \eta$ drops below the value of 12, as indicated in Table 10-1.1. The major effect of temperature is evident. Polymers must be above their glass-transition temperature to be molded into plastic products. The specific processing temperature depends on the pressures to be used.

Viscoelastic Deformation

Elastic strain, γ_e, and viscous flow, γ_f, occur simultaneously. In the simplest case, they are additive:

$$\gamma = \gamma_e + \gamma_f \qquad \text{(10-1.5)}$$

and we could plot strain versus time as the series model sketched in Fig. 10-1.7.†

FIG. 10-1.7

Viscoelasticity (Series Model). The total displacement is the sum of the elastic strain, γ_e and viscous flow, γ_f. Another common model places the two displacements in parallel. (See footnote following Eq. (10-1.5)).

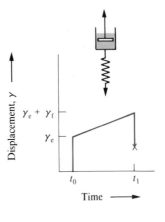

---

* Rearranged, this equation is

$$\log_{10} \eta = A + B/(52 + \Delta T) \qquad \text{(10-1.4b)}$$

where A and B are -5.5 and 910, respectively.

† An alternative model considers that elastic strain cannot occur without viscous movements; for example, the straightening of kinked elastomeric chains requires a relative flow within the material. Thus, γ_e and γ_f, proceed in parallel. More complicated models are required in other cases.

Figure 10–1.8(a) shows the shear stress required to produce 1 percent strain in polymethyl methacrylate (PMMA*) as a function of temperature. A marked change occurs slightly above 100° C. This temperature corresponds to the glass-transition temperature, T_g, of Fig. 10–1.8(b). Recall that the molecules have the freedom to kink and turn by thermal agitation above the glass temperature. Below that temperature, there is insufficient thermal agitation to permit rearrangements of molecules. Thus, this point represents a discontinuity in the thermal behavior of the material. Observe in Fig. 10–1.8(a) that the stress required for a deformation changes by more than two orders of magnitude at the glass temperature. Obviously, the glass temperature is important for polymer behavior.

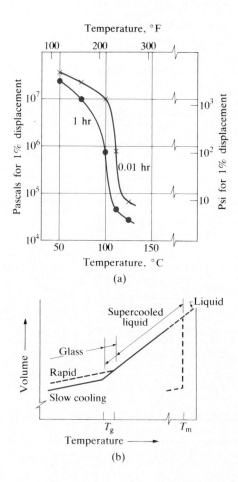

(a)

(b)

FIG. 10–1.8

Viscoelastic Deformation (Ordinate) Versus Temperature (Polymethyl Methacrylate, PMMA). (a) With more time at a given temperature, less stress is required for unit deformation. There is a major decrease in the modulus, M_{ve}, at the glass temperature. (b) The glass temperature is lower with slower cooling, because more time permits molecular adjustments to the stress.

---

* Lucite is one trade name. The composition of PMMA is given in Table 2–3.1.

The two curves of Fig. 10–1.8(a) indicate that less stress is required when the time of stressing is increased from 36 sec (0.01 hr) to 1 hr. The two curves also indicate that the glass temperature drops $\sim 10°$ C as the time is increased from 0.01 to 1.0 hr. This change is reflected in Fig. 10–1.8(b) as a drop in the glass-transition temperature with slower cooling. With slower cooling rates, or with longer times, the molecules can be rearranged at somewhat lower temperatures.

We can compare different molecular structures and their effect on deformation. In Fig. 10–1.9, the ordinate shows the *viscoelastic modulus* M_{ve}, where

$$M_{ve} = s/(\gamma_e + \gamma_f) \qquad (10-1.6)$$

As in Chapter 8, s is shear stress and γ is shear deformation, γ_e being elastic deformation and γ_f being displacement by viscous flow. The abscissa has been generalized. The right end includes higher temperatures and/or longer times, both of which introduce more deformation (and therefore lower values for M_{ve}). At the left end of Fig. 10–1.9(a) and below the glass temperature T_g, where only elastic deformation can occur, the material is comparatively *rigid;* a clear plastic tumbler

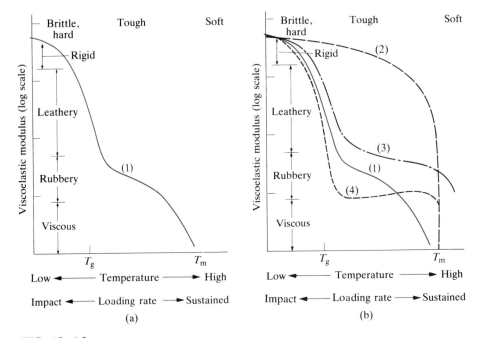

FIG. 10–1.9

Viscoelastic Modulus Versus Structure (Schematic). (1) Amorphous linear polymer. (2) Crystalline (100-percent) polymer. (3) Cross-linked polymer. (4) Elastomer (rubber).

used for soft drinks on air flights is an example. In the range of the glass temperature, the material is *leathery;* it can be deformed and even folded, but it does not spring back quickly to its original shape. In the *rubbery plateau,* polymers deform readily but quickly regain their previous shape if the stress is removed. A rubber ball and a polyethylene "squeeze" bottle serve as excellent examples of this behavior because they are soft and quickly elastic. At still higher temperatures, or under sustained loads, the polymer deforms extensively by *viscous flow.*

Figure 10-1.9(b) compares the deformation behavior for the different structural variants we mentioned earlier with the amorphous polymer just described. A highly (100-percent) *crystalline* polymeric material (curve 2 in Fig. 10-1.9b) does not have a glass temperature. Therefore, it softens more gradually as the temperature increases, until the melting temperature is approached, at which point fluid flow becomes significant. The higher-density polyethylenes (Table 7-5.1) lie between curves (1) and (2) of Fig. 10-1.9(b) because they possess approximately 50 percent crystallinity.

The behavior of *cross-linked* polymers is represented by curve (3) of Fig. 10-1.9. A vulcanized rubber, for example, is harder than is a nonvulcanized one. Curve (3) is raised more and more as a larger fraction of the possible cross-links are connected. Note that the effects of cross-linking carry beyond the melting point into the true liquid. In this respect, a network polymer such as phenolformaldehyde (Fig. 2-4.2) may be considered as an extreme example of cross-linking; it gains its thermoset characteristics from the fact that the three-dimensional amorphous structure carries well beyond an imaginable melting temperature.

Once the glass temperature is exceeded, *elastomeric* molecules can be rotated and unkinked to produce considerable strain. If the stress is removed, the molecules quickly snap back to their kinked conformations (Fig. 4-3.5). This rekinking tendency increases with the greater thermal agitation at higher temperatures. Therefore, the behavior curve (4) increases slightly to the right across the rubbery plateau (Fig. 10-1.9b). Of course, the elastomer finally reaches the temperature at which it becomes a true liquid, and flow proceeds rapidly.

Example 10-1.1

The glass-transition temperature of a thermoplastic polymer is 95° C. The viscosity at 110° C is four times too great for a particular molding process. (a) What temperature is required? Assume that the temperature cannot be controlled to closer than $\pm 1°$ C. (b) What viscosity variation might be expected?

Procedure An indicated for Eq. (10-1.4) and in Table 10-1.1, the viscosity drops sharply as the glass-transition temperature is exceeded. We shall solve the viscosity at 110° C. Then we shall determine the temperature at which the viscosity is reduced by a factor of four. Finally, we shall determine the viscosity variation as the temperature is raised or lowered 1° C.

Calculation

(a) At 110° C, $\Delta T = 15°$:

$\log_{10} \eta = 12 - (17.5 \times 15)/(52 + 15)$

$\log_{10} \eta = 8.1 \qquad \eta = 120 \text{ MPa} \cdot \text{s}$

Since 30 MPA·s is desired,

$\log_{10} 30 \times 10^6 = 12 - (17.5 \times \Delta T)/(52 + \Delta T)$

$\Delta T = 18° \qquad T = 113° \text{ C}$

(b) At 114° C, where $\Delta T = 19°$ C,

$\log_{10} \eta = 7.3 \qquad \eta = 20 \text{ MPa} \cdot \text{s}$

At 112° C, where $\Delta T = 17°$ C,

$\log_{10} \eta = 7.7 \qquad \eta = 50 \text{ MPa} \cdot \text{s}$

Comments As we indicated in the previous discussion, Eq. 10–1.4 is an approximation. Therefore, the relationship may vary, depending on the nature of the polymer. However, this calculation does illustrate the sensitivity of the viscosity to the temperature.

10–2

PROCESSING OF POLYMERIC PRODUCTS

Except for wood (Section 10–5) and fibers such as cotton and wool, most polymers are products of industrial chemistry. Following its synthesis, the typical polymeric raw material is compounded with appropriate additives and then shaped into a product.

Additives

The materials added to polymers may be incorporated for purposes of strengthening or toughening; or they may be included to produce flexibility. Thus, polyvinyl chloride (PVC) is widely used for floor tile; for this usage, an abrasion-resistant *filler* is added. The same polymer is used for raincoats, but with a *plasticizer* added to attain flexibility and to give it characteristics that are markedly different from those of floor tiles. Additives also may be used as *stabilizers,* to keep the polymer from deteriorating, or as *flame retardants.* Finally, additives may be included as *colorants,* for aesthetic purposes. Sometimes, certain materials are added to reduce the cost of the product, if at the same time other properties can be improved. For example, when a filler is added, a plastic becomes more rigid and costs less per unit volume than it would otherwise. Additives frequently serve multiple purposes. For example, carbon black strengthens rubber; it also absorbs ultraviolet light, so rubber offers more resistance to deterioration during service.

Fillers

We shall discuss those additives that are used in large proportions. Most of them are added to give strength or toughness to plastics. Thus, wood flour (a very fine sawdust) is commonly added to a PF (phenol-formaldehyde) plastic (Fig. 2-4.2) to increase strength (Fig. 10-2.1). More important is that 35 v/o wood flour more than doubles the toughness of a PF plastic. As further fringe benefits, wood flour is a replaceable resource that costs less than one-half of what an equal volume of PF would cost. Thus, the service properties of the product are improved at the same time as the cost is being reduced. This is real engineering! Note from Fig. 10-2.1 that the improvement in strength is not a result of the addition of the wood flour alone, because 100 v/o wood flour has nil strength. Rather, it is a result of the mutual interaction of the two components, just as a steel made of ferrite plus carbide is stronger than either one of the components alone.

Fillers may be of various types. Wood flour is a polymer—one containing cellulose. It has the advantage of being essentially of the same density as PF; therefore, the product retains a low specific gravity. Silica flour (finely ground SiO_2 made from quartz sand or quartzite rock) also is used (Example 10-2.1). It is appreciably harder than wood flour and therefore adds abrasion resistance to the plastic product. Furthermore, it neither burns nor softens at high temperatures; as a result, it adds thermal stability to the product. Of course, it does increase the product's total density, because the specific gravity of SiO_2 is about twice that of common resins. (See Appendix C.)

Fibrous fillers are especially effective for adding strength to a product. These fillers may be glass fibers, organic textile fibers, or fibers from mineral sources. They are often chopped into short lengths so that they can be mixed with the polymeric materials that are subsequently molded as fiber-reinforced plastics (FRP). In Section 10-4, we shall pay specific attention to composites in which a high volume fraction of continuous fibers reinforce polymers, often with considerable enhancement of properties.

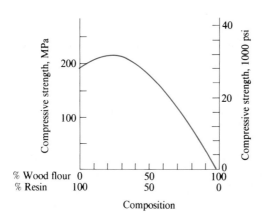

FIG. 10-2.1

Addition of a Filler to a Plastic (Wood Flour Added to Phenol-Formaldehyde). The mixture of the two is stronger than is either alone. (L. H. Van Vlack, *Materials for Engineers: Concepts and Applications,* Addison-Wesley, Reading, Mass., with permission.)

Plasticizers

At normal temperatures, the small molecules discussed in Section 2–2 and shown in Fig. 2–2.1 are generally liquids or gases. In contrast, the macromolecules that we have been describing here are solids, because they comprise long chains or networks. When small molecules are intimately mixed with macromolecules, they reduce the rigidity of the macromolecular — or polymeric — product. When small molecules surround large ones, the large molecules move more readily, in response to either thermal agitation or external forces. In brief, the small molecules *plasticize* the larger ones. Plastics that are normally stiff, such as polyvinyl chloride, can be made flexible, an obvious requirement if the product is to be used in film or sheet form. A plasticizer, in effect, lowers the glass-transition temperature T_g, so molecular movements and rearrangements can occur at room temperature in polymers that would otherwise be rigid.

A plasticizer must have certain characteristics. It should have a high boiling temperature (low vapor pressure), so that it will not readily evaporate. You can understand that a plasticized raincoat would become useless if the coat were to become stiff and brittle with time; this, of course, could happen if the plasticizer gradually evaporated, leaving a plastic that was below its glass temperature.

The plasticizer must not be soluble in the liquids with which it comes in contact. Although some plasticizers are nonvolatile, solvents such as petroleum products can dissolve a large variety of micromolecules. Thus, the inner surface layer of a plastic container may be embrittled as the plasticizer is depleted by solvents that are stored in the container.

A plasticizer must be "compatible" with the polymer. Although we leave the details to the polymer scientist, let us note that, in the case of some small molecules, the molecule may have greater attraction to another of its kind than to the surface of large molecules. Likewise, strong attraction between two adjacent large molecules may prevent the entry of the small molecules between them for plasticizing purposes. For a plasticizer to be suitable, the small molecules should be attracted to the surfaces of the large molecules; the small molecules should not segregate within the plastic. The choice of the most suitable plasticizer depends on the details of the characteristics of the molecular structure and usually involves extensive testing by the manufacturer.

Mixing

Uniformity is a necessity for quality materials. Even materials that are composed of multiple phases must have a "homogeneity in their heterogeneity." For example, in a steel it is necessary to have the hard carbides uniformly distributed throughout the tough ferrite matrix. Likewise, it is necessary to have an equal distribution of additives to a polymer. Slight gradients in colorants lead to obvious variations in products. Also, excess plasticizers or fillers at one point and deficiencies 1 mm away lead to suboptimal properties at both locations.

Mixing of additives into a plastic product is far from simple. Commonly, we are mixing unlikes. A silica flour filler has twice the density of many of the polymers with which it is combined. Some additives are in the form of liquids, others are solids, whereas polymers are either viscous melts or semisolids during the mixing. This latter characteristic is critical. Simple stirring does not suffice.

Most polymer mixing, or *compounding,* is performed as a batch process on an *open-roll mill* (also called a rubber mill). This is illustrated in Fig. 10–2.2. It has two rolls that can be internally heated or cooled. The *nip* of the rolls can be adjusted to different size gaps as required. They operate at different temperatures and speeds (3 to 4 sec/revolution). The velocity and viscosity gradients that result introduce a kneading action that is quite effective. Also, the temperature difference causes the sheet compound to invest one of the rolls; that is, the plastic mixture blankets one of the rolls and not the other. Typically, rubbers will coat the hot roll. Several hundred passes through the nip achieve a uniform mixing if the operator cuts the blanket and peels it from the coated roll several times, each time feeding it back into the nip to form a new blanket.

Polymers that are subject to oxidation at the mixing temperature must be blended in an internal mixer that is enclosed and excludes air. Rotors and blades perform the kneading action.

Shaping Processes

There are three general processing steps between the polymer mixture and the final product: (1) softening, (2) molding, and (3) hardening. The *softening* is commonly achieved by heating but may involve plasticization by solvents. The *molding* commonly involves the application of pressure, although there are certain exceptions. The *hardening* may occur by cooling, by a chemical reaction, or by the volatilization of a fugitive plasticizer. Typically, these three steps are sequenced within one process. Thus, as we examine a number of the forming processes, it will be convenient to identify these three steps.

A significant majority of polymers may have their thermal behavior categorized as (1) thermoplastic, or (2) thermosetting. The *thermoplasts* are *linear polymers* (Section 2–3) with only limited, if any, cross-linking or branching; therefore, they

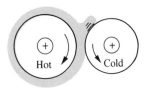

FIG. 10–2.2

Open-Roll Mill (Rubber Mill). The components of the plastic or rubber are blended by a shearing action in the nip region between the rolls. The two rolls rotate at slightly different speeds. They also are maintained at different temperatures. (L. H. Van Vlack, *Materials for Engineers: Concepts and Applications,* Addison-Wesley, Reading, Mass., with permission.)

soften at elevated temperatures. As the temperature rises, the molecules can respond to pressure by sliding past one another. Most of the vinyl compounds of Table 2–3.1 are linear and fall into this category. Of course, for thermoplasticity to be effective, the temperatures must be above the glass-transition temperature, T_g (Fig. 4–2.4).* Further, the normal temperatures of use must be in the range in which the plastic retains its shape. The *thermosets* are altered both chemically and structurally during thermal processing. They develop a three-dimensional structure, either a network structure (Fig. 2–4.2) or a cross-linked structure (Fig. 4–3.9). These structures are only partially completed before forming but become one big three-dimensional molecule when they are processed in the presence of heat and pressure. The *curing* within the heated mold completes the formation of the network. Therefore, a thermosetting polymer gains rigidity before the pressure and added temperature are removed. Thermoplastic polymers must be cooled in the mold (or on exit).

Finally, before describing the various forming processes, we should examine polymer viscosity more closely than we did in Section 10–1. When a polymer is under pressure, the relationship between viscosity, η, and the rate of strain, γ/t, is less ideal than that shown in Eq. (10–1.2).† Several factors contribute to this variance; among them is the "free space" that is present in the supercooled liquid (Example 7–5.2). This leads to a high compressibility that becomes evident as die swell (a volume expansion at the exit of the die) when the viscous polymer is extruded out of the die. Of course, this is undesirable when specific dimensions are required. It also leads to difficulties in predicting the rate of flow through feed channels *(sprues)* of closed dies. This can mean there will be incomplete filling of the dies during injection molding unless allowances are made.

Extrusion

The extrusion process is sketched in Fig. 10–2.3(a). Starting materials are commonly granules of a thermoplastic polymer that had been sized for easy flow. An *auger* (or screw) feeds the granules into a heated zone where the thermoplastic pellets soften. However, the heating does not come entirely from external heaters. There is considerable energy that goes into the polymer melt from the work of the auger as it compresses the melt and forces it through the die. For example, the motor turning the auger uses as much as 50 kW (~ 70 Hp) to drive a 10-cm screw. This power is absorbed by the plastic in the form of heat.

To control the temperature as closely as possible, we can segment the heater. The temperature of each zone is measured by thermocouples and is controlled

---

* T_g decreases somewhat when shear stresses are applied, since rearrangements of molecules do not depend solely on thermal energy.

† That is one reason we used an empirical η versus T relationship in Eq. (10–1.4), rather than the more orthodox viscosity Eq. (10–1.3c).

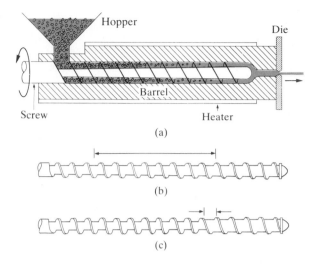

FIG. 10–2.3

Extrusion (Auger). (a) Schematic. Solid pellets are fed into the heated zone by an auger. The melted thermoplastic is extruded through an open-end die as bars, pipe, or sheet product. (b) Auger for polyethylene, which softens gradually. (c) Auger for nylon, which softens abruptly. The compression and metering section (arrows) is much shorter in the latter.

separately, so that the desired temperature profile is obtained along the barrel. This is important, because various thermoplastics have different softening characteristics. For example, low-density polyethylene, which is amorphous, softens progressively as the temperature increases. In contrast, nylon softens rather abruptly at a higher temperature because it possesses high crystallinity. Thus, the screw design of the auger must be modified depending on the product. As shown in Fig. 10–2.3, the feed into the heated zone and the compression to eliminate the pores is simultaneous and continuous for polyethylene (Fig. 10–2.3b). The compression is abrupt in Fig. 10–2.3(c) for nylon after the granules have passed well into the heated zones.

The die of an extruder may have a variety of geometries. Of course, the cross-section along the length of the product must remain constant. A circular orifice produces a solid rod, and a slit produces sheet or film products. It also is possible to manufacture longitudinal moldings of irregular cross-sections. Tubing and pipe must be extruded by mounting a *mandrel* (or "torpedo") in the center of the hole, so that extrusion is through the annulus. Alternatively, wire can be fed continuously into the center of the orifice to be coated with rubber insulation. Rates may approach 1 km/min. The extrusion process is adaptable to large-scale production.

The plastic must flow as it emerges through the die but must harden immediately on exit so as to retain its desired shape. The surface-to-volume ratio and the amount of mass to be cooled per sec dictate the choice of air or water for cooling.

Injection Molding

The open die of the extrusion process may be replaced with a closed die, as shown schematically in Fig. 10–2.4(a). The die is a split mold that contains the negative contours of the product to be made. The hot plastic is forced (injected) into the

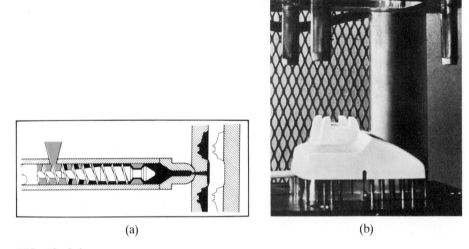

(a)   (b)

FIG. 10–2.4

Injection Molding (Telephone Receivers). (a) Extrusion into closed dies (schematic). The softened thermoplastic is forced through sprues to the cooled die cavity, where it hardens. (b) Formed product in the opened die (rotated 90° since the large auger in part (a) is operated horizontally). (Courtesy of Western Electric Co.)

mold by either an auger or a hydraulic plunger. The injection process provides more flexibility in product geometry than does extrusion, because the cross-section is not fixed longitudinally. Examples of injection-molded products are numerous, ranging from plastic ice-cream spoons (Fig. 2–3.4) to telephone receivers (Fig. 10–2.4b).

The injection molding of *thermoplastic* polymers generally uses water-cooled molds for hardening the product. This facilitates production, because the product becomes rigid almost immediately and can be removed, so there can be a sequel injection. However, this quick rigidity presents complications, since the viscous melt must enter through chilled sprues (feed channels) and mold cavities. This is illustrated in Fig. 10–2.5 where we see a rigid shell of chilled polymer that constricts the channel. Higher injection pressures are required to feed the far cavities and to avoid porosity from solidification shrinkage.

Thermosetting polymers may be injection molded more readily than they can be extruded, because the injection cycle may include time for *curing;* that is, for the completion of the polymerization reaction. It is not necessary to cool the mold inasmuch as the product "sets" rigidly while hot. Close control is required in the heating and feeding part of the cycle. The granular feed must be only partially polymerized, with an average of approximately two bonds per mer (Example 2–4.1). This gives it thermoplastic characteristics for molding. Further polymeri-

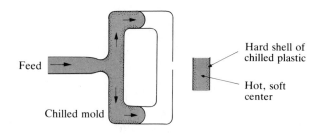

FIG. 10-2.5

Flow in Chilled Mold (Schematic). The mold is chilled to harden the plastic for stripping. However, this produces a rigid shell that constricts the flow channels. High injection pressures are required.

zation reactions occur during curing in the mold to produce a rigid network or cross-linked structure. If that second-stage polymerization were to start during the feeding step, the mold would not fill and the auger (or plunger) would bind.

Sheet Molding

In principle, sheet molding is the simplest forming process. An extruded sheet of thermoplastic material is clamped over the edges of a mold. The sheet is heated by infrared heaters. Gravity or a vacuum sags the sheet into the contour of the mold (Fig. 10-2.6). Variants of the process use air pressure or even incorporate a mating mold on the opposite side.

Sheet molding is an inexpensive process for products with suitable geometries. Applications range from raised-letter signs to the housings for automobile instrument panels.

Blow Molding

Bottles and related products that have a constricted neck cannot be molded by any of the previous processes. There would be no way to remove the interior mold. In the blow-molding process, which is an adaptation from the container glass industry, a soft plastic tube is extruded and is cut free from the die head (Fig. 10-2.7). The resulting "hollow drop," or *parison,* of soft plastic is surrounded by a mold. Air pressure is used to expand the parison until it takes on the shape of the mold. Blow molding, more than any other plastic-forming process, requires a knowledge of the relationships between viscosity, temperature, and viscoelastic behavior. Also not to be ignored is the accumulated empirical experience of the operator.

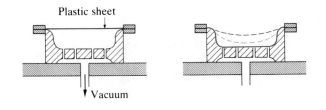

FIG. 10-2.6

Sheet Molding (Vacuum Molding). The molding pressure may also be created by air pressure or by a mating mold in the upper side.

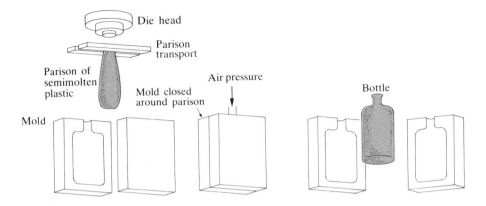

FIG. 10-2.7

Blow Molding. Air expands the viscous parison of plastic into the shape of the mold. Precise time and temperature controls are required in this process, which originated in the glass industry.

Spinning

Manufacturers make fibers by forcing the plastic through a multiple orifice *spinnerette.* This device contains as many as 50 or 100 holes, each less than 0.2 mm in diameter. In *melt spinning,* the thermoplastic polymers, such as nylon and polyesters, are heated to low viscosity. The slender, emerging fibers (Fig. 10-2.8a) cool quickly in a current of air before they travel over a take-up roll.

Rayon is made by a *dry-spinning* operation. It is dissolved in acetone to produce

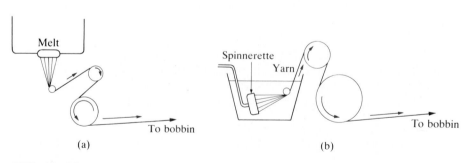

FIG. 10-2.8

Fiber Spinning. (a) Melt spinning (polyester fibers). The thermoplastic filaments solidify as they quickly cool. (b) Wet spinning (cellulose fibers). The extruded filaments complete their polymerization in the bath. The rolls move at different surface speeds (arrow lengths), so the filaments are stretched between the two take-up rolls, extending and orienting the molecules.

a thick solution that is extruded through the spinnerette. The acetone evaporates*
to permit the fibers to dry before they go over the take-up roll. In each of the above
spinning operations, the filaments travel at about 15 m/sec (3000 ft/min).

A third, slower process is *wet spinning.* It is used when the fibers must react with
the bath to complete the polymerization reaction after they leave the spinnerette
(Fig. 10–2.8b).

Molecular Orientation

We previously noted that, as linear molecules become uncoiled and straightened,
they align themselves with one another, and the crystallinity increases. Concurrently, the stress requirements increase for each additional increment of strain
(Fig. 10–1.3). Thus, the elastic modulus increases. Also the strength, the heat
resistance, and the moisture resistance increase. These facts are particularly useful
to the processor of fiber products. Figure 10–2.8 illustrates the procedure that is
used in fiber production.

The warm fibers emerge from the spinnerette as a viscous, but deformable,
supercooled liquid. They have been wrapped 360+° around each roll; the second
roll travels appreciably faster than does the first roll. Therefore, the fibers are
uniformly stretched several hundred percent, inducing crystallization. The
slender fibers cool rapidly to ambient temperatures while still under tension, thus
preserving the crystallization. Not only is the more fully crystalline product
stronger and more stable in service, but also the property changes that accompany
the glass-transition temperature are less abrupt, since only a minor fraction of the
product is amorphous. Essentially all fibrous products that are made artificially
are stretched in processing for molecular orientation and crystallization.

The molecules of film products also can be oriented; however, the process is
more complex, since the alignment must be bidirectional within the plane of the
film, rather than uniaxially along the length of a fiber.

The same principle could be used for a film product. Here, however, an undesirable anisotropy would be introduced. Specifically, the film would be strong in
one direction and weak at right angles to it.† A more desirable process includes
simultaneously stretching in the two coordinate directions, so that the molecular
orientation is biaxial. The *bubble method* (Fig. 10–2.9) is one such process. A
cylindrical film is extruded through an annular die. Air is blown through the
mandrel and expands the sleeve into a larger cylinder.‡ Both circumferential and
longitudinal stretching provide the biaxial orientation. The film may be slit if
desired, or it may be heat sealed into plastic bags.

---

* It is collected and reused.

† This is demonstrated by analogy with newsprint, in which the wood fibers are preferentially aligned
during the paper-making process. As a result, it is easy to tear newsprint straight in one direction, but
more difficult to do so in the perpendicular direction.

‡ Expansion ceases when the film temperature drops below T_g.

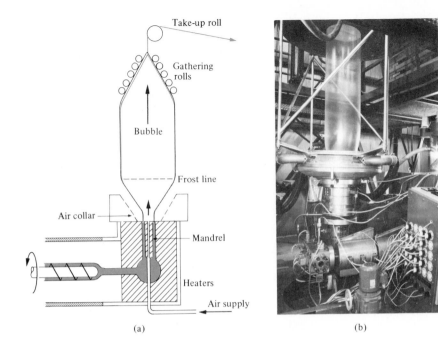

(a)   (b)

FIG. 10–2.9

Bubble Forming (Plastic Film). The cylindrical sheet is expanded simultaneously in two directions as it cools below the glass-transition temperature. Therefore, it develops biaxial strength in the two dimensions of the film. (a) Schematic. (L. H. Van Vlack, *Materials for Engineers: Concepts and Applications*, Addison-Wesley, Reading, Mass., with permission.) (b) Production of polyvinylidene chloride $(C_2H_2Cl_2)_n$. (Courtesy of Dow Chemical U.S.A.)

Example 10–2.1

Assume we mix 84 kg of silica flour (finely powdered SiO_2) with 100 kg of phenol-formaldehyde (PF). The densities are 2.65 and 1.3 g/cm³, respectively. What fraction of the volume is filler?

Procedure   Determine the true volume of each.

Calculation   Basis: 100 g PF and 84 g SiO_2.

$$100 \text{ g PF}/(1.3 \text{ g/cm}^3) = 77 \text{ cm}^3 \text{ PF} = 0.71$$
$$84 \text{ g SiO}_2/(2.65 \text{ g/cm}^3) = 32 \text{ cm}^3 \text{ SiO}_2 = 0.29$$

Comment   A filler is used not only because it is an inexpensive diluent for a more expensive polymer, but also because it gives additional thermal and dimensional stability to the plastic.

Example 10-2.2

Detect evidence of the ordering that accompanies the orientation of rubber molecules into longitudinal bundles when the rubber is stretched. Explain the results.

Procedure For this simple experiment, use a heavy but easily deformable rubber band. Your lip can serve as a sensitive detector of temperature changes. Place the band in contact with your lower lip. Stretch it rapidly; after a few seconds, quickly (without snapping) return the rubber band to its original length. Repeat this cycle several times.

Results If you are careful, you will be able to detect a temperature increase on stretching, and a temperature decrease on release.

Explanation (See discussion of conformational disorder in Section 4-3.) Stretching the rubber molecules produces longitudinal alignment, and introduces a greater amount of order to the structure (less entropy). In fact, crystallization can occur (Fig. 10-2.10). This releases the heat of fusion; but, since rubber is a very poor conductor and dissipater of heat, this heat raises the temperature of the rubber. Relaxing the rubber allows the oriented molecules to rekink, increasing the amount of disorder (entropy). This change requires energy, which must be drawn from the sensible heat, lowering the temperature.

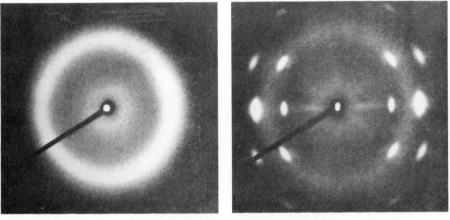

(a)   (b)

FIG. 10-2.10

Deformation Crystallization of Natural Rubber (Polyisoprene) Revealed by X-Ray Diffraction. (a) Unstretched. (b) Stretched. (S. D. Gehman, *Chemical Reviews,* vol. 26, page 203, by permission.)

10-3
POLYMERIC COMPOSITES

Composites contain two (or more) distinct materials as a unified combination. Thus, reinforced concrete is a composite of steel rods in a concrete matrix. Likewise, many sailboat hulls are made of fiber-reinforced plastics (FRP), in which the fiber is typically glass and the plastic is commonly a polyester. Other composites include products such as glass-coated steel for kitchen oven liners, and rocket motors with precisely aligned aramid fibers (Fig. 10-3.1) as the reinforcement.

Two-phase alloys generally are not considered to be a composite, because the two phases are not formulated separately, but rather originate within a single manufacturing process. Likewise, we previously considered polyblends on the basis of the microstructure that developed during processing (Section 7-5). Admittedly, a precise definition of a composite is arbitrary.

Most of the composite mixtures that will receive our attention are a combination of discontinuous phases and matrix phases. The former are commonly *fibers*, but also may be *particles*. The discontinuous component usually accounts for 10 percent or more of the total volume. The *matrix* is a polymer in many of the common composites, but it also may be metallic or ceramic, as it is in some of the more sophisticated materials that are being developed.

Particulate Composites

Particulate materials are essentially equiaxed; that is, there is no great disparity in their three spatial dimensions. They are seldom spheres or cubes, however, being instead generally angular, or partially rounded.

It is not uncommon for particulate materials to be used both for reinforcement and as an extender within a composite. In the latter case, they are called *fillers*.

FIG. 10-3.1

Fabrication of a Composite Rocket Motorcase. The polymer fibers are an aramid (see text) that will be bonded with an epoxy resin. (Courtesy of Morton Thiokol.)

Thus, the use of 25 percent wood flour with the phenol-formaldehyde resin in the material of Fig. 10–2.1 optimizes the strength, and reduces the cost, since the wood flour is much less expensive than is the resin. Fillers also may be added to increase fire resistance, to reduce shrinkage, or to add wear or skid resistance. Silica sand and silica flour can contribute to each of these goals, as well as saving costs (Example 10–2.1).

Rigid particles increase the hardness of a composite for the same reasons that spheroidized carbide particles strengthen a ferrite matrix in steels. That is, they provide plastic constraint (Section 9–4). Since fine ground particles possess more surface area per unit volume than do coarser particles, they increase the hardness of the composite more substantially for a given addition.

A foam could be called a composite of a material and a gas. Styrofoam can serve as our example. The near-zero density of the "particle" provides a light-weight product with low thermal conductivity. Of course, it does so at the expense of mechanical properties.

We commonly study and analyze properties of composites in terms of the properties of the contributing materials. We then formulate mixture rules, in which the properties are a function of the amounts and geometric distribution of each contributing material. The simplest mixture rules are for scalar properties, such as density or heat capacity. The density of the mixture is

$$\rho_m = f_1 \rho_1 + f_2 \rho_2 + \cdots \qquad (10\text{–}3.1a)$$

or, simply,

$$\rho_m = \Sigma f_i \rho_i \qquad (10\text{–}3.1b)$$

where f is the volume fraction of each contributor. Likewise, for heat capacity, c,

$$c_m = \Sigma f_i c_i \qquad (10\text{–}3.1c)$$

In this calculation, f is the volume fraction if the heat capacity is in $J\,°C^{-1}m^{-3}$. If the heat capacity is given in $J\,°C^{-1}kg^{-1}$, the f must be a mass fraction.

Fiber-Reinforced Plastics (FRPs)

Glass is the most widely used fiber for fiber-reinforced composites. It can be very strong, it is relatively inexpensive, and — of major importance — it has a higher elastic modulus than do the polymeric matrixes.

The E-glasses are used as fiber reinforcement. Designed originally for electrical applications, they are most commonly calcium aluminoborosilicate glasses that contain no Na^+ ions.* Fused-silica glasses have higher elastic moduli, but are too

---

* Sodium ions of a regular soda–lime–silica glass present a problem, because the glass fibers have a tremendously large surface-to-volume ratio. Ambient humidity can introduce ion exchange at the fiber surface:

$$H_2O_{air} + Na^+_{gl} \rightarrow H^+_{gl} + Na_2O_{surface} \qquad (10\text{–}3.2)$$

The Na_2O, which adsorbs additional moisture, can produce shorting across the surface in electric applications; it can also lead to loss of bonding between the fiber and matrix of a composite.

expensive for extensive use. All glass-fiber products must have their surfaces coated (1) to give protection from surface damage and loss of strength, and (2) to provide a transition bond between the fiber and the matrix.

Glass-fiber reinforced composites are useful for many applications. Although polymeric composites may have lower ultimate tensile strengths, S_u, than do many metals, their strength-to-density ratios, S_u/ρ, are high. Like strength, however, rigidity is a significant design factor, particularly in the presence of compressive or bending forces. Under those conditions, the elastic modulus–to-density ratio, E/ρ, is the figure of merit for making comparisons between materials in engineering design. Table 10–3.1 shows the E/ρ ratios for a number of common structural materials, including several of those in Appendix C. It is noteworthy that the majority have comparable values. The 50–50 glass–plastic composite has an E/ρ ratio of only 20,000 N·m/g. These lower values are typical for most artificial composites. A glass-reinforced composite would have to contain more than 70 v/o glass to approach the E/ρ ratio of metals. That percentage is difficult to produce inexpensively, because special efforts are required to align the fibers perfectly.

Higher E/ρ ratios must be obtained with fibers that have elastic moduli higher than that of E-glass (70 GPa, or 10,000,000 psi). This ratio can be raised by about 10 percent with the less dense, fused-silica fibers. However, as we noted previously, these fibers are too expensive to compensate for this marginal gain. Therefore, engineers are examining other materials as possible *high-performance* fibers. A number of these fibers are listed in Table 10–3.2.

TABLE 10–3.1 Modulus/Density (E/ρ) Ratios of Common Materials

MATERIAL	DENSITY, Mg/m³	YOUNG'S MODULUS, GPa	E/ρ N·m/g
Aluminum	2.7	70	26,000
Iron and steel	7.8	205	26,000
Magnesium	1.7	45	26,000
Glass (soda-lime)	2.5	70	28,000
Wood (spruce)	0.43	11	26,000
Wood (birch)	0.61	16	27,000
Polystyrene	1.05	2	2,700
Polyvinyl chloride	1.3	<4	<3,500
50 v/o glass-plastic*	~1.9	~37	~20,000
70 v/o glass-plastic*	~2.1	~50	~24,000

* Values for these composites vary, depending on the plastic; these values are typical of most glass-reinforced plastics.

TABLE 10-3.2 Modulus/Density (E/ρ) Ratios (Fibers for High-Performance Composites)*

FIBER	DENSITY, Mg/m^3	YOUNG'S MODULUS, GPa	E/ρ N·m/g
Alumina	3.9	400	100,000
Aramide	1.3	125	100,000
Boron	2.3	400	170,000
Beryllium	1.9	300	160,000
BeO	3.0	400	130,000
Carbon	2.3	700	300,000
Silicon carbide	3.2	500	160,000
Silicon nitride	3.2	400	120,000

* Because these fibers are significantly more expensive than are those in Table 10-3.1, their use is limited to specialized applications (see, for example, Fig. 10-3.2).

Carbon fibers are more expensive than are glass fibers, but have great attraction where the cost is justified. Originally made by carbonization of rayon fibers, almost all the carbon fibers now use polyacrylonitrile as a precursor. (This is the PAN polyvinyl of Table 2-3.1.) Nearly 90 percent of the structure of the Voyager aircraft was made with composites containing carbon fibers (Fig. 10-3.2).

FIG. 10-3.2

Application of High-Performance Composites (Voyager). Most of the structural members of the plane were made of graphite fibers embedded in an epoxy resin, giving an E/ρ ratio with high stiffness and a low weight. (Wide World Photos.)

Aramid fibers* are the only polymeric fibers that have the high-stiffness characteristics that are required in many advanced engineering designs. They have an
advantage over the other materials of Table 10–3.2 in that they undergo some
plastic deformation before failure; therefore, they introduce a greater toughness.
They are not as favorable in compression as are inorganic fibers, and they will
absorb moisture unless they are isolated from air and water.

10–4
PROPERTIES OF COMPOSITES

When improved strength is the major goal, the reinforcing component must have
a large *aspect ratio;* that is, its length-to-diameter ratio must be high, so that the
load is transferred across potential points of fracture. Thus, we place steel rods in a
concrete structure, and we combine glass fibers and polymers for fiber-reinforced
plastics (FRP).

It is obvious that the reinforcement must be the stronger component if it is to
carry the load. It may be less obvious that the reinforcement must have the higher
elastic modulus. Likewise, it may not be immediately apparent that the bond
between the matrix and the reinforcement is critical, since it is generally necessary
to transfer the load from the matrix to the fibers or rods if the reinforcement is to
serve its purpose.

Elastic Behavior

For the reinforcing materials to carry most of the load, *the reinforcement must
have a higher elastic modulus than does the matrix.* Consider Fig. 10–4.1, in
which a glass fiber reinforces polyvinyl chloride. When loaded in tension, the two
materials must deform together. Assume a strain of 0.002. With the two moduli
being $\sim 70{,}000$ MPa and ~ 350 MPa, respectively, and from Eq. (1–2.3), the glass
develops a stress of 140 MPa versus 0.7 MPa in the plastic.

$$e = 0.002 = (s_g/70{,}000 \text{ MPa}) = (s_{pvc}/350 \text{ MPa}) \qquad \text{(10–4.1)}$$

Because E_g is $\sim 200 \, (E_{pvc})$, we have $s_g \approx 200 \, (s_{pvc})$. Generally, in a two-component
composite,

$$s_1/s_2 = E_1/E_2 \qquad \text{(10–4.2)}$$

The modulus of elasticity (in tension) of a fiber-reinforced plastic, E_{FRP}, may be
estimated as the volume fraction average if all of the fibers are aligned parallel to
the direction of loading. Consistent with the derivation in Example 10–4.1,

$$E_{FRP} = \Sigma f_i E_i \qquad \text{(10–4.3)}$$

where f_i and E_i are the volume fractions and Young's moduli, respectively, for the
components. Consider a plastic reinforced with 50 v/o of parallel glass fibers

---

* Chemically, poly para-phenyleneterephthalamide. Do not memorize its name!

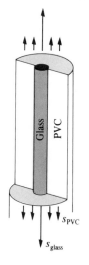

FIG. 10–4.1

**Stress in Composites
(Glass-Reinforced Plastic).**
The strain in the two must be
equal. Therefore, Eq. (10–4.2)
applies.

($E = 70,000$ MPa, or 10^7 psi) and the same fraction of a plastic that has a Young's
modulus of 4000 MPa (580,000 psi). The resulting composite has a Young's
modulus in its longitudinal direction of ~37,000 MPa (or 5.3×10^6 psi). (Of
course, the elastic moduli in the other two coordinate directions will be low and
close to that of the plastic, because they are at right angles to the reinforcement.)

If the same amount of glass is incorporated as a woven fabric into this plastic,
the composite gains two-way reinforcement. However, the values drop below that
of the mixture rule (Eq. 10–4.3). Furthermore, the 45° modulus is low (Fig.
10–4.2). It is now common to use matted glass fibers to avoid this anisotropy in

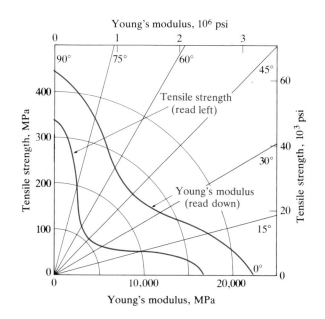

FIG. 10–4.2

**Directional Properties (Cross-
Laminated, Glass-Reinforced
Epoxy).** The reinforcing fibers are
woven in two right-angle
directions. Therefore, there is
high strength and good rigidity in
these directions. The corresponding values are low at 45°.
(Adapted from Broutman, *Modern Composite Materials,*
Addison-Wesley, Reading, Mass.)

composite sheet products. The fibers in the glass mats possess a sufficiently random distribution to give nearly uniform elastic moduli, and therefore uniform load distributions in the two dimensions.

Boundary Stresses

Civil engineers are aware that shear stresses develop at the interface between a reinforcing rod and concrete. For this reason, they specify "deformed" rods with merloned surfaces (ASTM, A305). Comparable shear stresses are encountered between fiber reinforcement and the surrounding plastic matrix in FRP. Here, however, the shear stresses are supported by a chemical rather than by a mechanical bond.

Interfacial shear stresses become particularly important if the fibers are not continuous. This is illustrated in Fig. 10–4.3. In this figure, s_f represents the stress that is carried by the fiber if there are no end effects (infinite length). This corresponds to the calculation in Problem 1041 and depends both on the volume fraction of the reinforcement and on the Young's moduli of the two materials. If the fiber is broken, however, its stress automatically drops to zero at the end of the fiber, and the load is transferred into the matrix (Fig. 10–4.3b). This transfer is made by shear stresses across the interface (Fig. 10–4.3c). The shear stress is very high near the fiber ends, and the weaker matrix must carry an overload (Fig. 10–4.4). This places a premium on long continuous fibers in load-bearing composites. It also favors greater numbers of small-diameter fibers rather than fewer

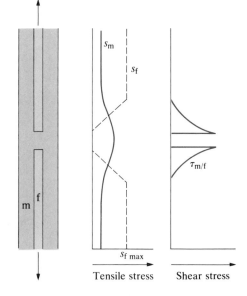

FIG. 10–4.3

Stress Distribution (at a Break in the Reinforcing Fiber). The fiber stress, s_f, drops from its normal value to zero. The load must be transferred across the interface from the fiber to the matrix by shear stresses, $\tau_{m/f}$, which is maximum at the end of the fiber. The matrix has to carry a higher stress s_m in the vicinity of the break (Fig. 10–4.4).

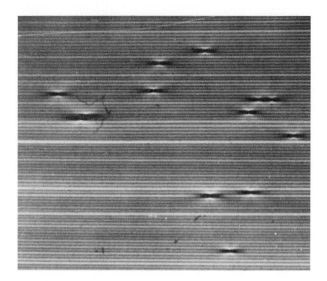

FIG. 10–4.4

Stress Redistribution (Fiber Breaks in a Composite). Polarized-light view of a monolayer, glass–epoxy composite under an axial tensile load shows the fiber breaks (at
the center of the dark bands). The adjacent unbroken fibers (brighter) must carry the
load. (Courtesy of B. W. Rosen, Materials Sciences Corp., Copyright ASTM. Reprinted
with permission.)

large fibers, since there is more interfacial area to support the shear loads, and less
chance that one broken fiber will introduce damaging flaws in the matrix. Finally,
a ductile matrix will adapt more readily to the stress concentrations at fiber ends
than will a brittle matrix.

Boundary stresses also develop during thermal expansion or contraction. As
described in Example 9–8.1, identical dimensional changes, $\Delta L/L$, must occur in
both components if the boundary is to remain intact:

$$(\Delta L/L)_1 = (\Delta L/L)_2 \qquad \text{(10–4.4a)}$$

so,

$$(\alpha \, \Delta T + e)_1 = (\alpha \, \Delta T + e)_2 \qquad \text{(10–4.4b)}$$

or,

$$(\alpha \, \Delta T + F/AE)_1 = (\alpha \, \Delta T + F/AE)_2 \qquad \text{(10–4.4c)}$$

This means that a tensile force, $+F$, develops on one side of the boundary, and a
compressive force, $-F$, develops on the other side, with a resulting shear stress
along the boundary to equalize the dimensional changes. These stresses can lead to
dejoining in nonductile composites.

Toughening and Fracture

We add fibers to composites to provide reinforcement. This may involve increasing the *strength,* so that larger loads can be carried. Reinforcement may also increase the *rigidity* through the use of larger fractions of fibers (Eq. 10–4.3), or through the introduction of fibers with a high value of E/ρ (Table 10–3.2). In engineering design, reinforcement also is used to increase the toughness of the product, so that more *energy* is required to initiate and propagate fracture. In fact, greater toughness is the prime incentive for the selection of many composites.

Fracture toughness is the measure of the energy required to cause failure under load (Section 8–4). The energy required for fracture can be increased in several ways. For brittle materials with linear elasticity, a doubling of the strength, quadruples the energy input before fracture (Fig. 9–8.5a). This increased toughness is applicable to thermosetting plastics with negligible ductility, and is one of the reasons for adding fillers to materials such as phenol-formaldehyde.

A more significant increase in toughness is realized when the microstructure requires the fracture path to traverse ductile phases. This traversal introduces two toughening factors. First, strain proceeds without fracture to increase the area under the stress–strain curve (Fig. 8–4.1b), and hence to increase the stress–strain product—$(N/m^2)(m/m)$ of Section 8–4. Second, ductility at the tip of the crack reduces stress concentrations (Fig. 8–4.5), because the stress concentration is an inverse function of the tip radius. This is the principal reason that most metals are tougher when the production process disperses the harder, reinforcing phase within the ductile matrix—for example, in precipitation hardening (Section 9–4), and tempering (Section 9–6). It is also the reason that domains of the elastomer polybutadiene can toughen ABS polyblends at temperatures below the T_g of a styrene–acrylonitrile copolymer (Section 7–5).

In composites, the energy for complete fracture can be increased by introducing *apparent strain* through *microcracking.* As discussed in the caption of Fig. 9–8.5(b), an apparent strain is developed with the numerous small displacements that occur as the affected area advances. These displacements all lead to a greater s–e product under the stress–strain curve, and, therefore, to greater energy requirements for final failure. The microcracking takes on various forms, which are identified in Fig. 10–4.5.

Fiber *debonding* (Fig. 10–4.5c) requires energy and permits microdisplacements. The materials engineer uses this and the following mechanisms to design tougher composites. We must be careful, however, to retain sufficient boundary cohesion that the strength of the composite is not adversely affected. Fiber *breakage* (Fig. 10–4.5d) also requires an expenditure of energy. Unlike fracture propagation in a homogeneous material, these cracks do not extend catastrophically, because the crack length (c of Eq. 8–4.1) is limited to the fiber dimension. Fiber *slippage* (Fig. 10–4.5e) also consumes energy. A nonductile fiber such as glass breaks statistically where flaws may be present. Thus, a fiber does not necessarily break exactly at the point of matrix failure (Fig. 10–4.6). The pullout after breakage requires a stress and introduces additional apparent strain. Each of the mecha-

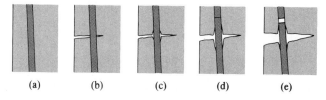

(a) (b) (c) (d) (e)

FIG. 10-4.5

Apparent Strain through Microcracking (Schematic). (a) Fiber in matrix. (b) Crack encounters fiber. (c) Debonding at fiber–matrix interface, which allows additional displacement. (d) Breakage in the fiber at nearby flaw, which is not in line with the progressing fracture through the composite. Energy is required. (e) Slippage, while pulling the fiber out of the matrix. Each of these steps adds to the apparent strain, and therefore increases the energy required for final failure.

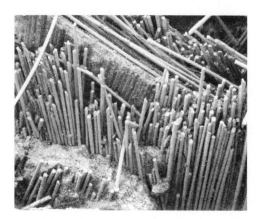

FIG. 10-4.6

Fiber-Matrix Debonding (FRP). Glass-fiber rovings were impregnated with an epoxy resin. During a fracture test, shear stresses debonded the fibers in the vicinity of the fracture. This adds to the energy required for failure. (×80) (Courtesy of General Motors Research Laboratories.)

nisms described in Fig. 10-4.5 delays the ultimate failure, as well as gives further opportunity for further plastic deformation of the matrix. Not only is greater toughness realized, but also opportunity is introduced for incorporating fail-safe features into engineering design. Such designs avoid catastrophic failure because opportunity becomes available to shut down for repairs, or to prevent further injuries or destruction.

Example 10-4.1

Formulate a mixture rule for Young's modulus (longitudinal) of a glass fishing rod. The rod contains longitudinally oriented glass fibers bonded with a polyester resin.

Procedure In longitudinal loading, the strain in the two components must be equal (if they are to deform elastically as a unit). [We shall ignore the lateral strains by assuming the two materials have the same Poisson's ratio.]

Derivation

$$e_{gl} = e_m = e_{pr}$$

$$\left[\frac{F/f\,A}{E}\right]_{gl} = \frac{F/A}{E_m} = \left[\frac{F/f\,A}{E}\right]_{pr}$$

where E_m is Young's modulus of the composite mixture.

$$F_{gl} = f_{gl}E_{gl}F/E_m$$

$$F_{pr} = f_{pr}E_{pr}F/E_m$$

Since $F = F_{gl} + F_{pr}$,

$$F = (f_{gl}E_{gl} + f_{pr}E_{pr})\,F/E_m$$

$$E_m = f_{gl}E_{gl} + f_{pr}E_{pr} \tag{10–4.5}$$

Comment This mixture rule applies only where strains are identical. A laminated composite with transverse loading will have a different mixture rule. (See Example 10–4.3.)

Example 10–4.2

An SAE 1060 steel wire (1 mm in diameter) is coated with copper (combined diameter 2 mm). What is the thermal expansion coefficient of the composite material?

Procedure In the composite, $(\Delta L/L)_{st} = (\Delta L/L)_{Cu}$, and, in the absence of an external load, $F_{Cu} = -F_{st}$. From Appendix C,

$$E_{st} = 205\ GPa \qquad\qquad E_{Cu} = 110\ GPa$$

$$\alpha_{st} = 11 \times 10^{-6}/°\ C \qquad \alpha_{Cu} = 17 \times 10^{-6}/°\ C$$

Basis for calculation, $\Delta T = +1°$ C.

Calculation

$$A_{st} = (\pi/4)(0.001\ m)^2 = 0.8 \times 10^{-6}\ m^2$$

$$A_{Cu} = (\pi/4)(0.002\ m)^2 - 0.8 \times 10^{-6}\ m^2 = 2.4 \times 10^{-6}\ m^2$$

$$(\Delta L/L)_{st} = (\Delta L/L)_{Cu}$$

$$[\alpha\,\Delta T + (F/A)/E]_{st} = [\alpha\,\Delta T + (F/A)/E]_{Cu} \tag{10–4.6}$$

$$(11 \times 10^{-6})(1) + \frac{F_{st}/0.8 \times 10^{-6}\ m^2}{205 \times 10^9\ N/m^2} = (17 \times 10^{-6})(1) + \frac{-F_{st}/2.4 \times 10^{-6}\ m^2}{110 \times 10^9\ N/m^2}$$

$$F_{st} \approx +0.61\ N \qquad \text{(tension on heating)}$$

$$F_{Cu} \approx -0.61\ N \qquad \text{(compression on heating)}$$

For $\Delta T = +1°$ C,

$$(\Delta L/L)_{Cu} = (17 \times 10^{-6}/°\ C)(1°\ C) + \frac{-0.61\ N/(2.4 \times 10^{-6}\ m^2)}{110 \times 10^9\ N/m^2}$$

$$\bar{\alpha} \approx 15 \times 10^{-6}/°\ C$$

Comment The thermal expansion varies slightly with the type of steel.

Example 10–4.3

Formulate a mixture rule for Young's modulus of a laminate, when the stress is applied perpendicular ($\perp$) to the "grain" of the laminate.

Derivation With transverse loading, $s_1 = s = s_2$.

$$\bar{E}_\perp = s_\perp/e = s/(e_1 f_1 + e_2 f_2)$$
$$= 1/(f_1/E_1 + f_2/E_2)$$
$$1/\bar{E}_\perp = f_1/E_1 + f_2/E_2 \tag{10–4.7}$$

Comment This derivation and the one for Eq. (10–4.5) assume that the Poisson ratios for the two components are comparable. Thus, there would be no secondary stresses because of differences in lateral strains.

Example 10–4.4

Refer to the composite of Example 10–4.2. The yield strength of the steel is 280 MPa (40,000 psi); that of the copper is 140 MPa (20,000 psi). (a) If this composite is loaded in tension, which metal will yield first? (b) How much load, F, can the composite carry in tension without plastic deformation? (c) What is Young's modulus for the composite?

Procedure From Example 10–4.2,

$$A_{st} = 0.8 \times 10^{-6} \text{ m}^2 \qquad A_{Cu} = 2.4 \times 10^{-6} \text{ m}^2$$
$$E_{st} = 205 \text{ GPa} \qquad E_{Cu} = 110 \text{ GPa}$$

The elastic strain must be equal in the two metals.

Calculation

$$(s/E)_{st} = e_{st} = e_{Cu} = (s/E)_{Cu}$$
$$s_{st} = s_{Cu} (205{,}000 \text{ MPa})/(110{,}000 \text{ MPa}) = 1.86 \, s_{Cu}$$

(a) With a 1.86 stress ratio, the steel is stressed 260 MPa when the copper is stressed 140 MPa. Therefore, the copper yields first.

(b) $F_{total} = F_{Cu} + F_{st}$
$$= (140 \times 10^6 \text{ N/m}^2)(2.4 \times 10^{-6} \text{ m}^2)$$
$$+ (260 \times 10^6 \text{ N/m}^2)(0.8 \times 10^{-6} \text{ m}^2)$$
$$= 540 \text{ N} \qquad\qquad \text{(or 55 kg on earth)}$$

(c) From Eq. (10–4.5),

$$\bar{E} = (fE)_{st} + (fE)_{Cu}$$
$$= 0.25 \, (205{,}000 \text{ MPa}) + 0.75 \, (110{,}000 \text{ MPa})$$
$$= 130{,}000 \text{ MPa}$$

Alternatively,

$$e = (s/E)_{st} = [F/(A_{st} + A_{Cu})]/\overline{E}$$

$$\overline{E} = (E/s)_{st}[540 \text{ N}/(3.2 \times 10^{-6} \text{ m}^2)]$$

$$= \left[\frac{205{,}000 \times 10^6 \text{ N/m}^2}{260 \times 10^6 \text{ N/m}^2} \right]\left[\frac{540 \text{ N}}{3.2 \times 10^{-6} \text{ m}^2} \right]$$

$$= 130 \times 10^9 \text{ N/m}^2 \qquad\qquad\qquad\qquad\qquad \text{(or 130,000 MPa)}$$

OPTIONAL ## 10-5
WOOD—A NATURAL COMPOSITE

We use more wood than we do any other engineering material, when measured on a volume basis. It has been available longer than has any other structural material, and it is a renewable resource, so it will continue to be important to us.

Wood has three noticeable characteristics: (a) It has a high strength-to-weight ratio; (b) many of its properties are anisotropic; and (c) it is easily processed to size, even on the job.

As with all materials, the properties of wood can be related to structure. Most woods exhibit a gross annual structure, called *grain,* that arises from the repeating growth cycles during the seasons of the year (Fig. 10–5.1a). All wood contains *biological cells* with cross-sectional dimensions near the visual limit of the eye. They commonly possess an elongated spindle shape (Fig. 10–5.1b). A closer examination of these cells reveals they are surrounded by *fibers* with submicron diameters. The two construct a natural composite (Fig. 10–5.1c). The principal molecular species in wood is *cellulose* (Fig. 10–5.1d). Its structure permits a significant amount of crystallinity within the fibers, thus adding to the strength.

Grain

The trunk of a tree grows by adding another ring every season. Usually, this is in the annual sequence of spring and summer growing seasons; however, in tropical regions, the rings may reflect a sequence of wet and dry periods. When the growth is fast, the biological cells are thin-walled and have a large central cavity. This growth is called the *earlywood* (Fig. 10–5.2). Later in the growing season, when the growth rate slows down, the cells have thicker walls and are denser. This is *latewood.* Because of its higher density, latewood is stronger than is earlywood. The cyclic pattern of the earlywood and latewood produces the typical ring pattern seen in the cross-section of the tree.

These alternating layers of earlywood and latewood and the biological cells that are oriented longitudinally in the tree are primarily responsible for the *anisotropy* of wood. They account for the easy longitudinal splitting, variation of elastic moduli, thermal expansion, conductivity, and so on that occur with direction.

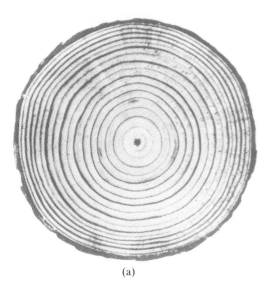

(a)

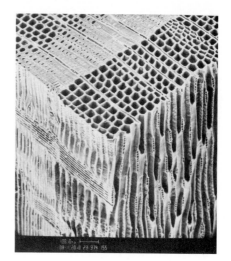

(b)

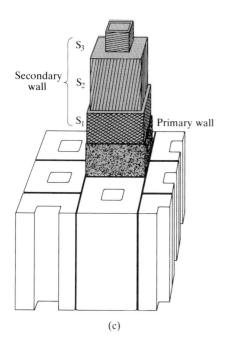

(c)

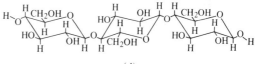

(d)

FIG. 10–5.1

Structures of Wood. (a) *Grain* formed by annual growth rings (American Forest Institute). (b) Biological cells (Douglas fir, ×50). The individual cells are hollow and spindle shaped. (Courtesy of W. A. Côtè, Jr., S.U.N.Y. College of Environmental Science and Forestry, Syracuse.) (c) Microfibrils in the cell wall (schematic). The three layers in the secondary wall provide reinforcement similar to that in tailored composites. (Courtesy of R. J. Thomas, North Carolina State University.) (d) Molecular (cellulose). It is a polymer, $+C_6H_{10}O_5+_n$.

FIG. 10-5.2

Microstructure of Wood. The units of structure
are biological cells. The cells that form with
early (spring) growth are larger but have thinner
walls than do those that form with late
(summer) growth. Cellulose (Fig. 10-5d) is the
major molecular constituent.

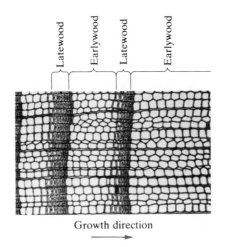

Latewood | Earlywood | Latewood | Earlywood

Growth direction

Because of these variations, wood technologists discuss properties in terms of the
longitudinal direction, 1, the *radial* direction, r, and the *tangential* direction, t
(Fig. 10-5.3a). Also, the lumberworkers distinguish between *quartersawed* wood
and *plainsawed* wood (Fig. 10-5.3b and c). Of course, the cut of the lumber may
have some intermediate orientation; however, it is sufficiently obvious to carpen-
ters that the properties of these two cuts differ, so they will intentionally select the
cut preferred for the job.

Finer Structures

The wood cells that show in Fig. 10-5.1(b) possess a structure within themselves.
They are made up of several layers of *microfibrils* that spiral at various angles

FIG. 10-5.3

Directions in Wood. (a) Log
(l, longitudinal; r, radial; t,
tangential). (b) Lumber
(quartersawed). (c) Lumber
(plainsawed).

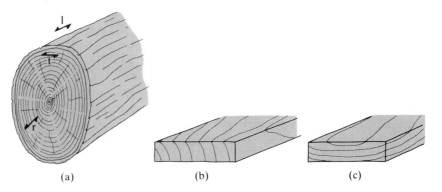

(a)   (b)   (c)

around the central hollow cavity. A typical microfibril has a cross-section of < 1 μm. The microfibrils are sketched schematically in Fig. 10–5.1(c). There are three secondary layers that account for 99 percent of the cell wall:

S_1	10–22% of wall thickness	50–70° from longitudinal
S_2	70–90% of wall thickness	10–30° from longitudinal
S_3	2–8% of wall thickness	60–90° from longitudinal

The multiple layers and the angles of pitch are nature's way of simultaneously giving strength and flexibility to the wood cell.

Cellulose is the principal molecular species in wood. It is a polymer with mers of $nC_6H_{10}O_5$ (Fig. 10–5.1d). These molecules make up the microfibrils of Fig. 10–5.1(c). Polycellulose possesses a high degree of crystallinity as a result of the presence of both —H and —OH attachments to the molecule (Fig. 10–5.1d). The degree of polymerization, n, sometimes exceeds 30,000. Cellulose accounts for ~42 percent (dry basis) of the mass of most wood. Wood in a tree contains a high percentage of water, and even seasoned wood contains moisture. The amount of moisture affects all properties. In extreme cases, it can more than double the density. Up to a certain point, it increases the volume and conductivity and lowers the strength and elasticity. The moisture content varies with the surrounding environment. The interior woodwork of a house may contain as little as 6 percent moisture in midwinter, yet increase to 15 percent moisture after a humid season; a wood piling on a boat dock can possess more than 100 percent water (kiln-dry basis) at the water line. Thus, before we consider properties, it will be helpful for us to learn how the water is contained within the wood.

Moisture in Wood

The reference point for moisture content is kiln-dried wood with 0 percent H_2O. Where wood is exposed to air, which always contains some water vapor, wood gains or loses moisture depending on the relative humidity of the air and the existing moisture status of the wood. The moisture content is always seeking an equilibrium. The equilibrium amount during midwinter is defined as 6 percent cited previously. This *ab*sorption into the wood is really an *ad*sorption onto the surface of the microfibrils sketched in Fig. 10–5.1(c). Warm humid summer air carries more moisture; therefore, more moisture is adsorbed onto the microfibrils. The microfibrils become saturated with moisture, and expansion ceases when about 30 to 35 percent H_2O (dry basis) is present in the wood. This is called the *fiber-saturation point,* FSP. The FSP is an important point, because shrinkage occurs and all strength properties increase as the moisture drops below this level. Moisture in excess of the FSP enters the cavity within the cells as free water (Fig. 10–5.1c). That does not occur from atmospheric moisture during normal lumber usage. The wood would need to be submerged in water or exposed to a continuous presence of water on its surface to fill these cavities.

Each *1 percent* of moisture (dry basis) below the FSP changes the dimensions of typical wood:

radial	~0.15 l/o per 1% moisture
tangential	~0.25 l/o per 1% moisture
longitudinal	~0.01 l/o per 1% moisture

Thus, there are constant but minor volume changes in finished wood in service related to prolonged moisture fluctuations of the atmosphere.

Density

Common woods in the United States have densities in the range of 0.4 to 0.7 g/cm^3. Among these woods, redwood and oak are near the limits (Table 10–5.1). However, other woods are more extreme, for example, balsa has a density of only 0.13 g/cm^3, and a few tropical woods have a density greater than water and, therefore, sink in water. The data of Table 10–5.1 are given on a 12 percent moisture basis.

TABLE 10–5.1 Properties of Common Woods (12% moisture)*

PROPERTY	WHITE OAK	PAPER BIRCH	DOUGLAS FIR	WHITE PINE	REDWOOD (OLD ROUGH GROWTH)
Density, g/cm³	0.68	0.55	0.50	0.38	0.40
Hardness (radial), N†	6000	4000	2900	1900	2100
Young's modulus, MPa	12,300	11,000	12,500	10,100	9200
Longitudinal‡ (10⁶ psi)	(1.78)	(1.60)	(1.81)	(1.46)	(1.33)
Strength MPa (psi)					
Tension (radial)§	5.4 (780)	—	2.4 (350)	—	1.7 (250)
Compression (long.)	51 (7400)	39 (5700)	51 (7400)	35 (5100)	42 (6100)
Compression (radial)	7.4 (1100)	4.1 (590)	5.2 (750)	3.2 (460)	4.8 (700)
Modulus of rupture	105 (15,200)	85 (12,300)	87 (12,600)	67 (9,700)	69 (10,000)

* Data extracted from Wood Handbook USDA, Forest Service, *Agric. Hdbk.,* No. 72.
† Force required to embed 0.444-in. (11.3-mm) ball to one-half of its diameter.
‡ See text for ranges of E_r and E_t.
§ Longitudinal strength in tension $S_{t(l)}$ is commonly 20 $S_{t(r)}$.

Elasticity and Strength

Selected mechanical properties of several woods are given in Table 10–5.1. In general, the Young's moduli lie near the centers of the following ranges:

longidutinal	7500 MPa – 16,000 MPa	(1,100,000 psi – 2,400,000 psi)
radial	500 MPa – 1000 MPa	(75,000 psi – 150,000 psi)
tangential	400 MPa – 700MPa	(60,000 psi – 100,000 psi)

The high values for longitudinal stressing are expected, since that is the alignment of the wood cells. The other two orientations are low because the cell cavity permits easy strain.

The longitudinal strength in tension, $S_{t(1)}$, is commonly 20 times (and sometimes as much as 40 times) the radial strength in tension, $S_{t(r)}$. This contrast will not be surprising to anyone who has worked with wood. Test specimens for mechanical properties must be specifically cut for each orientation (Fig. 10–5.4). Unlike that in ductile metals, the strength in compression of wood does not closely match the strength in tension. Longitudinal compression strengths, $S_{c(1)}$, are lower than in tension, $S_{t(1)}$, because the denser latewood buckles and permits shear to occur. Radial strengths in compression, $S_{c(r)}$, are higher than $S_{t(r)}$ (Table 10–5.1) because the tension failure occurs by splitting.

Example 10–5.1

A solid wood door has been made by gluing together edges of quartersawed boards of Douglas fir. That is, the width of the board is in the radial direction, and the thickness is in the tangential direction (and, of course, the length is the longitudinal direction of the wood). The door was initially trimmed to 761 mm by 2035 mm (to fit a 765 × 2050 mm opening) in late fall when the moisture content of the wood was 9 percent. Will the door "stick" when the moisture content increases to 14 percent?

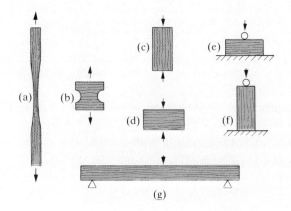

FIG. 10–5.4

Test Specimens of Wood. (a, b) Tension. (c, d) Compression. (e, f) Hardness. (g) Modulus of rupture. The modulus of rupture is the calculated stress at the lower (tension) surface.

Solution

radial change $= 0.0015\% \times [14 - 9\%] = +0.0075$

$\Delta L = (761 \text{ mm})(+0.0075) = +5.7 \text{ mm}$

It will stick, since $761 + 5.7$ mm is greater than 765 mm. The longitudinal change is negligible.

Comment   Suggest a design feature that could be used to minimize this problem.

Example 10–5.2

A plainsawed board of Douglas fir (width is tangential) is not permitted to expand its width as its moisture content increases from 5 to 10 percent. What pressure develops?

Procedure   Assume expansion is allowed, followed by compression ($-$) to the original dimension. Consider that $E = 550$ MPa, the midvalue of those values indicated in the text for the tangential modulus of elasticity.

Calculation

$\Delta L/L = (0.0025/\%)(10 - 5\%) = +0.0125$

$s = (-0.0125)(550 \text{ MPa})$
$= 7 \text{ MPa compression}$ (or 1000 psi)

Example 10–5.3

Cite examples of how wood can be processed to compensate for the following natural limitations: (a) *Strength and fracture anisotropy,* (b) *Dimensional instability under varying humidity,* (c) *Size and shape limitations,* and (d) *Biological degradation.*

Answers

(a) Plywood has a biaxial structure (the plies, or layers, run at right angles to one another) that capitalizes on the longitudinal stability, and reinforces the weakness in the radial direction.

(b) Resin impregnation is used to exclude moisture penetration. This procedure also reduces the anisotropy in properties.

(c) Wood is laminated to form large beams, curved seats, and so on.

(d) Treatment with penetrating chemicals inhibits degradation.

SUMMARY

Since polymers possess strong intramolecular bonds and weak intermolecular bonds, they respond somewhat differently to mechanical forces than do metals and crystalline ceramics.

1. Typical elastic behavior by bond stretching occurs below the glass-transition temperature, T_g. Above T_g, large elastic strains are possible in linear polymers, because the molecular bonds can be straightened from their kinked and coiled conformation. With straightening, the stress requirements increase and a nonlinear elastic modulus results. Elastomers have a positive dE/dT because higher temperatures promote the rekinking or recoiling.

Amorphous polymers flow viscously above T_g. As a result, elastic strain and flow are additive. The extent of viscous flow varies significantly with the molecular structure. It is inhibited by cross-linking, it decreases with added crystallinity, and, of course, it varies with time. Since viscosity decreases exponentially above T_g, temperature is an important factor both in process control and in applications.

2. Although most polymers originate as industrial chemicals, we looked first at the compounding of the polymer with appropriate additives. These include fillers, plasticizers, stabilizers, and flame retardants. Fillers, such as wood flour or other fine ground minerals, not only serve as diluents for the more expensive polymers, but also make a valuable contribution to strength and dimensional stability. Plasticizers generally are small molecules that dissolve into the amorphous structure among the larger polymer molecules. They facilitate molding processes, and add to the flexibility of the product. Thorough mixing is required for homogeneous distribution of all additives.

Shaping processes are numerous. The more common include extrusion, injection molding, sheet molding, blow molding, and spinning. All involve thermoplasticity. Molecular orientation is an important consequence of fiber and film processing, because aligned molecules lead to higher elastic moduli and to partial crystallization.

3. Composites contain two (or more) distinct materials as a unified combination. A common type of composite possesses a polymeric matrix with either particulate or fiber reinforcement. Figures of merit for many structural applications include the strength-to-density ratio, S_u/ρ, or the elastic modulus-to-density ratio, E/ρ. The latter is more important if structural rigidity is a design factor, as this requirement places a premium on fibers with high elastic moduli.

4. In the absence of flow, stress distributions are dictated by the elastic moduli (Eq. 10–4.2). This consequence leads to shear stresses along the phase boundaries. The toughness of composites is significantly increased by a microstructure that possesses elastomeric domains. (See the discussion of polyblends in Section 7–5). It also can be increased by the controlled development of microcracking. An apparent strain and therefore greater toughness is realized through the steps of debonding, fiber fracture, and fiber pullout.

5. Wood is a natural polymeric composite. Its anisotropic properties arise from its microstructure, which includes, by decreasing size, (a) grain, (b) biological cells, (c) microfibrils, and (d) polymeric molecules of cellulose. Moisture content has a major effect on wood's properties.

KEY WORDS

Additives

Blow molding

Boundary stresses

Composite

Debonding

Elastomers

Earlywood

Extrusion

Fiber debonding

Fiber-reinforced plastics (FRP)

Filler

Glass-transition temperature (T_g)

Grain (wood)

High-stiffness composite

Injection molding

Latewood	Reinforcement	Viscoelastic modulus
Matrix	Sheet molding	Viscoelasticity
Microcracking	Spinning	Viscosity
Microfibrils	Thermoplasts	Viscous flow
Molecular orientation	Thermosets	
Plasticizers	Viscoelastic deformation	

PRACTICE PROBLEMS

10–P11 Estimate the viscosity of polyvinyl chloride at 105° C

10–P12 The viscosity of water is 0.001 Pa·s at 20° C, and 0.00028 Pa·s at 100° C. (a) What is the activation energy for flow? (b) What is the viscosity at 0° C?

10–P21 Forty-five kg of silica flour (finely ground quartz sand) are mixed with each 100 kg of melamine-formaldehyde (mf). What is the volume fraction of filler?

10–P22 A 163-cm (64-in.) diameter outdoor sign is vacuum-formed into its final shape. What force would be required behind a die to provide the same force in a die-stamping operation as was obtainable by the vacuum?

10–P41 A steel-reinforced aluminum wire has the same dimensions as does the steel-reinforced copper wire of Example 10–4.2. What fraction of the load will be carried by each metal?

10–P42 Derive a mixture rule for the conductivity (electrical or thermal) of a laminate when the heat flow is parallel to the structure.

10–P43 Calculate the density of a glass-reinforced plastic fishing rod, in which the glass-fiber content is 15 w/o. (A borosilicate glass is used for the longitudinal fibers. The density of the plastic is 1.3 g/cm^3.)

10–P44 Estimate the thermal conductivity (longitudinal) of the composite in Problem 10–P43. ($k_{pl} = 0.00026$ W/mm·° C.)

10–P45 An AISI–SAE 1040 steel wire (cross-section 1 mm^2) has an aluminum coating such that the total cross-sectional area is 1.2 mm^2. (a) What fraction of a 450 N load (100 lb$_f$) will be carried by the steel? (b) What is the resistance of this composite wire per unit length?

10–P46 A glass-reinforced plastic rod (fishing pole) is made of 67 v/o borosilicate glass in a nylon matrix. What is the longitudinal thermal expansion coefficient?

10–P51 What is the mass of a mer of the polycellulose molecule?

10–P52 Birchwood veneer is impregnated with phenol-formaldehyde (PF) to ensure resistance to water and to increase the hardness and dimensional stability of the final product. Although this wood weighs only 0.56 g/cm^3, the true density of this cellulose-predominant material is 1.52 g/cm^3. (a) How many grams of PF are required to impregnate 10,000 mm^3 (0.6 in.^{3}) of dry birchwood? (b) What is the final density?

TEST PROBLEMS

1011 Estimate the temperature at which polystyrene will have a viscosity of 10^6 Pa·s.

1012 A window glass (Table 10–1.1) must have a viscosity of 10^6 Pa·s to be drawn into sheet product. Specify the temperature.

1021 Silica flour is used as filler for phenol-formaldehyde (PF). (a) What volume fraction is required to give a density of 1.7 g/cm³. (b) What is the weight fraction SiO_2?

1022 Sand is ground as a filler for a plastic. Which properties of the plastic will be affected by the amount of grinding—that is by the average size of the particles? Which properties will not be affected?

1023 "The shape of the aggregate affects the behavior of asphaltic concrete." Discuss this assertion.

1031 A 5-kW motor is used to power an auger extruding lucite rods (PMMA of Table 2–3.1). Assuming 80 percent of the power appears as temperature rise in the PMMA, by how much will the temperature rise if the through-put is 120 kg/hr? (c_p of PMMA is 1.5 J/ġ° C.)

1032 Suppose that you manufacture ball bearings, and you need to have 50 v/o of fine chopped glass fibers ($\rho = 2.4$ g/cm³) in the nylon ($\rho = 1.15$ g/cm³) that is to be used in these bearings. How much glass fiber should be batched with each kg of nylon in making this FRP product?

1041 Glass fibers provide longitudinal reinforcement for a nylon. The fiber diameters are 20 μm and the volume fraction is 0.45. (a) What fraction of the load is carried by the glass? (b) What is the stress in the glass when the average stress in the composite is 14 MPa (2000 psi)?

1042 Derive a mixture rule for the conductivity (electrical or thermal) of a laminate when the heat flow is perpendicular to the structure.

1043 A composite contains 50 v/o aramid fibers in a nylon matrix. The fibers are longitudinal. Using available data, calculate the E/ρ ratio of the composite.

1044 A 10-cm cube is made by laminating alternate sheets of vulcanized rubber and aluminum (0.5 mm and 0.75 mm thick, respectively). Calculate the thermal conductivity of the laminate (a) perpendicular to, and (b) parallel to the sheets. Either derive the appropriate mixture rules or refer to calculations in previous chapters.

1045 What is the glass–nylon interface area (cm²) per unit volume (cm³) in the composite of Problem 1041.

1046 What is the thermal expansion coefficient of the composite in Problem 1041?

1051 (a) What is the typical volume expansion as wood increases from 6 to 12 percent moisture? (b) The wood is dried from 12 to 6 percent moisture. What is the percent density change?

1052 The trimmed door in Example 10–5.1 was shut when the moisture content increased from 9 to 14 percent. What force will develop if the door is 30 mm thick? Assume that the doorway casing is rigid.

Chapter 11

CONDUCTING MATERIALS

In previous chapters, atoms and their arrangements have received the bulk of our attention. In this chapter, we shall focus on electrons and on their freedom to move among atoms.

Metals, with their weak hold on valence electrons, are good conductors of both electricity and heat. This conductivity occurs because little energy is required to activate delocalized electrons into conduction levels. In contrast, electrons must be raised across a large energy gap in an insulator. Semiconductors have small energy gaps, so a useful number can jump to the conduction band to transport charge. This leaves in the lower bands electron holes that serve as positive carriers of charge.

We can form simplified concepts of many devices based on an introductory understanding of energy gaps and junctions. We shall examine the basic principles of crystal growing and device preparation.

11–1

CHARGE CARRIERS

Various materials that are available to the engineer and scientist exhibit a wide range of conductivities (or resistivities, since $\sigma = 1/\rho$). As shown in Fig. 1–4.4 and again in Fig. 11–1.1, we commonly divide materials into three categories: *conductors, semiconductors,* and *insulators.* Metals fall in the first category, since they have delocalized electrons that are free to move throughout the structure (Sections

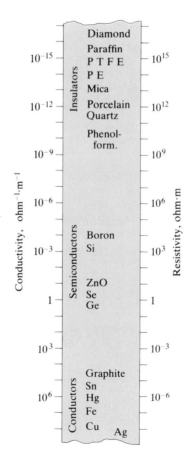

FIG. 11–1.1

Spectrum of Conductivity (and Resistivity). The values for commercial semiconductors lie between 10^{+4} and 10^{-4} ohm$^{-1}\cdot$m^{-1}.

2–2 and 2–4). Insulators include those ceramics and polymeric materials with strongly held electrons and nondiffusing ions. Their function is to isolate neighboring conductors. Not long ago, we considered only the two ends of the spectrum of Fig. 11–1.1 to be useful. Today, however, the middle, semiconducting category has become exceedingly important; it will be the chief subject of this chapter.

In those types of materials that conduct electricity, the charge is carried in modules of 0.16×10^{-18} coul, this being the charge on an individual electron. In metals, it is the individual electron that moves. In ionic materials, charge can be carried by diffusing ions. However, their charge is simply an integer number of electron charges ($-$ or $+$, for anions and cations, respectively). Thus, an SO_4^{2-} ion carries 0.32×10^{-18} coul of charge within a storage battery, and all Pb^{2+} ions have an absence of two electron charges as they move in the opposite direction.

Electrons and anions are *negative charge carriers.* In contrast, a cation such as Pb^{2+} is a *positive charge carrier* because, as we have just seen, it has an *absence* of electrons. There is another positive charge carrier that is important in semiconductors; namely, an *electron hole.* It is an absence of an electron within the energy band for the delocalized electrons discussed in Section 2–4. We shall come back to these electron holes in Section 11–4.

Conductivity σ and *resistivity ρ* ($=1/\sigma$) values for a material depend on the *number, n,* of charge carriers, the *charge, q,* on each, and the *mobility, μ,* of each:

$$\sigma = 1/\rho = nq\mu \qquad \text{(11–1.1)}$$

In this chapter, much of our attention shall be directed toward the number, n, of carriers per m^3 that are available for conduction. The value of q is fixed at 0.16×10^{-18} C (or 0.16×10^{-18} A·s) in electronic conductors. The mobility μ relates the drift velocity $\bar{v}$ to the electric field $\mathscr{E}$:

$$\mu = \bar{v}/\mathscr{E} \qquad \text{(11–1.2)}$$

Thus, the units for mobility are (m/s) per (V/m), or $m^2/V\cdot s$. As a result,

$$\sigma = (m^{-3})(A\cdot s)(m^2/V\cdot s) = ohm^{-1}\cdot m^{-1} \qquad \text{(11–1.3a)}$$

and

$$\rho = ohm\cdot m \qquad \text{(11–1.3b)}$$

Example 11–1.1

A semiconductor with 10^{21} charge carriers/m^3 has a resistivity of 0.1 ohm·m at 20° C. What is the drift velocity of the electrons if 1 amp of current is carried across a gradient of 0.15 volt/mm?

Procedure We know the resistivity (and thus conductivity) plus the number of carriers, and, of course, the charge q on each. We can calculate the mobility. The electric field also is available, so we can calculate the drift velocity.

Calculation

$\mu = \sigma/qn = (1/0.1 \text{ ohm} \cdot \text{m})/(0.16 \times 10^{-18} \text{ A} \cdot \text{s})(10^{21}/\text{m}^3)$
$= 0.0625 \text{ m}^2/\text{V} \cdot \text{s}$

$\bar{v} = \mu\mathscr{E} = (0.0625 \text{ m}^2/\text{V} \cdot \text{s})(0.15 \text{ V}/0.001 \text{ m})$
$= 9 \text{ m/s}$

Comment   Of course, the electrons move at a speed much faster than 9 m/s. This value is the *drift* or *net velocity* across the semiconductor.

Example 11–1.2

An alloy has a resistivity that varies linearly with temperature. It is 131×10^{-9} ohm·m at $0°$ C, and 162×10^{-9} ohm·m at $100°$ C. (a) What is its resistivity at $40°$ C? (b) What is the resistance per foot of an 0.022-in. wire at $40°$ C?

Procedure   Although we typically express resistivity in ohm·m or ohm·cm, we usually use English dimensions for wire products. Therefore, for (b), we must change to ohm·in.

Calculation

(a) 131×10^{-9} ohm·m $+ (162 - 131)(10^{-9}$ ohm·m$)(40°$ C$/100°$ C$)$
     $= 143 \times 10^{-9}$ ohm·m

(b) $(143 \times 10^{-9}$ ohm·m$)(39.37$ in./m$) = 5.6 \times 10^{-6}$ ohm·in. .

$$R = \rho L/A = \frac{(5.6 \times 10^{-6} \text{ ohm} \cdot \text{in.})(12 \text{ in./ft})}{(\pi/4)(0.022 \text{ in.})^2} = 0.18 \text{ ohm/ft}$$

11–2
METALLIC CONDUCTIVITY

We described the metallic bond in terms of *delocalized electrons* (Section 2–4). As such, the valence electrons are able to move throughout the metal as standing waves. Thus, there is no net charge transport in the absence of an electric field.

   If the metal is placed in an electrical circuit, the electrons moving toward the positive electrode *acquire more energy* and gain velocity in that direction. Conversely, those electrons moving toward the negative electrode *reduce their energy* and velocity. As a result, a *drift velocity* is developed. (See Example 11–1.1.)

Mean Free Path

Recall that waves move through periodic structures without interruption. A well-ordered crystal (Chapter 3) provides one of the most regular of the periodic structures available. Thus, a metallic crystal lattice provides an excellent medium for electron movements. However, any irregularity in the repetitive structures through which a wave travels may deflect the wave. Thus, if an electron had been traveling toward the positive electrode and was then deflected, it would no longer continue to gain velocity in that direction. The net effect is to *reduce the drift velocity,* even though the electric field has not been altered. In brief, irregularities in the lattice *decrease* the mobility described by Eq. (11-1.2); therefore, they *decrease* the conductivity and *increase* the resistivity (Eq. 11-1.1).

The average distance that an electron can travel in its wavelike pattern without deflection is called the *mean free path.* We will want to identify irregularities that deflect electron movements, because that will help us to understand why resistivities of metals are not all the same. These irregularities include thermal vibrations, solute atoms, and crystal damage.

Resistivity Versus Temperature

The resistivity of a metal increases with temperature (Fig. 11-2.1). To a first approximation, it is linear (except near absolute zero). We have no basis on which to conclude that the *n* of Eq. (11-1.1) decreases significantly with increased temperature in a metal;* rather, we must look at the mobility μ. Thermal agitation (Section 6-4) increases in intensity in proportion to increased temperature (except at very low temperatures). This increased agitation decreases the mean free path of the electrons by decreasing the regularity of the crystal and therefore decreases the mobility of electrons in a metal. The consequent change in resistivity is important to the engineer who is designing electrical equipment. In some cases, compensation must be introduced into a circuit to avoid an unwanted temperature sensitivity. In other cases, this temperature sensitivity provides a useful

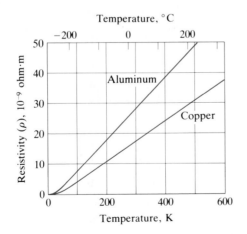

FIG. 11-2.1

Resistivity Versus Temperature (Metals). The resistivity of metals is linear with temperature in the 100 to 500 K range (−200 to 200° C). Except in superconductors, the "toes" of the curves are finite.

---

* The number of charge carriers increases with increased temperature in a semiconductor (Section 11-4).

"brake." Example 11–2.3 will point out this temperature effect in a familiar application (the toaster).

We can determine the ρ-versus-T relationship with a *temperature resistivity coefficient* y_T as follows:

$$\rho_T = \rho_{0°\,C}(1 + y_T\,\Delta T) \qquad\qquad \textbf{(11–2.1)*}$$

where $\rho_{0°\,C}$ is the resistivity at 0° C, and ΔT is $(T - 0°\,C)$. This value of this coefficient is approximately 0.004/° C for pure metals (Table 11–2.1). This suggests that the mean free path of electrons is reduced by a factor of two between 0° C and 250° C.

Resistivity in Solid Solutions

Another factor that can reduce the mean free path of electrons in a metal is the presence of solute atoms. A solid-solution alloy always has a higher resistivity than do its pure component metals.†

TABLE 11–2.1 Temperature Resistivity Coefficients

METAL	RESISTIVITY AT 0° C*, ohm·nm	TEMPERATURE RESISTIVITY COEFFICIENT, y_T, ° C^{-1}
Aluminum	27	0.0039
Copper	16	0.0039
Gold	23	0.0034
Iron	90	0.0045
Lead	190	0.0039
Magnesium	42	0.004
Nickel	69	0.006
Silver	15	0.0038
Tungsten	50	0.0045
Zinc	53	0.0037
Brass (Cu–Zn)	~60	0.002
Bronze (Cu–Sn)	~135	0.001
Constantan (Cu–Ni)	~500	0.00001
Monel (Ni–Cu)	~450	0.002
Nichrome (Ni–Cr)	~1000	0.0004

* These values do not agree with those in Appendix C, since they are based on different reference temperatures.

---

* The decreases in resistivity (increase in conductivity) we are describing are not related to the *superconductivity* that appears near absolute zero. That involves another phenomenon (quantum mechanical), which is beyond the scope of this book.

† Also, examine the data for metals and alloys in either Table 11–2.1 or Appendix C.

The reason for this generalization is that an electron encounters an irregularity in the local potential of the crystal lattice when it approaches an impurity atom. In the first place, the lattice is slightly distorted in an alloy such as brass, because the atomic radii differ a few percent; in addition, a zinc atom has 30 protons, whereas copper has 29. This discrepancy also alters the local field. Although these differences seem small, they deflect additional electrons and reduce the mean free path. Since brass (70 Cu – 30 Zn) has a resistivity three or four times as great as that of pure copper, we can assume that the mean free path for electrons is only 25 to 30 percent as long in brass as it is in pure copper. If high conductivity is paramount in a design, we should turn to pure metals (Fig. 11 – 2.2).

We can express empirically the increased resistivity (decreased conductivity) that arises through solid solution as

$$\rho_x = y_x \, x(1 - x) \qquad \textbf{(11-2.2)}$$

where x is the atom fraction of the solute and $(1 - x)$ is the atom fraction of the solid solvent. The *solution resistivity coefficient, y_X,* is specific for each binary alloy. The values for solid-solution alloys of copper are given in Table 11 – 2.2.

Resistivity from Strain Damage

Cold work increases the resistivity of metals (Fig. 11 – 2.3a). The increase arises in part from dislocations that are generated by plastic deformation (Section 9 – 3). These dislocations deflect the electrons from their wavelike movements, shortening the mean free path and decreasing the electrons' mobility. Annealing removes the strain damage within the crystals and restores the original conductivity.*

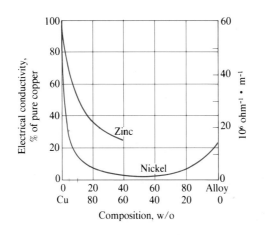

FIG. 11-2.2

Electrical Conductivity of Solid-Solution Alloys. The conductivity is reduced when nickel is added to copper, and when copper is added to nickel, because the impurity atoms shorten the paths of the electrons between deflections. (L. H. Van Vlack, *Materials for Engineers: Concepts and Applications,* Addison-Wesley, Reading, Mass., with permission.)

---

* The *recovery* of conductivity begins before the softening that accompanies recrystallization (Section 9 – 3), because part of the added resistivity results from vacancies and interstitials that develop during deformation and are eliminated at relatively low temperatures. The dislocations, which account for the bulk of the strain hardening, require more extensive crystalline rearrangements, and therefore must be eliminated by additional thermal activation — that is, at higher temperatures.

TABLE 11–2.2 Solution
Resistivity Coefficients (in
Copper at 20° C)*

SOLUTE	COEFFICIENT, ohm·m
Ag	$y_x = 0.2 \times 10^{-6}$
Al	$y_x = 0.8 \times 10^{-6}$
Ni	$y_x = 1.2 \times 10^{-6}$
Si	$y_x = 2.0 \times 10^{-6}$
Sn	$y_x = 2.5 \times 10^{-6}$
Zn	$y_x = 0.2 \times 10^{-6}$

* For dilute solutions only ($<$ 10 a/o).

The factors that contribute to the increases in metallic resistivity are additive:

$$\rho = \rho_T + \rho_x + \rho_e \qquad\qquad (11\text{--}2.3)$$

where the subscripts refer to temperature (T), solute content (x), and strain (e).
Figure 11–2.3(b) shows this relation for some copper–tin alloys.

We shall consider radiation damage under service performance in Chapter 14.

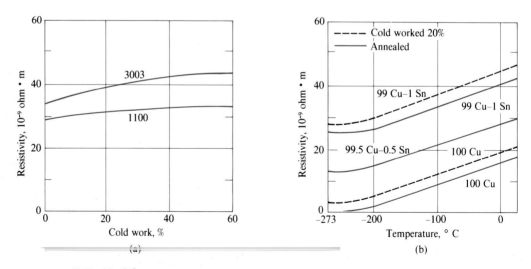

FIG. 11–2.3

Factors Affecting Resistivity. (a) Cold work (aluminum). The 1100 aluminum is
commercially pure; alloy 3003 contains 1.2 percent manganese. (b) Additive effects
(Cu–Sn alloys, a/o). The resistivity is the sum of the contributions from the temperature,
solution, and deformation (Eq. 11–2.3). (L. H. Van Vlack, *Materials Science for
Engineers,* Addison-Wesley, Reading, Mass., with permission.)

Thermal Conductivity

In a metal, electrons carry the majority of the energy for thermal conduction. Thus, there is a correspondence between the thermal and the electrical conductivities. (Compare Figs. 11–2.2 and 11–2.4.) In fact, it was pointed out some time ago that k/σ in most pure metals is about 7×10^{-6} watt·ohm/K at normal temperatures ($\sim 20°$ C) when thermal conductivity k and electrical conductivity σ are expressed in (watts/m²)/(K/m) and ohm^{-1}·m^{-1}, respectively (Fig. 11–2.5). This ratio, known as the Wiedemann–Franz (W–F) ratio, is proportional to temperature (in Kelvin). As a result, we can write

$$k/\sigma = LT = k\rho \qquad (11-2.4)$$

The proportionality constant L (called the *Lorenz number*) is 2.3×10^{-8} watts·ohm/K². The electrical resistivity, ρ, is the reciprocal of electrical conductivity, σ. Since data are readily available on the electrical resistivity values for metals at various temperatures, the Lorenz number provides a convenient rule of thumb for less readily available thermal-conductivity values.

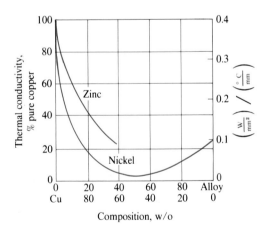

FIG. 11–2.4

Thermal Conductivity of Solid Solutions. These data correlate with the electrical conductivity of the same solid solutions (Fig. 11–2.2). In each case, the solute atoms reduce the mean free path of the electron movements to decrease the conductivities. (L. H. Van Vlack, *Materials for Engineers: Concepts and Applications,* Addison-Wesley, Reading, Mass., with permission.)

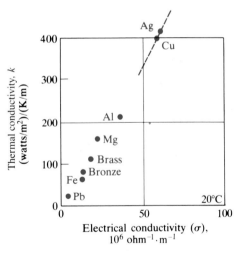

FIG. 11–2.5

Metal Conductivity (Thermal Versus Electrical at 20° C). Since electrons also transport thermal energy, a good electrical conductor is a good thermal conductor. (The k/σ ratio is proportional to absolute temperature. This relationship, known as the law of Wiedemann and Franz, applies best to pure metals. It does not include nonmetals.)

Example 11–2.1

Calculate the resistivity of silver at $-40°$ C.

Procedure Table 11–2.1 shows the resistivity at $0°$ C, and the temperature resistivity coefficient (i.e., the change per degree). Thus, this calculation is like one using a thermal expansion coefficient.

Calculation From Table 11–2.1, $\rho = 15$ ohm·nm at $0°$ C, and $y_T = 0.0038/°$ C.

$$\rho_{-40° \text{ C}} = 15 \text{ ohm·nm } [1 + (0.0038/° \text{ C})(-40° \text{ C})]$$
$$= 13 \text{ ohm·nm}$$

Example 11–2.2

What is the resistivity at $100°$ C of an alloy containing 99 Cu–1 Sn (weight basis)?

Procedure First, we change 99 Cu–1 Sn to atom percent so that we can calculate the *solution resistivity*, ρ_x. Then, we determine the resistivity of copper at $100°$ C by using the temperature resistivity coefficient, y_T from Table 11–2.1. The answer to the problem will be the sum of the two values:

$$\rho = \rho_x + \rho_T \tag{11–2.5}$$

Calculation Basis: 100 amu.

Sn: 1 amu/(118.7 amu/atom) = 0.0084 = 0.54 a/o
Cu: 99 amu/(63.54)   = 1.5581
                Total atoms = 1.5665

$$\rho = (2.5 \times 10^{-6} \text{ ohm·m})(0.0054)(0.995) + 16 \text{ ohm·nm } [1 + (0.0039/° \text{ C})(100° \text{ C})]$$
$$= 36 \text{ ohm·nm} \qquad \qquad (\text{or } 36 \times 10^{-9} \text{ ohm·m})$$

Comment The increase in resistivity is nearly linear with composition at low solute concentrations.

Example 11–2.3

A toaster uses 300 watts when it is in operation and the nichrome element is at $870°$ C. It operates off a 110-volt line. (a) How many amperes does it draw when it is hot? (b) How many does it draw when the switch is first snapped on?

Procedure Calculate the $870°$ C amperage and resistance from the basic relationships of $P = EI$ and $E = IR$. Since dimension changes are very minor (and partially compensating),

$$R_{20° \text{ C}}/R_{870° \text{ C}} = \rho_{20° \text{ C}}/\rho_{870° \text{ C}}$$

We can calculate $R_{870° \text{ C}}$ and $\rho_{20° \text{ C}}/\rho_{870° \text{ C}}$ from Eq. (11–2.1). Our answers let us determine the amperage at $20°$ C—again, by $E = IR$.

Calculation

(a) $I = 300 \text{ W}/110 \text{ V} = 2.7 \text{ amp}$

(b) $R_{870° \text{ C}} = 110 \text{ V}/2.7 \text{ A} = 40 \text{ ohm}$

$$\frac{R_{20}}{R_{870}} = \frac{\rho_0(1 + y_T 20° \text{ C})}{\rho_0(1 + y_T 870° \text{ C})}$$

$$R_{20} = 40 \text{ ohm } [1 + 0.0004(20)]/[1 + 0.0004(870)]$$
$$= 40 \text{ ohm } [1.008/1.35] = 30 \text{ ohm}$$

$$I_{20} = 110 \text{ V}/30 \ \Omega = 3.7 \text{ amp}$$

Comments   Had the element continued to draw 3.7 amperes, the temperature would continue to rise beyond 870° C, subjecting the heating element to faster oxidation and related service deterioration, even to melting.

Note that the temperature coefficients of resistivity of alloys are smaller than are those for pure metals, partly because the mean free path for the electron is already short and the resistivity is initially higher.

Example 11–2.4

A brass alloy is to be used in an application that will require a minimum ultimate strength of 275 MPa (40,000 psi) and an electrical resistivity of less than 50 ohm·nm. What percent zinc should the brass possess?

Procedure   Determine the "window" resulting from the two specifications. Data are available in Figs. 8–3.6(a) and 11–2.2.

Calculation $\rho_{Cu} = 17$ ohm·nm; therefore,

$\sigma_{Cu} = 60 \times 10^6$ ohm^{-1}·m^{-1}

$\rho_{Brass} < 50$ ohm·nm

and

$\sigma_{Brass} > 20 \times 10^6$ ohm^{-1}·m^{-1} and $> 0.33 \ \sigma_{Cu}$

Therefore, for ρ:

max. = 23% Zn

From Fig. 8–3.6, for S_u,

min. = 15% Zn

Use an 80 Cu–20 Zn brass.

Comment   The choice is on the high side of the "window," since zinc is less expensive than is copper; however, some margin is retained to allow for minor product and service variables.

11-3
ENERGY BANDS

Recall from Fig. 2-1.3 (resketched as Fig. 11-3.1a) that electrons of isolated atoms occupy only specific orbitals or energy levels, and that forbidden-energy gaps exist between these levels. In effect, the electrons establish standing waves around an individual atom. This pattern also is found in the inner or subvalence electrons of metals; however, the outer or valence electrons are delocalized when we have a large number of coordinated atoms. As a result, the valence orbitals form bands (Fig. 11-3.1b), and the standing waves are influenced by every atom that is involved. A consequence of this fact is that a band possesses as many standing-wave forms and discrete energy *levels* as there are atoms in the system. Since the number of energy levels is exceedingly great, and the energy bands are usually only a few electron volts wide, it follows that the energy levels within a band are separated by such infinitesimally small spaces that we can model the band as a continuum.

The Pauli exclusion principle states that only two electrons can occupy the same level (and that these two must be of opposite magnetic spin). Thus, with its multitude of levels, *a band may contain twice as many electrons as there are atoms.* As a result, a monovalent metal such as sodium has its energy band only half filled (Fig. 11-3.2a). Since aluminum has three valence electrons per atom, its first valence band is filled, and its second band is half filled (Fig. 11-3.2c).

The lower energy levels (states) of the energy band fill first. The uppermost filled level at 0° K is called the *Fermi energy, E_f*. At 0° K, none of the energy levels are occupied above E_f; meanwhile, all levels of the band are filled below E_f. We use the Fermi energy as our reference energy for understanding electronic conduction.

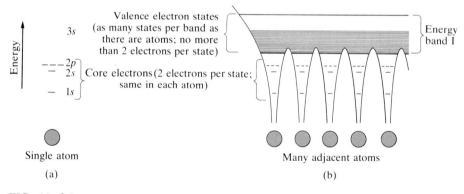

FIG. 11-3.1

Valence Electrons in Metal (Sodium). (a) Single atom (Fig. 2-1.3). (b) Multiatomic sodium metal. The valence electrons are delocalized into an energy band. These electrons are able to move throughout the metal and thus provide conductivity. The valence electrons fill only the bottom half of the band. Their average energy is lower than that of the 3s electrons with individual atoms.

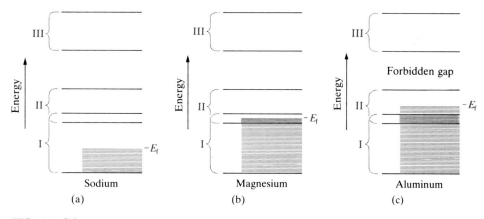

FIG. 11-3.2

Energy Bands. (a) Sodium. Since it has only one valence electron per atom, its first valence band (**I**) is only half filled. (b) Magnesium. Its first band would be full, except that its second band (**II**) overlaps, to contain a few electrons. (c) Aluminum. With three valence electrons, its first band is filled, and its second band is half full. At $0°$ K, no electrons are above the Fermi energy level, E_f; all levels below E_f are filled.

Definition of a Metallic Conductor

We characterized metals in Section 1-4 by their "ability to give up valence electrons," and pointed out that they were thus conductors. We now have a better definition of metallic conductors—namely, they are materials with only *partially filled energy bands.* Figure 11-3.2(a and c) shows this schematically for sodium and aluminum.

The empty energy levels within a band are important for conduction because they permit an electron to rise to a higher energy level when it moves toward the positive electrode. This would not be possible if the energy band were completely filled and an overlying forbidden-energy gap were present.

Magnesium, with its two valence electrons per atom, is expected to fill the first energy band. It so happens, however, that the first and second bands overlap (Fig. 11-3.2b). Thus, some of the $2N$ electrons (where N is the number of atoms) spill over into the second band, where there are plenty of vacant levels to receive the accelerating electrons. As a result, magnesium is metallic.

Semiconductors and Insulators

Silicon presents another story because its four valence electrons per atom completely fill the first two valence bands (Fig. 11-3.3). Furthermore, there is a forbidden-energy gap above the second band. Since electrons cannot be energized within the valence bands, they must be activated to levels across this gap in order to transport charge. Only a few electrons possess this much energy, so silicon is not a

FIG. 11–3.3

Energy Gap (Silicon). The four
valence electrons per atom of silicon
fill the first two energy bands. There
is a forbidden-energy gap between
the second and third bands.

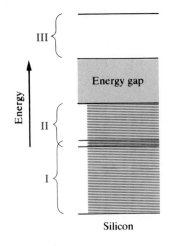

metallic conductor—it is only a *semiconductor.* (With pure materials at 20° C,
$\rho_{Si} = 2 \times 10^3$ ohm·m and $\rho_{Cu} = 17 \times 10^{-9}$ ohm·m—a ratio of about 10^{11}.)

The difference between semiconductors and insulators is related to the size of
the *energy gap, E_g,* that overlies the filled valence band (Fig. 11–3.4). Silicon is a
semiconductor because a usable number of electrons can "jump the gap" into the
conduction band (Section 11–4).

In terms of energy bands, an *insulator* is a material with a large energy gap
between the highest filled valence band and the next empty band (Fig. 11–3.4c).
The gap is so large that, in effect, we can state that electrons are trapped in the lower
valence band. Their number, n, in the conduction band for Eq. (11–1.1) is insignificant.

We commonly describe the valence electrons of insulators as being bound
within the negative ions (or in a covalent bond). Approximately 7 eV (=1.1 ×

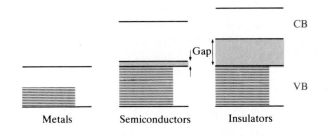

FIG. 11–3.4

Metals, Semiconductors, and Insulators. Metals have an unfilled energy band. Semiconductors have a narrow forbidden gap above the top filled valence band (**VB**). A few
electrons can "jump the gap" to the conduction band (**CB**). Insulators have a wide energy
gap, which is a barrier to the electrons.

10^{-18} J) of energy would be required to break an electron loose from the Cl⁻ ions in NaCl, and about 6 eV of energy would be needed to separate an electron from the covalent bond of diamond. These activation energies of 7 and 6 eV are also the dimensions of the energy gaps and can be compared to 1.1 and 0.7 eV for silicon and germanium, respectively. The physicist considers an energy gap of about 4 eV (=0.64×10^{-18} J) as an arbitrary distinction between semiconductors and insulators. Thus, NaCl and diamond are electronic insulators, whereas silicon and germanium are semiconductors* (Fig. 11–1.1).

Fermi Distribution

Equation (6–4.3) gave us a basis for the kinetic-energy distribution of molecules of gas. The distribution, an exponential function of temperature, T, is called the *Boltzmann distribution, B(E)*:

$$B(E) = 1/e^{(E - \bar{E})/kT} \qquad (11-3.1)$$

where $\bar{E}$ is the average energy of the molecules, and k is Boltzmann's constant, 13.8×10^{-24} J/ K, or 86.1×10^{-6} eV/ K. As in any thermally controlled situation, temperature must be expressed in Kelvin.

In the energy distribution within gases of Eq. (11–3.1), there is nothing to prevent several molecules from possessing the same energy. In the case of electron energies, however, only two electrons can occupy the same energy state (and they must be of opposite magnetic spins). This requirement alters the energy distribution only slightly, except near the Fermi energy level, E_f. Since we are interested in that region, we must make the necessary change in Eq. (11–3.1), to obtain

$$F(E) = 1/[1 + e^{(E - E_f)/kT}] \qquad (11-3.2)$$

for the *Fermi distribution, F(E)*.

This distribution, when plotted as shown in Fig. 11–3.5, indicates the probability that any particular energy level is occupied. Thus, far below E_f, there is full occupation; far above, E_f, there is negligible occupation. By definition, there is a 50-percent occupation at the Fermi energy. Calculations for 20° C (293 K) in the region of E_f give us the following probabilities:

$E_f + 0.5$ eV	$F(E) = 0.000\ 000\ 002$
$E_f + 0.2$ eV	$0.000\ 36$
$E_f + 0.1$ eV	0.019
E_f	0.5
$E_f - 0.1$ eV	0.981
$E_f - 0.2$ eV	$0.999\ 64$
$E_f - 0.5$ eV	$0.999\ 999\ 998$

---

* Diamond can be an electronic semiconductor if impurities are present; NaCl can be an ionic semiconductor if conditions are favorable for sodium *ion* diffusion.

FIG. 11–3.5

Fermi Distribution (Metal). (a) From Eq. (11–3.2) for 20° C and for 500° C. At E_f, there is a 50 percent probability of occupancy of the energy levels in the energy band. (b) Schematic. As electrons (•) are activated above the Fermi energy, electron holes (○) are opened below the Fermi energy.

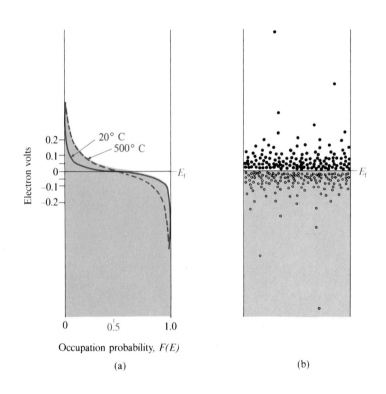

(a)   (b)

The symmetry is apparent when we compare the electron occupancy, $F(E)$, above E_f, with the presence of holes, $[1 - F(E)]$, below E_f.*

The effect of temperature on the probability of occupancy is shown in Fig. 11–3.6. This was obtained in Example 11–3.2 by plugging temperature and energy values in Eq. (11–3.2). Thus, the probability that there will be an electron at $(E_f + 0.2$ eV) is 0.000 60 at 40° C (313 K), and 0.000 20 at 0° C.

Example 11–3.1

A 0.1-mm film of polyethylene (PE) is used as a dielectric to separate two electrodes at 110 volts. Based on Fig. 11–1.1, what is the electron flux through the film?

Procedure Calculate R from $R = \rho L/A$, and the amperage from $E = IR$. Finally, the charge per electron (el) is 0.16×10^{-18} A·s.

---

* Although analogies are inexact, we can use one that involves the mean level of the sea. When the sea is placid and there is no wave energy, all the liquid water is below the surface; none is above. With the addition of energy, some water rises above the mean level in wave crests. Spray carries minor amounts to greater heights. Concurrently, the space below the mean level is not fully occupied, because of wave troughs and of bubbles entrained to greater depths.

With no thermal energy at 0° K, all the valence electrons are below the Fermi energy level; none are above. With the addition of thermal energy, some rise above E_f; a few rise far above it. Likewise, electron holes occur below this reference energy level.

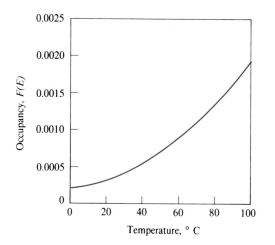

FIG. 11–3.6

F(E) **Versus Temperature (at $E_f +$ 0.2 eV).** The probability of occupancy increases with temperature according to Eq. (11–3.2). (See Example 11–3.2.)

Calculation Basis = 1 m² and 1 s.

$$R = \frac{(10^{14} \text{ ohm} \cdot \text{m})(10^{-4} \text{ m})}{(1 \text{ m}^2)} = 10^{10} \text{ ohm}$$

$$I = \frac{110 \text{ V}}{10^{10} \text{ ohm}} = 11 \times 10^{-9} \text{ amp}$$

$$\frac{(11 \times 10^{-9} \text{ A/m}^2)}{(0.16 \times 10^{-18} \text{ A} \cdot \text{s/el})} = 7 \times 10^{10} \text{ electrons/s} \cdot \text{m}^2$$

Comments This nanoampere current is small but measurable. Often, we must consider extraneous factors—such as impurities, pinhole porosity, and surface leakage—when we make measurements, because they may introduce larger currents.

Example 11–3.2

OPTIONAL

(a) Calculate and plot the probability that, between 0 and 100° C, an electron will occupy an energy level, *E*, that is at $(E_f + 0.2 \text{ eV})$. (Use 20° C temperature intervals.)

Procedure Solve for *F(E)* for the six temperatures using Eq. (11–3.2). ($k = 86.1 \times 10^{-6}$ eV/K for electron-volt energies.)

Calculation

(a) At 0° C: $F(E) = 1/[1 + e^{(0.2 \text{ eV})/k(273 \text{ K})}]$

$\qquad\qquad\qquad = 0.20 \times 10^{-3}$

20° C: $\qquad = 0.36 \times 10^{-3}$

40° C: $\qquad = 0.60 \times 10^{-3}$

60° C: $\qquad = 0.90 \times 10^{-3}$

80° C: $\qquad = 1.39 \times 10^{-3}$

100° C: $\qquad = 2.0 \times 10^{-3}$

(b) As the temperature increases through this temperature range, a greater probability exists that an electron will occupy an energy level that is 0.2 eV above the Fermi energy level. (See Fig. 11–3.6.)

Comment The value of $F(E)$ decreases with increasing temperature at energy levels below E_f.

_____

11–4

INTRINSIC SEMICONDUCTORS

Energy Gaps

Semiconductors and insulators are differentiated on the basis of the size of their forbidden-energy gaps (See Fig. 11–3.4). In a semiconductor, the energy gap is such that usable numbers of electrons are able to jump the gap from the filled valence band to the empty conduction band (Fig. 11–4.1). Those energized electrons can now carry a charge toward the positive electrode; furthermore, the resulting electron holes in the valence band become available for conduction, because electrons deeper in the band can move up into those vacated levels.

Figure 11–4.2 shows the energy gap schematically for C(diamond), Si, Ge, and Sn(gray). The gap is too large in diamond to provide a usable number of charge carriers, so diamond is categorized as an insulator (Table 11–4.1). With decreasing gap sizes, the number of carriers increases as we move down through Group IV of the periodic table to silicon, germanium, and tin; as a result, the conductivity increases, as shown in the accompanying Table 11–4.1. This conductivity is an inherent property of these materials and does not arise from impurities. Therefore, it is called *intrinsic semiconductivity.*

The crystal structure of diamond is repeated (from Chapters 2 and 3) in Fig. 11–4.3(a). Each carbon atom has a coordination number of 4, and each neighboring pair of atoms shares a pair of electrons (Section 2–2). Silicon, germanium, and

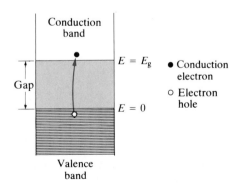

FIG. 11–4.1

Semiconduction. In semiconductors, a useful fraction of the valence electrons can jump the forbidden-energy gap. The electron is a negative carrier in the conduction band. The electron hole is a positive carrier in the valence band.

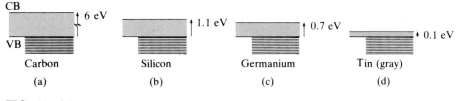

FIG. 11–4.2

Energy Gaps in Group IV Elements (Schematic). All these elements can have the same structure, and they all have filled bands. Because tin has the smallest energy gap, it has the most electrons in the conduction band (CB) at normal temperatures; therefore, it has the highest conductivity. (See Table 11–4.1.)

gray tin have the same structure.* Figure 11–4.4 uses germanium to represent schematically the mechanism of semiconductivity in these elements.

These four Group IV elements are the only elements that obtain semiconductivity from the structure of Fig. 11–4.3(a). However, a number of III-V *compounds* are based on the same structure (Fig. 11–4.3b). Atoms of elements from Group III of the periodic table (B, Al, Ga, In) alternate with atoms of elements from Group V of the periodic table (N, P, As, Sb). Most of the 16 III-V compounds that can form from these elements are semiconductors because every atom has four neighbors, and the average number of shared valence electrons is *four*. This matches exactly the situation for silicon and germanium, our predominant semiconductors.

Materials and electrical engineers also design electronic devices with *alloys of compounds;* for example, they use a solid solution of InAs and GaAs to give

TABLE 11–4.1 Energy Gaps in Semiconducting Elements

| | ENERGY GAP E_g | | AT 20° C (68° F) | |
| | | | FRACTION OF VALENCE ELECTRONS WITH | CONDUCTIVITY σ, |
ELEMENT	10^{-18} J	eV	ENERGY $> E_g$	ohm$^{-1}\cdot$m^{-1}
C(diamond)	0.96	~6	~$1/30 \times 10^{21}$	$< 10^{-16}$
Si	0.176	1.1	~$1/10^{13}$	5×10^{-4}
Ge	0.112	0.7	~$1/10^{10}$	2
Sn(gray)	0.016	0.1	~$1/5000$	10^6

---

* White tin is the more familiar polymorph. It is stable above 13° C (but may be supercooled to lower temperatures). White tin (bct) is denser than gray tin ($\rho_w = 7.2$ Mg/m^3 = 7.2 g/cm^3, while $\rho_g = 5.7$ Mg/m^3); therefore, the energy bands of white tin overlap, and this phase is a metallic conductor (see Fig. 11–3.2b).

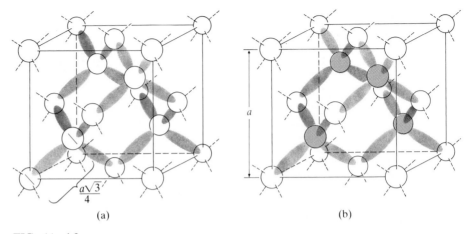

(a)   (b)

FIG. 11-4.3

Crystal Structures of Familiar Semiconductors. (a) Diamond, silicon, germanium, gray
tin. (b) ZnS, GaP, GaAs, InP, and so on. The two structures are similar, except that two
types of atoms are in alternate positions in the semiconducting compounds. All atoms
have CN = 4; each material has an average of four valence electrons per atom, and two
electrons per bond.

(In,Ga)As, or a solid solution of InP and InAs to give In(As,P). Furthermore,
four-component compounds of (In,Ga)(As,P) are possible. All such compounds
possess the same structure as do the individual III-V compounds (Fig. 11-4.3b).
The energy gaps of these III-V alloys are intermediate (but are not necessarily
linear) between the energy gaps of the end members (Fig. 11-4.5). Likewise, the

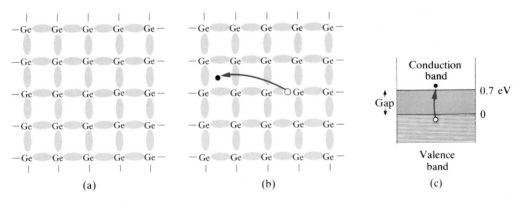

(a)   (b)   (c)

FIG. 11-4.4

Intrinsic Semiconductor (Germanium). (a) Schematic presentation showing electrons in
their covalent bonds. (b) Electron–hole pair. (Positive electrode at the left.) (c) Energy
gap, across which an electron must be raised to provide conduction. For each conduction
electron, there is a hole produced among the valence electrons.

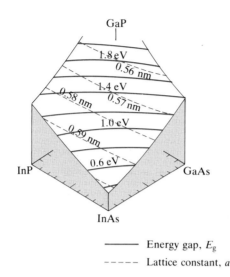

GaP

InP

InAs

GaAs

——— Energy gap, E_g
‑ ‑ ‑ ‑ ‑ Lattice constant, a

FIG. 11-4.5

Alloys of III-V compounds. By adjusting the In-Ga ratio and/or the P-As ratio, we can select combinations of energy gaps and lattice constants. (Adapted from C. J. Nuese, *J. Materials Education.*)

lattice constants are intermediate. The relationships of Fig. 11-4.5 are useful to the designer of complex devices in which it is necessary to match lattice size with specific energy-gap values.

Fermi Distribution in Intrinsic Semiconductors

Figure 11-4.4(c) shows only two charge carriers—one an electron in the conduction band, and the other an electron hole in the valence band. Of course, there are many more carriers in any electronic device. Their energy distribution in the energy bands is not random, but is dictated by the Fermi energy distribution. This distribution is graphed in Fig. 11-4.6 (a) as an overlay across the energy gap of one of the curves from Fig. 11-3.5. Several features are noticeable. First, the two "tails" of the distribution are inverted across the energy gap, as they must be if one hole is formed for each electron that has "jumped the gap." Second, the Fermi energy is in the center of the gap. This location is implied by our first preceding observation, even though no electrons reside in this forbidden-energy range. Third, there is an exponential decrease in the number of carriers above and below the edge of the energy gap. Figure 11-4.6(b) provides a schematic presentation of these three observations.

Charge Mobility

We must now modify our introductory equation on conductivity (Eq. 11-1.1) to match Fig. 11-4.1, since an intrinsic semiconductor has both negative and positive carriers.

The *electrons* that jump into the conduction band are the *negative*-type carriers. The conductivity they produce depends on their mobility μ_n through the conduction band of the semiconductor. The *electron holes* that are formed in the valence

FIG. 11–4.6

Fermi Distribution (Intrinsic Semiconductors). (a) Distribution across the energy gap. The Fermi energy, E_f, lies at the middle of the energy gap of an intrinsic semiconductor. (b) Schematic. The negative carriers are electrons (•); their numbers, n_n, decrease exponentially higher in the conduction band. The positive carriers are electron holes (○). Their numbers, n_p, decrease in a corresponding manner lower in the valence band.

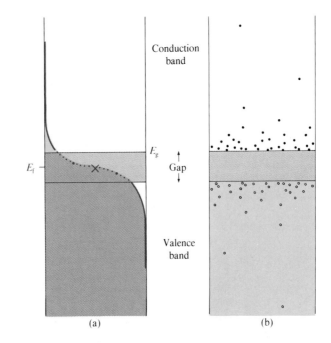

(a)   (b)

band are the *positive*-type carriers.* The conductivity they produce depends on their mobility μ_p through the valence band of the semiconductor. The total conductivity in an intrinsic semiconductor arises from both contributors

$$\sigma = n_n q \mu_n + n_p q \mu_p \qquad (11-4.1)$$

Of course, both the hole and the electron carry the same basic charge unit of 0.16×10^{-18} coul. In an intrinsic semiconductor, where there is a one-for-one formation of conduction electrons and electron holes, $n_n = n_p$; thus, we could simplify Eq. (11–4.1). However, we shall leave it in its present form, because n_n is not equal to n_p for *extrinsic* semiconductors (Section 11–5).

Table 11–4.2 summarizes the properties of a number of semiconductors. Note that we can make two generalizations:

1. The size of the energy gap commonly decreases as we move down in the periodic table: (C → Si → Ge → Sn), or (GaP → GaAs → GaSb), or (AlSb → GaSb → InSb).

2. The mobility of electrons within a given semiconductor is greater than is the mobility of electron holes in the same conductor.† The latter difference will be important when we consider the use of *n*-type semiconductors in contrast to *p*-type semiconductors.

---

* Comparably, anions, which have extra electrons, are negative-type, and cations, which are deficient in electrons, are positive-type.

† This relationship exists for all the semiconductors of Table 11–4.2, with the possible exception of AlSb, for which the mobility data have not been determined accurately.

TABLE 11-4.2 Properties of Common Semiconductors (20° C)*

| MATERIAL | ENERGY GAP E_g | | MOBILITIES, m²/V·s | | INTRINSIC CONDUCTIVITY, ohm⁻¹·m⁻¹ | LATTICE CONSTANT, a, nm |
	10^{-18} J	eV	ELECTRON, μ_n	HOLE, μ_p		
Elements						
C(diamond)	0.96	~6	0.17	0.12	$<10^{-16}$	0.357
Silicon	0.176	1.1	0.19	0.0425	5×10^{-4}	0.543
Germanium	0.112	0.7	0.36	0.23	2	0.566
Tin (gray)	0.016	0.1	0.20	0.10	10^6	0.649
Compounds						
AlSb	0.26	1.6	0.02	—	—	0.613
GaP	0.37	2.3	0.019	0.012	—	0.545
GaAs	0.22	1.4	0.88	0.04	10^{-6}	0.565
GaSb	0.11	0.7	0.60	0.08	—	0.612
InP	0.21	1.3	0.47	0.015	500	0.587
InAs	0.058	0.36	2.26	0.026	10^4	0.604
InSb	0.029	0.18	8.2	0.17	—	0.648
ZnS	0.59	3.7	0.014	0.0005	—	—
SiC (hex)	0.48	3	0.01	0.002	—	—

* Revised data collected by B. Mattes.

Semiconductivity (Intrinsic) Versus Temperature

Unlike metals, which have increased resistivity and decreased conductivity at higher temperatures (Fig. 11-2.1), the conductivity of intrinsic semiconductors *increases* at higher temperatures. The explanation is straightforward. Consider that the number of charge carriers, n, increases directly with the number of electrons that jump the gap (Fig. 11-4.1), and both increase exponentially with temperature. At 0° K, *no* electron would have the necessary energy to do this; however, as the temperature rises, the electrons receive energy, just as the atoms do. At 20° C, a useful fraction of the valence electrons in silicon, germanium, and tin have energy in excess of E_g, the energy gap (Table 11-4.1). The same situation obtains for compound semiconductors.

By analogy with Eq. (6-4.3a), the distribution of these thermally energetic electrons is

$$n_i \propto e^{-(E - \bar{E})/kT} \qquad \text{(11-4.2a)†}$$

---

* Equations (11-4.2a) and (6-4.3a) are the same at the upper end of the energy range only. In that range, the Pauli exclusion principle of two electrons per quantum state is not restrictive, since the probabilities of occupancy are very low. Of course, our interest is in this upper end of the range and in the initial electrons to jump the gap.

where n_i is the number of electrons per m^3 in the conduction band (and the number of holes per m^3 in the valence band). Within the forbidden-energy gap of an intrinsic semiconductor, the average energy $\overline{E}$ is at the middle of the energy gap, or $E_g/2$. Therefore,

$$n_i \propto e^{-E_g/2kT} \qquad \qquad \text{(11-4.2b)}$$

As in Chapter 6, T is the absolute temperature (K), and k is Boltzmann's constant, often expressed as 86.1×10^{-6} eV/K rather than as 13.8×10^{-24} J/K.

According to Eq. (11-1.1), conductivity σ is directly proportional to the number of carriers n; therefore,

$$\sigma = \sigma_0 e^{-E_g/2kT} \qquad \qquad \text{(11-4.3a)}$$

where σ_0 is the proportionality constant that includes, among other factors, both q and μ of Eq. (11-1.1). Admittedly, the mobility μ varies with temperature; however, its variation within the normal working range of most semiconductors is small compared with the exponential variation of the number of carriers n. Thus, we can rewrite Eq. (11-4.3a) in an Arrhenius form:

$$\ln \sigma = \ln \sigma_0 - E_g/2kT \qquad \qquad \text{(11-4.3b)}$$

If we measure the conductivity (or resistivity) of a semiconductor in the laboratory and plot $\ln \sigma$ versus $1/T$, we can calculate E_g from the slope of the curve; that is, slope $= -E_g/2k$. Conversely, from E_g and one known value of σ, we can calculate σ at a second temperature (Example 11-4.4).

Photoconduction

There is only a small probability that an electron in the valence band of silicon will be raised across the energy gap by thermal activation into the conduction band (~ 1 out of 10^{13}, according to Table 11-4.1). In contrast, a photon of red light (wavelength = 660 nm) has 1.9 eV of energy, which is more than enough to cause an electron to jump the 1.1-eV energy gap in silicon (Fig. 11-4.7). Thus, the conductivity of silicon increases markedly by photoactivation when this material is exposed to light.

Recombination

We can write the reaction that produces an *electron–hole pair,* as shown in Fig. 11-4.7 as

$$E \rightarrow n + p \qquad \qquad \text{(11-4.4a)}$$

where E is energy, n is the conduction electron, and p is the hole in the valence band. In this case, the energy came from light, but it could have come from other energy sources such as heat or fast-moving electrons.

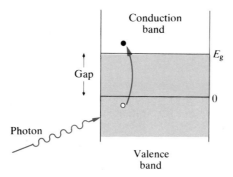

FIG. 11-4.7

Photoconduction. A photon (i.e., light energy) raises the electron across the energy gap, producing a "conduction electron + valence hole" pair, forming charge carriers. Recombination (Eq. 11-4.4b) occurs when the electron drops back to the valence band.

Since all materials are more stable when they reduce their energies, electron–hole pairs recombine sooner or later:

$$n + p \rightarrow E \qquad\qquad \textbf{(11-4.4b)}$$

In effect, the electron drops from the conduction band back to the valence band, just the reverse of the behavior shown in Fig. 11-4.4(c). If heat, light, or some other energy source did not continually produce additional electron–hole pairs, the conduction band would soon become depleted.

Luminescence

The energy released in Eq. (11-4.4b) may appear as heat. It also may appear as light; when it does, we speak of *luminescence* (Fig. 11-4.8). We can subdivide luminescence into several categories. *Photoluminescence* is the light emitted after electrons have been activated to the conduction band by light photons. *Chemoluminescence* is light emitted when the initial activation is due to chemical reactions. Probably, *electroluminescence* is best known, because this is what occurs in a television tube, where a stream of electrons (cathode rays) scans the screen, activating the electrons in the phosphor to their conduction band. Almost immediately, however, the electrons and holes recombine, emitting energy as visible light (Section 13-6).

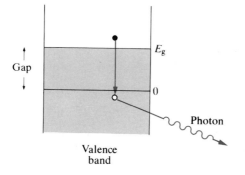

FIG. 11-4.8

Luminescence. During each microsecond, a fraction of the electrons energized to the conduction band return to the valence band. As an electron drops across the gap, the energy may be released as a photon of light.

Example 11–4.1

(a) What fraction of the charge in intrinsic silicon is carried by electrons? (b) What fraction is carried by electron holes?

Solution Recall that in an intrinsic semiconductor,

$$\sigma = \sigma_n + \sigma_p = (nq\mu)_n + (nq\mu)_p$$

With $n_n = n_p$,

$$\sigma_n/\sigma = \mu_n/(\mu_n + \mu_p)$$
$$= (0.19 \text{ m}^2/\text{V}\cdot\text{s})/(0.19 + 0.0425 \text{ m}^2/\text{V}\cdot\text{s}) = 0.82$$

$$\sigma_p/\sigma = 0.18$$

Comment Typically, conductivity is higher from electrons in the conduction band than it is from the electron holes in the valence band, because $\mu_n > \mu_p$. (See Table 11–4.2.)

Example 11–4.2

From Table 11–4.2, the compound gallium arsenide has an intrinsic conductivity of 10^{-6} ohm$^{-1}\cdot$m^{-1} at 20° C. How many electrons have jumped the energy gap?

Procedure Again, we use $\sigma = (nq\mu)_n + (nq\mu)_p = nq(\mu_n + \mu_p)$, and data from Table 11–4.2.

Calculation

$$n = (10^{-6} \text{ ohm}^{-1}\cdot\text{m}^{-1})/(0.16 \times 10^{-18} \text{ A}\cdot\text{s})(0.88 + 0.04 \text{ m}^2/\text{V}\cdot\text{s})$$
$$= 6.8 \times 10^{12}/\text{m}^3$$

Comment There will be 1.36×10^{13} carriers/m^3, since an electron hole remains for each electron activated across the energy gap.

Example 11–4.3

Each atom in gray tin has four valence electrons. The unit cell (uc) size (Fig. 11–4.3a) is 0.649 nm. Separate calculations indicate that there are 2×10^{25} conduction electrons per m^3. What fraction of the electrons has been activated to the conduction band?

Procedure The unit cell of tin has eight atoms (Fig. 11–4.3a); therefore, there are 32 valence electrons per a^3.

Calculation

$$\text{valence electrons/m}^3 = \frac{(8 \text{ atoms/uc})(4 \text{ el/atom})}{(0.649 \times 10^{-9} \text{ m})^3/\text{uc}} = 1.17 \times 10^{29}/\text{m}^3$$

$$\text{fraction activated} = \frac{2 \times 10^{25}}{1.17 \times 10^{29}} \approx 0.0002$$

Example 11-4.4

The resistivity of germanium at $20°$ C ($68°$ F) is 0.5 ohm·m. What is its resistivity at $40°$ C ($104°$ F)?

Procedure Two procedures are possible. The first is to solve for the $\rho_{20° C}/\rho_{40° C}$ ratio from $\rho_{20° C}$ and E_g. The second is to solve graphically (Fig. 11-4.9) from $\sigma_{20° C}$ and the slope, $E_g/2k$.

Solution Based on Eq. (11-4.3) and an energy gap of 0.7 eV (Table 11-4.2),

$$\frac{\rho_1}{\rho_2} = \frac{\sigma_2}{\sigma_1} = \frac{\sigma_0 e^{-E_g/2kT_2}}{\sigma_0 e^{-E_g/2kT_1}}$$

$$\ln\frac{\rho_1}{\rho_2} = \frac{E_g}{2k}\left[\frac{1}{T_1} - \frac{1}{T_2}\right] \tag{11-4.5}$$

$$\ln\frac{\rho_{20}}{\rho_{40}} = \frac{0.7\text{ eV}}{2(86.1 \times 10^{-6}\text{ eV/K})}\left[\frac{1}{293°\text{ K}} - \frac{1}{313\text{ K}}\right] = 0.9$$

$$\rho_{20}/\rho_{40} = {\sim}2.5$$

Thus, since $\rho_{20° C} = 0.5$ ohm·m, $\rho_{40° C} = 0.2$ ohm·m.

Alternate Solution From the slope (Fig. 11-4.9),

slope $= -0.7$ eV/2(86.1 $\times 10^{-6}$ eV/K) $= -4060$ K

or

slope $= -(0.112 \times 10^{-18}$ J)/2(13.8 $\times 10^{-24}$ J/K) $= -4060$ K
At $20°$ C (•),

$\ln \sigma = \ln (1/0.5$ ohm·m) $= 0.693$ (or $\log_{10} \sigma_{20° C} = 0.301$)

$1/T = 1/293$ K $= 0.00341$/K   (or $1000/T = 3.41$)

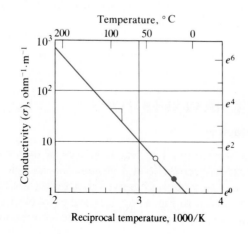

FIG. 11-4.9

Semiconduction Versus Temperature (Intrinsic Germanium). The slope is $-E_g/2k$ of Eq. (11-4.3) when the ordinate is ln σ and the abscissa is reciprocal temperature, K⁻¹. See Example 11-4.4.

At 40° C (∘),

$$1/T = 1/313 \text{ K} = 0.00319/\text{K}$$ (or $1000/T = 3.19$)

$$\text{slope} = -4060 \text{ K} = \frac{0.693 - \ln \sigma_{40° \text{ C}}}{(0.00341 - 0.00319)/\text{K}}$$

$$\ln \sigma_{40° \text{ C}} = 1.6$$ (or $\log_{10} \sigma_{40° \text{ C}} = 0.7$)

$$\sigma_{40° \text{ C}} = 5 \text{ ohm}^{-1} \cdot \text{m}^{-1}$$ (or $\rho = 0.2 \text{ ohm} \cdot \text{m}$)

Comments It is possible to measure resistance changes (and therefore resistivity changes) of less than 0.1 percent. Therefore, we can measure temperature changes of a small fraction of a degree. (See Problem 11–P61.)

OPTIONAL ### Example 11–4.5

An intrinsic semiconductor has an energy gap of 0.4 eV. Calculate and plot the probability between 0 and 100° C that the lowest level in the conduction band will be occupied. (Use 20° C intervals.)

Procedure In an intrinsic semiconductor, E_f is in the middle of the energy gap. Therefore, the energy, E, is at a level of $(E_f + 0.2 \text{ eV})$. Solve for $F(E)$ for the six temperatures using Eq. (11–3.2).

Calculation

At 0° C: $F(E) = 1/[1 + e^{(0.2 \text{ eV})/k(273 \text{ K})}]$
                $= 0.20 \times 10^{-3}$
                $\vdots$

This example duplicates Example 11–3.2. See Fig. 11–3.6 for the plot.

Comment Since the Fermi distribution curve is reversed across the energy gap in an intrinsic semiconductor, there is the same probability that an electron hole will be present at the top level in the valence band; that is, $(1.0 - F(E))$ at $(E_f - 0.2 \text{ eV})$. (See the discussion in the paragraph following Eq. 11–3.2.)

11–5
EXTRINSIC SEMICONDUCTORS

n-Type Semiconductors

Impurities alter the semiconducting characteristics of materials by introducing excess electrons or excess electron holes. Consider, for example, silicon containing an atom of phosphorus. Phosphorus has five valence electrons rather than the four that are found with silicon. In Fig. 11–5.1(a), the extra electron is present independently of the electron pairs that serve as bonds between neighboring atoms.

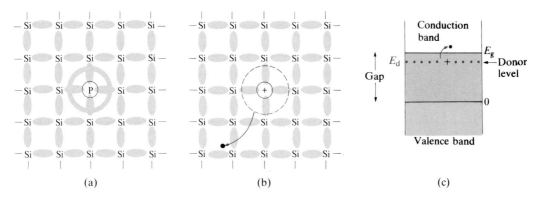

FIG. 11-5.1

Extrinsic Semiconductors (*n*-Type). A Group V atom has an extra valence electron beyond the average of four sketched in Fig. 11-4.3. This fifth electron can be pulled away from its parent atom with little added energy, and can be "donated" to the conduction band to become a charge carrier. We observe the donor energy level, E_d, as being just below the top of the energy gap. (a) An *n*-type impurity, such as phosphorous. (b) Ionized phosphorus atom. (Positive electrode at left.) (c) Band model.

When it is supplied with a small amount of energy, this electron can be pulled away from the phosphorus atom and can carry a charge toward the positive electrode (Fig. 11-5.1b). Alternatively, in Fig. 11-5.1(c), the extra electron—which cannot reside in the valence band because that is already full—is located near the top of the energy gap. From this position— called a *donor* level E_d—the extra electron can easily be donated to the conduction band. Regardless of which model is used (Fig. 11-5.1b or 11-5.1c), we can see that atoms from Group V (N, P, As, and Sb) of the periodic table (Fig. 2-1.1) can supply negative, or *n*-type, charge carriers to semiconductors.

p-Type Semiconductors

Group III elements (B, Al, Ga, and In) have only three valence electrons. Therefore, when such elements are added to silicon as impurities, electron holes are created. As shown in Fig. 11-5.2(a and b), each aluminum atom can accept one electron. In the process, a positive charge moves toward the negative electrode. Using the band model (Fig. 11-5.2c), we note that the energy difference for electrons to move from the valence band to the *acceptor level*, E_a, is much less than the full energy gap. Therefore, electrons are more readily activated into the acceptor sites than into the conduction band. The electron holes remaining in the valence band are available as positive carriers for *p*-type semiconduction. Therefore, with $(n_p)_{ex} = n_{III}$,

$$\sigma_{ex} = n_{III} q \mu_p \qquad \qquad (11\text{-}5.1)$$

where n_{III} is the number of acceptors per m³ (Group III atoms/m³).

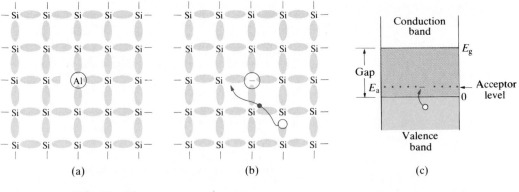

(a)   (b)   (c)

FIG. 11-5.2

Extrinsic Semiconductors (p-Type). A Group III atom has one less valence electron than the average of four sketched in Fig. 11-4.3. This atom can accept an electron from the valence band, thus leaving an electron hole as a charge carrier. The acceptor energy level, E_a, is just above the bottom of the energy gap. (a) A p-type impurity such as aluminum. (b) Ionized aluminum atom. (Negative electrode at right.) (c) Band model.

 OPTIONAL   ## Fermi Distribution in Extrinsic Semiconductors

The conduction band and the valence band are not balanced in an extrinsic semiconductor. In n-type semiconductors, the large majority of the electrons in the conduction band come from donors, rather than from the valence band (at normal temperatures). In p-type semiconductors, acceptors accommodate most of the electrons that open up electron holes in the valence band, and only a few electrons enter the conduction band. Thus, Fig. 11-5.3 indicates the distribution of the occupancies across the energy gaps. In addition to the asymmetry, we should note that the Fermi energy (50 percent occupancy) lies above the middle of the gap in a n-type semiconductor; it lies below the middle in a p-type one. These shifts are important in n-p junctions (Section 11-6). As there is with the intrinsic distribution, there is an exponential decrease in the numbers of carriers away from the edges of the gap.

Donor Exhaustion (and Acceptor Saturation)

Since donated electrons have only a small jump to the conduction band, they initiate *extrinsic* conductivity at relatively low temperatures. As the temperature is increased, the slope of the Arrhenius curve is $-(E_g - E_d)/k$, as shown at the right in Fig. 11-5.4.

If the donor impurities are limited in number (e.g., 10^{21} P/m^3 within silicon), essentially all the donated electrons have moved into the conduction band at temperatures below that of normal usage. This supply has been exhausted. In the

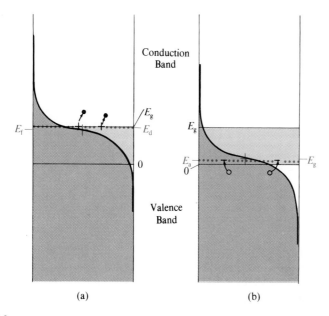

Conduction
Band

Valence
Band

(a)   (b)

FIG. 11–5.3

Fermi Distribution (Extrinsic Semiconductors at Normal Temperatures). (a) n-Type
semiconductor. Donors contribute the majority of the electrons in the conduction band.
(b) p-Type semiconductor. Acceptors receive electrons from the valence band, leaving
electron holes as charge carriers. (See Fig. 11–4.6, where electrons must be activated
across the energy gap to produce charge carriers.) At $0°$ K, $E_f > E_d$, and $E_f < E_a$. At
temperatures sufficiently high for electrons to jump to the conduction band the Fermi
energy, E_f, moves toward the center of the energy gap.

example of 10^{21} per phosphorus atoms per mm³, the extrinsic conductivity is

$$\sigma_{ex} = (10^{21}/m^3)(0.16 \times 10^{-18} \text{ A} \cdot \text{s})(0.19 \text{ m}^2/\text{V} \cdot \text{s})$$
$$= 30 \text{ ohm}^{-1} \cdot \text{m}^{-1}$$

The extrinsic conductivity will not continue to rise with further temperature
increases — there is a conductivity plateau.*

In the meantime, the *intrinsic* conductivity is very low in a semiconductor such
as silicon (5×10^{-4} ohm$^{-1} \cdot$ m^{-1} at $20°$ C, Table 11–4.1). Its Arrhenius curve is at
the left in Fig. 11–5.4 with the intrinsic slope of $-E_g/2k$; that is, -1.1 eV/$2k =$
6400 K. Only at elevated temperatures does the total conductivity increase above
the exhaustion plateau. (See Example 11–5.3.)

---

* With constant n, experiments can detect a slight decrease in μ that results from the shorter mean free
path that accompanies increased temperatures (Section 11–2).

FIG. 11–5.4

Donor Exhaustion. Intrinsic (left curve) and extrinsic (right curve) conductivities require energies of E_g and ($E_g - E_d$), respectively, to raise electrons into the conduction band (Fig. 11–5.1c). At lower temperatures, donor electrons provide most of the conductivity. Exhaustion occurs when all the donor electrons have entered the conduction band, and before the temperature is raised high enough for valence electrons to jump the energy gap. The conductivity is nearly constant in this temperature range.

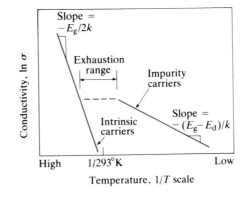

Donor exhaustion of n-type semiconductors has its parallel in *acceptor saturation* of p-type semiconductors. (You can paraphrase the previous paragraphs for the saturation analog.) Donor exhaustion and acceptor saturation are important to materials and electrical engineers, since these situations provide a region of essentially constant conductivity. This means that it is less necessary to compensate for temperature changes in electrical circuits than it would be if the log σ-versus-$1/T$ characteristics followed an ever-ascending line.

//**O P T I O N A L**//   **Hall Effect**

As charges move through a magnetic field, they are deflected by a 90° force according to the familiar right-hand rule of physics. This deflection induces a *Hall voltage,* $\mathcal{E}_H$, as shown in Fig. 11–5.5. The direction of the voltage depends on the sign of the charge carrier. Thus, the sign of the voltage identifies whether the carrier

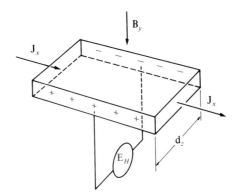

FIG. 11–5.5

Hall Voltage (n-Type). As the current flux, J_x, moves through a magnetic flux, B_y, the charge carriers are displaced to produce a Hall field, $\mathcal{E}_H$. (L. H. Van Vlack, *Materials Science for Engineers,* Addison-Wesley, Reading, Mass., with permission.)

is an electron (negative) or an electron hole (positive), and distinguishes between *n*- and *p*-type semiconductors.

The Hall voltage also provides a basis for determining the mobility, μ, of the carriers. The deflecting force, F, on the electron is

$$F = -qB_y \bar{v}_x \tag{11-5.2}$$

where B_y is the magnetic flux density in $N/A \cdot m$ (or webers/m^2), and $\bar{v}_x$ is the drift velocity through the semiconductor. In accordance with Eq. (11–1.2), we can express this velocity as $\mathscr{E}_x\mu$, where $\mathscr{E}_x$ is the voltage gradient for conduction, and μ is the charge mobility:

$$F = -qB_y\mathscr{E}_x\mu \tag{11-5.3}$$

This force is balanced by an equal force, $q\mathscr{E}_H$, on the charge carrier from the induced, or Hall, field $\mathscr{E}_H$:

$$F_z = 0 = -qB_y\mathscr{E}_x\mu - q\mathscr{E}_H \tag{11-5.4}$$

As a result, we can determine the mobility from the formula

$$\mu = -\mathscr{E}_H/\mathscr{E}_x B_y \tag{11-5.4*}$$

where the three values on the right can all be determined experimentally.

Defect Semiconductors

The iron oxide of Fig. 4–5.2 possessed Fe^{3+} ions in addition to the regular Fe^{2+} ions. A similar situation occurs in Fig. 11–5.6(a), when NiO is oxidized to give some Ni^{3+} ions; in fact, the situation is relatively common among transition-metal oxides that have multiple valences. In nickel oxide, three Ni^{2+} are replaced by two Ni^{3+} and a vacancy, $\square$. This replacement maintains the charge balance; it also permits easier diffusion and therefore some ionic conductivity. More important, however, electrons can hop from an Ni^{2+} ion into acceptor sites in Ni^{3+} ions. Conversely, an electron hole moves from one nickel ion to another as it migrates toward the negative electrode. Nickel oxide and other oxides with $M_{1-x}O$ defect structures are *p*-type semiconductors.

There are also *n*-type oxides. Zinc oxide, when exposed to zinc vapors, produces $Zn_{1+y}O$. A zinc ion moves into an interstitial position (Fig. 11–5.6b). The Zn^+ ions that balance the charge have one electron more than do the bulk of the Zn^{2+} ions. The former ions can donate electrons to the conduction band for *n*-type semiconductivity.

---

* The units are $m^2/V \cdot s = [V/m]/(V/m)(N/A \cdot m] = A \cdot m/N$, or $V \cdot A \cdot s = N \cdot m$.

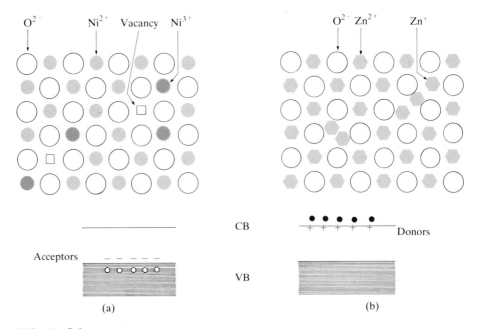

FIG. 11–5.6

Defect Semiconductors. (a) $Ni_{1-x}O$. The Ni^{3+} ions serve as electron acceptors, so holes (∘) form in the valence band. (b) $Zn_{1+y}O$ (schematic). The Zn^+ ions donate electrons (•) to the conduction band for *n*-type semiconduction.

Example 11–5.1

Silicon, according to Table 11–4.2, has a conductivity of 5×10^{-4} ohm$^{-1} \cdot$ m^{-1} when pure. Specifications call for a conductivity of 200 ohm$^{-1} \cdot$ m^{-1} when the silicon contains aluminum as an impurity. How many aluminum atoms are required per m^3 of silicon?

Procedure The total conductivity, σ, is the sum of $\sigma_{in} + \sigma_{ex}$. However, from Table 11–4.2, $\sigma_{in} = 5 \times 10^{-4}$ ohm$^{-1} \cdot$ m$^{-1} \ll 200$ ohm$^{-1} \cdot$ m^{-1}, as required here. Therefore, we can use Eq. (11–5.1) directly.

Calculation

$$n_{III} = n_p = (200 \text{ ohm}^{-1} \cdot \text{m}^{-1})/(0.16 \times 10^{-18} \text{ A} \cdot \text{s})(0.0425 \text{ m}^2/\text{V} \cdot \text{s}).$$
$$= 3 \times 10^{22}/\text{m}^3$$

Comments Each aluminum atom contributes one acceptor site and hence one electron hole. Therefore, 3×10^{22} aluminum atoms are required per m^3. This is a large number; however, it is still small (0.6 ppm) when compared with the number of silicon atoms per m^3. (See Problem 11–P51.)

Example 11–5.2

Early transistors used germanium with an extrinsic resistivity of 200 ohm $\cdot$ m and a conduction-electron concentration of $0.87 \times 10^{17}/m^3$. (a) What is the mobility of the electrons in the germanium? (b) What impurity elements could be added to the germanium to donate the conduction electrons?

Procedure   We can write an equation for *n*-type semiconduction analogous to Eq. (11–5.1):

$$\sigma_{ex} = n_v q \mu_n = 1/\rho_{ex} \qquad\qquad (11\text{–}5.5)$$

since the extrinsic conduction arises from electron donors.

Calculation

   (a) $\mu_n = 1/(200 \text{ ohm} \cdot \text{m})(0.87 \times 10^{17}/m^3)(0.16 \times 10^{-18} \text{ A} \cdot \text{s})$
        $= 0.36 \text{ m}^2/\text{V} \cdot \text{s}$

   (b)   Group V elements: N, P, As, Sb.

Comments   Note that the electron mobility does not depend on which of these Group V elements is added, since the electron, once in the conduction band, moves through the germanium structure independent of its donor.

    Group VI elements also could be added. Since they have a second additional electron (beyond the four necessary for bonding), it would take only $0.4 \times 10^{17}/m^3$ of these atoms to supply $0.8 \times 10^{17}/m^3$ conduction electrons.

Example 11–5.3

The residual phosphorus content of purified silicon is 0.1 part per billion (by weight). Will the resulting conductivity exceed the intrinsic conductivity of silicon?

Solution   Based on density (Appendix B): 1 m^3 silicon $= 2.33 \times 10^6$ g   Si $= 2.33 \times 10^{-4}$ g P.

   $(2.33 \times 10^{-4} \text{ g/m}^3)/(30.97 \text{ g}/0.6 \times 10^{24}) = 4.5 \times 10^{18}/m^3$

$$\sigma_{ex} = (4.5 \times 10^{18}/m^3)(0.16 \times 10^{-18} \text{ A} \cdot \text{s})(0.19 \text{ m}^2/\text{V} \cdot \text{s})$$
$$= 0.14 \text{ ohm}^{-1} \cdot m^{-1}$$

versus

   $\sigma_{in} = 5 \times 10^{-4} \text{ ohm}^{-1} \cdot m^{-1}$ (from Table 11–4.2).

Comment   To achieve the low impurity levels required for semiconduction production, engineers had to develop entirely new processing procedures.

Example 11–5.4

There are 10^{22} Al/m^3 in silicon to produce a *p*-type semiconductor. At what temperature will the intrinsic conductivity of silicon be equal to the maximum extrinsic conductivity?

Procedure Saturation is required for maximum extrinsic conductivity; therefore, $n_p =$ n_{III}. Of course, μ_p (only) for silicon must be used for this extrinsic conductivity.

From Table 11–4.2, $\sigma_{in} = 5 \times 10^{-4}$ ohm$^{-1}\cdot$m^{-1} at 20° C and $E_g = 1.1$ eV. Calculate the temperature at which $\sigma_{in} = \sigma_{ex}$.

Calculation

$$\sigma_{ex} = (10^{22}/\text{m}^3)(0.16 \times 10^{-18} \text{ A}\cdot\text{s})(0.0425 \text{ m}^2/\text{V}\cdot\text{s})$$
$$= 68 \text{ ohm}^{-1}\cdot\text{m}^{-1}$$

$$\frac{\sigma_{20°C}}{\sigma_T} = \frac{5 \times 10^{-4} \text{ ohm}^{-1}\cdot\text{m}^{-1}}{68 \text{ ohm}^{-1}\cdot\text{m}^{-1}} = \frac{\sigma_0 e^{-1.1/2k(293 \text{ K})}}{\sigma_0 e^{-1.1/2kT}}$$

$$\ln \frac{5 \times 10^{-4}}{68} = -11.8 = \frac{-1.1 \text{ eV}}{2(86.1 \times 10^{-6} \text{ eV/K})}\left[\frac{1}{293 \text{ K}} - \frac{1}{T}\right]$$

$$T = 640 \text{ K} \qquad\qquad (\text{or } 367° \text{ C})$$

Comments The most general equation for conductivity in semiconductors is

$$\sigma = \sigma_{in} + (\sigma_n)_{ex} + (\sigma_p)_{ex}$$
$$= (n_{in}q)(\mu_n + \mu_p) + (n_n q\mu_n)_{ex} + (n_p q\mu_p)_{ex} \qquad (11\text{–}5.6)$$

Typically, only one of the three terms is significant at a time, and the others can be dropped. In this example, only $(\sigma_p)_{ex}$ is significant at 20° C.

Example 11–5.5

Donors in an *n*-type silicon possess energies that position the Fermi energy at 0.1 eV below the top of the energy gap at 20° C. What is the ratio of the occupancy probability at the bottom edge of the conduction band to the probability of an electron hole being present at the top edge of the valence band?

Procedure Refer to Fig. 11–5.3(a). We are comparing the number of possible carriers at each edge of the energy gap, which is 1.1 eV for silicon (Table 11–4.2). We want to compare $F(E)$ at $(E_f + 0.1)$ with $[1 - F(E)]$ at $(E_f - 1.0)$. Although the conduction band (cb) and the valence band (vb) do not mirror each other in extrinsic semiconductors, the Fermi distribution curve is inverted around E_f. Therefore, $1 - F(E)_{vb} = F(E)_{cb}$ at $(E_f + 1.0)$.

Calculation

At $(E_f + 0.1)$: $F(E)_{cb} = [1/(1 + e^{0.1 \text{ eV}/k(293 \text{ K})})] = 18 \times 10^{-3}$

At $(E_f + 1.0)$: $1 - F(E)_{vb} \doteq F(E)_{cb}$ at $(E_f + 1.0)$.
$$= [1/(1 + e^{1.0 \text{ eV}/k(293 \text{ K})})] = 6 \times 10^{-18}$$

$$\frac{F(E)_{cb}}{1 - F(E)_{vb}} = \frac{18 \times 10^{-3}}{6 \times 10^{-18}} = 3 \times 10^{15}$$

Comment For all practical purposes, we can ignore the number electron holes for this extrinsic silicon at 20° C.

11-6

SEMICONDUCTOR DEVICES

There are many electronic devices that use semiconductors. We shall consider but a few.

Conduction and Resistance Devices

We have already seen that the conductivity of a *photoconductor* varies directly with the amount of incident light. This capability leads to *light-sensing* devices. The radiation does not have to be visible—it also may be ultraviolet or infrared, provided that the photons have energy comparable to or greater than the energy gap.

A second device is a *thermistor.* It is simply a semiconductor that has had its resistance calibrated against temperature. If the energy gap is large, such that the ln σ versus $1/T$ curve is steep, it is possible to design a thermistor that will detect temperature changes of 10^{-4} °C.*

Because many semiconducting materials have low packing factors, they have a high compressibility. Experiments show that, as the volume is compressed, the size of the energy gap is measurably reduced; this, of course, increases the number of electrons that can jump the energy gap. Thus, pressure can be calibrated against resistance for *pressure gages.*

A *photomultiplier* device makes use of electron activation, first by photons *and* then by the electrons themselves. Assume, for example, that a very weak light source—even just one photon—were to hit a valence electron. Our eye would not be able to detect it. However, if that electron were raised to the conduction band, and simultaneously the semiconductor was within a very strong electric field, that electron would be accelerated to high velocities and high energies. In turn, it could activate one or more additional electrons that also would respond to the very strong field. We can use this multiplying effect to advantage. A very weak light signal can be amplified. With appropriate focusing, the image in nearly complete darkness can be brought into visible display.

Junction Devices (Diodes)

A number of devices use junctions between *n*- and *p*-type semiconductors. The most familiar of these junctions is the *light-emitting diode* (LED). We see it used in the digital displays (red) of many hand-held calculators and clocks. An LED operates on the principle shown schematically in Fig. 11-6.1. The charge carriers

---

* In technical practice, thermistors are better than are other types of thermometers for measuring small temperature *changes.* However, thermocouples and similar devices are more convenient for measuring the temperature itself. The measurement of temperature changes is important in microcalorimetric studies involving chemical or biological reactions.

FIG. 11–6.1

Light-Emitting Diode (LED; Schematic). (a) An LED is a junction device between *n*- and *p*-type semiconductors. (b) When a forward bias is placed across the junction, carriers of both types cross the junction; there they recombine emitting a photon (Eq. (11–6.1) and Fig. 11–4.8).

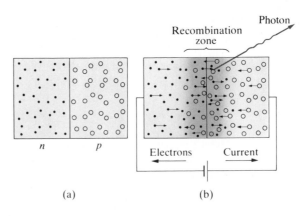

(a)   (b)

on the *n*- and *p*-sides of the junction are electrons and holes, respectively. If a current is passed through the device in the direction shown, the holes of the valence band move through the junction, into the *n*-type materials; conversely, the electrons of the conduction band cross into the *p*-type material. Adjacent to the junction, there are excess carriers that recombine and produce luminescence:

$$n + p \rightarrow \text{photon} \tag{11–6.1}$$

When GaAs is used, the photons emitted in the recombination zone are red; a phosphorus substitution to give Ga(As,P) produces green photons.

The junction of Fig. 11–6.1 also can serve as a *rectifier;* that is, it can function as an electrical "check valve" that lets current pass one way and not the other. With the *forward bias* of Fig. 11–6.2(a), current can pass because carriers—both electrons and holes—move through the junction. With a *reverse bias* of Fig. 11–6.2(b), the carriers are pulled away from each side of the junction to leave a

FIG. 11–6.2

Rectifier (Schematic). (a) Current flows with a forward bias because charge carriers pass the junction. (b) With a reverse bias, charge carriers are depleted from the junction region. Extrinsic conductivity disappears from the junction region, and only a small amount of intrinsic conductivity remains.

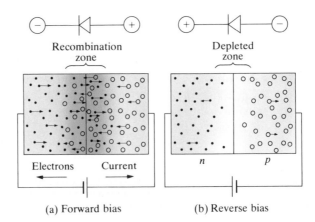

(a) Forward bias   (b) Reverse bias

carrier-depleted "insulating zone" at the junction. If a greater voltage is applied, this *depletion zone* is simply widened. Usable current passes with only the forward bias.

This behavior holds over a wide range of reverse voltages. There is a point, however, where a surge of current is allowed to pass because there is an electrical breakdown in the depleted zone. Specifically, those few carriers that are in the depleted zone are accelerated to high velocities by the steep potential drop. As occurs in the photomultiplier device that we described, these energetic electrons can knock loose other electrons. An *avalanche* develops that leads to a high current. In effect, we have a "safety valve" that opens at a definite voltage. Lightening arrestors operate on this principle.

As an alternate to Fig. 11–6.2, we can explain the behavior of the *n-p* junction on the basis of the Fermi distribution. We shall describe that perspective because it can help us to visualize more complex devices.

Figure 11–6.3(a) shows the *n-* and *p*-distributions when the two types are separated, and the energy gaps are equated. When the two are in contact, the gaps must shift so as to equalize the Fermi energy, as shown in Fig. 11–6.3(b). (This

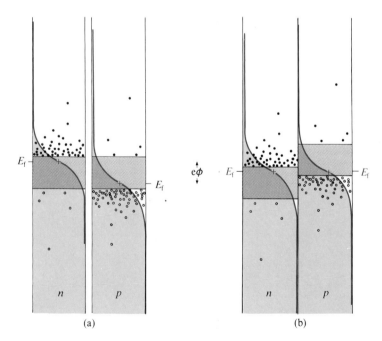

(a)   (b)

FIG. 11–6.3

n–p **Junction (Schematic).** (a) Separate *n-* and *p*-distributions. (b) Distributions in contact. The Fermi distributions must equalize. This introduces a contact potential, Φ, and requires the corresponding energy, $e\Phi$, equal to the initial ΔE_f.

also equalizes the distribution probabilities at *all* energy levels within the two sides of the junction.) A contact potential, Φ, and the resulting energy, $e\Phi$, are established because electrons were moved from the valence band of the *n*-type material to the valence band of the *p*-type material.

As presented in Fig. 11–6.4(a), the conductivity through the junction is low because few electrons are available above the top of the combined gaps (and few electron holes are available below the lower side). If a *forward bias* is placed across the junction (Fig. 11–6.4b), however, then there is an exponential increase in the number of carriers with the increased voltage. Once across the junction into the *p*-side, the electrons recombine with the electron holes according to Eq. (11–6.1). The analogous situation occurs with the electron holes as they move into the *n*-side. The resulting energy may produce light of a wavelength that corresponds to the gap width.

Current flows with the forward bias. A reverse bias, however, reduces the carriers to a negligible number (Fig. 11–6.4c). The *n–p* junction serves as a rectifier (Fig. 11–6.4c). With appropriate circuitry, rectifiers can change ac current into dc current.

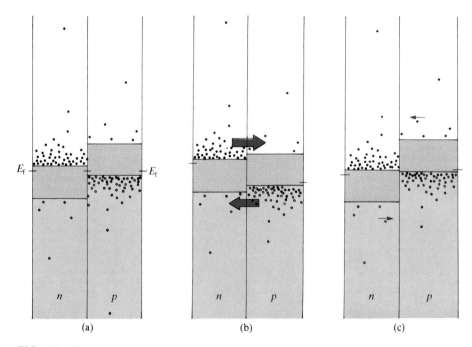

(a)   (b)   (c)

FIG. 11–6.4

Rectifying Junction. (a) *n–p* Junction, no bias. (See Fig. 11–6.3b) (b) Forward bias. The numbers of carriers in both bands increase exponentially with the applied voltage. (c) Reverse bias. The numbers of carriers crossing the junction become negligible.

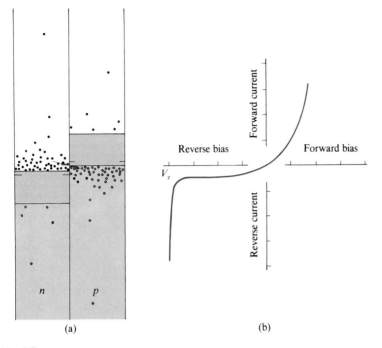

FIG. 11-6.5

Zener Diode. A forward bias readily increases the current; a reverse bias produces little reverse current initially. When the reverse voltage is sufficiently great, V_z, electrons can "tunnel" through the junction (a), causing a dramatic flow of reverse-biased current (b).

The Zener diode, which conducts current from a large reverse bias, is a consequence of an extreme shift, as shown in Fig. 11-6.5. Electrons and holes can "tunnel" across the junction to transport the current. Since we can choose various doping levels in the semiconducting materials, we can tailor the threshold voltage for triggering these diodes to required specifications.

Transistors

Transistors are junction devices that amplify weak signals to make stronger, usable outputs.

The simplest type of transistor,* the *field-effect transistor* (FET), makes use of a carrier-depleted zone to amplify the output of a circuit. In Fig. 11-6.6, the *p-n*

---

* The term *transistor* is a contraction of "transfer resistor"; the early rationale for this device's performance was based on that concept of operation.

FIG. 11–6.6

Transistor (Field-Effect). A single n–p junction is used. A voltage change across the gate, which is reverse biased, changes the width of the depleted zone and, hence, the cross-section of the conduction channel between the source and the drain. A small variation in the input voltage produces a major change in the resulting current through the conduction channel.

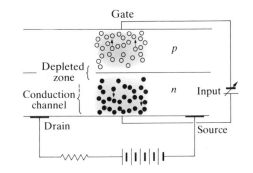

junction is reverse-biased by the input signal. As the signal varies, the carrier-depleted zone varies in size and thus alters the resistance between the *source* and the *drain.* In turn, the current going to the output varies in a controlled manner. A small signal produces a significant current fluctuation.

A second type of transistor called a *junction transistor,* has two junctions in series. The transistor thus can be either p–n–p or n–p–n. The former has been used somewhat more in the past; however, we shall consider the n–p–n transistor, since it is easier to visualize the movements of electrons, rather than the movements of holes. The principles behind each type are the same.

Before we describe the makeup of the transistor, recall that as holes move across the junction with a forward bias (Fig. 11–6.2a), they recombine with the electrons in the n-type material according to Eq. (11–4.4b). Likewise, the electrons combine with the holes as the electrons move beyond the junction and into the p-type material. However, the reaction of Eq. (11–4.4b) does not occur immediately. In fact, an excess number of positive and negative carriers may move considerable distances beyond the junction. The number of excess, unrecombined carriers is an exponential function of the applied voltage; it is important in transistor operation.

A transistor consists of an *emitter,* a *base,* and a *collector* (Fig. 11–6.7). For the moment, consider only the *emitter junction,* which is biased such that electrons move into the base (and toward the collector). As we said, the number of electrons that cross this junction and move into the p-type material is an exponential function of the emitter voltage, V_e. Of course, these electrons at once start to combine with the holes in the base; however, if the base is narrow, or if the recombination time is long (τ of Section 13–6), the electrons keep moving on through the thickness of the base. Once they are at the second junction, the *collector junction,* the electrons have free sailing, because the collector is an n-type semiconductor. The total current that moves through the collector is controlled by the emitter voltage, V_e. As the emitter voltage fluctuates, the collector current, I_c, changes exponentially. Written logarithmically,

$$\ln I_c \approx \ln I_0 + V_e/B \tag{11–6.2a}$$

or

$$I_c = I_0 e^{V_e/B} \tag{11–6.2b}$$

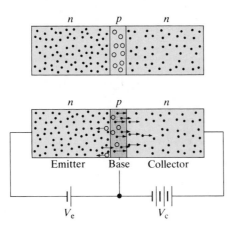

FIG. 11-6.7

Transistor (n-p-n). The number of electrons crossing from the emitter-base junction is highly sensitive to the emitter voltage. If the base is narrow, these carriers move to the base-collector junction, and beyond, before recombination. The total current flux, emitter to base, is highly magnified, or amplified, by fluctuations in the voltage of the emitter.

where I_0 and B are constants for any given temperature. Thus, if the voltage in the emitter is increased even slightly, the amount of current is increased markedly. It is because of these relationships that a transistor serves as an amplifier.

Example 11-6.1

Zinc sulfide is used as a thermistor. To what fraction sensitivity, δ, must the resistance be measured to detect a $0.001°$ C change at $20°$ C?

Procedure We want to know $\delta(=\Delta R/R)$ for $\pm 0.001°$ C. The shape of the thermistor is fixed; therefore, $R_2/R_1 = \rho_2/\rho_1 = \sigma_1/\sigma_2$.

Calculation

$$\delta = \frac{R_2 - R_1}{R_1} = \frac{\rho_2 - \rho_1}{\rho_1} = \frac{\rho_2}{\rho_1} - 1 = \frac{\sigma_1}{\sigma_2} - 1$$

$$1 + \delta = \frac{\sigma_1}{\sigma_2} = \left[\frac{\sigma_0 e^{-3.7eV/2kT}}{\sigma_0 e^{-3.7eV/2k(T\pm0.001)}}\right]$$

$$\ln(1+\delta) = \frac{-3.7 \text{ eV}}{2(86.1 \times 10^{-6} \text{ eV/K})}\left[\frac{1}{T} - \frac{1}{T \pm 10^{-3}}\right]$$

Because

$$\frac{1}{T} - \frac{1}{T \pm 10^{-3}} = \frac{(T \pm 10^{-3}) - T}{T(T \pm 10^{-3})} \approx \pm\frac{10^{-3}}{T^2}$$

$$\ln(1+\delta) = (-21,500 \text{ K})(\pm 10^{-3} \text{ K}/[293 \text{ K}]^2)$$
$$= \pm 2.5 \times 10^{-4}$$

$$\delta = \pm 0.00025 \qquad\qquad (\text{or } \pm 0.025\%)$$

Comment A bridge-type instrument would be required.

Example 11–6.2

A transistor has a collector current of 4.7 milliamperes when the emitter voltage is 17 mV. At 28 mV, the current is 27.5 milliamperes. Given that the emitter voltage is 39 mV, estimate the current.

Solution   Based on Eq. (11–6.2),

$$\ln 4.7 \approx \ln I_0 + 17/B = 1.55$$
$$\ln 27.5 \approx \ln I_0 + 28/B = 3.31$$

Solving simultaneously, using *milli*units, we have

$$\ln I_0 \approx -1.17 \quad \text{and} \quad B \approx 6.25 \text{ mV}$$

At 39 mV,

$$\ln I_c \approx -1.17 + 39/6.25 \approx 5.07$$
$$I_c \approx 160 \text{ milliamp}$$

Comments   The electrical engineer modifies Eq. (11–6.2) to take care of added current effects. These effects, however, do not change the basic relationship: The variation of the collector current is much greater than is the variation of the signal voltage.

11–7

SEMICONDUCTOR PROCESSING

The specifications for semiconductors are very critical. Impurities must be kept to the absolute minimum (ppm or less). Crystals must be essentially defect-free. The circuitry of devices must be positioned within microns, and the compositions precisely defined.

Purification

Some impurities introduce donors and negative (*n*-type) carriers; others introduce acceptors and positive (*p*-type) carriers. Even when these *dopants* are desired, their amounts must be closely controlled. Therefore, it is common practice to purify the silicon (or other semiconducting material) to the highest level possible, and then to make precise additions of the required dopant.

The *zone-refining* process is commonly used for the purification of silicon. The phase diagram is basic to the process. Consider silicon that contains 1 percent aluminum (Fig. 5–3.3). If 100 g of such a material is equilibrated at 1300° C (Fig. 11–7.1), the result is 5 g of liquid containing 20 percent Al, and 95 g of β containing approximately 0.02 percent Al. (This value is obtained from interpolation along the solidus.) If the liquid is removed and discarded, the impurities in the remaining silicon will have been reduced by a factor of approximately 50. Now

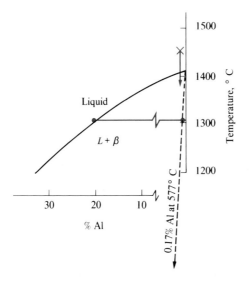

FIG. 11-7.1

Purification of Silicon. (See Fig. 5-3.3.) Assume 1 percent Al, and 1300° C. Liquid (20 Al-80 Si) may equilibrate with solid (~0.02 percent Al, balance Si) to give a 5-95 liquid-solid ratio. If we discarded that liquid, and equilibrated the retained material (0.02 percent Al) at 1400° C, the resulting solid would have approximately 99.998 percent Si (~0.002 percent Al). We could then once again discard the accompanying liquid (2 percent Si) and could repeat the process still further. Zone refining is a practical alternative that makes use of this principle (see Fig. 11-7.2).

repeat the equilibration at 1400° C. The precise concentrations become difficult to read from the phase diagram; however, we again find a major decrease in impurity content.

In theory, the separation processes can be repeated several times until the desired impurity reduction is attained. In practice, the minor amounts of liquid cannot be separated cleanly. Zone refining provides a viable alternative. One end of a rod of impure silicon is melted by radio-frequency (r-f) heating, as shown in Fig. 11-7.2. The r-f coil, and therefore the molten *zone,* is moved slowly (<mm/ min) along the rod, solidifying at the bottom. The compositions of the solidifying solid and of the overlying liquid are defined by the solidus and the liquidus compositions, respectively. Thus, as the zone travels along the rod, the impurity content of the solidifying phase is always less than that in the liquid by a factor of C_S/C_L, or the purification ratio is C_L/C_S. Since the solid-liquid interface has a planar front, there is no problem of liquid entrapment within the solid. Several passes are commonly made, so the impurity level may be decreased to ppb.

The zone-refining process has an added advantage when it is set up vertically. The molten zone can be contained by the adjacent solids with neither a crucible nor furnace walls, either of which could introduce impurities into the liquid.

FIG. 11–7.2

Zone Refining. The r-f coil is moved slowly up
the silicon rod, producing a molten zone. The
liquid–solid composition ratio is C_L/C_S from
Fig. 5–3.3. Thus, the impurity content is
segregated into the liquid, leaving a more pure
solid. The procedure is repeated several times,
reducing the impurity content to less than 1
ppm. (The liquid is held in place by surface
tension.)

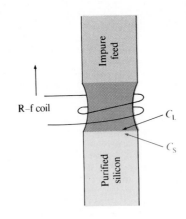

Crystal Growing

Single crystals are required for the large majority of semiconductor applications,
since grain boundaries reduce carrier mobility and decrease the recombination
time of excess carriers. The reduction of recombination time affects the performance of many junction devices. Single-crystal growth generally uses one of two
techniques in semiconductor technology — crystal-pulling and floating-zone
methods (Fig. 11–7.3).

For *crystal pulling,* the semiconducting material is first melted; then a single-
crystal *seed crystal* is touched to the surface and is slowly pulled away (~ 1 mm/
min) as it is rotated (~ 1/s). If the liquid is only slightly above its melting temperature, it will solidify on the seed crystal as the seed is pulled upward. The solidifying
atoms continue the crystal structure of the seed. Dopants of Group III or V
compounds can be added to the liquids in the amounts ($\sim 10^{-6}$ a/o) required to
make *p*- and *n*-type products.

The crystal-pulling technique (Fig. 11–7.3a) is satisfactory for germanium and
other materials that melt below 1000° C. However, it is not satisfactory for silicon

FIG. 11–7.3

**Single-Crystal Growth (Semiconduc-
tors).** (a) Crystal-pulling method.
The seed — a single crystal — is
slowly pulled upward. The liquid
crystallizes on its lower surface. (b)
Floating-zone procedure. The molten
zone is raised along the semiconductor bar, solidifying the lower edge as
a single crystal. The liquid is held in
position by surface tension and does
not come into contact with a
container.

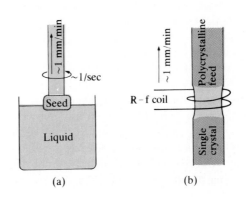

for a couple of reasons. Silicon melts above 1400° C. Not only does it readily pick up contaminants at that temperature, but also the dopants more rapidly vaporize, so compositional control becomes more difficult. Therefore, the *floating-zone* process is used for silicon. The method starts with a rod (diameter > 5 cm) of purified polycrystalline silicon sitting on a disc of a previously prepared *single* crystal. The rod and disk are melted where they are in contact by r-f heating. The r-f coil is then slowly raised (Fig. 11–7.3b) to move the molten zone upward. The polycrystalline solid melts with the upward movement and feeds the molten zone. The initial single crystal grows upward with the movement, crystallizing at the lower side of the molten zone. As occurs in the crystal-pulling process, the upward movement of the molten zone advances at a rate of ~ 1 mm/min.

Thin (0.25-mm) wafers are cut from the bar. These are polished and chemically cleaned. The diameters of commercially prepared single crystals have increased almost without limits, since a large wafer leads to more efficient device preparation; diameters of more than 200 mm (8 in.) are common (Fig. 7–1.1).

Device Preparation

Figure 11–7.4 shows a processed wafer that holds dozens of chips, each with literally thousands of circuit elements. These chips have a *planar* configuration; that is, the preparation and contacts are on one flat surface (Fig. 11–7.5) rather than at the ends and sides (Fig. 11–6.7). The thicker substrate, an *n*-type silicon base that was crystallized by the floating-zone process, is doped to have low resistivity and increased conduction. Following the masking procedure, boron is implanted into the surface of the silicon, changing the selected regions from *n*- to

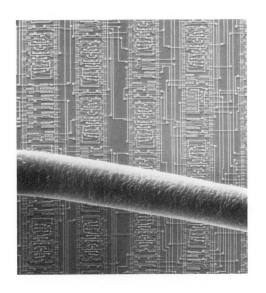

FIG. 11–7.4

Processed Wafer (Silicon Chip and Human Hair). The circuits of the chip are in planar configuration, made with sequential maskings and overlays. See Fig. 11–7.5 for cross-sections. (Courtesy of General Electric Co.)

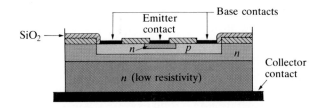

FIG. 11–7.5

Planar Transistor (Junction-Type). The substrate is a low-resistivity, *n*-type, single-crystal wafer of silicon. It is covered with an epitaxial layer of higher-resistivity silicon that serves as the collector. Boron is diffused through the openings of an SiO₂ mask to produce the *p*-type base. A second masking admits phosphorus for the *n*-type emitter. Aluminum contacts are added.

p-type. The latter becomes the base of the transistor. Ion implantation is favored because it can provide the required sharp *n–p* junction. A second masking and implantation step introduces a Group V element to produce the *n*-type emitter. Final maskings, etchings, and coating establish a passive silica (SiO₂) surface and contacts for circuit leads.

The thickness of the base in Fig. 11–7.5 may be only 0.5 μm. With this short distance between the two junctions, any recombination can be minimized, as described in Section 11–6. The dimensions and dopant concentrations are controlled by diffusion time and temperature, and by gas composition. The production process is highly reproducible.

Example 11–7.1

The Al–Si diagram (Fig. 5–3.3) has the solubility limit exaggerated for Al in β because, at the eutectic temperature (577° C), β will hold a maximum of only 0.17 w/o Al (and 99.83 w/o Si).

(a) Assume a linear solid solubility curve; what is the aluminum analysis of β at 1300° C?

(b) An alloy containing 98 Si–2 Al is equilibrated at 1300° C, and the liquid removed. What fraction of the aluminum is removed?

(c) The remaining solid is remelted at 1450° C. At what temperature will the new solid start to form during cooling?

(d) What is the new solid's chemical analysis?

Procedure and Calculations

(a) Being linear, and using ratios,

$$\frac{0.17\%}{(1410 - 577° \text{ C})} = \frac{C_\beta}{(1410 - 1300° \text{ C})}$$

$$C_\beta = 0.023 \text{ w/o Al}$$

(b) From Fig. 5–3.3, $C_L = 20$ w/o Al. Using 100 g,

$L = [100 \text{ g}][(2 - 0.023)/(20 - 0.023)] = 9.9 \text{ g}$

$\beta = 90.1 \text{ g}$

The fraction of aluminum removed is

$$\frac{(9.9 \text{ g})(0.20)}{(100 \text{ g})(0.02)} = 0.99$$

(c) The upper end of the liquidus curve of Fig. 5–3.3 is approximately linear. Therefore, by using ratios again,

$$\frac{20}{(1410 - 1300° \text{ C})} = \frac{0.023}{(1410 - T° \text{ C})}$$

$$T = 1410 - 0.127° \text{ C}$$

(d) Repeating (a),

$$\frac{0.17\%}{(1410 - 577° \text{ C})} = \frac{C'_\beta}{0.127° \text{ C}}$$

$$C'_\beta = 0.00003 \text{ w/o Al}$$

Comment By modifying this type of procedure into the *zone-refining* process, the materials engineer is able to produce silicon with $< 1/10^9$ aluminum for controlled-purity semiconductor materials.

11–8
SUPERCONDUCTIVITY

Since the early twentieth century, we have known that certain metals and a significant number of intermetallic compounds possess *superconductivity;* that is, they have zero resistance and undetectable magnetic permeability below their critical temperature, T_C. The change from normal conductivity to superconductivity is abrupt; it occurs as a function of temperature and magnetic field (Fig. 11–8.1).

Only about one-half of the metals are superconductive. Thus, researchers have continued to give attention to intermetallic compounds in an effort to find materials with higher critical temperatures. Of course, this goal is of not only scientific but also technic interest, because designs and products presently are limited for very low-temperature applications. The progress toward developing materials with higher critical temperatures had been steady (at about 0.3 K/yr), but was not encouraging, because the highest T_C in 1986 was only 23 K (for Nb_3Ge), thus requiring liquid-helium environments.

With that background, it is not surprising that, when the ceramic "high-temperature" superconductors were discovered in early 1987 that had critical temperatures of 90 K (and even higher), the technical community was excited. These

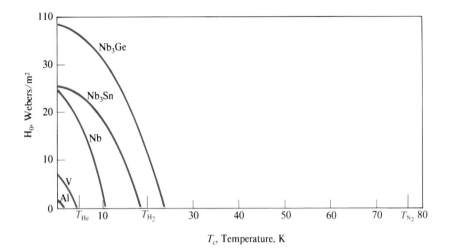

T_c, Temperature, K

FIG. 11–8.1

Conditions for Superconductivity. Superconductivity occurs below the critical temperature, T_C, and critical magnetic field, H_0, for each material. For metals, the T_C is below
10 K; intermetallic compounds have been developed that have a T_C as high 23 K, but
that still require liquid hydrogen for cooling. The recent advent of high-temperature
ceramic superconductors is a breakthrough because it should permit the use of liquid
nitrogen and thereby simplify design and applications.

materials would permit the use of inexpensive and widely available liquid nitrogen
as a coolant (77 K), and would simplify the requirements for cryogenic designs.

The ceramic that received the most initial attention may be described as follows: Align three unit cells of $CaTiO_3$ (Fig. 3–2.9b) in series, with the origin
located at the Ti ion. Replace the titanium with copper, and the three Ca with two
Ba and one Y (Fig. 11–8.2). Not all the oxygen ions are required to balance the
charges. Scientists initially investigated the substitution of various rare-earth ions
for some of the yttrium ions. This research was followed by the substitution of
other alkaline-earths ions for the barium ions.

When we make these Ba, Y, and Cu insertions, and use copper as Cu^{2+}, we
create the compound of $Ba_2YCu_3O_{6.5}\square_{2.5}$. Experiments indicate, however, that
fewer oxygen vacancies, $\square$, are more desirable, so we make $Ba_2YCu_3O_{7\pm x}\square_{2\pm x}$.*
The variable oxygen content is possible because some of the Cu^{2+} may be replaced
by Cu^{3+}, depending on the oxygen pressure. The nonstoichiometry is important
because the superconductivity requires an optimum number of vacancies in the
oxygen part of the lattice. The resulting structure is shown in the high-resolution

---

* This description is overly simplified, because the oxygen vacancies introduce some distortion of the
(010) planes adjacent to the yttrium atoms, which in turn produces an orthorhombic unit cell, rather
than the cubic cell that is possessed by $CaTiO_3$ (Fig. 3–2.9).

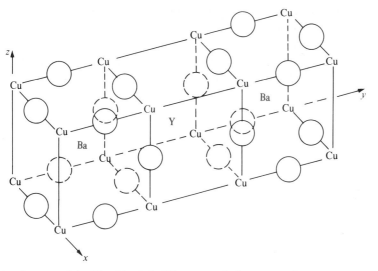

FIG. 11–8.2

Ceramic Superconductor
$(Ba_2YCu_3O_{7\pm x}\square_{2\pm x})$. The
structure was determined
by X-ray diffraction.
Nonstoichiometry varies
with the oxygen pressure
during processing. This
structure may be described
in terms of the tripling of
the $CaTiO_3$ unit cell. (See
the text.)

electron micrograph in Fig. 11–8.3. We currently interpret the superconductivity
as occurring principally along the [010] and [001] directions—directions in which
there are missing oxygen ions.

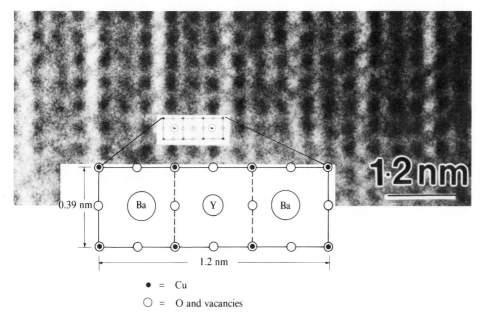

FIG. 11–8.3

Ceramic Superconductor (High-Resolution Electron Micrograph). The dark spots locate
the heavier Ba and Y ions as sketched in Fig. 11–8.2. (Courtesy of T. M. Shaw and
R. B. Beyers, International Business Machines Corporation.)

The initial technical applications will involve thin-film technology for use in computer circuits, where low temperatures and zero resistance will increase computation speed. Development will be slower in products that require magnetic coils and similar wire applications. The brittle nature of ceramic phases precludes the simple processing that is available to make wire out of metallic materials.

S U M M A R Y

In this chapter, we looked at metallic conductivity, then examined the concepts of energy bands and of energy gaps in solids. These concepts give us a basis for understanding semiconduction and simple semiconductor devices.

1. The electron $(0.16 \times 10^{-18} \text{ A·s})$ is the basis of charge transport within solid materials. *Electrons* are negative carriers, as are *anions*, because of their excess electron(s). Positive ions *(cations)* carry a positive charge by virtue of their missing electron(s). Likewise, a missing electron in the valence band, called an *electron hole*, can transfer a positive charge.

2. We can view *metals* as possessing delocalized electrons that are capable of transporting charge in a valence band. Any lattice irregularity—such as thermal vibrations, solute atoms, or dislocations from plastic strain—shortens the *mean free path* for an electron, and therefore (i) reduces the *drift velocity*, (ii) decreases the *mobility*, (iii) lowers the *conductivity*, and (iv) raises the *resistivity*.
Thermal conductivity is proportional to electrical conductivity in metals (only), because the electrons transport the bulk of the thermal energy.

3. Metals possess a partially filled valence band. The valence bands of insulators and semiconductors are filled. In *semiconductors*, however, the overlying energy gap is small; thus, there is some opportunity for conduction. The large energy gap in *insulators* precludes conductivity. The most energetic electrons obey the *Fermi distribution*, which defines the occupation probabilities in the energy bands.

4. A *semiconductor* has a sufficiently small energy gap that a usable number of electrons can be activated from the valence band to the conduction band by heat, by light, or by other electrons. *Intrinsic semiconductors* have equal numbers of negative carriers (electrons) and positive carriers (electron holes). The III-V compounds have a structure that is similar to those of diamond, silicon, germanium, and gray tin. They all have filled valence bands as a result of their covalent bonds. The numbers of electrons to "jump the gap," and therefore to introduce conductivity into semiconductors, follows the Arrhenius relationship with temperature; that is, *ln n and ln σ are linear with 1/T.* In general, the *mobility* of the conduction electrons is greater than that of the electron holes.

5. *Extrinsic semiconduction* originates from impurities. *Donors* provide electrons to the conduction band. *Acceptors* receive electrons from the valence band. Thus, the donors and acceptors produce negative and positive carriers, respectively. *Defect semiconduction* exists in nonstoichiometric compounds of those elements that have more than one valence (e.g., Fe^{2+} and Fe^{3+}).

6. There are two main categories of semiconducting devices. The first includes thermistors, photoconductors, pressure gages, and photomultipliers, all of which depend on the *conduction* (or resistance) variations of the semiconducting material. The second category, which involves *junctions* between n- and p-type semiconductors, includes light-emitting diodes (LEDs), rectifiers, Zener diodes, and transistors.

7. The major steps of semiconduction processing include materials purfication, single-crystal growth, and device preparation. *Zone refining* is the principal puri-

The *Study Guide* that accompanies this text has three study sets—*Metallic Conductivity, Intrinsic Semiconduction,* and *Extrinsic Semiconduction*—available for the student who favors a more detailed approach. The Fermi distribution is used on an optional basis to describe common devices.

fication procedure for silicon. *Crystal-pulling* and *floating-zone* processes are the more common for crystal growth. The latter has the advantage of avoiding contact with containers, a source of impurities. Device preparation involves *masking, etching,* and *ion implantation* of the configurations for planar circuits.

8. Since the higher-temperature superconductors have critical temperatures in the range of liquid nitrogen, they possess fascinating potential applications. Their initial use probably will be in computers. These superconductors are ceramics; thus, they pose manufacturing challenges.

KEY WORDS

Acceptor
Acceptor saturation
Band, conduction (CB)
Band, valence (VB)
Charge carriers
Compounds, III-V
Conductivity (electrical)
Crystal pulling
Defect semiconductor
Depleted zone
Donor
Donor exhaustion
Drift velocity ($\bar{v}$)
Electric field ($\mathscr{E}$)
Electron charge (q)
Electron hole (p)
Electron–hole pair

Energy band
Energy gap (E_g)
Fermi distribution, $F(E)$
Fermi energy (E_f)
Floating zone
Forward bias
Hall effect
Insulator
Junction (n–p)
Light-emitting diode (LED)
Luminescence
Metal
Mobility (μ)
Photoconduction
Photomultiplier
Photon
Photovoltaic

Recombination
Rectifier
Resistivity (ρ)
Resistivity coefficient, solution (y_x)
Resistivity coefficient, thermal (y_T)
Semiconductor
Semiconductor, extrinsic
Semiconductor, intrinsic
Semiconductor, n-type
Semiconductor, p-type
Superconductivity
Thermistor
Transistor
Zener diode
Zone refining

PRACTICE PROBLEMS

11–P11 (I^2R review) A dry cell (1.5 V) is connected across a 10-ohm circuit. (a) What is the resulting current? (b) What is the wattage or energy rate? (c) How much energy is used per hour?

11–P12 (I^2R review) An 18-m copper wire connects two terminals of a dry cell (1.5 V). (a) What is the current density in the wire? (b) How many watts go through the wire, if it has a 3-mm² cross-section? (c) What diameter must be used to limit the current to 1 ampere?

11–P13 A wire must have a diameter of less than 1 mm, and a resistance of less than 0.1 ohm/m. Which of the materials in Appendix C are suitable?

11–P14 Seventy mV are placed across the 0.5-mm dimension of a semiconductor with the carrier mobility of 0.23 m²/v·s. What drift velocity develops?

11–P15 Charge carriers within a semiconductor, which has the resistivity of 0.0313 ohm·m, possess a drift velocity of 6.7 m/sec when a 1.2-V differential is placed across a 9-mm piece. How many carriers are there per m³?

11–P16 Laboratory measurements indicate that the drift velocity of electrons in a semiconductor is 149 m/sec when the voltage gradient is 15 V/mm. The resistivity is 0.7 ohm·m. What is the carrier concentration?

11–P17 List and compare the various types of charge carriers in solids.

11–P21 Based on Fig. 11–2.2, what is the electrical resistivity of an annealed 70–30 brass?

11–P22 Estimate the resistivity of a tungsten wire at 1500° C.

11–P23 A 6 percent variation (maximum) is permitted in the resistance between 0 and 25° C. Which metals of Table 11–2.1 meet this specification?

11–P24 Based on data of Table 11–2.1 and of Appendix C, estimate the effect of copper additions on the resistivity of silver (ohm·nm per a/o). (Why do the data of Table 11–2.2 not apply?)

11–P25 Using the data of Appendix C, determine the ratios of thermal to electrical conductivities, k/σ, for each metal. (b) Based on your calculations, make a generalization about the relationship between electrical and thermal conductivities.

11–P26 Estimate the thermal conductivity of copper at 40° C from its electrical conductivity.

11–P27 A brass must have a resistivity of <50 ohm·nm, and a hardness of > 50 R_F. (a) What range of zinc content is permissible? (b) What is the *best* choice for a composition within that range?

11–P28 A brass wire must carry a load of 45 N (10 lb) without yielding and must have a resistance of less than 0.033 ohm/m (0.01 ohm/ft). Calculate the diameter of the smallest wire that can be used, if the wire is made of (a) 60–40 brass, (b) 80–20 brass, and (c) 100 percent Cu.

11–P29 A certain application requires a piece of metal that has a yield strength greater than 100 MPa and a thermal conductivity greater than 0.04 W/m·° C. Specify either an annealed brass or an annealed Cu–Ni alloy that will meet the requirements.

11–P31 Calculate the fraction of the energy states in a metal are occupied at $E = E_f + 0.05$ eV (a) at 100° C, (b) at 200° C, (c) at 400° C, and (d) at 800° C.

11–P32 (a) Calculate and plot the Fermi distribution for 20° C in increments of 0.05 eV from $E = E_f - 0.15$ eV to $E = E_f + 0.15$ eV. (b) Explain what this distribution means.

11–P33 Explain to a classmate why there are energies that are forbidden (a) to electrons that are associated with individual atoms; (b) to delocalized electrons that are in metallic structures.

11–P34 Differentiate among *metallic conductors, semiconductors,* and *insulators* on the basis of energy bands.

11–P41 (a) What fraction of the charge is carried by electron holes in intrinsic gallium arsenide (GaAs)? (b) What fraction is carried by the electrons?

11–P42 The resistivity of a semiconductor that possesses $10^{21}/m^3$ of negative carriers (plus a negligible number of positive carriers) is 0.016 ohm·m. (a) What is the conductivity? (b) What is the electron mobility? (c) What is the drift velocity when the voltage gradient is 5 mV/mm? (d) What is the drift velocity when the voltage gradient is 0.5 V/m?

11–P43 Based on the data in Table 11–4.2, which is larger—the conductivity from electrons in intrinsic InP, or the conductivity from electron holes in intrinsic InAs?

11–P44 Pure silicon has 32 valence electrons per unit cell (eight atoms with four valence electrons each). Its resistivity is 2000 ohm·m. What fraction of the valence electrons are available for conduction?

11–P45 The mobility of electrons in silicon is 0.19 $m^2/V·s$. (a) What voltage is required across a 2-mm chip of Si to produce a drift velocity of 0.7 m/s? (b) What electron concentration must be in the conduction band to produce a conductivity from negative carriers of 20 $ohm^{-1}·m^{-1}$? (c) What would be the total conductivity for pure silicon?

11–P46 The conductivity for silicon is 5×10^{-4} $ohm^{-1}·m^{-1}$ at 20° C. Estimate the conductivity at 30° C.

11–P47 An intrinsic semiconductor has a conductivity of 390 $ohm^{-1}·m^{-1}$ at 5° C and of 1010 $ohm^{-1}·m^{-1}$ at 25° C. (a) What is the size of the energy gap? (b) What is the conductivity at 15° C?

11–P51 Silicon has a density of 2.33 g/cm^3. (a) What is the concentration of silicon atoms per m^3? (b) Phosphorus is added to the silicon to make it an *n*-type semiconductor of 100 $ohm^{-1}·m^{-1}$. What is the concentration of donor electrons per m^3?

11–P52 How many silicon atoms are there for each aluminum atom in Example 11–5.1?

11–P53 Aluminum is a critical impurity when present in silicon during processing for semiconductors. Assume only 10 ppb (0.000 001 a/o) Al remains in the final product. Will the resulting extrinsic semiconductivity be greater than or less than the intrinsic conductivity of silicon at 20° C?

11–P54 Differentiate between *acceptor* and *donor* impurities.

11–P55 Refer to Problem 11–P53. At what temperature will 1 percent of the conductivity be intrinsic?

11–P56 Detail the mechanism of acceptor saturation in terms of Fig. 11–5.4

11–P57 In an extrinsic semiconductor, the Fermi energy moves toward the center of the energy gap as the temperature is increased. Why does this happen?

11–P58 Refer to Example 4–5.1. (a) Is the oxide *n*- or *p*-type? (b) How many charge carriers are there per mm^3?

11–P59 Silicon has one atom of gallium replacing one atom of silicon per 10^6 unit cells. (a) Will the material be *n*- or *p*-type? (b) On the basis of the data in Table 11–4.2, what is the resulting conductivity?

11–P61 The resistance of some silicon for a thermistor is 1031 ohm at 25.1° C. With no change in the measurement procedure, the resistance decreases to 1029 ohm. What is the change in temperature?

11–P62 To what temperature must you raise GaP to make its resistivity 50 percent of the value at 0° C?

11–P63 Refer to Example 11–6.2. If the emitter voltage is increased 70 percent from 17 to 29 mV, by what factor is the collector current increased?

TEST PROBLEMS

1111 The resistivity of iron is 100 ohm · nm. What is the resistance per 1 m of an iron wire that has a 1.25-mm diameter?

1112 A flashlight bulb has a resistance of 8 ohms when it is used in a three-cell flashlight battery. Assume 4.5 V. How many electrons move through the filament per min?

1113 A material has a resistivity of 0.1 ohm · m. There are 10^{20} conduction electrons/m^3. What is the conductivity? (b) What is the charge mobility? (c) What electric field is required for a drift velocity of 1 m/s?

1114 A material with a charge mobility of 0.43 $m^2/V \cdot s$ develops a conductivity of 75 $ohm^{-1} \cdot m^{-1}$. (a) How many charge carriers are involved? (b) What is the voltage gradient when the drift velocity of the charge carriers is 19 m/s?

1115 The mobility of electrons in a conductor that has a resistivity of 0.0043 ohm · m was determined to be 145 $m^2/V \cdot sec$. (a) What is the carrier concentration? (b) What voltage gradient will produce a drift velocity of 10 m/s?

1121 A copper wire has a resistance of 0.5 ohm per 100 m. Your company is considering using a 80–20 brass instead of copper. What would be the resistance of the brass wire if the size were the same?

1122 The electric power loss in a low-voltage copper power line is 1.01 percent per km on a 0° C day. What is the power loss at 30° C (86° F)?

1123 The resistivity of copper doubles between 20 and 300° C. At what temperature is the resistivity of gold equal to the higher value of copper?

1124 (a) Estimate the resistivity of a 95 Cu–5 Sn (weight basis) bronze at 0° C. (b) Estimate the resistivity at 50° C.

1125 Estimate the thermal conductivity of gold at 0° C from the data in Table 11–2.1.

1126 You need a copper alloy (either brass or cupronickel) with an ultimate strength of at least 245 MPa (35,000 psi) and a resistivity less than 50×10^{-9} ohm · m. (a) Select a suitable alloy, bearing in mind that $\$_{Ni} > \$_{Cu} > \$_{Zn}$. (b) What is the ductility of your alloy?

1127 A wire has a 1-mm diameter and a resistance of 0.33 ohm/m. It is cold drawn to a 0.8-mm diameter, and then has a resistance of 0.66 ohm/m. What fraction of the original conductivity remains after cold working?

1128 The design engineer has a choice of either a brass alloy or a Cu–Ni alloy to meet room-temperature specifications of hardness $>80 R_f$, $S_u > 275$ MPa, ($>40,000$ psi), and a thermal conductivity, $k > 40$ W/m·°C (>10 percent k_{Cu}). Pick a suitable composition.

1129 When used in a steam system, a brass plate must conduct more than 2.7 cal/mm²·sec (or 11.3 W/mm²) when its two faces have a temperature difference of 125°C. It also must have an ultimate strength–thickness product of more than 0.55 MPa·m; that is, $(S_u \cdot t) > 0.55$ MPa·m. Consider 90–10, 80–20, 70–30 brasses, and select the alloy that will meet the requirements least expensively.

1131 At what temperature will 65 percent of the electron energy states be occupied at the ($E = E_f - 0.04$ eV) energy level?

1132 (a) Calculate and plot the Fermi distribution at $E = E_f + 0.05$ eV for 20°C temperature intervals from 0 to 100°C. (b) Explain what this distribution means.

1133 (a) Discuss the various factors that affect the drift velocity of electrons in metals. (b) Why does the conductivity of metals decrease (resistivity increase) at higher temperatures?

1134 From Eq. (11–3.2), show that $F(E)$ at $[E_f + x$ eV] is equal to $[1 - F(E)]$ at $[E_f - x$ eV].

1141 The intrinsic conductivity of germanium is 2 ohm⁻¹·m⁻¹ at 20°C. (a) What is the conductivity from the positive carriers? (b) What is it from the negative carriers?

1142 At what temperature will the conductivity of silicon be one-half the conductivity at 30°C?

1143 Using the data and figures in Section 11–4, predict whether boron nitride, BN, will be an insulator or a semiconductor.

1144 What energy gap is required to permit a semiconductor to double its conductivity between 15 and 25°C?

1145 At an elevated temperature, one of every 10^{12} valence electrons in intrinsic silicon is in the conduction band. (a) What is the conductivity? (b) What is the temperature?

1146 By what factor does the conductivity of germanium decrease when the temperature is dropped from $+20$ to $-20°$ C?

1147 To what temperature must gray tin be reduced to have only one of every 10,000 valence electrons in the conduction band? (Gray tin has the same structure as silicon, but with $a = 0.649$ nm.)

1151 Boron is to be aded to silicon to produce a conductivity of 1 ohm⁻¹·m⁻¹. (a) How much boron is required? (Give your answer in silicon unit cells per boron atom, and in ppb.) (b) What is the average distance between the boron atoms in the silicon?

1152 Some extrinsic germanium, formerly used for transistors, has a resistivity of 0.017 ohm·m and an electron hole concentration of 1.6×10^{21}/m³. (a) What is the mobility of the holes in this semiconductor? (b) What impurity atoms could have been added to create the holes?

1153 There are 10^{12} magnesium atoms, which replace an equal number of silicon atoms, in a 10 mm³ silicon wafer. How will this replacement affect the conductivity compared to silicon with an impurity content of 10^{12} aluminum atoms?

1154 Extrinsic silicon is produced by melting 1 g of a master (Si-Sb) alloy that contains 3.22 mg of antimony (Sb) with 1 kg of pure silicon. (a) Will the product be n- or p-type? (b) Calculate the concentration of antimony (in atoms/cm³) in the silicon product.

1155 Three g of n-type silicon that had been doped with antimony to produce a conductivity of 600 ohm⁻¹·m⁻¹ are melted with 3 g of p-type silicon that had been doped with boron to produce a conductivity of 600 ohm⁻¹·m⁻¹. (a) What will be the resulting conductivity? (b) Will the product be n-type, p-type, or intrinsic?

1156 Gallium arsenide is made extrinsic by adding 0.000 001 a/o phosphorus (and retaining a stoichiometric Ga/As ratio). Calculate the extrinsic conductivity at exhaustion.

1157 Rework Example 11–5.5, but assume that the doping is different, such that E_f resides 0.3 eV above the bottom of the energy gap.

1158 Refer to Problem 1151. At what temperature will the intrinsic conductivity, σ_{in}, of the silicon be 5 percent of the extrinsic conductivity, σ_{ex} (i.e., when $\sigma_{in} = 0.05\ \sigma_{ex}$)?

1159 Why is the slope for the intrinsic semiconduction steeper in Fig. 11–5.4 than is the slope for the *n*-type extrinsic semiconduction?

115.10 Indium phosphide, InP, has gallium atoms replacing phosphorus atoms by the amount of one for every 10^6 unit cells. (a) Will the InP be *n*- or *p*-type? (b) On the basis of Table 11–4.2, what is the conductivity?

115.11 Experiments with $Fe_{<1}O$ indicate that 99 percent of the charge is carried by electrons hopping from cation to cation, thus moving electron holes in the opposite direction. (The remaining 1 percent of the charge is carried by the diffusion of cations.) Assume that the oxide of Problem 11–P58 has a conductivity of 93 ohm$^{-1}\cdot$m^{-1}. What is the mobility of the electrons?

115.12 The copper ions of Cu_2O are predominantly Cu^+, but some Cu^{2+} ions are present. Will this oxide be *n*- or *p*-type?

1161 Eight watts are required of a solar cell that produces 60 A/m² at 0.45 V. What area is required?

1162 Intrinsic silicon is to be used as a thermometer. How sensitive (in plus or minus percent) should the resistance instrument be to detect an 0.02° C temperature change at the temperature of the human body (37° C, or 98.6° F)? (*Hint:* Change $[1/T_1 - 1/T_2]$ to $[T_2 - T_1]/T_2 T_1$.)

1163 A transistor operates between 10 and 120 mV across the emitter. At the lower voltage, the collector current is 3 mA; at the higher voltage, it is 300 mA. Estimate the current when the voltage is 30 mV.

1171 (See Problem 6–P23.) Assuming fully efficient operation, how many passes by an r-f induction coil are required to refine a rod of 90 Si–10 Al material to "8-nines" purity, that is, 99.999999 percent Si at the starting end of the rod? (*Hint:* Near the melting temperature of silicon (Fig. 5–3.3), the ratio C_S/C_L, of aluminum contents in the solid and in the liquid is ~0.001.)

1172 What is the conductivity of the refined silicon of the Problem 1171 where there is "8-nines" purity?

Chapter 12

MAGNETIC PROPERTIES OF CERAMICS AND METALS

From lodestone initially, and from iron at a later date, the technical progress in magnetic materials has been continuous. Cold-worked steel provided a "hard," or permanent, magnet. The development of Alnico alloys in the beginning of the twentieth century increased that permanency by a factor of more than 10. Developments of rare earth–cobalt magnets in the 1960s introduced another factor-of-10 improvement in the 1970s (Fig. 1–2.3). Now even further increases are developing, and these with the use of less strategic materials, such as iron and boron. These advances help us to reduce motor sizes, for example, and affect engineering design considerations in numerous ways.

Commonly, we consider magnets to be metals; however, ceramic magnets are critical for many applications, particularly in high-frequency circuits where eddy current losses become severe in metals.

The technology of high-performance magnets depends significantly on the internal structure that is achieved—grains, phases, boundaries, dislocations, and solid solutions, plus the domains that we shall introduce in this chapter. Thus, processing methods become important.

12-1
MAGNETIC MATERIALS

The best known magnets contain metallic iron. Several other elements also can show magnetism, and not all magnets are metallic. Modern technology uses both metallic and ceramic magnets. It also makes use of other elements to enhance magnetic capabilities.

Soft and Hard Magnets

Magnets are commonly categorized as "soft" or "hard." A *hard magnet* attracts other magnetized materials to it. It retains this obvious magnetism more or less permanently. A *soft magnet* can become magnetized and be drawn to another magnet; however, it has obvious magnetism only when it is in a magnetic field. It is not permanently magnetized.

The differences between permanent, or hard, magnets and soft magnets, are best shown with the *hysteresis loop* used in physics courses (Fig. 12-1.1). We shall explore the reasons for these differences.

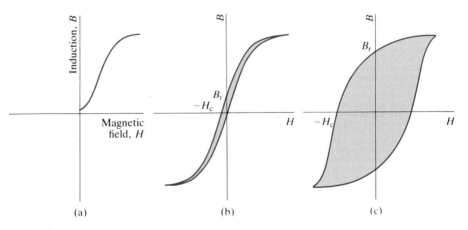

(a)   (b)   (c)

FIG. 12-1.1

Magnetization Curves. (a) Initial induction B versus magnetic field H. (b) Hysteresis loop (soft magnet). (c) Hysteresis loop (hard magnet). Both the remanent induction (flux density) B_r, and coercive field, $-H_c$, are large for hard magnets. The BH product is a measure of energy for demagnetization.

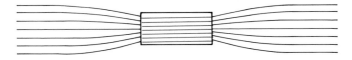

FIG. 12–1.2

Flux Density (Magnetic Induction). Lines of magnetic force become concentrated within magnetic materials. The induction is greater when the atoms and electrons are able to align their magnetic moments with the magnetic field. The alignment is retained in a permanent magnet, but becomes disordered in a soft magnet after the external field is removed.

When a magnetic material is inserted into a *magnetic field, H,* the adjacent "lines of force" are collected into the material to increase the *flux density* (Fig. 12–1.2). In more technical terms, increased *magnetic induction, B,* results. Of course, the amount of induction depends on the magnetic field and on the nature of the material. However, the increased induction is not linear; rather, it follows a *B–H* relationship similar to that of Fig. 12–1.1(a), jumping to a high level, and then remaining nearly constant in still stronger magnetic fields.

In a *soft magnet* (Fig. 12–1.1b), there is a near-perfect backtrack as the magnetic field is removed. A reversed magnetic field provides a symmetrical curve in the third quadrant.

The hysteresis curve of a *permanent magnet* differs significantly. When the magnetic field is removed, most of the induction is retained to give a *remanent induction, B_r.* A reversed field, called the *coercive field, $-H_c$,* is necessary before the induction drops to zero. In common with that of a soft magnet, the completed loop of a permanent magnet possesses 180° symmetry.

Since the product of the magnetic field (A/m) and the induction ($V \cdot s/m^2$) is energy per unit volume, the integrated area within the hysteresis loop is the energy required to complete one magnetization cycle from 0 to $+H$ to $-H$ to 0. Soft magnets require negligible energy; hard magnets require sufficient energy that, under ambient conditions, demagnetization is precluded. The magnetization is permanent.

Properties of Permanent Magnets

The magnetic permanence can be indexed by the coercive field, $-H_c$, that is required to return the induction to zero (Table 12–1.1). A value of $-H_c = 1000$ A/m is sometimes used to separate soft and hard (permanent) magnets. The *maximum* instantaneous energy product, BH_{max}, is a better measure, because it represents the critical energy barrier that must be exceeded for demagnetization. The advantage of the *BH* product is apparent in Fig. 12–1.3, where the second-quadrant profiles for several permanent magnets are shown. $BaFe_{12}O_{19}$ has a large

TABLE 12–1.1 Properties of Selected Hard Magnets*

MAGNETIC MATERIAL	REMANENCE, B_r V·s/m²	COERCIVE FIELD, $-H_c$ kA/m	MAXIMUM DEMAGNETIZING PRODUCT, BH_{max}, kJ/m³
Plain-carbon steel	1.0	4	1
Alnico V	1.2	55	34
Ferroxdur ($BaFe_{12}O_{19}$)	0.4	150	20
RE–Co†	1.0	700	200
$Nd_2Fe_{14}B$		1600	

* Data from various sources.
† Rare earth–cobalt, particularly samarium.

$-H_c$ value when compared to Alnico V,* but it has only a moderately high BH_{max}, because its remanent induction, B_r, is lower. As a result, the Alnico V material is somewhat more permanent.

You may have sensed the magnetic "strength" of an ordinary steel magnet. For comparison, refer to Fig. 1–2.3, which graphs the data of Table 12–1.1. That figure shows the dramatic improvements that have been made possible by those scientists and engineers who have developed newer magnetic materials, such as the rare earth–cobalt magnets, and the still more recent boron-containing, iron-based magnets.

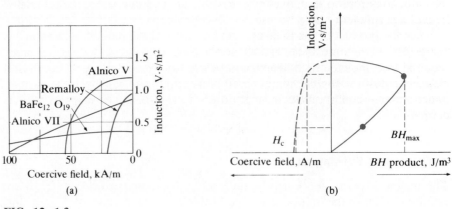

FIG. 12–1.3

Demagnetization Curves (Second Quadrant). (a) Hard magnets. (b) *BH* products. The maximum value is commonly used as an index of magnetic permanency.

---

* The Alnico magnets are iron-based alloys containing **aluminum**, **nickel**, and **cobalt**.

Properties of Soft Magnets

Soft magnets are the obvious choice for applications requiring ac power and high-frequency operation, since they must be magnetized and demagnetized many times per second. Among the more critical specifications for soft magnets are saturation induction (high), coercive field (low), and maximum permeability (high). Selected data are shown in Table 12–1.2; contrast them with the data of Table 12–1.1 (for permanent magnets). The *permeability* is the B/H ratio. However, since the BH behavior is nonlinear, the engineer commonly specifies the maximum B/H ratio. This *maximum relative permeability, μ_r,* is the secant of the initial magnetization curve of Fig. 12–1.4. A high value of μ_r implies easy magnetization, because only a small magnetic field produces a high flux density (induction).

TABLE 12–1.2 Properties of Selected Soft Magnets*

MAGNETIC MATERIAL	SATURATION INDUCTION, B_s V·s/m²	COERCIVE FIELD, $-H_c$ A/m	MAXIMUM RELATIVE PERMEABILITY, μ_r(max)
Pure iron (bcc)	2.2	80	5,000
Silicon ferrite transformer sheet (oriented)	2.0	40	15,000
Permalloy, Ni–Fe	1.6	10	2,000
Superpermalloy, Ni–Fe–Mo	0.2	0.2	100,000
Ferroxcube A, (Mn, Zn)Fe₂O₄	0.4	30	1,200
Ferroxcube B, (Ni, Zn)Fe₂O₄	0.3	30	700
Metallic glass (Fe₈₁B₁₃.₅Si₃.₅C₂)	1.5	1	100,000+

* Data from various sources.

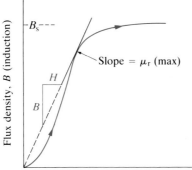

FIG. 12–1.4

Magnetic Permeability, μ_r. This is the ratio of flux density, or magnetic induction, B, to the magnetic field, H. Induction saturation, B_s, occurs when all the domains are aligned in one direction. (See Table 12–1.2.)

Example 12–1.1

Estimate the maximum energy product, BH_{max}, for Alnico VII in Fig. 12–1.3.

Procedure At $H = 0$ and at $H = -H_c$, the BH product is zero. The maximum is some-where in the mid-range. It will be quicker to take several pairs of data from the graph than to formulate and differentiate an equation for the curve.

Estimate

At $H = 50,000$ A/m	$B \approx 0.5$ V·s/m²	$BH = $ **25,000** J/m³
75,000	≈ 0.3	22,500
55,000	≈ 0.45	24,750
45,000	≈ 0.55	24,750

$$BH_{max} \approx 25,000 \text{ J/m}^2$$

Comparison The maximum energy product for pure iron, a soft magnetic material, is ≈ 50 J/m².

12–2
MAGNETIC DOMAINS

Magnetic materials contain *domains.* These are regions within crystals in which all the unit cells have a common magnetic orientation. Parts of three domains are shown schematically in Fig. 12–2.1(a).

Of course, all the unit cells within a phase have the same arrangements of atoms. At elevated temperatures, the magnetic dipoles of individual atoms are oriented randomly. During cooling past the *Curie temperature, T_c,* a coupling develops between the magnetic dipoles of adjacent atoms. This coupling produces the same magnetic orientations across many unit cells to develop a domain.

If a crystal is placed in an external magnetic field, those domains that match that polarity will grow at the expense of the adjacent, unfavorably oriented domains (Fig. 12–2.1b). This growth increases the flux density within the material and accounts for the big jump in the B–H curve of Fig. 12–1.4. With saturation, B_s, essentially all the unfavorably oriented domains have disappeared. Any further increase in the induction is achieved only through rotation of the N–S dipoles into better compliance with the applied field. This increase is very slight, as shown at the upper end of the curve in Fig. 12–1.4.

A single, large domain would appear to be favored, since there would be parallel magnetic orientation throughout. Actually, such a domain is unstable when the external magnetic field is removed, because the "lines of force" would have to be completed outside the material. To avoid that situation, the domains adjust to

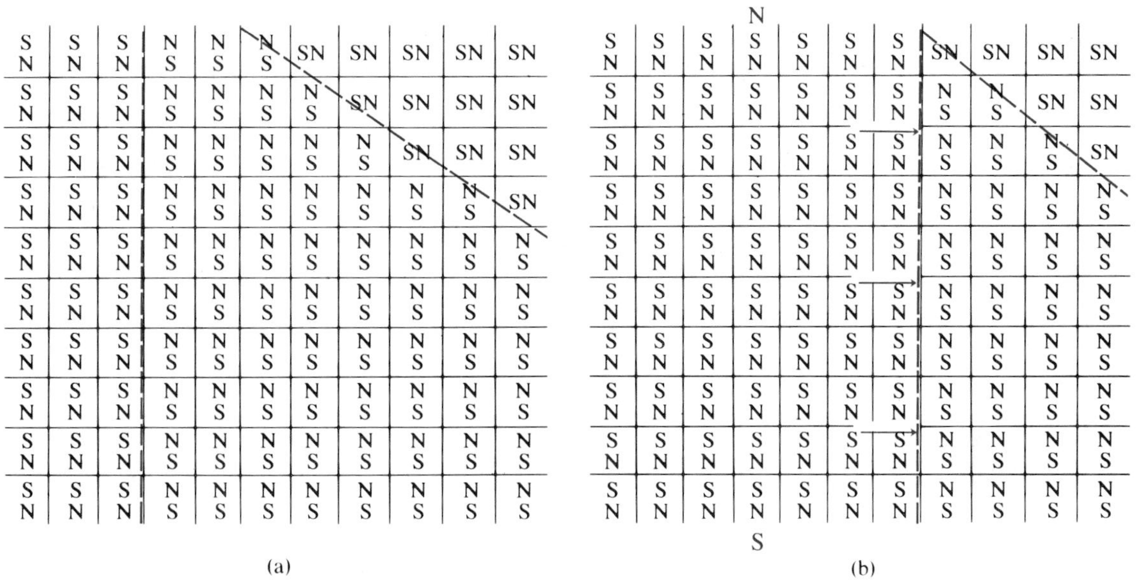

(a)

S/N	S/N	S/N	N/S	N/S	N/S	SN	SN	SN	SN	SN
S/N	S/N	S/N	N/S	N/S	N/S	N/S	SN	SN	SN	SN
S/N	S/N	S/N	N/S	N/S	N/S	N/S	N/S	SN	SN	SN
S/N	S/N	S/N	N/S	N/S	N/S	N/S	N/S	N/S	N/S	SN
S/N	S/N	S/N	N/S	N/S	N/S	N/S	N/S	N/S	N/S	N/S
S/N	S/N	S/N	N/S	N/S	N/S	N/S	N/S	N/S	N/S	N/S
S/N	S/N	S/N	N/S	N/S	N/S	N/S	N/S	N/S	N/S	N/S
S/N	S/N	S/N	N/S	N/S	N/S	N/S	N/S	N/S	N/S	N/S
S/N	S/N	S/N	N/S	N/S	N/S	N/S	N/S	N/S	N/S	N/S
S/N	S/N	S/N	N/S	N/S	N/S	N/S	N/S	N/S	N/S	N/S

(b)

N (above column 4)

S/N	S/N	S/N	S/N	S/N	S/N	S/N	SN	SN	SN	SN
S/N	S/N	S/N	S/N	S/N	S/N	S/N	N/S	N/S	SN	SN
S/N	S/N	S/N	S/N	S/N	S/S	S/S	N/S	N/S	N/S	SN
S/N	S/N	S/N	S/N	S/N	S/N	S/N	N/S	N/S	N/S	N/S
S/N	S/N	S/N	S/N	S/N	S/N	S/N	N/S	N/S	N/S	N/S
S/N	S/N	S/N	S/N	S/N	S/S	S/S	N/S	N/S	N/S	N/S
S/N	S/N	S/N	S/N	S/N	S/N	S/N	N/S	N/S	N/S	N/S
S/N	S/N	S/N	S/N	S/N	S/N	S/N	N/S	N/S	N/S	N/S
S/N	S/N	S/N	S/N	S/N	S/S	S/S	N/S	N/S	N/S	N/S
S/N	S/N	S/N	S/N	S/N	S/N	S/N	N/S	N/S	N/S	N/S

S (below column 6)

FIG. 12-2.1

Domains. Adjacent unit cells develop identical magnetic alignments to produce a domain. (a) Schematic. Adjacent domains cancel the external effects, unless a magnetic field is encountered that moves the boundaries, enlarging favored domains and restricting unfavorably oriented domains. (b) Domains of a permanent magnet retain this magnetization after the external field is removed.

retain the magnetic flux within the material, as shown in Fig. 12-2.2. The return to $B = 0$ requires that the vector sum of all magnetic dipoles be equal to zero. In soft magnetic materials, this randomization occurs readily. In hard magnetic materials, it does not, and a coercive field must be introduced.

Domain-Boundary Movements (and Pinning)

For the domains to grow, the domain boundaries* must move to expand the favorably oriented domains, as occurred between the sketches in parts (a) and (b) in Fig. 12-2.1. These boundaries move quite readily in perfect single crystals; however, they do not move across grain or phase boundaries. Likewise, dislocations will pin the domain boundaries and prevent their movements. Thus, the best soft magnetic materials must be single phase, annealed, and coarse grained (Fig. 12-2.3) to give easy boundary movements for high permeabilities (saturation

---

* These boundaries are sometimes called *Bloch walls.* Actually, the reorientation of the polarity between domains is not abrupt; rather, rotation occurs gradually over many layers of atoms.

FIG. 12–2.2

Domain Closures (Single Crystal of Iron). Within an external magnetic field, the lines of flux are completed outside the material. If that field is removed, the domains will attempt to realign to be completed within the material. The completion can occur in a soft magnetic material (colored lines), but not in a hard magnetic material. In the latter, the external lines of force can attract other magnetizable materials. Photomicrograph by H. J. Williams. (From *Ferromagnetism* by R. M. Bozorth. Copyright by Litton Educational Publishing Company, Inc. Reproduced by permission of Van Nostrand Reinhold Publishing Company.)

with a small magnetic field). Likewise, easy domain-boundary movements are required to provide the low remanance and coercive field necessary for soft magnets (Fig. 12–1.1b).

Conversely, hard magnets are designed with microstructures that impede domain-boundary movements. The oldest process was the cold working of steels, in which the strain hardening introduced dislocations that immobilized (pinned) the domain boundaries. Thus, the remanence, B_r, remained high with H = zero, and a major coercive field, $-H_c$, was required to remove the induction.[#] The Alnico magnets attain their permanence from their very fine, two-phase microstructure (Fig. 12–2.4), which allows them to arrest the domain boundary movements for

---

[#] The term *hard magnet* was associated with a hardened steel. Conversely, an annealed, carbide-free steel is mechanically soft, as well as being magnetically "soft."

FIG. 12-2.3

Soft Magnetic Steel (Stator Core of a Large Electrical Generator). Easy demagnetization requires unimpeded domain-boundary movements. Therefore, to avoid pinning, we use steel with (i) negligible carbon to eliminate carbides, (ii) large grains to minimize grain boundaries, and (iii) annealing to remove dislocations. (Courtesy of General Electric Co.)

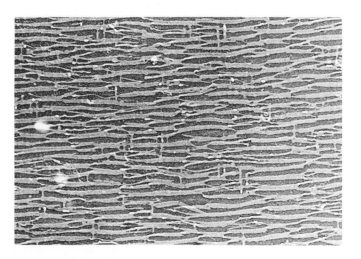

FIG. 12-2.4

Alnico Magnet (×50,000). This permanent magnet, which contains **aluminum, nickel,** and **cobalt** in an iron-base alloy is magnetized as a single phase, then is heat treated to produce a very fine-grained, two-phase microstructure. The extensive phase-boundary area that results pins the domain boundaries to produce a high *BH* product. (B. D. Cullity, *Introduction to Magnetic Materials,* Addison-Wesley, Reading, Mass., with permission.)

demagnetization. The ceramic magnets made of $BaFe_{12}O_{19}$ and the RE–Co magnets are magnetically hard because they do not have cubic unit cells. As such, the passage of the domain boundary is not simply a reorientation of the N–S polarity, but also requires atom displacements. Of course, those displacements also require an energy input.

Paramagnetism

Above the Curie temperature, thermal energy destroys the coupling between the magnetic atoms. Without that coupling, there are no domains, and the large flux densities are not realized. The increase in flux density with an increase in the magnetic field is relatively low and linear. We use the term *paramagnetism* to describe this situation. Similar low-permeability paramagnetism exists in materials that have too few magnetic atoms to permit coupling.

12–3
CERAMIC MAGNETS

Magnetization

Each electron of an atom possesses a *magnetic moment, p_m*. For this reason, the physicist speaks of the *spin* of an electron. The value of this magnetic moment, called a Bohr *magneton,* is 9.27×10^{-24} A · m^2. Electrons normally pair off in their orbitals with as many spins "up" as "down." Thus, any external evidence of these moments is masked. An atom will appear magnetic only when there is an unbalance in the spins of the electrons. For the most part, this rules out nearly one-half of the elements from having magnetic characteristics, since 50 percent of the elements have an even number of effective valence electrons. It also rules out the majority of ions, since an ion has either shed or accepted electrons to achieve completed shells of eight electrons. Finally, covalent bonds always involve pairs of electrons (of opposite spins). The consequence is that few elements have atoms with unbalanced electron spins to achieve a net magnetic moment. Those that qualify are those transition elements with unfilled subvalent shells (Fig. 12–3.1).

	K	Ca	Sc	Ti	V	Cr	Mn	Fe	Co	Ni	Cu	Zn
$4s$	↑	↑↓	↑↓	↑↓	↑↓	↑	↑↓	↑↓	↑↓	↑↓	↑	↑↓
$3d$ { ↑	—	—	1	2	3	5	5	5	5	5	5	5
↓	—	—	—	—	—	—	—	1	2	3	5	5
Net $(3d)$	—	—	1	2	3	5	5	4	3	2	0	0

FIG. 12–3.1

Unbalanced Magnetic Spins (Isolated Atoms). The $4s$ orbital fills before the $3d$ orbitals when each additional proton is introduced to the nucleus.

Iron, cobalt, and nickel are most familiar among these; in addition, gadolinium of the rare-earth series (unfilled 4-f shells) has pronounced magnetism. The total flux density, B, depends on the magnetic field, H, and on the magnetization, M:

$$B = \mu_0 (H + M) \qquad (12\text{-}3.1)$$

where μ_0 is the magnetic permeability of a vacuum. The *magnetization* of a material is the net sum of the magnetic moments per unit volume:

$$M = \Sigma \, p_m / V \qquad (12\text{-}3.2)$$

Since magnetization supplements the external field, H, that can be applied in circuits to produce higher flux densities (Eq. 12–3.1), this property is important in the development and choice of magnetic materials. We can illustrate this choice through the calculation of the magnetization of certain ceramic magnets.

Ferrites

The most common soft ceramic magnets are *ferrites* with the general composition of $[RFe_2O_4]_8$, where the indicated iron is Fe^{3+}, and **R** may be Fe^{2+}, Ni^{2+}, Co^{2+}, Mn^{2+}, Zn^{2+}, and so on. Each ferrous ion has lost two electrons. These are the two $4s$ electrons; the *six* $3d$ electrons remain to give four unpaired electrons (Fig. 12–3.2). A ferric ion has lost the two $4s$ electrons and *one* $3d$ electron; thus, *five* unpaired electrons remain.

Nature's original magnet is the mineral magnetite (Fe_3O_4), formerly called lodestone. Magnetite is a natural ceramic phase with an fcc lattice of O^{2-} ions. The iron ions are in both four-fold and six-fold interstitial sites (Fig. 12–3.3). More specifically, the Fe^{2+} ions are in 6-f sites; and the Fe^{3+} ions are equally divided between 6-f and 4-f sites. The resulting structure is called a *spinel*.

The unit cell of this structure is magnetic because the magnetic moments of the ions in the 6-f sites are all aligned in the same direction, and those in the 4-f sites are all aligned in the opposite direction. We can make an accounting in the compound Fe_3O_4, or $[Fe^{2+}Fe_2^{3+}O_4]_8$, since there are eight formula weights per unit cell. For every 32 oxygen atoms, which are not magnetic, there are eight Fe^{2+} ions and eight Fe^{3+} ions with alignment in the $\uparrow$ direction. There are eight Fe^{3+} ions with alignment in the $\downarrow$ direction. From this plus the information in the previous two paragraphs, we obtain the following table.

INTERSTITIAL SITE	SPIN ALIGNMENT	Fe^{2+} (4β)	Fe^{3+} (5β)	β	MAGNETIC MOMENT
6-f	↑	8		+32	$+8(4)(9.27 \times 10^{-24} \, A \cdot m^2)$
6-f	↑		8	+40	$+8(5)(9.27 \times 10^{-24} \, A \cdot m^2)$
4-f	↓		8	−40	$-8(5)(9.27 \times 10^{-24} \, A \cdot m^2)$
		Net:		+32	$+32(9.27 \times 10^{-24} \, A \cdot m^2)$

FIG. 12–3.2

Unbalanced Magnetic Spins (Iron). An Fe^{2+} ion has a mismatch of four electron moments ($\beta = 4$); for Fe^{3+} ions, $\beta = 5$. Thus, the magnetic moment of an Fe^{3+} ion is $5(9.27 \times 10^{-24}$ A$\cdot$m^2). (The $4s$ electrons are lost first with ionization, since protons are not removed.)

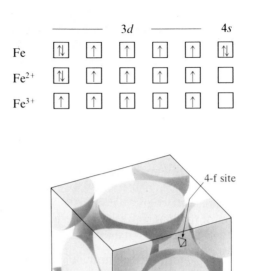

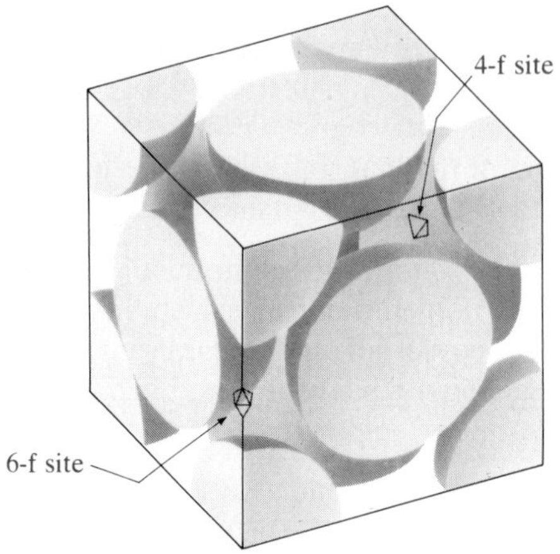

FIG. 12–3.3

Interstitial Sites (fcc). The 6-f site is among six neighboring atoms. There are four such sites per unit cell. The 4-f site is among four neighboring atoms. There are eight of these sites per unit cell.

We can calculate the magnetization, M, of magnetite from Eq. (12–3.2). The unit cell of $[Fe^{2+}Fe_2^{3+}O_4]_8$ is cubic and has a lattice constant, a, of 0.837 nm. Therefore, the *saturation* (maximum) *magnetization, M_s,* for magnetite is

$$32(9.27 \times 10^{-24} \text{ A}\cdot\text{m}^2)/(0.837 \times 10^{-9} \text{ m})^3 = 0.5 \times 10^6 \text{ A/m}$$

This compares well with the 0.53×10^6 A/m determined in laboratory experiments.

Nickel ferrite $[Ni^{2+}Fe_2^{3+}O_4]_8$ has the same structure as magnetite; however, Ni^{2+} ions have only two unpaired electrons. (Compare Figs. 12–3.1 and 12–3.2). Therefore, nickel ferrite has a magnetic moment of only $(+8)(2)(9.27 \times 10^{-24}$ A$\cdot$m^2) per unit cell. Its unit-cell dimensions are close to that of magnetite; thus, its saturation magnetization is about one-half that of Fe_3O_4.*

The ceramic magnets just described are *ferrimagnetic.* That is, they possess an opposing, but unbalanced, alignment of the magnetic atoms; thus, they have a net

---

* These two magnetic compounds, $[Fe_3O_4]_8$ and $[NiFe_2O_4]_8$, are called *ferrites* on the basis of their chemistry. They possess 32 O^{2-}, 16 Fe^{3+}, and 8 Fe^{2+} (or 8 Ni^{2+}) per unit cell. The lattice constant must be twice that shown in Fig. 12–3.3 in order to obtain a matching of occupied interstices with full unit-cell translations.

magnetization. Some ceramic compounds—for example, manganous oxide (MnO) and nickel oxide (NiO)—are *antiferromagnetic.* In NiO, which has the NaCl structure (Fig. 12–3.4), alternate planes of Ni^{2+} ions have antiparallel (opposite) orientations. Thus, although each Ni^{2+} ion has two unpaired electrons, the alternate planes lead to zero net magnetization.

Example 12–3.1

Calculate the theoretical magnetization of $Mn^{2+}Fe_2O_4$, which has the same structure as $Ni^{2+}Fe_2O_4$ and magnetite. There are eight $MnFe_2O_4$ formula weights per unit cell (cubic). The lattice constant is 0.85 nm.

Procedure   This problem is nearly identical to the calculation for the magnetization of magnetite. (See discussion of ferrites in the previous section). Since Mn^{2+} has lost its two $4s$ electrons by ionization, it possesses five Bohr magnetons (net of Fig. 12–3.1). The magnetization is $\Sigma\, p_m/V$.

Calculation

$\uparrow$:　　$(8Mn^{2+})(5\beta/Mn^{2+})(9.27 \times 10^{-24}\ A\cdot m^2)$

$\uparrow$:　　$(8Fe^{3+})(5\beta/Fe^{3+})(9.27 \times 10^{-24}\ A\cdot m^2)$ 　　$\Big\}$ net $= 3.7 \times 10^{-22}\ A\cdot m^2$

$\downarrow$:　$-(8Fe^{3+})(5\beta/Fe^{3+})(9.27 \times 10^{-24}\ A\cdot m^2)$

$\qquad\quad M = 3.7 \times 10^{-22}\ A\cdot m^2/(0.85 \times 10^{-9}\ m)^3 = 600{,}000\ A/m$

Comment   This compares with 570,000 A/m by experiment.

Example 12–3.2

A ceramic magnet has a true density of 5.41 Mg/m^3 (=5.41 g/cm^3). A poorly sintered sample weighs 3.79 g dry, and 3.84 g when saturated with water. The saturated sample

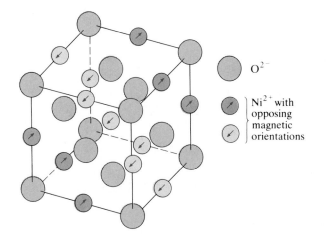

O^{2-}

Ni^{2+} with opposing magnetic orientations

FIG. 12–3.4

Antiferromagnetism (NiO). Magnetism exists; however, there is a balance of magnetic moments in the two opposing directions on alternate planes. Therefore, special laboratory procedures are required to detect the presence of the magnetism.

weighs 3.08 g when suspended in water. (a) What is its *true volume?* (b) What is its *bulk volume* (total volume)? (c) What is its *apparent (open) porosity?* (d) What is its *total porosity?*

Background There are three measurements of volumes:

$$\text{true volume} = \text{total (bulk) volume} - \text{total pore volume}$$

$$\text{apparent volume} = \text{total volume} - \text{open-pore volume}$$
$$= \text{true volume} + \text{closed-pore volume}$$

$$\text{bulk (total) volume} = \text{true volume} + \text{total pore volume}$$

Since $\rho = m/V$, there are also three densities.

Recall from Archimedes' principle that an object is buoyed up by the weight of the displaced fluid.

Calculation

(a) true volume $= m/\rho_{tr} = 3.79 \text{ g}/(5.41 \text{ g/cm}^3) = 0.70 \text{ cm}^3$

(b) From Archimedes' principle,

water displaced by bulk sample = buoyancy $= 3.84 \text{ g} - 3.08 \text{ g} = 0.76 \text{ g}$

$\therefore$ total volume $= 0.76 \text{ g}/(1 \text{ g/cm}^3) = 0.76 \text{ cm}^3$

This value includes material plus all pore space.

(c) apparent porosity = (open-pore volume)/(total volume)
$= [(3.84 - 3.79 \text{ g})/(1 \text{ g/cm}^3)]/0.76 \text{ cm}^3$
$= 0.066$ (or 6.6 v/o, bulk basis)

(d)   total porosity $= (V_{total} - V_{true})/V_{total}$
$= (0.76 - 0.70 \text{ cm}^3)/0.76 \text{ cm}^3$
$= 0.079$ (or 7.9 v/o bulk basis)

Comment The *closed porosity* is 1.3 v/o. The *bulk density* is 3.79 g/0.76 cm^3 = 5.0 g/cm^3. We also may speak of an *apparent density:*

mass/(bulk volume $-$ open-pore volume)

This density is

$(3.79 \text{ g})/(0.76 \text{ cm}^3)(1 - 0.066) = 5.34 \text{ g/cm}^3$ (or 5.34 Mg/m^3)

12–4

METALLIC MAGNETS

Bcc iron is the most common metallic magnetic material. There are other metallic magnets, as indicated in Tables 12–1.1 and 12–1.2. In particular, note the very high maximum permeability, μ_{max}, of superpermalloy, and the large coercive field, $-H_c$, of the newer boron-containing, iron-based materials.

Transformer Sheet

Many tons of magnetic-iron products are made and used each year as sheets for transformer cores and motor parts (Fig. 12–2.3). In these applications, the magnets must be soft in order to respond to the 60-Hz power sources. They also must have high resistivity to reduce eddy current losses. Typically, these steels contain 1 to 4 percent silicon in solid solution. The resistivity increases according to Eqs. (11–2.2 and 11–2.3). Silicon also increases the permeability. The carbon specification is near zero in these steels to avoid the presence of iron carbide, which interferes with domain-boundary movements during each half cycle. Likewise, steels are annealed to eliminate dislocations, and the grain size is coarsened.

Steels for power transformers commonly are processed to provide a preferred orientation of the grains. As shown in Fig. 12–4.1(a), the ⟨100⟩ directions have greater permeability than do the other directions. Therefore, when the grains are oriented to near common ⟨100⟩ orientation, a greater transformer efficiency is realized. The increase is from ~97 to 98 percent. At first glance, this 1 percent increase appears minor; however, we must consider that the distribution of gigawatts of power requires a step up to a high voltage for cross-country transmission, then several step-down stages to the 120 to 220V level for homes and industry. A 1 percent savings on *each* step totals gigadollars of energy savings each year!

Metallic magnets pose a major disadvantage in high-frequency circuits because the rapidly changing magnetic field induces current flow and I^2R losses in the core

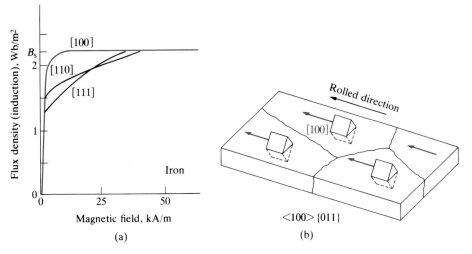

FIG. 12–4.1

Magnetic Anisotropy (bcc Iron). (a) Permeability, (B/H) versus crystal direction. (b) Processed to give cube-on-edge. This processing permits the ⟨100⟩ directions to be predominant for easier magnetization of the transformer sheet. An alternate cube-on-face orientation is slightly more efficient in a transformer, but is more difficult to process.

(E^2/R losses). This is why thin laminates are used in transformers (to increase the resistance). At radio frequencies, however, the heating effect is excessive, so it is necessary to switch to ceramic magnets, which have appreciably higher resistivities.

Ferromagnetism

We explained *ferri*magnetism and *antiferro*magnetism on the basis of the polarities of the magnetic atoms (Section 12–3). The *ferro*magnetism of metals must be explained in terms of energy bands. As is true in all metals, the valence bands are not filled.* Since each energy level within the band contains two electrons of opposite magnetic spins, Fig. 12–4.2(a) presents the filled levels with two spin directions that have a common Fermi energy. When an external magnetic field is added, the energies of one spin direction are reduced, and those of the other are increased (Fig. 12–4.2b). Thus, some of the spins realign into lower energy states, and the Fermi energies are again equal (Fig. 12–4.2c).

Example 12–4.1

The magnetization of pure iron is 1.7×10^6 A/m. What is the magnetic moment per atom?

Procedure Since magnetization is the total magnetic moment per unit volume (Eq. 12–3.2), we must determine the number of iron atoms per unit volume. We can do this calculation on the basis of either the unit cell or the density. In this example, we'll choose the latter.

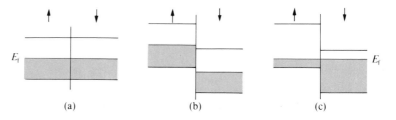

(a)   (b)   (c)

FIG. 12–4.2

Valence Band (Ferromagnetic Metal). (a) Equal spins in opposing directions ($3d$ and $4s$ electrons). (b) An external magnetic field lowers the energy of those electrons with a matching alignment, and raises the energies of those with an opposing alignment. (c) Some electrons realign their spins into lower energy levels, producing a net magnetization.

---

* The transition metals have overlapping $3d$ and $4s$ bands, both of which are partially filled. (See magnesium in Fig. 11–3.2.)

Calculation

$(7.87 \times 10^6 \text{ g/m}^3)/[(55.85 \text{ g}/(0.6 \times 10^{24} \text{ atoms})] = 8.5 \times 10^{28} \text{ atoms/m}^3$

$(1.7 \times 10^6 \text{ A} \cdot \text{m}^2/\text{m}^3)/(8.5 \times 10^{28} \text{ atoms/m}^3) = 20 \times 10^{-24} \text{ A} \cdot \text{m}^2 \text{ per atom}$

Since the magnetic moment per electron (Bohr magneton, β) is 9.27×10^{-24} A$\cdot$m^2, we can calculate the net number of electron spins there are per atom:

$(20 \times 10^{-24} \text{ A} \cdot \text{m}^2)/(9.27 \times 10^{-24} \text{ A} \cdot \text{m}^2/\beta) = 2.2 \ \beta/\text{atom}$

Comment The magnetic coupling in metallic iron aligns a net of 2^+ more electrons per atom in one direction than are aligned in the opposing direction (Fig. 12-4.2).

12-5

DIAMAGNETISM

An external magnetic field increases the flux density (induction) in all magnetic materials (Fig. 12-1.2). *Diamagnetic* materials (nonmagnetic) reduce the flux density, as shown schematically in Fig. 12-5.1(a). In almost every diamagnetic material, however, the decrease is appreciably less than 1 percent. Thus, to date, diamagnetism has had almost no technological importance.

Superconductors

Superconductors (Fig. 12-5.1b), on the other hand, do have properties of potential importance. The flux density drops to zero when these materials are below the

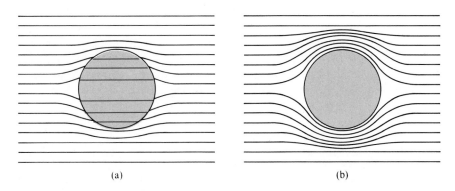

(a)   (b)

FIG. 12-5.1

Diamagnetism. (a) Nonmagnetic materials. The flux density in the material is less than that in a vacuum. This decrease is insignificant in almost all materials. (b) Superconductors. The flux density is zero when these materials are in the superconducting range. This zero flux density can be useful in many potential design applications, because the material is repelled by a magnetic field.

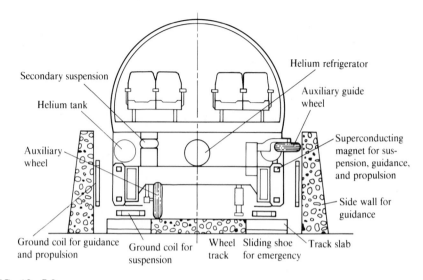

FIG. 12–5.2

Magnetically Levitated Train. Because superconducting materials are completely diamagnetic (Fig. 12–5.1), they are repelled by a magnetic field. This design makes use of that property to levitate a train over ground coils. Friction is eliminated, so only a small propulsion system is required. (Design by the Japanese National Railway. Reprinted with permission from the *Encyclopedia of Materials Science and Engineering,* Pergamon Books Ltd.)

critical values of temperature, T_c, and magnetic field, H_0, that introduce superconductivity (Fig. 11–8.1). Thus, a superconducting material is *repelled* by a magnet. This property could have significant technical importance for future applications of "high-temperature" superconductors (Section 11–8). For example, a train with a superconducting base could be levitated above a magnetic track (Fig. 12–5.2). Because there is no solid-to-solid contact, the train could be propelled quietly and smoothly with only a small, specially designed motor.

S U M M A R Y

1. The major properties of permanent (hard) magnets are a high *remanence,* a large *coercive field,* and a large *BH* product. The last measure is an index of the energy required for demagnetization. A soft magnet requires low values for these properties and a high relative *permeability,* a measure that indicates the ease of magnetization.

2. Energy is required for domain-boundary movement. In hard magnets, *domain boundaries* become pinned to (i) dislocations, (ii) grain boundaries, and (iii) phase boundaries; also, they have reduced mobility in anisotropic materials.

   Thermal energy destroys the coupling between

magnetic atoms above the *Curie temperature* to produce paramagnetism.

3. Ceramic magnets are generally *ferrimagnetic*. We can calculate their magnetization from the magnetic moments of the ions. They can be used in high-frequency circuits, because their high resistivity limits induced eddy currents.

4. Metallic magnets include tonnage quantities for

transformer sheets and for other related applications requiring very soft magnetic characteristics. Magnetic alloys and intermetallic compounds have been developed with *BH* products that are 10^4 times higher than that of the familiar cold-worked steel.

5. A *diamagnetic* material repels the magnetic flux lines. Typically, this effect is sufficiently insignificant that it is ignored. A superconductor however, completely rejects all the magnetic induction; thus, these materials have fascinating design potentials.

KEY WORDS

Antiferromagnetism
BH product
Coercive field (magnetic, $-H_c$)
Curie temperature (magnetic, T_c)
Diamagnetism
Domain (magnetic)

Domain boundary
Ferrimagnetism
Ferrites
Ferromagnetism
Induction (magnetic, B)
Induction (remanent, B_r)

Magnet, permanent (hard)
Magnet, soft
Magneton, Bohr (β)
Paramagnetism
Permeability (magnetic, μ)

PRACTICE PROBLEMS

12–P11 Estimate the maximum *BH* product of Alnico V on the basis of data in Fig. 12–1.3.

12–P21 Why are strain-hardened steels also harder magnetically, and why are annealed ones softer?

12–P22 (a) What magnetic characteristics are required for a soft magnet? (b) What structural features optimize these behaviors?

12–P23 Estimate the amount of energy required to demagnetize an Alnico V magnet that has a volume of 2.1 cm³.

12–P31 The ceramic magnetic material nickel ferrite has eight [$NiFe_2O_4$] formula weights per unit cell, which is cubic with $a = 0.834$ nm. Assume that all the unit cells have the same magnetic orientation. What is the saturation magnetization? (The Ni^{2+} ion lost both $4s$ electrons, but retained all $3d$ electrons, when it ionized.)

12–P32 Electron hopping occurs from Fe^{2+} to Fe^{3+} ions in iron oxide (Fig. 4–5.3 and Problem 115.11). Consider Fe_3O_4, of [$Fe^{2+}Fe_2^{3+}O_4$]₈, in which one electron per unit cell has hopped from an Fe^{2+} ion to an Fe^{3+} ion that is in a 4-f site. (a) What is the resulting magnetization per unit cell? (b) What would be the effect if it had hopped from an Fe^{2+} ion to an Fe^{3+} ion that was in a 6-f site?

12–P33 Refer to the calculations for the magnetization of Fe_3O_4 in Section 12–3. Al^{3+} ions can substitute for Fe^{3+} ions in the 6-f sites of the magnetite to give [$Fe(Fe,Al)_2O_4$]₈ as a solid solution. If one-half of these Fe^{3+} ions are replaced by Al^{3+} ions, what is the resulting magnetization? (Assume that the Fe^{3+} ions in the 4-f sites are unaffected.)

12–P41 Metallic cobalt has a saturation magnetization of 1.43×10^6 A·m²/m³. Calculate the magnetic moment per atom (a) in A·m², and (b) in Bohr magnetons.

12–P42 (a) Distinguish between *ferrimagnetism* and *antiferromagnetism*. (b) Distinguish between *paramagnetism* and *ferromagnetism*.

12–P43 Refer to Example 12–3.1. Assume that 40 percent of the domains have their N—S polarity reversed 180° after the magnetic field H_c is dropped to zero. What is the remanent induction, B_r?

TEST PROBLEMS

1211 Assume a linear B–H curve for the second quadrant of magnetization for a rare earth–cobalt magnet (Table 12–1.1). (a) What is the maximum BH product? (b) Will your answer be higher or lower than an experimentally determined energy requirement? Explain why there will be a difference.

1221 The metallic glass of Table 12–1.2 has a very small coercive field, $-H_c$. Provide a rationale for this fact.

1222 (a) What magnetic properties are required for a hard magnet? (b) What structural characteristics optimize these properties?

1223 A "square-loop" magnet is preferred for magnetic memory. That is, the second (and fourth) quadrant of the hysteresis loop should be as square-shaped as possible. Why is this characteristic desirable?

1231 You are considering using cobalt ferrite, $CoFe_2O_4$, as a soft magnet. What is this material's magnetic saturation? (It is a spinel that has the same structure as does nickel ferrite and magnetite.)

1232 By measurement, the magnetization of $[Li_{0.5}Fe_{2.5}O_4]_8$ is 2×10^{-22} A·m² per unit cell of 32 oxygens. Is this value consistent with the material's structure, which is like that of magnetite, but with the eight Fe^{2+} ions in each unit cell replaced by four Li^+ and four Fe^{3+} ions?

1233 A solid solution of zinc ferrite and nickel ferrite can occur to produce a unit cell of approximately $(Zn_3,Fe_5)Fe_{16}O_{32}$. However, the Zn^{2+} ions choose 4-f sites preferentially and force Fe^{3+} ions into 6-f sites. What is the saturation magnetization?

1234 Figure 12–3.4 shows that atoms of NiO possess a lattice very similar to those of NaCl (Fig. 3–1.1). However, alternate {111} planes of cations have opposing magnetic orientations. Thus, the *magnetic unit cell* is not identical to the cubic NaCl structure. (a) What is the magnetic unit cell's lattice constant? (b) How many nickel ions are there in the magnetic unit cell? (Recall from Chapter 3 that positions related by a full unit-cell translation must be identical in *every* respect.)

1241 The saturation magnetization of metallic nickel is 0.48×10^6 A/m. How does the number of Bohr magnetons per atom compare with that of iron?

1242 Two textures can enhance the permeability of transformer sheet steel. One is the cube-on-edge texture shown in Fig. 12–4.1(b). The second is a cube-on-face texture. Make a sketch of the second texture.

1243 The rare-earth metal gadolinium is ferromagnetic below 16° C with 7.1 β per atom. (a) What is the magnetic moment per g? (b) What is the saturation magnetization? (The atomic weight of Gd is 157.26 amu, and $\rho = 7.8$ g/cm³.)

Chapter 13

DIELECTRIC AND OPTICAL PROPERTIES OF CERAMICS AND POLYMERS

Although dielectrics are nonconducting and therefore are insulators, they may be more than just isolators. Their polarization can be used for functional purposes in an electrical circuit. Polarization is associated with the atomic and molecular structures, and the resulting displacements to produce electrical dipoles. Applications include sensors that transpose pressure and touch into electric signals that can be used as feedback in automated equipment.

The use of optical materials as waveguides for information transmission introduces stringent requirements for transparency and light sources. This is another of the frontiers for high-performance materials.

13–1

DIELECTRIC MATERIALS

Dielectric materials separate two electrical conductors without a passage of current. Thus, metals cannot be dielectrics; but many (not all) ceramics and polymers fall in this category. In the simplest situation, a dielectric is an *insulator,* playing an inert role in the electrical circuit. The prime property for an insulator is *dielectric "strength,"* which is the potential gradient value, V/mm, that the designer can use and still expect to avoid an electrical breakdown. Table 13–1.1 lists values of the dielectric strengths of various insulators, with their resistivities. A good insulator has high values of each; however, no correlation exists between the two because eventual electrical breakdown generally is a consequence of impurities, cracks, flaws, and other imperfections, rather than of the inherent electrical characteristics of the material. This also accounts for the fact that the dielectric strength is a function of thickness.

Dielectric materials do not conduct electricity. They are not, however, entirely inert to an electrical field. The electrons and the proton-containing atomic nuclei will shift their positions in response to the field. For example, the mean position of the electrons will be on the side of the atom nearer the positive electrode, whereas the atomic nucleus itself, which contains the protons, will shift slightly toward the

TABLE 13–1.1 Typical Properties of Electrical Insulators

MATERIAL	RESISTIVITY (20° C) ohm·m	DIELECTRIC STRENGTH,* V/mm
Ceramic Materials		
Soda-lime glass	10^{13}	10,000
Pyrex glass	10^{14}	14,000
Fused silica	10^{17}	10,000
Mica	10^{11}	40,000
Steatite porcelain	10^{12}	12,000
Mullite porcelain	10^{11}	12,000
Polymeric Materials		
Polyethylene	$10^{13} - 10^{16}$	20,000
Polystyrene	10^{16}	20,000
Polyvinyl chloride	10^{14}	40,000
Natural rubber	—	16,000–24,000
Polybutadiene	—	16,000–24,000
Phenol-formaldehyde	10^{10}	12,000

* Not constant with thickness.

negative electrode (Fig. 13–1.1). We call this shift *polarization.* If an ac field is applied, these internal charges "dance" back and forth in step with the frequency of the alternating field.

Electronic Polarization

Polarization can be categorized into several different types in terms of the displaced units. We have just described *electronic polarization.* Being small, the electrons have a very high natural frequency ($\sim 10^{16}$ Hz) as they form their standing waves around that atoms. Thus, this polarization can occur not only in 60-Hz circuits and at radio frequencies, but also in response to light frequencies ($\sim 10^{15}$ Hz).

Ionic Polarization

The displacement of negative and positive ions toward the positive and negative electrodes, respectively, is called *ionic polarization.* Like electronic polarization, ionic polarization is *induced,* since net displacements occur only when an external field is present.

Because they are more massive than are electrons, the ions cannot become polarized as rapidly. Ionic polarization is limited to a maximum frequency of approximately 10^{13} Hz. This value is below the frequency of visible light. Therefore, incident light cannot produce ionic polarization, as it does electronic polarization.

Molecular Polarization

Molecular polarization occurs when polar molecules are in an electric field. In polar molecules, the center of gravity for the positive charges and that for the negative charges are not coincident. A small molecular *dipole* is present. An

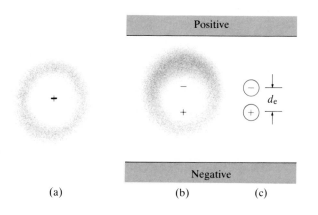

(a)   (b)   (c)

FIG. 13–1.1

Electronic Polarization (Schematic). (a) No external field. The electrons are equally distributed around the nucleus in the shaded area. (b) External field. The positive nucleus shifts toward the negative electrode. The electrons shift in the other direction to separate the center of negative charge $\ominus$ from the center of positive charge $\oplus$ by the distance d_e. (c) The dipole moment, p_e, is the product, Qd_e (Q being the charge).

example is methyl chloride (CH_3Cl, Fig. 2–2.4b). The chlorine atom has a complement of 17 electrons, whereas each of the three hydrogen atoms is an exposed proton at the far end of a covalent bond. As shown schematically in Fig. 13–1.2, the centers of positive and negative charges are separated by a distance d_e.

Molecular polarization is permanent, since it is inherent in the molecular structure. These dipoles can be oriented with the field. Furthermore, the molecule "flips" each half cycle of an ac field. Since the masses involved depend on the size of the molecule, the maximum frequency of reponse varies significantly from material to material. It is, however, always less than that for electronic and ionic polarization. Furthermore, it is highly sensitive to temperature; for example, polarization of polyvinyl chloride $+C_2H_3Cl+_n$ does not reverse below the glass temperature T_g, because molecular movements are restricted.

Space Charge

A *space charge* (also called *interfacial polarization*) develops when there is local conduction within a dielectric. For example, if an Al_2O_3 product, a nonconductor, contained small particles of metallic aluminum, the conduction electrons in the latter will be shifted toward the positive electrode in an ac field. However, they are contained inside the metallic particles (Fig. 13–1.3). We use this example of metallic particles for illustrative purposes; it is not typical of engineering materials. However, it is not uncommon for some ceramic phases to contain particles of semiconducting oxides (e.g., Ti_2O_3 in TiO_2). Generally, this situation is to be avoided, because it leads to dielectric losses in high-frequency circuits.

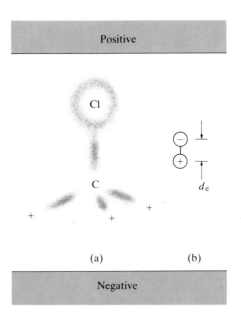

FIG. 13–1.2

Molecular Polarization (Schematic with CH_3Cl). Asymmetric molecules possess positive and negative ends and orient themselves in an electric field. (a) Electron regions are shaded. Hydrogen atoms are protons at the end of a covalent bond. (b) Electric dipole. (Center of positive charges, ⊕; negative charges, ⊖.)

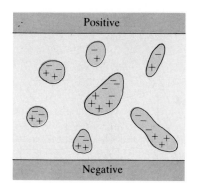

FIG. 13-1.3

Space Charge. When conducting particles are present within a dielectric, polarization can occur. Each particle develops a positive side and a negative side in an electric field.

13-2
POLARIZATION CALCULATIONS

Dipole Moments

The electric *dipole moment* p_e of a polarized molecule (or atom or unit cell) is the product of the charge Q and the distance d_e between the centers of positive and negative charges:

$$p_e = Qd_e \tag{13-2.1}$$

This is illustrated in Fig. 13-1.2, where CH_3Cl has 26 electrons, each with 0.16×10^{-18} C. If the electric field produces an average displacement $\bar{d}_e$ of 0.01 nm between the center of positive charge and the center of negative charge, the dipole moment per molecule is $26(0.16 \times 10^{-18} \text{ C})(10^{-11} \text{ m}) = 40 \times 10^{-30}$ C·m.

Polarization is measured as the sum of the dipole moment per unit volume:

$$\mathcal{P} = \Sigma p_e/V = \Sigma Qd_e/V \tag{13-2.2}$$

Thus, if there are 10^{20} methyl chloride molecules per cubic meter, the polarization of the CH_3Cl is

$$10^{20} \, (40 \times 10^{-30} \text{ C·m})/(1 \text{ m}^3) = 40 \times 10^{-10} \text{ C/m}^2$$

Dielectric Constant

An electric field can induce electronic and ionic polarization and can orient permanently polarized molecules. Conversely, polarization leads to an increased charge density on a capacitor. We can show this by separating two capacitor plates by a distance d and applying a voltage E between them (Fig. 13-2.1). The *electric field, $\mathcal{E}$,* is the voltage gradient:

$$\mathcal{E} = E/d \tag{13-2.3}$$

Under these conditions, when there is nothing between the plates (Fig. 13–2.1a), the *charge density* $\mathcal{D}_0$ on each plate is proportional to field $\mathcal{E}$, with a proportionality constant, ϵ_0, of 8.85×10^{-12} C/V·m:

$$\mathcal{D}_0 = \epsilon_0 \mathcal{E} = (8.85 \times 10^{-12} \text{ C/V·m})\mathcal{E} \qquad (13\text{–}2.4)^*$$

Thus, if the voltage gradient is 1 V/m, there will be 8.85×10^{-12} coulombs/m² on the electrodes. With each electron carrying a charge of 0.16×10^{-18} C, the electron density on the electrodes will be 55×10^6/m² (or 55/mm²). The voltage gradients are normally much higher in electrical circuits, and thus the charge densities are significantly greater.

If any material is placed between the capacitor plates in Fig. 13–2.1(b), the charge density on the plates is increased because of the polarization within the material. In effect, the upward shift of the negative charges in the material (and the downward shift of the positive charges) increases the value of $\mathcal{D}$ in Eq. (13–2.4) by a constant factor, κ:

$$\mathcal{D}_m = \kappa \epsilon_0 \mathcal{E} \qquad (13\text{–}2.5)$$

The factor κ, called the relative *dielectric constant,* is the ratio $\mathcal{D}_m/\mathcal{D}_0$, the charge density with and without the spacer material of Fig. 13–2.1. The relative dielectric constant is a property of the material that is used as the dielectric. Table 13–2.1 lists the relative dielectric constants for a number of common dielectrics. Note that they may be sensitive to temperature and to frequency.

We also can view polarization as the added charge density that originates from the dielectric,

$$\mathcal{P} = \mathcal{D}_m - \mathcal{D}_0 \qquad (13\text{–}2.6a)$$

and, from Eqs. (13–2.4 and 13–2.5),

$$\mathcal{P} = (\kappa - 1)\epsilon_0 \mathcal{E} \qquad (13\text{–}2.6b)$$

This is illustrated schematically in Fig. 13–2.2.

FIG. 13–2.1

Charge Density. The charge density (coulombs/m²) on the electrodes depends on the electric field, $\mathcal{E}$, and the polarization of material that is present.

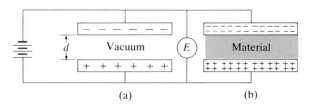

(a)   (b)

---

* ε_0 is the *permittivity* of free space.

TABLE 13–2.1 Relative Dielectric Constants*

MATERIAL	AT 60 Hz	AT 10^6 Hz
Soda-lime glass	7	7
Pyrex glass	4.3	4
"E" glass	4.2	4
Fused silica	4	3.8
Porcelain	6	5
Alumina	9	9
TiO_2 (film)		20–50
ZrO_2		12
$BaTiO_3$ (max)		5000+
Nylon 6/6	4	3.5
Polyethylene $+C_2H_4+_n$	2.3	2.3
Teflon, $+C_2F_4+_n$	2.1	2.1
Polystyrene, PS	2.5	2.5
Polyvinyl chloride, PVC		
plasticized ($T_g \approx 0°$ C)	7.0	3.4
rigid ($T_g = 85°$ C)	3.4	3.4
Rubber (12 w/o S, $T_g \approx 0°$ C)		
$-25°$ C	2.6	2.6
$+25°$ C	4.0	2.7
$+50°$ C	3.8	3.2

* 20° C, unless stated otherwise.

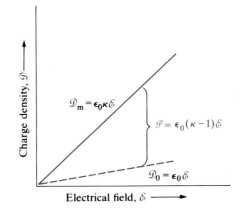

FIG. 13–2.2

Charge Density Versus Electric Field. Polarization, $\mathscr{P}$, is that part of the charge density resulting from the presence of the material.

Example 13-2.1

Polystyrene (PS) has a relative dielectric constant of 2.5 when subjected to a dc field. What is the polarization within the PS when a 0.5-mm sheet separates 100 V?

Solution The field is 2×10^5 V/m. From Eq. (13-2.6b),

$$\mathscr{P} = (2.5 - 1)(8.85 \times 10^{-12} \text{ C/V} \cdot \text{m})(2 \times 10^5 \text{ V/m})$$
$$= 2.7 \times 10^{-6} \text{ C/m}^2$$

Comment Since the electron charge is 0.16×10^{-18} C, this polarization is equivalent to 1.7×10^7 el/mm².

Example 13-2.2

A capacitor (Fig. 13-2.1a) with two parallel plates 1 cm $\times$ 2 cm each, receives a 2.25-V potential difference between the electrodes. (a) How far apart must the plates be to produce a charge density of 10^{-7} coul/m²? (b) How many electrons accumulate on the negative plate under these conditions? No dielectric insulator is placed between these plates.

Calculation

(a) From the first paragraph of the preceding section *(Dielectric Constant)*,

$$\mathscr{D}_0 = (8.85 \times 10^{-12})\mathscr{E} \tag{13-2.4}$$

$$10^{-7} \text{ coul/m}^2 = (8.85 \times 10^{-12} \text{ coul/volt} \cdot \text{m})(2.25 \text{ volt}/d)$$

$$d = 0.0002 \text{ m} \qquad \text{(or 0.2 mm)}$$

(b) no. of electrons $= \dfrac{1 \text{ cm} \times 2 \text{ cm} \times 10^{-11} \text{ coul/cm}^2}{0.16 \times 10^{-18} \text{ coul/electron}}$

$$= 125 \times 10^6 \text{ electrons.}$$

Example 13-2.3

The relative dielectric constants κ of a glass and of a plastic are 3.9 and 2.1, respectively. What voltage should be applied between electrodes that are separated with 0.13 cm of glass if we want the same charge density $\mathscr{D}_m$ as would develop on another set of electrodes separated by 0.42 cm of plastic and 210 V?

Calculation

$$\kappa_{gl}\mathscr{D}_0 = \mathscr{D}_{gl} = \mathscr{D}_{pl} = \kappa_{pl}\mathscr{D}_0$$

Since

$$\mathscr{D}_0 = (8.85 \times 10^{-12} \text{ coul/V} \cdot \text{m})\mathscr{E}$$

$$[\kappa(8.85 \times 10^{-12})(E/d)]_{gl} = [\kappa(8.85 \times 10^{-12})(E/d)]_{pl}$$

$$3.9 \, E/0.13 \text{ cm} = 2.1(210 \text{ V})/(0.42 \text{ cm})$$

$$E = 35 \text{ V}$$

13–3
POLYMERIC DIELECTRICS

Temperature and Frequency

The dielectric constant of polymers originates from both electronic polarization and molecular orientation. As revealed in Example 13–3.1, the dielectric constant is less than 3.0 above 10^8 Hz with *only* electronic polarization present. Molecular orientation may become a contributing factor below ~ 10^8 Hz. The protons of the side hydrogen atoms and the polar groups such as $>C=O$, $-C\equiv N$, and $-Cl$ respond to the electric field. Of course, this is possible only above the glass temperature, T_g.

Figure 13–3.1 shows schematically the effect of polarization, $\mathscr{P}$, on the relative dielectric constant κ as a function of frequency. Below T_g, however, molecular movements cannot contribute to the dielectric constant. As soon as T_g is exceeded, the dielectric constant increases markedly, as shown in Fig. 13–3.2. However, at still higher temperatures, we see a drop in the dielectric constant, because the more intense thermal agitation destroys the orientation of the molecular dipoles.

The curves of Fig. 13–3.2 also illustrate another point. The glass temperature T_g varies with frequency. Just as in Fig. 10–1.8, longer times (in this case, milliseconds for 60 Hz versus microseconds for 10^6 Hz) permit molecular movements at a somewhat lower temperature. A dc field has a still lower T_g because extended time is available for molecular orientation.

Liquid Crystals

Just as mechanical stressing can orient linear molecules, polar polymeric molecules (Section 2–3) can be oriented by the forces of an electric field. Possessing the same orientation, these long molecules can then mesh together into a crystalline array. When the electric field is removed, the molecules become disordered by thermal motions and lose their crystal periodicity. Materials of this type are called *liquid crystals* because they can possess some of the qualities of both liquids and

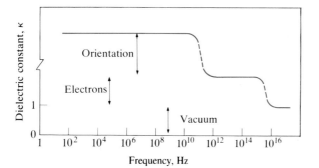

FIG. 13–3.1

Relative Dielectric Constant Versus Frequency (Schematic). Electrons respond to the alternating electric field below ~ 10^{15} Hz. Molecular dipoles can respond below ~ 10^{10} Hz.

FIG. 13–3.2

Dielectric Constant Versus Temperature. Below the glass temperature, T_g, the molecular dipoles cannot respond to the alternating electric fields. (The glass temperature is lower with low frequencies—see Fig. 10–1.8b.) The dielectric constant is greatest just above T_g. At higher temperatures, more vigorous thermal agitation destroys the polarization and therefore reduces the dielectric constant κ.

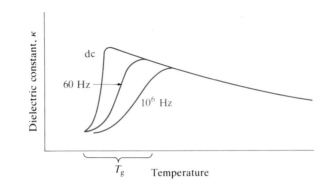

crystals. However, the properties of the crystalline and those of the liquid variants differ. For example, they have differences in light reflection and refraction. Thus, it is possible to design digital displays, such as those found in watches and calculators. Transparent electrodes—for example, SnO—are used to polarize and crystallize the regions of a thin film that correspond to the required digit. Essentially no current is required; however, incident light must be available.

Example 13–3.1

The relative dielectric constants for polyvinyl chloride (PVC) and polytetrafluoroethylene (PTFE) are as follows:

FREQUENCY, Hz	PVC	PTFE
10^2	6.5	2.1
10^3	5.6	2.1
10^4	4.7	2.1
10^5	3.9	2.1
10^6	3.3	2.1
10^7	2.9	2.1
10^8	2.8	2.1
10^9	2.6	2.1
10^{10}	2.6	2.1

(a) Plot the capacitance-versus-frequency curves for three capacitors with 3.1 cm × 102 cm effective area separated by 0.025 mm of (1) vacuum, (2) PVC, and (3) PTFE.

(b) Account for the decrease in the relative dielectric constant of PVC with increased frequency, and for the constancy in the relative dielectric constant of PTFE.

Procedure From physics, capacitance is given in farads, F, or coulombs per volt, $A \cdot s/V$. Rearranging the units of Eqs. (13–2.3 and 13–2.4), we obtain

$$A \cdot s/m^2 = \kappa\epsilon_0(V/m) \quad \text{to} \quad A \cdot s/V = \kappa\epsilon_0(m^2/m)$$

$$C = \kappa\epsilon_0 A/d = \kappa(8.85 \times 10^{-12}\ A \cdot s/V \cdot m)A/d \qquad \textbf{(13–3.1)}$$

Sample calculation

(a) At 10^2 cycles per sec,

$$C_{vac} = (1)(8.85 \times 10^{-12}\ A \cdot s/V \cdot m)(0.031\ m)(1.02\ m)/(25 \times 10^{-6}\ m) = 0.0112\ \mu F$$

$$C_{pvc} = (6.5)(0.0112\ \mu F) = 0.073\ \mu F$$

See Fig. 13–3.3(a) for the remainder of the results.

(b) The relative dielectric constant of PVC is high at low frequencies because PVC has an asymmetric mer with a large dipole moment (Fig. 13–3.3b). At high frequencies, these dipoles cannot maintain alignment with the alternating field, so only electronic polarization exists. On the other hand, PTFE has a symmetric mer, and therefore its polarization is only electronic. Although the dipoles in PTFE are weaker, they can be oscillated at the frequencies of Fig. 13–3.3.

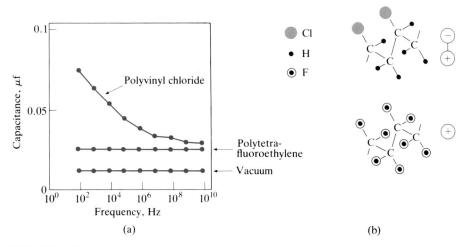

(a)

(b)

FIG. 13–3.3

Capacitance Versus Frequency. (a) See Example 13–3.1. (b) Symmetry comparison of polyvinyl chloride and polytetrafluorethylene mers.

13-4
CERAMIC DIELECTRICS

Ceramics are widely used as electrical insulators, and as substrates (Fig. 13–4.1). Pertinent properties are high resistivities and dielectric strengths (Table 13–1.1).

Thermal conductivity also is important in those applications where heat is generated and must be removed. For example, an automobile spark is hot enough that it would melt the metal electrode of a spark plug, were it not for the fact that the Al_2O_3 insulator conducts the heat to the metal shell of the plug. Likewise, a substrate for computer circuitry must dissipate the heat that is generated by the multitude of operating elements. Alumina is particularly suited for these roles,

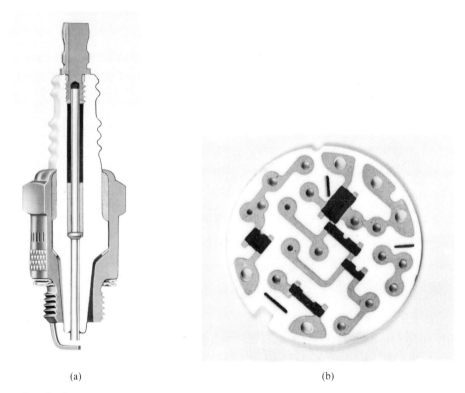

(a)   (b)

FIG. 13–4.1

Al_2O_3 Ceramics. (a) Sparkplug insulator. (Courtesy of Champion Spark Plug Co.)
(b) Substrate for a printed circuit. (Courtesy of AC Spark Plug Division, General Motors.) The principal material in these electrical insulators is Al_2O_3. Each Al^{3+} ion has lost its three valence electrons, and each O^{2-} ion contains two extra electrons, which it holds tightly, so they are not available for electrical or thermal conduction. The cations (3+) and anions (2−) are strongly attracted to each other, as evidenced by the high melting temperature ($\sim 2020°$ C, or 3670° F), and by the great hardness of Al_2O_3. (One name for Al_2O_3 is emery.)

because both elements in Al_2O_3 have a small mass, and therefore have a high vibration frequency for rapid thermal conduction.

Commonly, the dielectric constants for ceramics are higher than are those for polymers. More important, they are more constant over a range of frequencies (Table 13–2.1). However, plastic films generally are more adaptable for the production of capacitors and similar electronic units.

Piezoelectric Materials

We saw in Section 13–3 that certain molecules possess a permanent dipole; they are polar molecules. Likewise, certain crystals possess a permanent electrical dipole because their centers of positive and negative charges are not at the centers of the unit cells. These unit cells are polarized. This polarization is illustrated by $BaTiO_3$. Recall from Fig. 3–3.6 that $BaTiO_3$ is cubic—above 120° C. Below that temperature, called the *ferroelectric Curie point,* there is a slight but important shift in the ions. The central Ti^{4+} ion shifts about 0.006 nm with respect to the corner Ba^{2+} ions. The O^{2-} ions shift in the opposite direction, as indicated in Fig. 13–4.2.* The center of positive charge and the center of negative charge are separated by the dipole length, d.

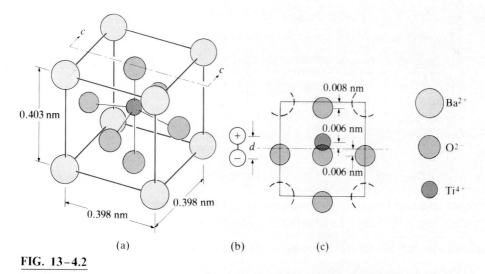

(a)   (b)   (c)

FIG. 13–4.2

Tetragonal $BaTiO_3$. Above 120° C, $BaTiO_3$ is cubic (see $CaTiO_3$, Fig. 3–2.9). (a) At ambient temperatures, the unit cell is noncubic. (c) The ions have shifted with respect to the corner Ba^{2+} ions (dashed). (b) Since the Ti^{4+} and O^{2-} ions shift in opposite directions, the centers of positive and negative charges are separated. This separation introduces an electric dipole of length d, and with a charge, Q, of $6(0.16 \times 10^{-18}$ A·s).

---

* The shift of the Ti^{4+} in Fig. 13–4.2 is upward as drawn. Actually, the shift could be in any one of the six coordinate directions. In any event, the O^{2-} ions shift in the opposite direction.

A material such as $BaTiO_3$ changes its dimensions in an electric field because the negative charges are pulled toward the positive electrode, and the positive charges are pulled toward the negative electrode, thus increasing the dipole length, d. This displacement also increases the dipole moment Qd, and the polarization $\mathcal{P}$, since the latter is the total of the dipole moments $\Sigma\, Qd$ per unit volume V (Eq. 13–2.2).

This sequence of effects provides a means of changing mechanical energy into electrical energy and vice versa. To understand this possibility, refer to Fig. 13–4.3(a). The cooperative alignment of the dipole moments of the many unit cells gives a polarization that collects positive charges at one end of the crystal and negative charges at the other end. Now consider two alternatives, as shown in Fig. 13–4.3(b and d). (1) Compress (or pull) the crystal with a stress s. There is a strain e, which is dictated by the elastic modulus. This strain changes the dipole length d and directly affects the polarization ($=\Sigma\, Qd/V$) since Q and V remain essentially constant. With a smaller polarization (from compression), there is excess of charge density on the two ends of the crystal. If the two ends are isolated, a voltage differential develops (Fig. 13–4.3b). If they are in electrical contact, electrons will shift from one end to the other (Fig. 13–4.3c). (2) No pressure is applied in Fig. 13–4.3(d); rather, a voltage is applied, which increases the charge density at the

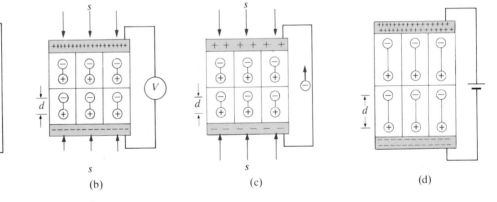

(a)   (b)   (c)   (d)

FIG. 13–4.3

Piezoelectric Material (Schematic). (a) No external field: Centers of positive and negative charges are not coincident. (b) Pressure applied: Compression leads to voltage differential. (c) Pressure applied and electrical path present: Compression leads to charge transfer. (d) External potential applied: Dipole lengthening produces dimensional change.

---

* The most widely used piezoelectric ceramics are $PbZrO_3$–$PbTiO_3$ solid solutions called PZTs. They have the structure of $BaTiO_3$ in Fig. 13–4.2. However, their ferroelectric Curie points are higher than that of $BaTiO_3$, and they have a greater range of design applications.

two ends. The negative charges within the $BaTiO_3$ are pulled one way, and the positive charges are pulled in the opposite direction, thus changing not only the dipole length, d, but also that dimension of the crystal.

These two situations indicate how mechanical forces and dimensions can be interchanged with electrical charges and voltages. Devices with these capabilities are called *transducers.* Materials with these characteristics are called *piezoelectric* (i.e., pressure-electric). Crystals of the $BaTiO_3$ type* are used for pressure gages, for tactile sensors of robotic manipulators, for phonograph cartridges, and for high-frequency sound generators. Let us examine a ceramic phonograph cartridge as an example. The stylus, or needle, follows the groove on the record. A small transducer is in contact with the stylus, which detects the vibrational pattern recorded in the groove. Both the frequency and the amplitude can be sensed as voltage changes (Fig. 13-4.3b). Although the voltage signal is small, it can be amplified through electronic circuitry until it is capable of driving a speaker.

Quartz crystals (SiO_2) also are piezoelectric. They are produced (Fig. 13-4.4) for special applications in circuits requiring frequency control. Once they are cut to a selected geometry, the elastic vibrational frequency is constant to one part in 10^8! Thus, in resonance, the crystal can control the frequency of an electronic circuit to that same precision. Fine watch circuits and control circuits for radio broadcasting are but two applications that use the piezoelectric characteristics of quartz crystals.

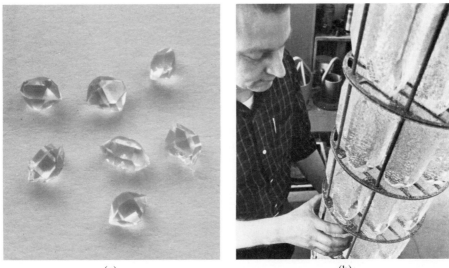

(a)   (b)

FIG. 13-4.4

Quartz Crystals (SiO_2). (a) Natural crystals. (b) Artificial crystals being removed from the hydrothermal cell. They will be used for the frequency controls of radio circuits. (Courtesy of Western Electric.)

Ferroelectric Materials

The polarity of quartz that we discussed in the previous section is permanently established by the atom arrangements of the unit cells. The unit cell cannot be reoriented. One end of the crystal is always positive and the other is always negative. This is not true for the $BaTiO_3$ of Fig. 13–4.2. As indicated in Fig. 13–4.5(b), a small displacement of the Ti^{4+} ions downward (and of the O^{2-} ions upward) produces a mirror image, which has opposite polarity. This reversal can be induced by an electric field—either one externally applied, or one created locally by the unit cell next door. This reversibility is called *ferroelectricity.** As a result of it, domains form that are conceptually similar to the domains in a magnet (Fig. 13–4.6).

The presence of domains leads to hysteresis in an ac field (Fig. 13–4.7). Consider a ferroelectric material containing many domains, with all the six possible orientations preferred. (Refer to Fig. 13–4.5; the six are the $\langle 100 \rangle$ directions, both positive and negative.) If an external electric field is applied, the boundaries of the domains move such that the more favorably oriented domains expand and the less

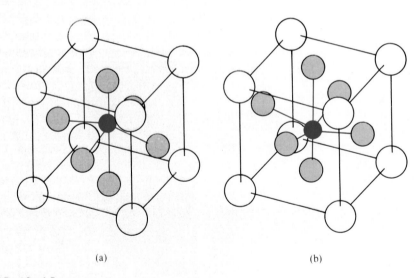

(a)   (b)

FIG. 13–4.5

Ferroelectric Crystal ($BaTiO_3$). (a) Same polarity as Fig. 13–4.2. (b) Reversed polarity. The Ti^{4+} ion is displaced downward and the O^{2-} ions are displaced upward. Either polarity is stable until reversed by an external field.

---

* All ferroelectric materials are piezoelectric; not all piezoelectric materials are ferroelectric. For example, quartz cannot be reoriented by the simple atom displacements effective for $BaTiO_3$ (Fig. 13–4.5). Rather, covalent bonds would have to be broken to produce a reversal.

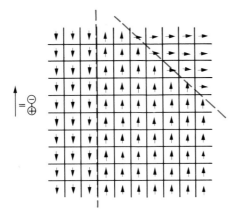

FIG. 13-4.6

Ferroelectric Domains. Adjacent unit cells interact to give similar polarity. Domain walls, ------, can be shifted by external electric fields. (See Fig. 12-2.1.)

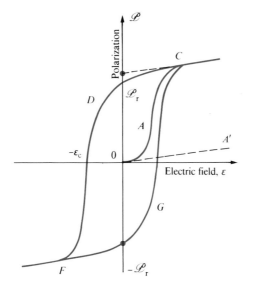

FIG. 13-4.7

Ferroelectric Hysteresis. Starting from 0, an electric field, $\mathscr{E}$, polarizes the material until saturation is reached at C. If the field is dropped to $\mathscr{E} = 0$, a remanent polarization, $\mathscr{P}_r$, remains. It takes a coercive field of $-\mathscr{E}_c$ to drop the net polarization to zero. The hysteresis loop, CDFGC, is followed for each succeeding cycle. (L. H. Van Vlack. *Materials for Engineers: Concepts and Applications,* Addison-Wesley, Reading, Mass., with permission.)

favorably oriented domains contract in volume. This reorientation gives a net polarization, which increases rapidly, as shown in the O-to-A part of the $\mathscr{P} - \mathscr{E}$ diagram of Fig. 13-4.7.

Eventually, the alignment approaches saturation, and a further increase in the electric field causes only a slight increase in the polarization. Removing that external electric field does not remove the polarization, so a remanent polarization, $\mathscr{P}_r$, is maintained. Not until a coercive field, $-\mathscr{E}_c$, of opposite orientation is applied does the material lose its net polarization. Cyclic fields produce a hysteresis loop along the CDFGC path of Fig. 13–4.7, which resembles the magnetic hysteresis loop.*

Example 13–4.1

A piezoelectric material has a Young's modulus of 72,000 MPa (10,400,000 psi). What stress is required to change its polarization from 640 $C \cdot m/m^3$ to 645 $C \cdot m/m^3$?

Procedure The increase in polarization must originate from a strain that increases d of the dipole moment. The stress must be positive (tension).

Calculation $\Delta \mathscr{P}/\mathscr{P} = (645 - 640)/640 = +0.0078 = (\Delta Qd/V)/(Qd/V)$. However, Q and V do not change; therefore,

$$\Delta d/d = +0.0078 = e = s/E$$
$$s = +0.0078 \ (72,000 \text{ MPa})$$
$$= +560 \text{ MPa}$$

or

$$s = +81,000 \text{ psi} \qquad \text{(tension)}$$

Additional Information The charge density changes 5 C/m^2 with this pressure. Therefore, if the two electrodes of Fig. 13–4.3(c) are connected, there will be a transfer of

$$\text{electrons} = 5 \ C/m^2/(0.16 \times 10^{-18} \ C/el)$$
$$= 3 \times 10^{19} \text{ el/m}^2 \qquad \text{(or } 3 \times 10^{13} \text{ el/mm}^2\text{)}$$

Example 13–4.2

Calculate the polarization, $\mathscr{P}$, of $BaTiO_3$, based on Fig. 13–4.2 and or the information in this section.

Procedure Calculate the dipole moment for each atom of the unit cell. Use the middle of the unit cell as the fulcrum for the moment arm.

Calculation

ION		Q, coul	d, m	Qd, coul·m
Ba^{2+}	(reference)	$+2(0.16 \times 10^{-18})$	0	0
Ti^{4+}		$+4(0.16 \times 10^{-18})$	$+0.006(10^{-9})$	3.84×10^{-30}
$2 \ O^{2-}$	(side of cell)	$-4(0.16 \times 10^{-18})$	$-0.006(10^{-9})$	3.84×10^{-30}
O^{2-}	(top and bottom)	$-2(0.16 \times 10^{-18})$	$-0.008(10^{-9})$	2.56×10^{-30}
				$\Sigma = 10.24 \times 10^{-30}$

---

* The name *ferroelectric* originated because of the direct comparison between the $\mathscr{P}-\mathscr{E}$ hysteresis loop and the $B-H$ loop for ferromagnetic materials. The name is not fully appropriate, since ferroelectric materials seldom, if ever, contain iron ions.

From Eq. (13–2.2)

$$\mathcal{P} = \Sigma \; Qd/V$$
$$= (10.24 \times 10^{-30} \; C \cdot m)/(0.403 \times 0.398^2 \times 10^{-27} \; m^3)$$
$$= 0.16 \; coul/m^2$$

Additional information   This means that polarized $BaTiO_3$ can possess a charge density of $0.16 \; coul/m^2$, equivalent to 10^{12} electrons/mm^2.

___

13–5
TRANSPARENT SOLIDS

The great majority of materials that are used for optical purposes are assigned to light-transmission roles. Filtering action (selective transmission) is sometimes specified. In the most familiar applications, such as architectural and automotive windows, the optical requirement is transmission without distortion. Distortion-free transmission calls for flat, parallel surfaces with the absence of internal flaws. In selected applications, infrared or ultraviolet radiation must be filtered out.

Optical *lenses* have the added purpose of refracting the light along an optical path (Fig. 13–5.1). In lenses for eyesight, the refraction is generally controlled by grinding for surface curvature. The index of refraction, which is a property of the material, is a second factor in lens specification and must be considered in all optical systems.

Optical fibers for long-distance communication channels are a recent innovation. Obviously, the fibers must have near-zero light absorption. Somewhat surprisingly, the index of refraction also is a factor.

Indices of Refraction

The index of refraction, n, of a material is the ratio of the light velocity in a vacuum, v_0, to the velocity in that material, v_m:

$$n = v_0/v_m \qquad\qquad (13\text{–}5.1)$$

FIG. 13–5.1

Optical-Glass (Prisms, Lenses, and Mirrors). Stringent requirements for composition and homogeneity are mandatory to assure clarity and precise optical paths for high resolution. (Courtesy of American Optical.)

When light passes from one material to another, the ray is refracted according to Snell's law. When we incorporate that law with Eq. (13–5.1), we have

$$n_1/n_2 = v_2/v_1 = \sin \phi_2/\sin \phi_1 \qquad (13\text{–}5.2)$$

with the angles and refraction being those of Fig. 13–5.2(a).

The index of refraction is related to the polarization we discussed in Section 13–1. Since only electronic polarization is operative at light frequencies (10^{15} Hz), the relationship is between the index, n, and the electronic dielectric constant, κ_e:

$$n = \sqrt{\kappa_e} \qquad (13\text{–}5.3)$$

The value of this dielectric constant depends on the number of electrons that are encountered per unit length and their polarizability. There are two consequences of this relationship. First, a denser phase of a given material has a larger index of refraction. Thus, in Table 13–5.1, quartz ($\rho = 2.65$ g/cm³) shows a higher index of refraction than does fused silica ($\rho = 2.2$ g/cm³). Both are SiO_2; therefore, the change in the index is a function of $\rho^{-1/3}$. Second, in noncubic materials, the electronic polarization varies with the direction of light vibration. Thus, the index of refraction is anisotropic. For example, calcite ($CaCO_3$) has its $(CO_3)^{2-}$ ions oriented such that the four atoms lie in the (0001) plane of the hexagonal crystal (Fig. 13–5.3). The electronic polarization is 24 percent greater for light vibrating in this plane (**A**), which is perpendicular to the vertical axis, than for light vibrating in the plane parallel to the vertical axis (**B**). This produces two indices of refraction — 1.66 and 1.49, respectively. This difference is called *birefringence* and appears as double refraction. Glasses and materials with cubic crystals do not have birefringence because the electronic polarization is isotropic.

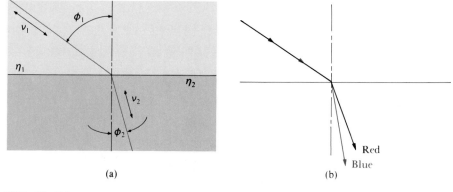

(a)   (b)

FIG. 13–5.2

Refraction. (a) The index of refraction is inversely proportional to the velocity of light and is measured by the refraction angles (Eq. 13–5.2). (b) Dispersion. The indices of refraction vary with the frequency (and wavelength) of the light. Red light travels slightly faster than does light with a shorter wavelength; therefore, it has a lower index.

TABLE 13-5.1 Selected Indices of Refraction

MATERIAL	INDEX (20°)	
	$\lambda = 589.3$ nm	656.3 nm
Air	1.00028	
Water	1.333	
Optical glass (heavy flint)	1.650	1.644
Optical glass (crown)	1.517	1.514
Fused silica	1.458	1.457
Quartz	$\begin{cases} 1.544 \\ 1.553 \end{cases}$	
Calcite	$\begin{cases} 1.658 \\ 1.486 \end{cases}$	
Diamond	2.438	2.426
Polytetrafluoroethylene	1.4	
Polypropylene	1.47	
Polyethylene		
(low density)	1.51	
(high density)	1.54	
Polymethyl methacrylate	1.49	
Polystyrene	1.60	

The electronic polarization increases slightly with frequency; therefore, the index of refraction also increases with frequency. The spread in index between blue and red is called *dispersion* (Fig. 13-5.2b). The shorter wavelengths have higher indices (slower velocities). This is of importance in lens systems where it is mandatory that all colors have the same plane of focus. It also means that different

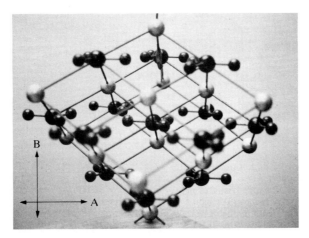

FIG. 13-5.3

Optical Anisotropy ($CaCO_3$). Light vibration **A** perpendicular to the vertical axis—parallel to (0001). Light vibration **B** parallel to the vertical axis—perpendicular to (0001). The electronic polarization from the transmitted light is 24 percent greater in direction **A** than in direction **B**; therefore, the index of refraction is greater for the vibration direction of **A**. This difference produces birefringence, giving two indices of refraction.

colors of light will get out of phase as they travel through optical waveguides. Table 13–5.1 lists the indices for a few selected materials. Two wavelengths (yellow and red) are included for two optical glasses and for fused silica (used for optic fibers). The high index and dispersion of diamond produces its "brilliance."

Internal Reflection

Light is bent away from the perpendicular as it travels from a denser phase into a less dense material; for example, from glass into air. As ϕ_2 of Fig. 13–5.2(a) is increased for an "upward" traveling ray, the refracted angle, ϕ_1, approaches 90°. If ϕ_2 exceeds a *critical* angle, the light undergoes *internal reflection* and does not leave the denser material (Fig. 13–5.4a). That redirection occurs sharply at the surface. The reflection can be made gradually if the composition near the surface is *graded* so that the change in density, and therefore the index of refraction, is not abrupt (Fig. 13–5.4b).

A transparent material becomes *translucent* if grain and phase boundaries are present to scatter the light by refraction or by internal reflections. Thus, the maximum transparency dictates the use of glasses, amorphous plastics, or single crystals.

Optic Waveguides

The recently developed multichannel, modulated-light communication systems require optical fibers that will transmit light over great distances (Fig. 13–5.5b). The engineer must give special attention to the compositions, structure, and

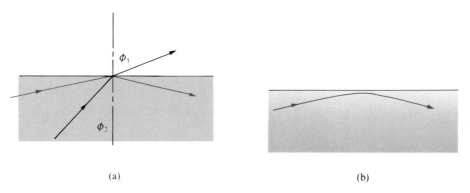

(a)   (b)

FIG. 13–5.4

Internal Reflection. (a) As ϕ_2 exceeds the critical angle that makes ϕ_1 equal to 90° according to Eq. (13–5.2), the light is internally reflected. (b) Graded index. A gradual change in composition at the surface, and therefore a change in the index of refraction, can produce a gradual redirection of the internally reflected light. This procedure is used in fiber optics.

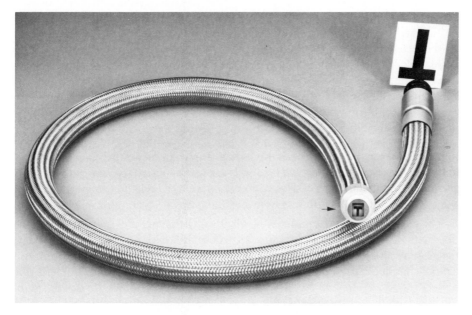

(a)

(b)

FIG. 13–5.5

Glass-Fiber Optics. (a) Imagescope. The bundle of 27,000 glass fibers, each 60 μm in diameter, permits the viewing of inaccessible and hazardous locations. (Courtesy of Galileo Electro-Optics Corp.) (b) Optic waveguides. Optical fibers for communication systems require kilometers of transparency, plus coherent light signals. Thus, the material must have not only extremely high purity, but also a controlled profile for the index of refraction (Fig. 13–5.6). (Courtesy of American Information Technologies.)

processing of the fibers. The fibers must be free of all oxides that absorb light — in particular, they must contain no transition metal oxides, such as FeO.

We depend on total internal reflections to keep the light within the optical waveguide. None of the light that is within a few degrees of the axis of the fiber can escape; rather, it is channeled along the optic fiber. Thus, in principle, the optic waveguide is simple. There are, however, two other factors that we must consider. Surface flaws, no matter how minute, can lead to scattering of light — and any handling of the fiber will introduce flaws. These flaws can be avoided if the light–transmitting glass fiber is clad with a second glass of lower index. The light is internally reflected at the interface between the two glasses, which cannot be damaged. We can make this dual fiber by jacketing a glass rod with a lower-index glass tube, heating the two, and drawing the composite into a cored filament (Fig. 13–5.6a).

The second factor is the path length for the rays that are precisely parallel to the fiber axis *versus* that for the rays that are internally reflected (Fig. 13–5.6b). The light signals along these two paths get out of phase over the length of the communication channel unless the fiber is extremely thin. To avoid this problem, the processing procedure can be altered. A fused-silica glass tube is vapor-coated on the *inside* with increasing concentrations of a glass-forming material, such as GeO_2, which has a higher index of refraction. The index is graded. The tube is heated and collapsed into a rod, which is then drawn into a fiber, as sketched in Fig. 13–5.6(a). Because the core has a graded index, which increases toward the axis, the velocity of the light increases away from the center line of the final fiber. Thus, the rays along the shorter axial route do not advance ahead of the reflected rays.

Example 13–5.1

Water has an index of refraction of 1.331. (a) What is the critical angle for internal reflection of light at an air–water surface? (b) Will the index for ice be higher or lower than 1.331?

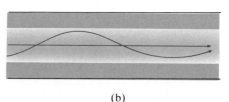

FIG. 13–5.6

Optical Waveguides (Longitudinal Sections).
(a) Dual fiber. A low-index glass (outer) and normal-index glass (center) are hot drawn together. (b) Clad, graded-index fiber. The center of the fiber has a higher index, so the paths have the same forward velocity. (See Fig. 13–5.4b.)

(a)

(b)

Procedure (a) The refracted angle in air will be 90°, which we can use in Eq. (13–5.2). Consider that the index of refraction for air is 1.0. (b) The molecules of ice and water are the same kind, H_2O, but are not present in equal numbers per unit volume. Which will have the greater polarization? (Compare densites, ρ.)

Answers

(a) $1.331/1.0 = \sin 90°/\sin \phi_{H_2O}$:

$\qquad \phi_{H_2O} = 48.7°$

(b) With $\rho_{ice} < \rho_{water}$, $(\Sigma \, Qd/V)_{ice} < \mathscr{P}_{water}$. Therefore, κ and n are less for ice than for water (Eqs. 13–2.6b and 13–5.3).

Comment The indices of air and of ice are 1.0003 and 1.31, respectively.

Example 13–5.2

Sodium light has a frequency of 5.09×10^{14} Hz. (a) What is its wavelength in outer space? (b) In glass with $n = 1.56$?

Procedure The velocity changes with the index; but the frequency v must remain constant. From physics, $\lambda = c/v$.

Calculation

(a) $\lambda = (3 \times 10^8 \text{ m/s})/(5.09 \times 10^{14}/\text{s})$
$\qquad = 589 \times 10^{-9}$ m $\hfill$ (or 589 nm)

(b) $\lambda = [(3 \times 10^8 \text{ m/s})/1.56]/(5.09 \times 10^{14}/\text{s})$
$\qquad = 378 \times 10^{-9}$ m $\hfill$ (or 378 nm)

Example 13–5.3

The attenuation of light in a glass fiber is purported to be -2.3 db per km. (a) What fraction of the initial intensity remains after 1 km? (b) After 8 km?

Procedure A "bel" is the exponent (base 10) of I_2/I_1;

$\qquad \therefore 1 \text{ db} = 0.1 \text{ bel} = 0.1 \log_{10}(I_2/I_1) \hfill$ **(13–5.4)**

Calculation

(a) $-2.3 \text{ db} = -0.23 \text{ b} = \log_{10}(I_2/I_1)$;

$\qquad I_2/I_1 = 10^{-0.23} = 0.59$

(b) $(-2.3 \text{ db/km})(8 \text{ km}) = -18.4 \text{ db} = -1.84 \text{ b}$;

$\qquad\qquad I_3/I_1 = 10^{-1.84} = 0.014$

Comment Ordinary window glass has an attenuation in excess of -2.3 db/m.

13-6

LIGHT-EMITTING SOLIDS

Hot solids emit light. Our examples could include coals in a campfire, and the tungsten filament of an incandescent light bulb. In this section, however, we shall limit our attention to fluorescent materials and to the stimulated emission of light (lasers).

Fluorescence

We discussed luminescence in Section 11–4. The *recombination* of excess electrons and electron holes from the conduction band and from the valence band, respectively, provided the energy for the emitted light (Eq. 11–4.4b). In a light-emitting diode (LED), the excess electrons and holes were introduced by a current that moved these charge carriers across an $n-p$ junction (Section 11–6). Excess electrons also can be made available by exposure of the material to light or heat, or by the activation from other electrons. Luminescence exists not only in semiconductors, but also in ionic solids where electrons can be energized to higher energy states associated with individual atoms. Photons are emitted as these electrons drop back to their ground state.

Luminescent materials in which the recombination is noticeably delayed are labeled *phosphorescent.* If the recombination rate is sufficiently rapid that we do not observe an afterglow, we speak of *fluorescence.* This distinction is significant on the screen of a television set or video display terminal. The luminescence from the stream of electrons must not interfere with the following scan.

The *recombination rate* varies from material to material. However, it follows a regular pattern because, within a specific material, every conduction electron has the same probability of recombining within the next second (or minute). This equal probability leads to the relationship:

$$N = N_0 e^{-t/\tau} \tag{13-6.1a}$$

which we usually rearrange to

$$\ln(N_0/N) = t/\tau \tag{13-6.1b}*$$

In these equations, N_0 is the number of electrons in the conduction band at a particular time (say, after the light has been turned off for photoluminescence, or

---

* Equation 13–6.1(b) can be derived through calculus (by those who wish to do so) from the information stated here. We use $1/\tau$ as the proportionality constant:

$$dN/dt = -N/\tau \tag{13-6.2}$$

Rearranging,

$$dN/N = -dt/\tau$$

then integrating,

$$\ln(N/N_0) = -t/\tau \tag{13-6.1c}$$

the electron beam has scanned for electroluminescence). After an additional time, t, the number of remaining electrons is N. The term, τ, is a characteristic of the material. It is called the *recombination time* (or *relaxation time*).

Since the recombination rate is proportional to the number of activated electrons, the intensity, I, of luminescence also follows Eq. (13–6.1b):

$$\ln(I_0/I) = t/\tau \qquad (13\text{–}6.3)$$

Thus, the decay curve for the intensity is that shown in Fig. 13–6.1. At $t = \tau$, $I = I_0/e$, and so on.

For a television tube, the engineer chooses a phosphor with a relaxation time such that the light continues to be emitted as the next scan comes across. Thus, our eyes do not see the light–dark flickering. However, the light intensity from the previous raster should be weak enough that it does not compete with the new scan that follows $\frac{1}{30}$ sec later. (See Example 13–6.1.)

Lasers

A laser (**L**ight **A**mplification by **S**timulated **E**mission of **R**adiation) provides a coherent light source; that is, all the light is in phase. As such, it must be monochromatic.

The original laser material (it is still used) is a single crystal rod of ruby, which is Al_2O_3 containing ~ 0.5 percent Cr^{3+} ions in solid solution. Numerous types of materials are currently used, including gases, liquids, solids, glasses, and semiconductors. In each case, electrons of the atoms are excited into a higher, or upper, energy state. In the ruby laser, the subvalence electrons of the Cr^{3+} ions are energized. The *pumping* of the electrons to the upper energy state commonly is achieved by photons from a flash lamp.

As they do in luminescence, the activated electrons will return to their ground state in a random, statistical manner. In the process, they release energy as photons of light. The rate of relaxation is normally defined by the decay equation (13–6.2). However, this return to the lower energy state may be forced to occur prematurely if the electron is *stimulated* by a photon whose energy is equal to the release energy.

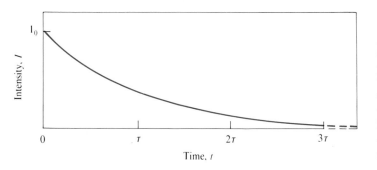

FIG. 13–6.1

Decay Curve (Luminescence). The relaxation time, τ, is a characteristic of the material. After $t = \tau$, $I = I_0/e$; and at $t = 2\tau$, $I = I_0/e^2$; and so on. A television screen works best when the relaxation time is approximately 15 msec.

Thus, one photon can release another (and another) in a chain reaction, as long as the photons remain in the material. Furthermore, the succeeding photons are released in phase with the triggering photon.

The procedure to retain photons for repeated use is to polish and silver the ends of the laser to provide reflection (Fig. 13–6.2). The pulse of energy is released in nanoseconds, and power levels (J/s) in the watts-to-megawatt range because of the short time.

As a device (Fig. 11–6.2), a laser has a light pump, which may be a gas discharge tube, surrounded by a reflector to concentrate the activating energy into the ruby rod (or other laser material). The emerging beam is monochromatic, coherent, and parallel, so that it can be focused onto a "pinpoint." This latter attribute permits diverse applications, ranging from eye surgery for detached retinas to surface hardening of steel.

Semiconductor lasers can also produce coherent light. A $p-n$ junction is made of a material such as gallium arsenide (GaAs). Referring to Fig. 13–6.3, an applied current carries electrons and electron holes across the junction to produce excesses that quickly recombine, and emit a monochromatic light. The emitted light is reflected at the two sides, returning to stimulate the recombination of additional, new electron–hole pairs that are being continually formed by the current. The light that exits the half-mirrored side can be used to feed an optic waveguide or other electro-optic device. The intensity of emitted light, and hence the number of recombining electron–hole pairs, can be controlled by the applied current.

Example 13–6.1

The scanning beam of a television tube covers the screen with 30 frames per second. What must the relaxation time for the activated electrons of the phosphor be if only 20 percent of the intensity is to remain when the following frame is scanned?

FIG. 13–6.2

Laser (Light Amplification by Stimulated Emission of Radiation) Device. Photon energy is absorbed by the ruby (Al_2O_3 plus Cr_2O_3) rod and remitted as an intense, coherent, and monochromatic light beam. (See Fig. 13–6.3.)

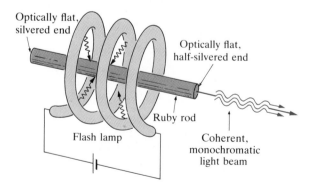

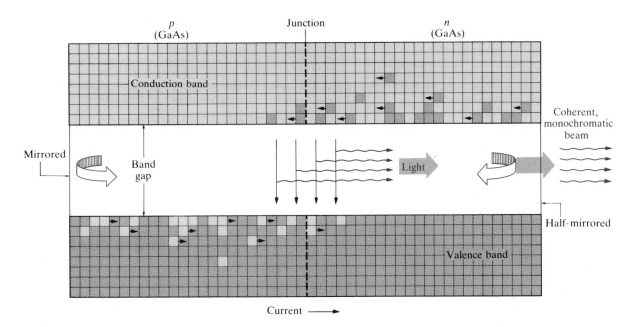

FIG. 13-6.3

Semiconductor Laser (Schematic). Photons of light are emitted when electrons combine with electron holes at the $p-n$ junction of properly doped gallium arsenide or comparable semiconductors. The emitted light is reflected at the two ends, returning to stimulate the combination of additional electron–hole pairs. These pairs build up an intense coherent beam that exits into an optical train for a variety of sophisticated applications.

Procedure Use Eq. (13–6.3). The time t in question is $\frac{1}{30}$ s.

Calculation

$$\ln(1.00/0.20) = (0.033 \text{ s})/\tau$$
$$\tau = 0.02 \text{ s}$$

Comments We use the term *fluorescence* when the relaxation time is short compared to the time of our visual perception (~ 25 ms). If the luminescence has a noticeable afterglow, we use the term *phosphorescence.*

SUMMARY

1. *Dielectric "strength"* is the resistance of an insulator to electrical breakdown. Polarization occurs within insulators by electronic and ionic displacements, and by the orientation of polar molecules.

2. *Polarization, $\mathscr{P}$,* is the sum of the dipole moments per unit volume. It also is equal to the excess *charge density, $\mathscr{D}_m - \mathscr{D}_0$,* developed in a capacitor. The relative *dielectric constant κ* is the ratio of the charge densities on a capacitor with and without a material present.

3. Polymer dielectrics have only electronic polarization below the glass temperature, T_g, and at high frequencies ($> 10^{10}$ Hz). Molecular polarization becomes operative above T_g, so the dielectric constant takes a big jump, particularly for dc and low-frequency ac circuits. At still higher temperatures, thermal agitation destroys the molecular orientation with the electric field, so the dielectric constant decreases.

4. Crystalline materials without a center of symmetry are *piezoelectric.* As such, they can be used for *trans-ducers,* which can be made to convert electric signals to mechanical signals, and vice versa. Materials with reversible polarity are *ferroelectric,* and produce a hysteresis loop in an alternating electric field.

5. The *index of refraction* is the inverse measure of light velocity and wavelengths within materials. It depends on the electronic polarization, and it varies slightly with wavelength. *Optical fibers* for communication transmission must have exceptionally low light absorption. Use is made of lower-index (and graded) coatings to keep the light path away from the fiber surfaces, which can be damaged to permit light leakages.

6. Photons are released in *lasers* when activated electrons return to their lower energy levels. These photons stimulate additional emission. A chain reaction of released photons produces a power burst of coherent (inphase), monochromatic light. Semiconducting lasers are used to feed fiber-optic systems.

KEY TERMS

Attenuation (optical)
Birefringence
Charge density ($\mathscr{D}$)
Coercive field (electric, $\mathscr{E}$)
Curie point (electric)
Dielectric
Dielectric constant, relative (κ)
Dielectric strength
Dipole moment (p_e)
Dispersion

Domain (electric)
Domain boundary
Electric field ($\mathscr{E}$)
Ferroelectric
Fluorescence
Index of refraction (n)
Internal reflection
Laser
Luminescence
Permittivity (ϵ)

Phosphorescence
Piezoelectric
Polarization, dielectric ($\mathscr{P}$)
Polarization, electronic (p_e)
Polarization, ionic (p_i)
Polarization, molecular (p_m)
Polarization, remanent ($\mathscr{P}_r$)
Space charge
Transducer

PRACTICE PROBLEMS

13–P21 Two capacitor plates (20 mm × 30 mm each) are parallel and 2.2 mm apart, with nothing between them. What voltage is required to develop a charge of 0.24×10^{-10} C on the electrodes?

13–P22 What is the electron density on the electrodes in Problem 13–P21?

13–P23 A 2.2-mm sheet of polystyrene is inserted in

the space between the plates of Problems 13–P21. With 10 V, the charge is 0.24×10^{-10} C without the polymer sheet, and 0.6×10^{-10} C with the sheet. What is the relative dielectric constant of the plastic?

13–P31 Dielectric constants of polymers are greatest at intermediate temperatures. Why is this so? How does the dielectric constant differ between ac and dc?

13–P32 A plate capacitor must have a capacitance of $0.25 \, \mu f$. What should its area be if the 0.0005-in. (0.0131-mm) mylar film that is used as a spacer has a dielectric constant of 3.0? (See Example 13–3.1.)

13–P41 A piezoelectric crystal has an elastic modulus of 130 GPa (19,000,000 psi). What stress will reduce its polarization from 560 to 557 C/m^2?

13–P42 Refer to Example 13–4.2. What is the distance between the centers of positive and negative charges in each unit cell?

13–P43 Lead zirconate is cubic in one of its polymorphs. The unit cell can be chosen such that each corner has a Zr^{4+} ion; the center of an edge, an O^{2-} ion; and the center of the unit cell, a Pb^{2+} ion. (a) What is the chemical formula? (b) Relocate the unit cell so that the Zr^{4+} ion is in the center of the unit cell. (c) Speculate on the ferroelectric possibilities of lead zirconate.

13–P44 Explain basis for the ferroelectric Curie temperature in terms of crystal structure.

13–P45 Microwave heating involves molecular polarization. Explain this statement.

13–P51 The critical angle for internal reflection at a glass–air surface is 41°. (a) What is the index of the glass? (Assume that $n_{air} = n_{vac}$.) (b) What is the velocity of light in the glass?

13–P52 Light retains 60 percent of its intensity after traveling 600 m through a glass fiber. What is its attenuation in db/km?

13–P53 Two spectral lines have wavelengths of 589.3 and 656.3 nm. Their indices of refraction in an optical glass are 1.516 and 1.501, respectively. How much will their refracted angles differ, if the incident ray of each is 19°30′ from the normal as they enter the glass from the air?

13–P61 A phosphorescent material is exposed to ultraviolet light. The intensity of the reemitted light decreases by 20 percent in the first 37 min after the ultraviolet light is removed. (a) How long will it be after the ultraviolet light has been removed before the light has only 20 percent of its original intensity (a decrease of 80 percent)? (b) How long before it has only 1 percent?

13–P62 A phosphorescent material must have an intensity of 50 (arbitrary units) after 24 hr, and of 20 after 48 hr. Based on these figures, what initial intensity is required? (Solve *without* using a calculator.)

TEST PROBLEMS

1311 Draw a sketch that shows why the $-\overset{\displaystyle Cl}{\underset{\displaystyle H}{C}}-$ portion of polyvinyl chloride is a polar group.

1312 Distinguish between *dielectric constant* and *dielectric strength*.

1321 What is the dipole moment per mm^3 of the polystyrene in Problem 13–P23?

1331 Discuss the relationships among *viscoelastic modulus, glass temperature,* and *dielectric constant.*

1332 What voltage difference (dc) is required to develop a charge density of 2×10^{-6} C/m^2 on the metalized surfaces of a 0.2-mm polyethylene film?

1333 The polarization of the polyethylene in Problem 1332 contributed how many electrons per mm^2 in the metal coating?

1334 A radio frequency (10^6-Hz) capacitor has been made with an 0.5-mm plasticized PVC spacer. The PVC spacer is replaced with an 0.33-mm polyethylene spacer. How much will the charge density be affected, assuming that other factors remain unchanged?

1335 (a) What is the polarization for the teflon (PTFE) of Example 13–3.1 when 4 V are applied to the capacitor? (b) What will be the electron density on the capacitor plate? (c) Explain why the curve for PFTE is lower than the curve for PVC, and why it is frequency insensitive in the range of 10^2 to 10^8 Hz.

1341 Calculate whether a capacitor using a 0.125 mm thick ribbon of "E" glass will have greater or lower capacitance than will another capacitor having the same area, but using a 110-μm film of nylon 6/6 (a) in a 60-Hz circuit, and (b) in a 1-kHz circuit.

1342 A quartz crystal ($E = 300$ GPa, or 43,000,000 psi) is piezoelectric and produces a polarization of 43.5 C/m² in an electric circuit. It is compressed with a stress of 1200 MPa (175,000 psi). (a) How much (in percentage) will the polarization be changed? (b) Will it be increased or decreased?

1343 A compressive stress of 1200 MPa is applied to a piezoelectric material. The polarization changes from 805 C/m² to 800 C/m². What is the elastic modulus for the material?

1344 Refer to Example 13–4.2. A 2.0-cm cube (8 cm³) of BaTiO₃ is compressed with a linear strain of 0.5 percent. The two ends receiving the pressure are connected electrically. How many electrons travel from the negative end to the positive end?

1345 Why can all ferroelectric materials be piezoelectric, whereas not all piezoelectric materials can be ferroelectric?

1346 A number of compounds have the same structure as does BaTiO₃. Among these is KNbO₃, with K^+ and Nb^{5+} as the cations. (a) Make a plausible sketch of the unit cell. (b) Assume that the dimensions of KNbO₃ and BaTiO₃ have the same lattice constants. What is the distance between the center of positive charge and the center of negative charge in each unit cell? (c) What is the dipole moment, $p = Qd$, of the unit cell? (d) What polarization is possible?

1347 Barium titanate is ferroelectric as well as piezoelectric. The polarization saturation, $\mathscr{P}_s$, is

0.16 C/m², as calculated in Example 13–4.2. After the electric field is partially reversed, depolarization is not complete; rather, $\mathscr{P}_r$ is equal to 0.14 C/m². (See Fig. 13–4.7.) What fraction of unit cells have realigned?

1348 BaTiO₃ is ferroelectric (and piezoelectric) below 120° C, but it is not ferroelectric above that temperature. Explain why the change occurs.

1349 The term, *ferroelectric* was coined because of the similarities between these material's hysteresis loops and ferromagnetic hysteresis loops. Identify the similarities. Also identify the differences between ferroelectric and ferromagnetic materials and between their properties.

1351 The index of refraction increases in proportion to the linear density of electrons along the path of the light. Vitreous silica (SiO₂ glass), with a density of 2.2 g/cm³, has an index of refraction of 1.46. Estimate the index of refraction for cristobalite (crystalline SiO₂) with a density of 2.32 g/cm³.

1352 An optic glass fiber ($n = 1.4581$) is to be clad with a second glass to ensure internal reflection that will contain all light traveling within 1° of the fiber axis. What maximum index of refraction is required for the cladding?

1353 Light is attenuated in glass by -0.15 db per km. How long can an optic fiber be and still retain 10 percent of its initial intensity?

1354 Visible light can produce electronic polarization in a material, but it cannot produce ionic polarization. Why is this statement true?

1361 (a) What is the minimum frequency of light required to supply all the necessary energy for an electron to be raised across the energy gap in gallium arsenide (GaAs)? (b) What is the wavelength? (c) Will visible light supply enough energy? (Draw on your knowledge of physics for the relationships among energy, frequency, and wavelength.)

1362 The relaxation time for a phosphor is 100 ms. How much time is required before the light intensity is 50 percent of its initial value? (b) What relaxation time would be required to have 50 percent intensity in 50 ms?

Chapter 14

PERFORMANCE OF MATERIALS IN SERVICE

Ideally, every product retains its initial characteristics after it enters service. However, service conditions may modify materials; and materials may fail unless their alteration is anticipated in the original design. Every material that fails in service has had its structure altered. For example, plastics may have been fractured by overload, aluminum alloys weakened by excessive heat, rubbers degraded by ultraviolet light, steels corroded by seawater, and so on. The scientist and engineer must identify and foresee the changes that can occur in service to ensure that these failures are avoided.

   In this chapter, we shall examine several types of failure: (1) failure by *corrosion;* (2) *delayed fracture,* commonly called *fatigue;* (3) failure at *elevated temperatures,* or by *reaction* with the surrounding environment; and (4) failure due to *radiation* that damages a material atom by atom, and electron by electron.

501

Of course, there are other failure modes, but these four will suffice to point out the correlation of performance with structural changes, and the resulting property modifications.

14–1

SERVICE PERFORMANCE

If engineers are to adapt materials and energy for society's needs, as we stated in Chapter 1, they must consider the performance of the product during a period of extended service. Materials in an automobile, a bridge, a computer, or a petroleum refinery must be selected for performance of 1 year, 10 years, or more, and not simply on the basis of design calculations for the newly delivered product. For example, the shaft shown in Fig. 14–1.1 performed for several years before it fractured. At no time was it under a load greater than that anticipated in design. Obviously, allowance was not made for extended service, during which time changes occurred within the material. Likewise, a cutting tool may be specified and manufactured with a desired microstructure of tempered martensite (Fig. 14–1.2a), and with the required hardness. However, the temporary absence of a cutting fluid, or the use of excessive cutting speeds, can produce heating at the cutting tip that changes the microstructure and therefore softens the tool. In turn, the tool tip dulls and fails (Fig. 14–1.2b). The total sequence can occur within seconds.

FIG. 14–1.1

Fatigue Fracture [14-cm (5½-in.) Steel Shaft]. Fracture progressed slowly from the set-screw hole at the top through nearly 90 percent of the cross-section before the final rapid fracture (bottom). (Courtesy of H. Mindlin, Battelle Memorial Institute.)

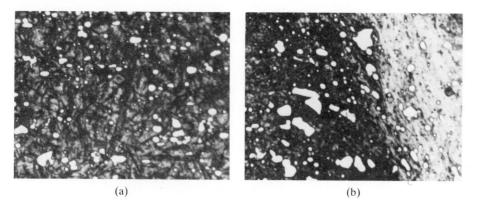

(a)   (b)

FIG. 14-1.2

Structural Change During Service (High-Speed Tool Steel, ×500). (a) Microstructure
before use. (b) Microstructure after cutting at excessive speeds. The temperature of the tip
increased sufficiently to permit overtempering and softening, which lead to failure.
(Courtesy of Crucible Research Center.)

Failure

Product failure need not necessarily be catastrophic, or involve fracture, leakage,
or excessive wear. Failure is simply the point at which *the product is no longer
capable of fulfilling its intended purpose.* Thus, failures range from the fracture of a
welded bridge beam to the dulling of a rear-view mirror in an automobile; from
potholes in streets to the oxidized contact points of a microswitch in a home
thermostat; and from a punctured tire to a faded color print. In some products,
failures are nearly instantaneous; the intended functions of other products are lost
gradually, extending over a period of months or years.

Since failure of a product or device occurs within the material, it is natural to
assume that there was a deficiency in that material. However, the failure of some
products arises from improper design judgments or unanticipated service conditions; in others, it results from misuse or from deficient maintenance. Insufficient
quality control during the manufacturing process also may be a factor leading to
failure. In any event, design and manufacturing engineers must anticipate the
effects of service conditions on their products, and on the materials that constitute
those products. The more the engineer knows about the nature of delayed failure,
the easier it is for her to control such failure, or to design around it.

We shall examine four types of delayed failure that arise from the alteration of
materials during service. They are failures that develop from chemical reactions,
from cyclic or continuing stresses, from elevated temperatures, and from radiation
damage.

FIG. 14–1.3

Stainless-Steel Alternative (1936 Car, with Original Body). Materials availability, international political factors, processing complications, and the customer's willingness to pay the cost differential all affect decisions concerning materials selection. (Courtesy of Allegheny Ludlum Steel Corp.)

Design Considerations

Many factors affect the design engineer's choice of material for a product. Let us examine several examples:

1. The strongest material may not withstand impact loads because high-strength materials generally are not tough materials. A tradeoff is necessary.

2. A material with an ultimate corrosion resistance (e.g., gold) is too expensive to use to solve the automotive corrosion problem. Stainless steel could solve the problem (Fig. 14–1.3) but it is not practical, because the local supply of the necessary chromium is too limited and the material must be imported. As a result, economic and political factors influence the choice of material.

3. Aluminum is less expensive than copper, and can be used as an electrical conductor, but it requires special techniques for splicing and terminal connectors to avoid high resistances and possible melting. Should aluminum conductors be used in residential products? In commercial installations?

4. Age-hardened aluminum is widely used in airplane construction because it has one of the highest strength-to-weight ratios. The engineer must know its overaging characteristics and the service conditions before choosing aluminum for a structural component of a space shuttle.

These examples, as well as all other technical designs, show that we must consider multiple factors before choosing a material. As noted, not all the decisions are based on technical considerations.

14–2
CORROSION REACTIONS

Corrosion is the process of surface deterioration of metals and related materials. The alteration occurs by the metal, M, losing electrons and becoming a positive ion:

$$M^0 \rightarrow M^{m+} + m\,e^- \qquad\qquad (14\text{–}2.1)$$

For the reaction to proceed, both the electrons and the metal ions must be removed.* The electrons are accepted by a nonmetallic element, N, or by other metallic ions, M^+:

$$N^0 + n\,e^- \rightarrow N^{n-} \qquad \text{(14–2.2a)}$$

or

$$M^{m+} + m\,e^- \rightarrow M^0 \qquad \text{(14–2.2b)}$$

The metal ions of Eq. (14–2.1) either dissolve in the surrounding electrolyte, or combine with nonmetallic ions to form a surface deposit.

Figure 14–2.1 shows the corrosion reaction schematically. The location of the electron-loss (oxidation) reaction is called the *anode*. That of the electron-gain (reduction) reaction is called the *cathode*. The two reactions may proceed adjacently, as in Fig. 14–2.1, or they may be separated by a considerable distance, provided they have a low-resistance electrical connection (for electron transfer). However, the two reactions must occur simultaneously if corrosion is to proceed. Neither can continue alone. Corrosion will stop if (1) the electrical connection is interrupted, (2) the cathode reactants are depleted, or (3) the anode products ($M^{m+} + m\,e^-$) are saturated. We shall examine these three considerations to increase our understanding of corrosion and its control.

Anode Reactions

All metals are subject to the oxidation reaction of Eq. (14–2.1). However, the tendency to ionize differs from metal to metal. For example, we can write Eq. (14–2.1) for zinc and copper:

$$Zn \longrightarrow Zn^{2+} + 2\,e^- \qquad \text{(14–2.1}_{Zn}\text{)}$$

and

$$Cu \rightarrow Cu^{2+} + 2\,e^- \qquad \text{(14–2.1}_{Cu}\text{)}$$

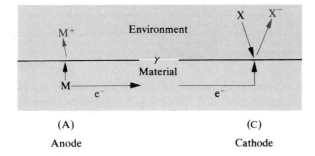

(A) Anode

(C) Cathode

FIG. 14–2.1

Corrosion (Schematic). (A) Anode (Eq. 14–2.1). (C) Cathode ($X + e^- \rightarrow X^-$). The material (commonly metal) must provide a continuous electrical path between the anode and cathode. The environment (commonly a liquid electrolyte) receives the corrosion product and supplies the reactant to the cathode.

---

* Otherwise, the reverse reaction becomes significant and equilibrium is established.

Both release electrons and lead to an electric potential, the value of which *cannot* be measured in isolation. However, a standard voltage difference between the two reactions can be measured when the metals are in 1-molar solutions of their own salts at 25° C (Fig. 14–2.2). In an open circuit (black switch), that voltage difference is 1.1 V. If electrical contact is made (color switch), the voltage difference is removed because electrons move from the zinc to the copper. (The "current" is in the opposite direction.) Thus, reaction ($14–2.1_{Zn}$) proceeds with zinc becoming the anode. Reaction ($14–2.1_{Cu}$) is reversed to reaction (14–2.2b) and copper atoms are reduced from the 1-molar solution of Cu^{2+} ions. Zinc is corroded at the anode; copper is plated onto the cathode.

We identify the electrodes as *anode* and *cathode* to avoid the confusion that accompanies the positive and negative labels for electrodes. The *anode supplies* electrons to the *external* circuit. Conversely, the *cathode receives* electrons from the *external* circuit. This definition is appropriate for corrosion, batteries, electroplating, cathode-ray tubes, and all other electronic devices in which current flows. All corrosion occurs at the *anode*. [Electroplating onto the cathode is "corrosion in reverse" (Example 14–2.3).]

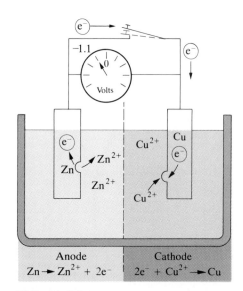

FIG. 14–2.2

Galvanic Cell (Zn–Cu). With the switch closed, zinc serves as an anode to provide electrons through the external circuit to copper. A 1.1-V potential difference develops when the circuit is opened. (Table 14–2.1, with molar solutions.)

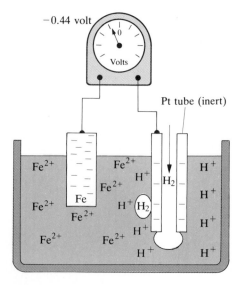

FIG. 14–2.3

Potential Difference, Fe Versus H_2. Iron produces a lower electron potential than does H_2 (see Table 14–2.1). Therefore, iron will be the anode and hydrogen the cathode when electrical contact is made. (Platinum is not the cathodic element because there are no Pt^{4+} ions present to receive electrons.)

The voltage differences could be measured between all possible pairs of metals in the manner just described. However, the more common practice is to use the $H_2 \rightarrow 2 H^+ + 2 e^-$ reaction as a reference for other anode (electron-releasing) reactions. As shown schematically in Fig. 14-2.3, iron is more reactive (more anodic) than is hydrogen to the extent that the potential difference is 0.44 V. It is arbitrary whether we call this difference $+0.44$ V, or -0.44 V; therefore, we will use -0.44 V for the reason cited in the footnote of Table 14-2.1.

The equations of Table 14-2.1 are for anode reactions; that is, reactions that release electrons. A reversal of any of these reactions is electron-consuming, and therefore is cathodic. Since neither type of reaction can occur in isolation, but must occur as an anode-cathode reaction pair, the equations of Table 14-2.1 are called *half-cell reactions*.

TABLE 14-2.1 Electrode Potentials (25°C; 1-molar solutions)

ANODE HALF-CELL REACTION*	ELECTRODE POTENTIAL USED BY ELECTROCHEMISTS AND CORROSION ENGINEERS,† V	
$Au \rightarrow Au^{3+} + 3 e^-$	$+1.50$	↑
$2 H_2O \rightarrow O_2 + 4 H^+ + 4 e^-$	$+1.23$	
$Pt \rightarrow Pt^{4+} + 4 e^-$	$+1.20$	Cathodic (noble)
$Ag \rightarrow Ag^+ + e^-$	$+0.80$	
$Fe^{2+} \rightarrow Fe^{3+} + e^-$	$+0.77$	
$4(OH)^- \rightarrow O_2 + 2 H_2O + 4 e^-$	$+0.40$	
$Cu \rightarrow Cu^{2+} + 2 e^-$	$+0.34$	
$H_2 \rightarrow 2 H^+ + 2 e^-$	0.000	Reference
$Pb \rightarrow Pb^{2+} + 2e^-$	-0.13	
$Sn \rightarrow Sn^{2+} + 2 e^-$	-0.14	
$Ni \rightarrow Ni^{2+} + 2 e^-$	-0.25	
$Fe \rightarrow Fe^{2+} + 2 e^-$	-0.44	
$Cr \rightarrow Cr^{2+} + 2 e^-$	-0.56	
$Zn \rightarrow Zn^{2+} + 2 e^-$	-0.76	Anodic (active)
$Al \rightarrow Al^{3+} + 3 e^-$	-1.66	
$Mg \rightarrow Mg^{2+} + 2 e^-$	-2.36	
$Na \rightarrow Na^+ + e^-$	-2.71	
$K \rightarrow K^+ + e^-$	-2.92	
$Li \rightarrow Li^+ + e^-$	-2.96	↓

* The arrows are reversed for the cathode half-cell reaction.

† The convention used by certain technical specialties is to interchange the $(+)$ and $(-)$ signs of these electrode potentials. (As noted in the text, the choice is arbitrary.) IUPAC recommends the convention used in this table.

As we noted, the standard electrode potentials of Table 14–2.1 require an electrolyte that is a 1-molar solution (25° C). The actual potential varies with concentration and temperature according to the *Nernst relationship,*

$$\phi = \phi 298^0 + (kT/n)(\ln C) \qquad \text{(14–2.3a)}$$

where $\phi 298^0$ is the standard electrode potential at 298 K (25 °C), as listed in Table 14–2.1. Concentration C is expressed as moles per liter (i.e., molar). Since the constant k is 86.1×10^{-6} eV/K, Eq. (14–2.3a) can be condensed to

$$\phi = \phi 298^0 + (0.0257/n)\ln C \qquad \text{(14–2.3b)}$$

for use at ambient temperatures (25° C or 298 K). The term n is the number of electrons released per atom. For example, $n = 3$ for $Al \rightarrow Al^{3+} + 3\ e^-$.

Most electrolytes that pertain to corrosion are dilute, with $C \ll 1$. Therefore, the values of Table 14–2.1 are commonly shifted in the anodic direction. For example, with $C_{Fe^{2+}} = 0.1$ M, ϕ is -0.47V, rather than -0.44V.

Cathode Reactions

The cathode reaction of Fig. 14–2.2 was the reduction of copper from the electrolyte containing a 1-molar Cu^{2+} solution. Of course, that reaction cannot occur if the electrolyte contains no copper ions. Under such conditions, the zinc (of Fig. 14–2.2) will still be able to corrode if the electrons can be consumed in other reactions. Possibilities include (1) other metals that are cathodic to zinc—that is, that are above zinc in Table 14–2.1; (2) evolution of hydrogen (since it is cathodic to zinc); (3) hydroxyl formation; and (4) water formation:

electroplating:	$M^{m+} + m\ e^- \rightarrow M^0$	(14–2.2b)
hydrogen evolution:	$2\ H^+ + 2\ e^- \rightarrow H_2 \uparrow$	(14–2.4)
hydroxyl formation:	$O_2 + 2\ H_2O + 4\ e^- \rightarrow 4(OH)^-$	(14–2.5)
water formation:	$O_2 + 4\ H^+ + 4\ e^- \rightarrow 2\ H_2O$	(14–2.6)

Each of these cathode half-cell reactions consumes electrons; each is the reverse of one of the anode half-cell reactions in Table 14–2.1. Based on Eq. (14–2.3), greater concentrations shift the electrode potential to become more cathodic. Thus, a more acid solution (higher hydrogen-ion concentrations) increases the potential difference between Eqs. (14–2.4 and 14–2.1$_{Zn}$) for the corrosion of zinc.

Probably the most ubiquitous cathode reaction is Eq. (14–2.5) that forms hydroxyl ions, because the reactants include oxygen and water. Under the majority of situations, special efforts are required to avoid the presence of these components. We readily see the corrosion product when iron corrodes and OH^- ions are formed, because the iron ions combine with the OH ions to produce $Fe(OH)_3$, which is ordinary red iron rust.

Galvanic Corrosion Cells

We can categorize corrosion couples, called galvanic cells, in three separate groups: (1) composition cells, (2) stress cells, and (3) concentration cells.

A *composition* cell may be established between any two *dissimilar* metals. In each case, the metal lower in the electromotive series, as listed in Table 14–2.1, acts as the anode. For example, on a sheet of *galvanized* steel (Fig. 14–2.4), the zinc coating acts as an anode and protects the underlying iron even if the surface is not completely covered, because the exposed iron is the cathode and does not corrode. Any corrosion that does occur is on the anodic zinc surface. As long as zinc remains, it provides protection to the adjacent exposed iron.

Conversely, a *tin* coating on sheet iron or steel provides protection only for as long as the surface of the steel is completely covered. If the surface coating is punctured, the tin becomes the cathode with respect to iron, which acts as the anode (Fig. 14–2.5). The galvanic couple that results produces corrosion of the iron. Since the small anodic area must supply electrons to a large cathode surface, very rapid localized corrosion can result.

Other examples of galvanic couples often encountered are (1) steel screws in brass marine hardware, (2) Pb–Sn solder around copper wire, (3) a steel propeller shaft in bronze bearings, and (4) steel pipe connected to copper plumbing. Each of these is a possible galvanic cell unless it is protected from a corrosive environment. Too many engineers fail to realize that the contact of dissimilar metals is a potential source of galvanic corrosion. In an actual engineering application, a brass bearing was once used on a hydraulic steering mechanism made of steel. Even in an oil environment, the steel acted as an anode and corroded sufficiently to permit leakage of oil through the close-fitting connection.

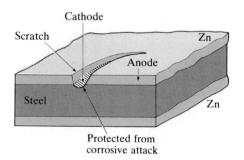

FIG. 14–2.4

Galvanized Steel (Cross-Section). Zinc serves as the anode; the iron of the steel serves as the cathode. Therefore, the iron is protected even though it is exposed where the zinc is scraped off.

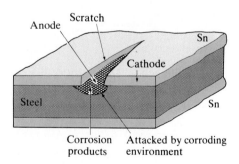

FIG. 14–2.5

Tinplate (Cross-Section). The tin protects the iron while the coating is continuous. When the coating is broken, the iron of the steel becomes the anode and is subject to accelerated corrosion.

Galvanic cells can be microscopic in dimension, because each phase has its individual composition and structure; therefore, each possesses its own electrode potential. As a result, galvanic cells can be set up in two-phase alloys when those metals are exposed to an electrolyte. For example, the pearlite of Fig. 7–3.2 reveals the carbide lamellae because the carbide was the anode in the electrolyte that was used as an etch.* Figure 14–2.6 shows the microstructure of an Al–Si casting alloy. Again, we depend on corrosion to reveal the two phases. One is the anode, the other is the cathode.

Heat treatment may affect the corrosion rate by altering the microstructure of the metal. Figure 14–2.7 shows the effect of tempering on the corrosion of a previously quenched steel. Prior to tempering reactions, the steel contains a single phase—martensite. The tempering of the martensite produces many galvanic cells and grain boundaries of ferrite and carbide, and the corrosion rate is increased. At higher temperatures, the coalescence of the carbides reduces the number of galvanic cells and the number of grain boundaries, which decreases the corrosion rate markedly.

When only a single phase is present, the corrosion rate of an age-hardenable aluminum alloy is low (Fig. 14–2.8), but the corrosion rate is significantly increased with precipitation of the second phase. Still greater agglomeration of the precipitate once again decreases the rate, but never to as low a level as in the single-phase alloy. The maximum corrosion rate occurs in the overaged alloy.

Stress cells do not involve compositional differences; rather, they involve dislocations, grain boundaries, and highly stressed regions. As shown in Fig. 4–1.9, where the grain boundaries had been etched (i.e., corroded), the atoms at the boundaries between the grains have an electrode potential different from that of

FIG. 14–2.6

Galvanic Microcells (Al–Si Alloy). Any two-phase alloy is more subject to corrosion than is a single-phase alloy. A two-phase alloy provides anodes and cathodes. (Courtesy of Aluminum Company of America.)

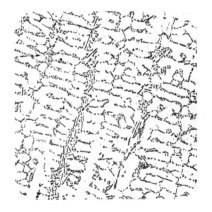

---

* The etch was 4 percent picral. The carbides are darkened because a corrosion reaction product remains on the surface. The electrode potentials of ferrite and carbide are sufficiently close that, with other electrolytes, their cathodic and anodic roles may be interchanged.

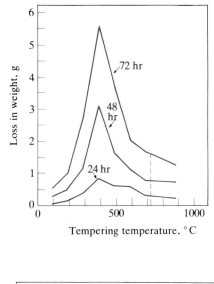

FIG. 14-2.7

Microcells and Corrosion. After quenching, only martensite exists. After intermediate-temperature tempering, many small galvanic cells exist as a result of the fine (α + carbide) structure in tempered martensite. After high-temperature tempering, the carbide is agglomerated and fewer galvanic cells are present. (Adapted from F. N. Speller, *Corrosion: Causes and Prevention*, McGraw-Hill.)

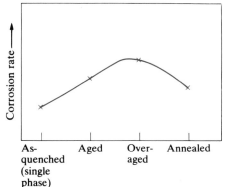

FIG. 14-2.8

Age Hardening and Corrosion (Schematic). The single-phase, quenched alloy has a lower corrosion rate than do the subsequent two-phase modifications.

the atoms within the grains; thus, an anode and a cathode have developed (Fig. 14-2.9). The grain-boundary zone may be considered to be stressed, since the atoms are not at their positions of lowest energy.

The effect of internal stress on corrosion also is evident after a metal has been *cold worked*. A simple example is shown in Fig. 14-2.10(a), where strain hardening exists at the bend of an otherwise annealed wire. The highly cold-worked metal serves as the anode, and the unchanged metal serves as the cathode.*

---

* *Corrosion in Action,* published by the International Nickel Company, demonstrates (with illustrative experiments) the effect of cold work on galvanic corrosion.

FIG. 14–2.9

Grain-Boundary Corrosion. The grain
boundaries served as the anode because the
boundary atoms have a higher energy. (See Fig.
4–1.9.)

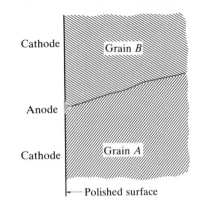

FIG. 14–2.10

Stress Cells. In these two
examples of strain hardening,
the anodes are in the more
highly cold-worked areas.
The electrode potential of a
deformed metal is more
anodic than is that of an
annealed metal."

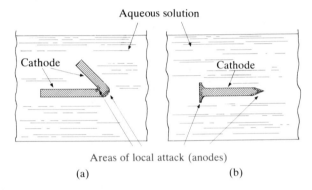

Areas of local attack (anodes)

(a)   (b)

The engineering importance of the effects of stress on corrosion is plain. When
engineering components must be used in a corrosive environment, the presence of
stress may significantly accelerate the corrosion rate.

Concentration cells arise from differences in electrolyte compositions. According to the Nernst equation (14–2.3), an electrode in a dilute electrolyte is anodic
with respect to a similar electrode in a concentrated electrolyte. We can view this
relation in terms of Fig. 14–2.11 and Eq. (14–2.7):

$$Cu^0 \underset{Conc.}{\overset{Dilute}{\rightleftarrows}} Cu^{2+} + 2\ e^-$$

$$(14\text{–}2.7)$$

The metal on side (D) of Fig. 14–2.11 is in the more dilute Cu^{2+} solution. Therefore, reaction (14–2.7) readily proceeds to the right. The metal on side (C) is in a
solution with a higher concentration of Cu^{2+}. Therefore, reaction (14–2.7) more
readily plates copper on that electrode. The electrode in the concentrated electrolyte is protected and becomes the cathode; the electrode in the dilute electrolyte
undergoes further corrosion and becomes the anode.

*The concentration cell accentuates corrosion, but it does so where the concentra-
tion of the electrolyte is lower.*

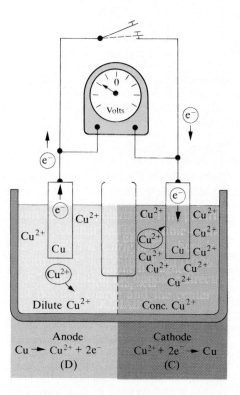

FIG. 14–2.11

Concentration Cell. When the electrolyte is not homogeneous, the less concentrated area becomes the anode.

Dilute Cu^{2+} Conc. Cu^{2+}

Anode	Cathode
Cu $\rightarrow$ Cu^{2+} + 2e$^-$	Cu^{2+} + 2e$^-$ $\rightarrow$ Cu
(D)	(C)

Concentration cells of this type are encountered frequently in chemical plants, and also under certain flow-corrosion conditions. However, in general, they are of less widespread importance than are *oxidation-type concentration cells.* When oxygen in the air has access to a moist metal surface, corrosion is promoted. However, the most marked corrosion occurs in the part of the cell with an oxygen deficiency.

We can explain this apparent anomaly on the basis of the reactions at the cathode surface, where electrons are consumed. We can restate Eq. (14–2.5) to indicate the role of O$_2$ (at the cathode) in promoting corrosion in oxygen-free (anode) areas:

$$2 \, H_2O + O_2 + 4 \, e^- \rightleftarrows 4(OH)^-$$

Since this cathode reaction, which requires the presence of oxygen, removes electrons from the metal, more electrons must be supplied by adjacent areas that do not have as much oxygen. The areas with less oxygen thus serve as anodes.

The oxidation cell accentuates corrosion, but it does so where the oxygen concentration is lower. This generalization is significant. Corrosion may be accelerated in apparently inaccessible places, such as in cracks or crevices, and under accumulations of dirt or other surface contaminations (Fig. 14–2.12), because these oxygen-deficient areas serve as anodes.

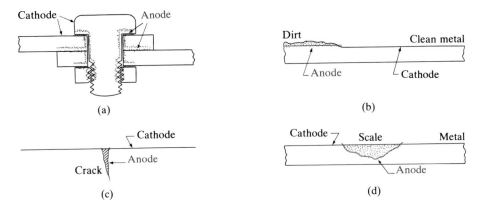

FIG. 14–2.12

Oxidation Cells. Inaccessible locations with low oxygen concentrations become anodic. This situation arises because the mobility of electrons and of the metal ions is greater than that of oxygen or oxygen ions.

Local Corrosion

Corrosion seldom progresses uniformly into a metal surface. Rather, it acts locally. In part, this local action is due to microscopic-sized anodes and cathodes in what appears to be a homogeneous material. For example, the grain boundary is anodic to the grain proper (Fig. 14–2.9). Likewise, carbides and ferrite form microscopic anode–cathode pairs within an otherwise uniform steel product.

These initial heterogeneities can lead to accelerated corrosion. If the microspot that temporarily served as an anode becomes coated with an insoluble corrosion product, that local area now has less oxygen. As a result, it remains as the anode and is corroded further, establishing a permanent corrosion pit that "bores" its way into a metal (Fig. 14–2.13). The more exposed areas become the cathode. *Pit corrosion* of this origin is a common mode of failure in sheet-aluminum products, because the holes eventually penetrate the full thickness of the metal.

The examples of Fig. 14–2.12 also illustrate localized corrosion, because an oxidation cell develops its anode in the regions that are inaccessible to O_2. In this condition, called *crevice corrosion*, the smaller metal ions can diffuse out of the crevice to a surface to combine with the oxide (or OH^- ions). Most commonly, the rust spots that we see on car fenders that have not been otherwise damaged originate in this manner. Often, they start on the underside of the fender adjacent to weld spots or metal bends that collect dirt and locally exclude oxygen.

Example 14–2.1

Determine the electrode potential (with respect to hydrogen) for a chromium electrode in a solution containing 2 g of Cr^{2+} ions per liter (25° C).

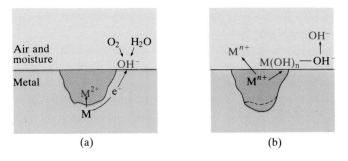

FIG. 14-2.13

Pit Corrosion (Schematic). If initiated by any type of nonuniformity, the corroded spot becomes more anodic. (a) Electrons can move to the available oxygen and water (Eq. 14-2.5). (b) The corrosion continues in the oxygen-free area. The metal ions and the hydroxyl ions either diffuse into the electrolyte or combine to form a hydroxide (e.g., iron rust). The debris-filled pit continues to deepen by localized corrosion.

Procedure   Table 14-2.1 applies to 1-molar electrolyte at 25° C. This solution has (2/52.0) moles of Cr/liter. We use the Nernst equation (14-2.3) to adjust for the concentration.

Calculation

$$\phi = -0.56 \text{ V} + (0.0257/n)\ln (2/52.0)$$
$$= -0.60 \text{ V}$$

Example 14-2.2

You are informed that the voltage difference is 1.2 V when silver and cadmium are tested in a standard cell similar to that in Fig. 14-2.2 (1-molar solutions of Ag^{2+} and Cd^{2+}). You do not know whether the value is +1.2 or -1.2 V. Where should the $Cd \rightarrow Cd^{2+} + 2 e^-$ half-cell reaction be placed in Table 14-2.1?

Analysis   Since $Ag \rightarrow Ag^{2+} + 2 e^-$ is +0.80 V, the $Cd \rightarrow Cd^{2+} + 2 e^-$ reaction must be

$$+0.80 + 1.20 = +2.00 \text{ V} \quad \leftarrow or \rightarrow \quad +0.80 - 1.20 = -0.40 \text{ V}$$

The +2.0 value would make cadmium more noble than is gold (+1.5 V), an unexpected result. The -0.4 V value appears more realistic. An alternate check would be to couple Cd with another metal, say nickel, and to see whether the standard potential difference is 0.15 V (or 2.25 V).

Example 14-2.3

Electroplating is described as "corrosion in reverse." How thick will a nickel-plated layer be if a current density of 500 A/m² is used for 1 hr on a 16 cm × 43 cm part?

Procedure Determine the charge and the number of electrons involved. From this, calculate the g per hr of nickel, and the thickness. Use 1 cm^2 as the basis.

Calculation

$$(500 \text{ A/m}^2)(3600 \text{ sec})/(10^4 \text{ cm}^2/\text{m}^2) = 180 \text{ C/cm}^2$$
$$(180 \text{ A} \cdot \text{s/cm}^2)(1 \text{ Ni}^{2+}/2 \text{ el})/(0.16 \times 10^{-18} \text{ A} \cdot \text{s/el}) = 5.6 \times 10^{20} \text{ Ni}^{2+}/\text{cm}^2$$
$$(5.6 \times 10^{20} \text{ Ni}^{2+}/\text{cm}^2)(58.7 \text{ g}/0.6 \times 10^{24} \text{ Ni}) = 0.055 \text{ g/cm}^2$$
$$(0.055 \text{ g/cm}^2)/(8.9 \text{ g/cm}^3) = 0.006 \text{ cm} \qquad \text{(or 60 } \mu\text{m)}$$

Comment The 16 cm $\times$ 43 cm surface will receive 38 g of Ni.

14–3
CORROSION CONTROL

We can minimize corrosion by isolating the metal surface from its environment. A layer of paint, a nickel-plated surface, and a vitreous enamel coating separate the metal from the electrolyte that supplies the cathode reactants and receives the anode products (Fig. 14–2.1). The choice of protective coatings depends on the application, as summarized in Table 14–3.1.

Corrosion can also be restricted if galvanic couples are avoided. It is the responsibility of the design engineer to ascertain that unlike metals are not in contact in places where corrosion is critical. Alternatively, unlike metals, which of course have different electrode potentials, may be insulated from each other. For example, residential plumbing codes commonly require that a teflon $+ C_2F_4 +_n$ washer be used in the coupling where copper plumbing joins steel pipes, in order to separate the two metals.

TABLE 14–3.1 Comparison of Inert Protective Coatings

TYPE	EXAMPLE	ADVANTAGES	DISADVANTAGES
Organic	Baked "enamel" paints	Flexible Easily applied Cheap	Oxidizes Soft (relatively) Temperature limitations
Metal	Noble metal electro-plates	Deformable Insoluble in organic solutions Thermally conductive	Establishes galvanic cell if ruptured
Ceramic	Vitreous enamel, oxide coatings	Temperature resistant Harder Does not produce cell with base	Brittle Thermal insulators

Passive Surface Films

Corrosion rates are typically more rapid for those metals at the anodic end of the electrode potential scale (Table 14–2.1), particularly if they are coupled with a cathode reaction at the other end of the scale. However, there are major exceptions. Aluminum, for example, appears to be relatively inert to many service conditions. We even use it for boat hulls. Also, stainless steels are "stainless" because of their chromium content. Yet, Cr, like Al, is more anodic in Table 14–2.1 than is iron.

When aluminum is the metal of Fig. 14–2.1, it readily forms Al^{3+} ions. Any oxygen in the environment first reacts with the released electrons to produce O^{2-} ions; also the Al^{3+} and O^{2-} ions react with each other at the metal surface to form a thin, invisible surface layer of Al_2O_3. The Al_2O_3 is protective because it has tightly bonded, stable Al^{3+}-to-O^{2-} bonds, and also the Al_2O_3 is *coherent* with the underlying metal. That is, the crystal structure of the Al_2O_3 nearly matches the {111} planes in the aluminum (Fig. 14–3.1). An Al_2O_3 film not only is protective, but also reforms almost immediately if it is ruptured. This *passive* (nonreactive) film loses its protection in alkaline solutions that dissolve Al_2O_3.

Anodizing is a commercial process that uses the electrical current to force the formation of a thicker Al_2O_3 film than occurs naturally. It can be built up to provide protection from saline solutions, such as seawater, which can destroy the thinner, natural Al_2O_3 film. Furthermore, when the appropriate ions are introduced into the processing electrolyte, the Al_2O_3 (sapphire) film can be colored blue, red, or other color, to meet any appearance requirements.

Stainless steels are stainless because their surfaces are passivated by a thin chromium oxide film. In Table 14–2.1, the chromium (12 to 25 percent of the stainless steel) reacts to form Cr ions before the iron reacts. If oxygen is present, it immediately claims the released electrons and forms an extremely thin ($\sim$ monolayer) chromium-oxide film that halts all further reaction. The reaction, called *passivation,* gives a high degree of protection.

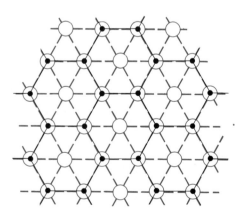

FIG. 14–3.1

Al–Al_2O_3 Coherency. Color—pattern of aluminum on {111} planes of the metal. Black—pattern of aluminum on the (0001) plane of the oxide. The Al-to-Al distances differ only slightly. Therefore, the boundary atoms can be part of both structures.

If, however, the service conditions are reducing and no oxygen is present, this passive layer can be destroyed. Without that protection, the steel becomes active and the corrosion of Cr-bearing steels proceeds rapidly. Table 14–3.2 presents a *galvanic series* of common alloys that indicates the alloys' relative position in the cathode–anode spectrum. Observe that the Cr-containing steels are listed twice, (A) in their activated and (P) in their passivated condition. Obviously, the designer must anticipate the possibility that these stainless steels may encounter reducing (low-oxygen) conditions in service. When this occurs, the steels are no longer stainless.

Inhibitors provide another variant of a passive surface film. Best known of these are the rust inhibitors that are used in automobile radiators (commonly as an additive with the antifreeze). These inhibitors contain chromate or similar highly oxidized ions that are adsorbed onto the metal surface. The protection received is similar to that afforded by the passive film on stainless steel, except for the fact that it cannot be regenerated from the metal. New inhibitor must be added if the film is flushed away.

TABLE 14–3.2 Galvanic Series of Common Alloys*

Graphite	Cathodic	Nickel—A
Silver	↑	Tin
12% Ni, 18% Cr, 3% Mo steel—P		Lead
20% Ni, 25% Cr steel—P		Lead-tin solder
23 to 30% Cr steel—P		12% Ni, 18% Cr, 3% Mo steel—A
14% Ni, 23% Cr steel—P		20% Ni, 25% Cr steel—A
8% Ni, 18% Cr steel—P		14% Ni, 23% Cr steel—A
7% Ni, 17% Cr steel—P		8% Ni, 18% Cr steel—A
16% to 18% Cr steel—P		7% Ni, 17% Cr steel—A
12 to 14% Cr steel—P		Ni-resist
80% Ni, 20% Cr—P		23 to 30% Cr steel—A
Inconel—P		16 to 18% Cr steel—A
60% Ni, 15% Cr—P		12 to 14% Cr steel—A
Nickel—P		4 to 6% Cr steel—A
Monel metal		Cast iron
Copper–nickel		Copper steel
Nickel–silver		Carbon steel
Bronzes		Aluminum alloy 2017-T
Copper		Cadmium
Brasses		Aluminum, 1100
80% Ni, 20% Cr—A		Zinc
Inconel—A	↓	Magnesium alloys
60% Ni, 15% Cr—A	Anodic	Magnesium

* Adapted from C. A. Zapffe, *Stainless Steels,* American Society for Metals. A—active; P—passivated.

Galvanic Protection

We can achieve service protection against corrosion in some applications by arranging for the product to be the cathode. The galvanized steel of Fig. 14–2.4 provided a preview for us. That steel was coated with zinc, which is anodic to iron. Therefore, the zinc anode corrodes, and the steel, being the cathode, is protected. Several adaptations of these *sacrificial anodes* are shown in Fig. 14–3.2. They can be replaced after the anode has been spent.

Galvanic protection also can be provided with a dc current that feeds electrons into the metal that must be protected against corrosion (Fig. 14–3.3). In a pipe-line, for example, the *impressed current* stops, and even reverses, the corrosion reaction,

$$M^0 \rightleftarrows M^{n+} + n\,e^- \qquad\qquad (14\text{–}2.1)$$

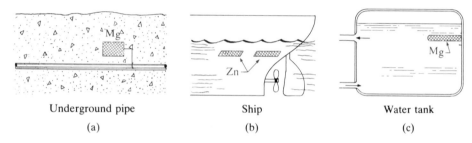

| Underground pipe | Ship | Water tank |
| (a) | (b) | (c) |

FIG. 14–3.2

Sacrificial Anodes. (a) Buried magnesium plates along a pipeline. (b) Zinc plates on ship hulls. (c) Magnesium bar in an industrial hot-water tank. Each of these sacrificial anodes can be replaced easily. They cause the *equipment* to become a cathode.

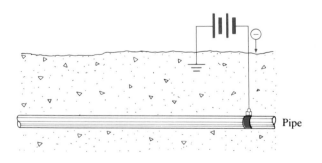

Pipe

FIG. 14–3.3

Impressed Voltage. A small dc voltage will provide sufficient electrons to make the equipment a cathode.

Example 14–3.1

Aluminum oxide, Al_2O_3, is hexagonal with the Al^{3+} ions spaced 0.267 mm apart. (See Fig. 14–3.1). Compare this distance with these center-to-center distances on the {111} planes of metallic aluminum. For fcc Al, $a = 0.4049$ nm.

Procedure Neighboring atoms "touch" on the {111} planes of an fcc metal (Fig. 3–7.7). Thus,

$$d = 2 R = a\sqrt{2}/2$$
$$= 0.286 \text{ nm}$$
$$\Delta d/d = (0.286 - 0.267 \text{ nm})/0.286 \text{ nm}$$
$$= 0.066$$

Comment With less than 7 percent difference, the aluminum atoms at the Al_2O_3–Al interface can be common to both structures. This coherency provides a tight bond between the two phases.

14–4
DELAYED FRACTURE

There are many documented examples of eventual failure by fracture of equipment that had previously performed satisfactorily for long periods of time. Figure 14–1.1 showed one such fracture of a rotating shaft that underwent cyclic stressing; static loading also can lead to delayed fracturing in moist or otherwise reactive environments.

Delayed fracture resulting from extended service is called *fatigue,* because people formerly assumed that the material got "tired" and failed through "weariness." We now know that fatigue fracturing progresses through a material via changes within the material at the tip of a crack, where there is a high stress intensity (Section 8–4). We shall look at several situations: (1) cyclic fatigue, (2) fatigue of plastics, and (3) stress corrosion and static fatigue. The last two are closely related in concept.

Cyclic Fatigue

Many mechanical designs require cyclic loading. Commonly, the stresses cycle from tension to compression, as occurs in a loaded rotating shaft (Figs. 14–4.1a and 14–4.2a). Fatigue also can occur with fluctuating stresses of the same sign, as occur in a leaf spring or in a diving board (Fig. 14–4.1b).

Figure 14–4.2 shows three examples of cyclic loading. The axle of a train has many sinusoidal stress cycles (tension to compression). Examples of low-cycle stresses are found in the rotor of a generator that is used as a "topping" unit to meet peak demand for electricity. Likewise, an airplane fuselage has tension stresses

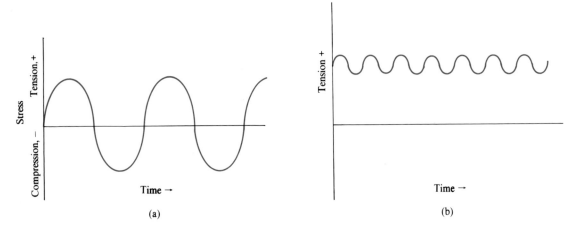

FIG. 14-4.1

Cyclic Loading. (a) Tension–compression (+ to −). (b) Tension–tension (+ to +).
Design stresses must be lower than are those for static loading.

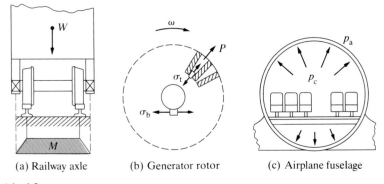

(a) Railway axle   (b) Generator rotor   (c) Airplane fuselage

FIG. 14-4.2

Examples of Cyclic Loading. (a) Axle of rail car. (b) Rotor of generator during starting
and stopping. (c) Pressurization and depressurization of airplane. The latter may
encounter only once every few thousand cycles; however, as indicated by Fig. 14-4.3,
the yield strength cannot be used by designers. (Courtesy of *ASTM*, R. E. Peterson,
"Fatigue in Metals," *Materials Research and Standards.*)

imposed during pressurization following each takeoff; these stresses are relaxed
each time the plane returns to the ground, producing tension-to-tension cycling.
These two examples may involve only a few cycles per day; however, during the
period of a few years of service, that number can significantly influence engineering design considerations (Fig. 14-4.3).

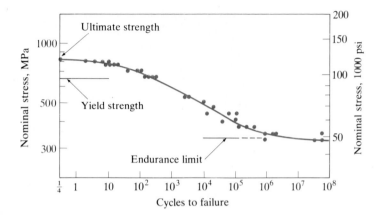

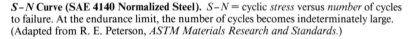

FIG. 14–4.3

S–N **Curve (SAE 4140 Normalized Steel).** *S–N* = cyclic *stress* versus *number* of cycles to failure. At the endurance limit, the number of cycles becomes indeterminately large. (Adapted from R. E. Peterson, *ASTM Materials Research and Standards.*)

The number of cycles, *N*, that a material will endure decreases with increased stress, *S*, particularly when the stresses involve both tension and compression. Figure 14–4.3 presents a typical *S–N* curve for fatigue fracture of a steel. (See Fig. 14–4.6 for curves of two plastics.) The yield and ultimate strengths that we considered in Chapters 8 and 9 can be used as a guide in design only for structures that are in service under static loading. For cyclic loading, the allowable stresses must be reduced, as shown in Fig. 14–4.3. Some materials possess a *fatigue limit,* or *endurance limit,* below which permissible cycling is nearly unrestricted. It is important to note that other materials, such as nonferrous metals, do not possess such a limit.

Fatigue cracks start at points of high stress intensity. This was the case in the shaft of Fig. 14–1.1, where the crack was initiated at a hole for a set screw. Such high-stress points must be avoided in good engineering design. Avoiding them may involve the use of generous fillets (Fig. 14–4.4). Finite element analysis is now a helpful tool in designing machine components. However, in critical situations, even the surface finish must be considered, as indicated by the data in Table 14–4.1.

The precrack nucleation of fatigue reveals that microscopic and irreversible slip occurs within individual metal grains. This slip causes *extrusions* and *intrusions* on the external surface of the grains (Fig. 14–4.5). There is a gradual reduction in the ductility along these slip planes, causing microscopic cracks to form early in the cycling process. The cracks progress slowly but steadily during the remaining cycles. Eventually, a crack exceeds the critical size to raise the stress intensity to K_{Ic}, and final catastrophic failure occurs (Fig. 14–1.1). Typically, the crack-growth period accounts for 75 to 90 percent of the fatigue life of the part. The precrack nucleation period is relatively short.

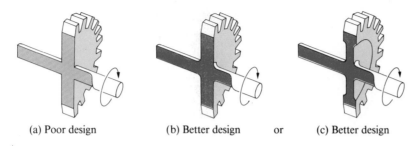

    (a) Poor design   (b) Better design   or   (c) Better design

FIG. 14–4.4

Design of Fillet. The use of generous fillets is recommended in mechanical engineering design. Observe that (c) is a better design than (a), even with some additional material removed. Of course, if too much metal is removed, failure may occur by mechanisms other than fatigue.

Fatigue of Plastics

Stress-cycled polymers, like metals, exhibit a reduced strength. For polymethyl methacrylate (PMM of Table 2–3.1), there is an endurance limit (Fig. 14–4.6); however, nylon-6 and most unreinforced amorphous polymers exhibit no such limit. For them, design calculations are based on the *fatigue strength,* which is the stress corresponding to the number of loading cycles during the expected lifetime of the plastic.

    The viscoelasticity of a polymer (Section 10–1) complicates that polymer's fatigue characteristics, because there can be incremental flow during each cycle. Therefore, *hysteretic heating* develops if the stress-relaxation time is close to the cycling period.* Since polymers are poor thermal conductors, the heat is not

TABLE 14–4.1 Surface Finish Versus Endurance Limit (SAE 4063 Steel, Quenched and Tempered to 44R$_C$)*

TYPE OF FINISH	SURFACE ROUGHNESS		ENDURANCE LIMIT	
	μm	μin.	MPa	psi
Circumferential grind	0.4–0.6	16–25	630	91,300
Machine lapped	0.3–0.5	12–20	720	104,700
Longitudinal grind	0.2–0.3	8–12	770	112,000
Superfinished (polished)	0.08–0.15	3–6	785	114,000
Superfinished (polished)	0.01–0.05	0.5–2	805	116,750

* Adapted from M. F. Garwood, H. H. Zurburg, and M. A. Erickson, "Correlation of Laboratory Tests and Service Performance," *Interpretation of Tests and Correlation with Service,* Amer. Soc. Metals.

* The same principle operates microwave ovens, except that the frequency is higher.

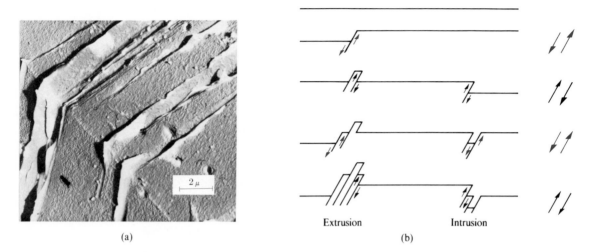

(a)   (b)

FIG. 14–4.5

Intrusions and Extrusions (×4500). (a) Electron-microscope replica after cyclic
straining. (A. H. Cottrell and D. Hull, *Proc. Royal Soc.* (London), **A242,** with permission.) (b) Formation of intrusions and extrusions. Progressive irreversible slip by cyclic
shear stresses can form surface irregularities that produce stress concentrations to
nucleate cracks. (L. H. Van Vlack, *Materials Science for Engineers,* Addison-Wesley,
Reading, Mass., with permission.)

readily dissipated and softening occurs. Strength and rigidity are therefore reduced.

Because of these complications, fiber reinforcement is almost mandatory when
plastics are used in applications with cyclic stresses. Although microcracking (Fig.
10–4.5) can be progressive, composites are more likely to exhibit an endurance
limit than are nonreinforced plastics. Also, the greater the elastic modulus of the
reinforcing fiber, the more nearly the fatigue strength approaches the static tensile
strength. These two factors provide a more rational basis for design calculations.

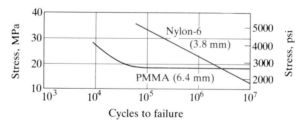

FIG. 14–4.6

S–N **Curves (Nylon-6 and Polymethylmethacrylate).** Design specifications must call for
lower stresses (S) when a greater number (N) of stress reversals are encountered.
(Adapted from M. N. Riddell et al., *Polymer Science and Engineering.*)

Stress Corrosion and Static Fatigue

All materials may exhibit delayed fracture in a corrosive environment. Conversely, materials corrode more rapidly when stressed. Figure 14–4.7 shows the rupture time of a plastic as a function of stress, with and without the presence of an organic solvent. Glass is subject to *static fatigue,* in which delayed fracture may occur at constant loads in the presence of water vapor. Finally, *stress corrosion* is well known in metals that are loaded by tension in corrosive environments (Fig. 14–4.8).

Stress corrosion proceeds through a metal as nonductile cracks (even in ductile metals). Higher stresses shorten the time for fracture (Fig. 14–4.9). The corrosive

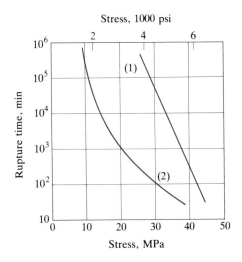

FIG. 14–4.7

Time for Failure (Polymers, Typical). Curve (1): rupture time in a dry environment; curve (2): rupture time in an organic solvent. (Adapted from Alfrey and Gurnee, *Organic Polymers,* Prentice-Hall.)

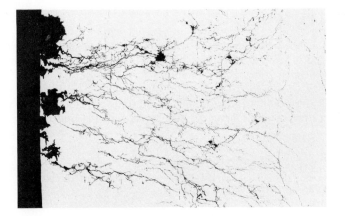

FIG. 14–4.8

Stress-Corrosion Cracking (316 Stainless Steel, Boiling NH₄Cl). This type of failure is common when metals are stressed in the presence of an electrolyte. (×150.) (Courtesy of A. W. Loginow, U.S. Steel Corp.)

FIG. 14–4.9

Stress Corrosion (Metal in Electrolyte). Failure is accelerated at higher stresses. The testing procedure is shown in Fig. 14–4.10(b).

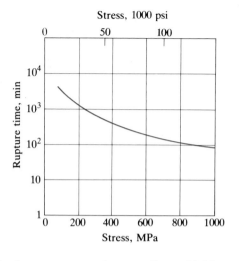

environments that lead to stress corrosion usually are highly specific; for example, electrolytes with Cl^-, NO_3^-, and OH^- ions generally produce fracture across the grains of the metal (Fig. 14–4.8).

The mechanism of stress corrosion is not fully understood. One possibility is that corrosion is extremely rapid at the tip of a flaw, where the stress intensity factor is high. Thus, the crack penetrates faster than it would if the tip were blunted by plastic deformation.

The fracture toughness, K_{Ic}, of a metal drops with time in a corrosive medium, as shown in Fig. 14–4.10(a). The long-term value is labeled K_{Iscc}, the stress-intensity factor for **Stress Corrosion Cracking**. Figure 14–4.10(b) shows schematically the procedure for simultaneously testing of corrosion and fracture toughness.

Design engineers must specify stress levels far below the static yield strength if their products are to be used in corrosive environments. Naturally, the sensitivity to stress corrosion varies from material to material.

Example 14–4.1

The fracture toughness for the steel of Example 8–4.2 drops by 40 percent when the material is tested in seawater. (See K_{Iscc} in Fig. 14–4.10(a).) What crack length can be tolerated, with yielding as the mode of failure?

Procedure The stress cannot exceed 1100 MPa; however, any crack must be short enough to stay under a stress intensity factor of 54 MPa·√m—that is, under 60 percent of K_{Ic}.

Calculation

$$c\pi = [K_{Iscc}/(S_y)\alpha]^2$$
$$c = [54 \text{ MPa} \cdot \sqrt{m}/(1100 \text{ MPa})(1.1)]^2/\pi$$
$$= 0.6 \text{ mm}$$

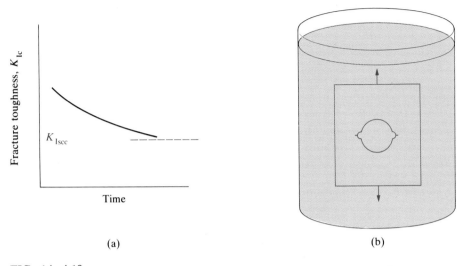

(a)

(b)

FIG. 14–4.10

Fracture Toughness (in an Electrolyte). (a) The value of K_I drops with time in the electrolyte (schematic). The design limit is K_{Iscc}, the critical stress intensity factor for stress corrosion cracking. (b) Test method (schematic). Cracks are initiated to known depths before loading the specimen in the chosen electrolyte.

14–5
PERFORMANCE OF METALS AT HIGH TEMPERATURES

In general, materials deform more readily with increased temperatures. They do so because plastic deformation proceeds most commonly through dislocation movements that involve the sequential displacements of atoms to new neighbors (Fig. 8–3.4). These movements are accommodated by thermal energy, and lead to reduced strength.

In this section, we shall consider an *elevated temperature* as a temperature at which a material undergoes structural changes during service. Of course, structural changes affect properties and require design considerations. Since different materials respond differently to heat, an elevated temperature for one material may not affect another material. In addition, *time* is a factor in structural change. Recrystallization (Section 9–3) provided an example of time–temperature relationships. A strain-hardened metal should not be specified if the service conditions are to be above the recrystallization temperature.

Property deterioration can occur at elevated temperatures, with a variety of structural responses, including creep, oxidation, and spalling, as well as overaging (Section 9–4), and overtempering (Figs. 9–6.7 and 14–1.2).

Creep

At *low temperatures,* fine-grained metals generally are stronger than are coarser-grained ones, because grain boundaries interfere with the progression of the dislocation movements. At elevated temperatures, the situation is reversed. Above temperatures at which atom diffusion is significant, the grain boundaries are a source of weakness in a material. We can understand this better by examining Fig. 14–5.1, which shows schematically several grains loaded vertically in tension. With tension in one direction, there is a perpendicular contraction. (See Poisson's ratio, Section 8–1.) Thus, atoms along vertically oriented boundaries are crowded; atoms along horizontally oriented boundaries have an increased amount of space. This situation induces diffusion from the vertical boundaries to the horizontal boundaries, with the net effect that there is a gradual change in the shape of the metal.*

The strain just described is one of several mechanisms of *creep.* With smaller grains and, therefore, more grain-boundary area, creep is more rapid. There are more "sinks" for atoms along horizontal boundaries, and more "sources" of atoms from vertical boundaries (Fig. 14–5.1). Equally important is that the diffusion distances are shorter in fine-grained materials.

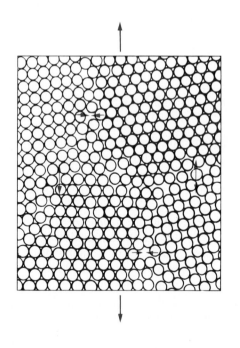

FIG. 14–5.1

Grain Boundaries and Deformation (Schematic). Under tension, vertical boundaries are crowded and lose atoms by diffusion; the horizontal boundaries receive relocated atoms because they develop extra space. The resulting creep becomes a significant factor in the use of materials at high temperatures, where the atoms diffuse readily. (Base sketch from Fig. 4–1.7.)

---

* Compressive forces induce similar, but opposite, shape changes.

Of course, this mechanism of creep does not occur at low temperatures where there are negligible atom movements, but it does increase exponentially with the increases in the values of diffusivity, D, for self-diffusion (Section 6–5). As with recrystallization, the temperature for the reversal of these grain-size effects is a function of time, bond strength, impurities, and so on. Controlling grain size and shape is a means of minimizing creep (Fig. 14–5.2).

As implied by its name, creep is a slow mechanism of strain. The rates range from a few percent per hour at high stress levels, or at high temperatures, down to less than 10^{-4} %/hr* (Fig. 14–5.3). The latter rate may seem insignificant; however, consider its importance in a design for a steam power plant or nuclear

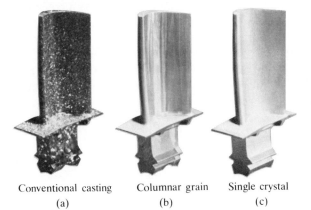

Conventional casting   Columnar grain   Single crystal

(a)   (b)   (c)

FIG. 14–5.2

Advances in Turbine Airfoil Materials (Cast Turbine Blades). (a) Since grain boundaries permit strain at high temperatures by creep, conventional processing specifies coarse grains (and reduced grain-boundary area). (b) This more sophisticated process not only directionally solidifies the metal, but also lowers the area of the boundaries further (and aligns them parallel to the centrifugal forces). (c) Advanced (and more difficult) procedures eliminate all grain boundaries by producing a blade out of a single crystal of metal. (Courtesy of M. Gell, D. H. Duhl, and A. F. Giamei, Pratt and Whitney Aircraft, *Superalloys 1980*, ASM.)

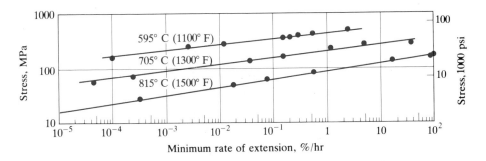

FIG. 14–5.3

Creep Data (Type 316 Stainless Steel). Higher stresses and higher temperatures increase the creep rate. (From Shank, in McClintock and Argon, *Mechanical Behavior of Metals*, Addison-Wesley, Reading, Mass., with permission.)

---

* ~1 percent/yr.

reactor, which must be in high-temperature service for many years. Likewise, creep becomes important in gas turbines and in equipment that must be operated without change in dimensions at high stresses and elevated temperature to maximize energy-conversion efficiencies.

We can plot strain as a function of time for various stresses and temperatures. Figure 14–5.4 is schematic. When a material is stressed, it undergoes immediate elastic deformation, which is greater when either the stress or the temperature is high. In the first short period (stage 1), the material makes additional, relatively rapid, plastic adjustments at points of stress concentration along grain boundaries and at internal flaws. These initial plastic adjustments give way to a slow, nearly steady rate of strain that we define as the *creep rate ė*. This second stage of *steady creep* continues over an extended period, until sufficient strain has developed that necking down and area reduction starts. With this change in area at a constant load, the rate of strain accelerates (stage 3) until terminated by rupture. If the load could be adjusted to match the reduction in area, and thus a constant stress could be maintained, the creep rate of stage 2 would continue until rupture.

In Fig. 14–5.4, the following relationships are shown schematically: (1) the steady creep rate increases with both increased temperature and stress, (2) the total strain at rupture also increases with these variables, and (3) the time before eventual failure by *stress rupture* is decreased as the temperatures and applied stresses are increased (Fig. 14–5.5).

Oxidation of Metals

Although metals oxidize at all temperatures, the rate of oxidation increases from negligible at ambient temperatures to rapid at elevated temperatures. In simplest terms, the oxidation reaction is

$$\text{metal} + O_2 \rightarrow \text{metal oxide} \tag{14–5.1}$$

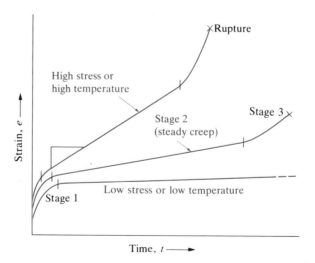

FIG. 14–5.4

Creep. The steady rate of creep in the second stage determines the useful life of the material.

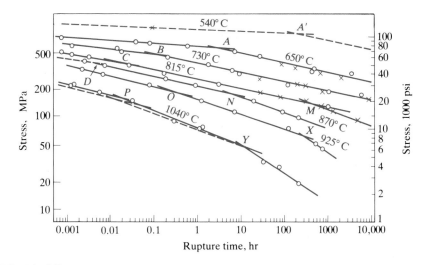

FIG. 14–5.5

Stress-Rupture Data. The time to rupture is shorter with higher stresses and higher temperatures. (After N. J. Grant, American Society for Metals.)

The actual reaction is somewhat more complex, because an oxide scale soon separates the metal from the air that supplies the oxygen. For oxidation to continue, either the metal must diffuse outward or the oxygen must diffuse inward through the scale. This mechanism is illustrated in Fig. 14–5.6 for iron. At high temperatures, the first oxygen molecules react to give

$$2Fe + O_2 \rightarrow 2FeO \qquad (14\text{–}5.2)$$

Additional reaction requires that Fe^{2+} ions diffuse through the scale to meet the oxygen. This diffusion occurs much more rapidly than does any possible diffusion of oxygen in the opposite direction, because (1) the Fe^{2+} ions ($r = 0.074$ nm) are much smaller than are the O_2 molecules (or for O^{2-} ions, $R = 0.140$ nm); and (2) FeO is iron deficient—specifically, $Fe_{1-x}O$—with an Fe/O ratio of 0.96 in the

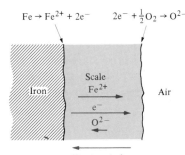

FIG. 14–5.6

Scaling Mechanism. Fe^{2+} ions and electrons diffuse more readily through the scale than do O^{2-} ions. As a result, the reaction $Fe^{2+} + O^{2-} \rightarrow FeO$ is predominant at the scale–air interface. (L. H. Van Vlack, *Materials for Engineers: Concepts and Applications,* Addison-Wesley, Reading, Mass., with permission.)

scale adjacent to the metal (Fig. 4–5.2). This deficiency introduces Fe vacancies to permit easy diffusion of the Fe^{2+} ions. Electrons must go through the scale, too. Figure 4–5.3 reveals the mechanism, which is by electron hops from Fe^{2+} ions to Fe^{3+} ions. The former Fe^{2+} ion is now an Fe^{3+} ion and may receive still another electron to continue the transfer.

Oxidation of the type shown in Fig. 14–5.6 has a parabolic rate; that is, the thickness, x, increases with the square root of time, t.

$$x = k\sqrt{t} \qquad\qquad (14\text{–}5.3)$$

This relationship can be predicted by metallurgists, since the rate of this type of corrosion is controlled by the thickening scale layer. If, however, the scale flakes off, or if the scale occupies less volume than does the original metal,* the rate of oxidation will be more rapid than is indicated in Eq. (14–5.3) and will approach

$$x = k't \qquad\qquad (14\text{–}5.4)$$

This more rapid oxidation is particularly characteristic of the alkali metals (Li, Na, etc.), and of the adjacent alkaline earth metals (Mg, Ca, etc.).

The energies released by aluminum oxidization to Al_2O_3 are great. However, the oxide layer that forms nearly halts all further oxidation. [The k of Eq. (14–5.3) is extremely small.] As a result, the scale that develops is nearly invisible and provides excellent protection from further oxidizing. Two factors contribute to this protection: (1) more so than many other metal oxides, the Al^{3+} and O^{2-} ions are very tightly bonded, so the aluminum ions cannot readily diffuse through the oxide layer toward the surface, and (2) the crystal structures of Al_2O_3 and aluminum are *coherent* — that is, the crystal structures of the two phases match dimensions. There is, thus, a strong bond between the scale and metal (Fig. 14–3.1).

We considered stainless steels in Section 14–3 because of their corrosion resistance to corrosive liquids. Those characteristics that make them passive in liquid electrolytes also make them oxidation resistant in hostile gaseous atmospheres. Like the aluminum of the previous paragraph, a chromium-containing steel forms a coherent chromium oxide scale. Thus, stainless steels can be used at higher temperatures than can either low-alloy or plain-carbon steels.

Decarburization

The two components of an alloy seldom have identical oxidation tendencies. Thus, we see chromium oxidizing preferentially before iron to passivate stainless steel. Carbon oxidizes readily at the surface of plain-carbon steels. This reaction is not restricted by a scale, since the product is carbon monoxide:

$$2C + O_2 \rightarrow 2CO\uparrow \qquad\qquad (14\text{–}5.5)$$

---

* See Example 14–5.1.

FIG. 14-5.7

Decarburization (×100). The 1040 steel is softened as the carbon is preferentially oxidized from the surface. (L. H. Van Vlack, *Materials for Engineers: Concepts and Applications,* Addison-Wesley, Reading, Mass., with permission.)

Surface

Ferrite + pearlite   Ferrite only

Carbon also diffuses from the interior of the steel, and the CO disappears as a gas. This leaves a decarburized zone behind the surface (Fig. 14-5.7). Of course, a decarburized surface is very soft compared to the balance of the steel.

Example 14-5.1

We oxidize 1 cm³ magnesium ($\rho = 1.74$ g/cm³) to MgO ($\rho = 3.6$ g/cm³). What is the volume of the resulting oxide?

Calculation

$$1 \text{ cm}^3 = 1.74 \text{ g Mg}$$
$$\text{g MgO} = 1.74 \text{ g (40.31 amu/MgO)/(24.31 amu/Mg)}$$
$$= 2.9 \text{ g MgO}$$
$$\text{volume of MgO} = 2.9 \text{ g/(3.6 g/cm}^3)$$
$$= 0.8 \text{ cm}^3 \text{ MgO}$$

Comments The Mg-O bond is sufficiently stronger than is the Mg-Mg bond in the metal, so there is a volume contraction with the addition of oxygen. [*Note:* $T_{Mg} = 650°$ C; and $T_{MgO} = 2800°$ C.] The volume contraction cracks the scale and permits more rapid oxidation than is found in aluminum and in other metals.

14-6
SERVICE PERFORMANCE OF POLYMERS

OPTIONAL

Heat Resistance

Polymers are rigid below their glass-transition temperature, T_g. As the temperature is raised beyond the glassy range and into the viscoelastic range (Fig. 10-1.8), creep and distortion begin to be significant under applied loads. The rate of strain is, of course, dependent on the temperature and the nature of the polymer. For this reason, standardized *heat-distortion* tests have been devised. For example, ASTM standard D648 heats a sample bar, 5 in. × 0.5 in. × 0.5 in., at the rate of 2° C per minute while the bar is loaded in flexure. The temperature at which a deflection of 0.010 in. (0.25 mm) develops is called the heat-distortion temperature. Of course, this temperature varies with the load. Therefore, a calculated "outer fiber" stress of either 66 psi (0.455 MPa) or 264 psi (1.82 MPa) is commonly chosen. This value gives a reproducible distortion temperature for any given polymer that, as expected, varies significantly among products, even those of the same type, because fillers, plasticizers, molecular weight, and so on are additional factors.

Stress Relaxation

In the discussion of Fig. 10–1.9, we assumed constant stress and increasing strain. In other situations, strain is constant, and since viscous flow proceeds, the stress is reduced. You have undoubtedly observed this phenomenon if you have removed a stretched rubber band from a bundle of papers after an extended period. The rubber band does not return to its original length. For this reason, it has not been holding the papers as tightly as it did initially. Some of the stress has disappeared.

The stress decreases in a viscoelastic material that is under constant strain because the molecules can gradually flow by one another. The rate of stress decrease $(-ds/dt)$ is proportional to the stress level, s:*

$$(-ds/dt)\tau = s \tag{14–6.1}$$

Rearranging and integrating,

$$ds/s = -dt/\tau$$

$$\ln s/s_0 = -t/\tau \tag{14–6.2a}$$

or

$$s = s_0 e^{-t/\tau} \tag{14–6.2b}$$

The stress ratio, s/s_0, relates the stress at time t to the original stress s_0 at t_0. The proportionality constant τ of Eq. (14–6.1) must have the units of time; it is called the *relaxation time*. When $t = \tau$, $s/s_0 = 1/e = 0.37$.

Stress relaxation is a result of molecular movements; therefore, we find that temperature affects stress relaxation in much the same manner as it affects diffusion. Since the relaxation time is the reciprocal of a rate,

$$1/\tau \propto e^{-E/kT} \tag{14–6.3}$$

or, as in Eq. (6–5.5),

$$\ln 1/\tau = \ln 1/\tau_0 - E/(13.8 \times 10^{-24} \text{ J/K})(T) \tag{14–6.4}$$

---

* In this section, shear stress is represented by s rather than by τ used in Eq. (8–1.3); τ will be used for relaxation time (Eq. 14–6.1ff).

In these equations, E, k, and T have the same meanings and units as in Eq. (6–5.5); $1/\tau$, like D, contains $\sec^{-1}$ in its units, since both involve rates.

Degradation

Although viscoelastic deformation is accelerated as the temperature exceeds T_g, this softening does not break the primary covalent bonds within the molecule. Under more severe conditions, however, these bonds may be ruptured. Of course, any resulting change in structure affects the properties. Extreme heat can degrade a polymer by breaking of bonds and by oxidation.

The most obvious *degradation* of polymers is charring. If the side groups or the hydrogen atoms of a vinyl polymer (Table 2–3.1) are literally torn loose by thermal agitation, only the backbone of carbon atoms remains. We also see this occur with the carbohydrates of our morning toast or with charred wood. It is generally to be avoided in a commercial product.* Carbonization is accelerated in the presence of air, because oxygen reacts with the hydrogen atoms along the side of the polymer chain.

The degradation just described is accentuated by oxygen. In addition, oxygen can have other effects. For example, many rubbers are vulcanized with only 5 to 20 percent of the possible positions anchored by sulfur cross-links. This permits the rubber to remain deformable and "elastic." Over a period of time, the rubber may undergo further cross-linking by oxygen of the air. The result is identical to Fig. 4–3.10, except that oxygen rather than sulfur is the connecting link (Fig. 14–6.1). Naturally, the rubber becomes harder and less deformable with an increased number of cross-links.

Several factors accelerate the oxidation reaction just described:

1. Oxygen in the form of ozone, O_3, is much more reactive than is normal O_2

2. Ultraviolet light can provide the energy to break existing bonds, so that the oxidation reaction can proceed

3. The existing bonds are broken more readily when the molecules are stressed

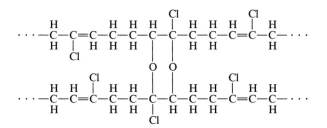

FIG. 14–6.1

Vulcanization of Rubber in the Presence of Oxygen. Like sulfur (Fig. 4–3.10), oxygen from the air can cross-link rubber molecules (chloroprene), which hardens the rubber. Still further oxidation would degrade the chloroprene to micromolecules. (L. H. Van Vlack, *Materials for Engineers: Concepts and Applications,* Addison-Wesley, Reading, Mass., with permission.)

---

* Under controlled conditions, a graphite fiber can be formed from a polymer fiber. Such fibers hold considerable promise as a high-temperature reinforcement for composites.

Because of these features, tires commonly contain carbon black or similar light absorbers to decrease the oxidation rate. Applying the same principle, accelerated testing procedures for product stability commonly expose the polymer to ozone and/or ultraviolet light.

Swelling

Micromolecules are intentionally added to polymeric materials to make the latter more flexible (Section 10–2). This is not always desirable, however. Consider a polyvinyl alcohol, $+C_2H_3R+_n$, with the **R** being —OH (Table 2–3.1). Water molecules can be absorbed among the vinyl chains. This absorption leads to a weakening and a swelling of the polymer. Likewise, unless the necessary structural adaptations are made, petroleum molecules can be absorbed into the rubber of a gasoline hose, which produces swelling and reduces the usefulness of the hose.

Swelling (Fig. 14–6.2) is not compatible with engineering specifications, of course, so the materials engineer looks for ways to prevent it. Cross-linking reduces swelling, since the molecules are linked together. Also, crystallized plastics are less subject to swelling than are amorphous polymers because crystallized plastics have more closely intermeshed molecular structures. Finally, the engineer can choose

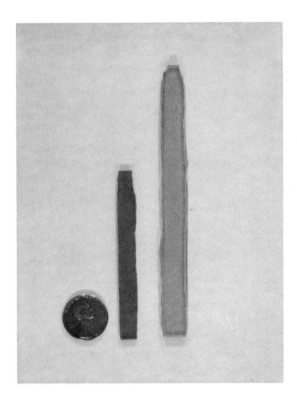

FIG. 14–6.2

Swelling. Micromolecules of benzene are absorbed between large molecules of isoprene rubber in the sample at the right, expanding it 60 l/o. Rubber, if it is not sufficiently cross-linked, is particularly subject to this type of deterioration. (Courtesy of G. S. Y. Yeh, The University of Michigan.)

among a number of polymers in order to avoid swelling. Small molecules distribute themselves among macromolecules more readily when the two types are chemically similar. For example, the polyvinyl alcohol, C_2H_3—OH, and water, H—OH, just described are closely related. Therefore, water is readily absorbed among these vinyl molecules. Likewise, hydrocarbon petroleum fluids are absorbed into hydrocarbon rubbers. Where swelling is critical, such similarities should be avoided through judicious selection of polymers.

Example 14-6.1

A stress of 11 MPa (1600 psi) is required to stretch a 100-mm rubber band to 140 mm. After 42 days at 20° C in the same stretched position, the band exerts a stress of only 5.5 MPa (800 psi). (a) What is the relaxation time? (b) What stress would be exerted by the band in the same stretched position after 90 days?

Procedure (a) The stress drops to 50 percent of its initial value. Therefore, $42/\tau = \ln 2.0$. (b) Here, $90/\tau = \ln (s_{90}/s_0)$, or we can use Eq. (14-6.2b).

Calculation

(a) From Eq. (14-6.2a),

$$\ln \frac{5.5}{11} = -\frac{42}{\tau}$$

$$\tau = 61 \text{ days}$$

(b) From Eq. (14-6.2b),

$$s_{90} = 11e^{-90/61}$$
$$= 2.5 \text{ MPa} \qquad\qquad\qquad\qquad\qquad\qquad \text{(or 360 psi)}$$

An alternative answer for (b), with 48 *additional* days, is

$$s_{48} = 5.5e^{-48/61} = 2.5 \text{ MPa}$$

Example 14-6.2

The relaxation time at 25° C is 50 days for the rubber band in Example 14-6.1. What will be the stress ratio, s/s_0, after 38 days at 30° C?

Procedure Observe in Example 14-6.1 that $\tau_{20^\circ} = 61$ days. With this and $\tau_{25^\circ} = 50$ days, we use Eq. (14-6.4) to solve simultaneously for E and $\ln (1/\tau_0)$. With these values, we can calculate τ_{30°, and hence the remaining stress.

Calculation At 20° C,

$$\ln 1/\tau_0 = \ln 1/61 + E/(13.8 \times 10^{-24} \text{ J/K})(293 \text{ K})$$

At 25° C,

$$\ln 1/\tau_0 = \ln 1/50 + E/(13.8 \times 10^{-24} \text{ J/K})(298 \text{ K})$$

Solving simultaneously,

$$E = 4.8 \times 10^{-20} \text{ J}$$

$$\ln 1/\tau_0 = 7.76$$

At 30° C,

$$\ln 1/\tau = 7.76 - 4.8 \times 10^{-20} \text{ J}/(13.8 \times 10^{-24} \text{ J/K})(303 \text{ K})$$

$$\tau = 41 \text{ days}$$

$$s_{38} = s_0 e^{-38/41}$$

$$s_{38}/s_0 = 0.4$$

Comments The relaxation time shortens at higher temperatures.

A rubber is more subject to oxidation when it is under stress. When this occurs, the structure is modified, and we observe other changes, such as hardening and cracking.

Example 14–6.3

A nylon absorbs 1.66 w/o water (dry basis). Because of differences in density, this means that 100 cm³ of dry nylon absorbs 1.9 cm³ of H_2O to produce 101.5 cm³ of saturated nylon. Account for the missing 0.4 cm³.

Solution In this case, where absorption occurs, the attraction between unlike molecules is greater than is the attraction between like molecules; that is, nylon–water attractions are greater than are the average of nylon–nylon and water–water attractions. A contraction in volume results.

Comments If the attraction between like molecules had been greater than that between unlike nylon–water pairs, the water and nylon would have become segregated and absorption would have been limited.

14–7

PERFORMANCE OF CERAMICS AT HIGH TEMPERATURES

As we stated in Section 9–8, many ceramics exhibit an excellent thermal stability because of their strong internal bonding. Even so, with sufficiently severe thermal conditions, they are subject to alteration and failure. *Fluxing* and *thermal stresses* have been the major high-temperature failure mechanisms. Unlike polymers and metals, creep had not been a major factor in the higher-temperature service of ceramics in the past. However, the prospective use of ceramics in "heat engines" does require enhanced *creep resistance* to meet the demands of extremely high rotating stresses.

Refractories are ceramics that are used in exceptionally high-temperature applications, such as for the linings of furnaces that produce glass, generate power for steam, or melt and refine metals. In steel furnaces, for example, temperatures up to and exceeding 1650° C (3000° F) are encountered (Fig. 14–7.1). However, the environment involves more than temperature alone. For example, the 2800° C melting temperature of MgO can be lowered to less than 1600° C as more and more iron oxide is encountered (Fig. 5–6.6). Likewise, the eutectic temperature between MgO and SiO_2 is 1540° C (2800° F). However, when FeO and SiO_2 are both present (plus other oxides that are always contained in molten slags), the MgO refractories of the furnace walls are fluxed by the more complex eutectic liquids at temperatures as low as 1400° C (2550° F)—significantly below the melting temperature of the molten steel. To understand and minimize this chemical attack, the engineer must draw on a thorough knowledge of the pertinent polycomponent phase diagrams. We shall not explore those phase diagrams in this text.

Spalling is thermal cracking, resulting chiefly from sharp temperature gradients that develop during heating or cooling. Several factors contribute to this type of failure. First, most ceramic materials have relatively low thermal conductivities. Thus, a sharp thermal gradient can be established, and the thermal contraction of the surface of a ceramic will introduce severe tensile stresses in that region during cooling.* Furthermore, these stresses can be higher in ceramics for a given thermal strain than in other materials, because ceramic materials generally have high elastic moduli. Finally, ceramics are generally nonductile, and therefore possess a

FIG. 14–7.1

Refractories (Roof of Electric Steel-Melting Furnace). The roof has just been removed after tapping steel at 1600° C (2900° F). We see the underside. In general, ceramics possess higher melting temperatures than do metals; furthermore, they react less with the surrounding air. Thus, the higher-melting-temperature ceramics (refractories) are used for high-temperature applications. (Courtesy of Harbison–Walker Refractories.)

---

* The tempering of glass (Section 9–8) does not produce these tensile stresses, because the temperature of the interior glass is above T_g and will adjust. As illustrated in Fig. 9–8.3, the deferred contraction of the interior produces compressive stresses in the surface.

low critical stress-intensity factor, K_{Ic}. For these reasons, cautious heating and cooling have been the norm in the application of ceramics.

Silicon nitride, Si_3N_4, and *silicon carbide,* SiC, are ceramics that have been developed on the basis of technology rather than of the availability of specific raw materials. They have potential for dynamic high-temperature service, such as in rotors in gas turbines (Fig. 14–7.2). Their success depends on our ability to process them with an essentially flaw-free microstructure to avoid stress raisers.

These two compounds (and Al_2O_3) also benefit from having a higher thermal conductivity than do most other ceramics, particularly glass (Fig. 14–7.3). Thus, they are less subject to thermal spalling, for the reasons cited previously.*

FIG. 14–7.2

Turbine Rotor (Silicon Nitride). With the development of this ceramic material, the operating temperature for a gas turbine could be increased above that of metals (Fig. 14–5.2). (Courtesy of T. Whalen, Ford Motor Co.)

FIG. 14–7.3

Thermal Conductivity (Selected Ceramics). Higher conductivities reduce spalling. Crystalline compounds of lightweight elements have higher conductivities, other factors being equal.

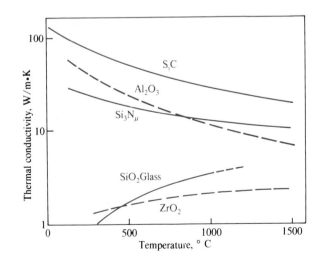

---

* These materials have high thermal conductivities, because their low atomic weights and their relatively simple crystal structures facilitate phonon movements—that is, energy transfer by atomic vibrations. Glass has no crystal structure, and therefore has extensive phonon scattering.

With gas-turbine applications near $1300°$ C ($2400°$ F), and rotational speeds of 80,000 rpm, the centrifugal stresses can lead to grain-boundary creep if the boundaries contain any glass or other amorphous phase. Improvements can be made if (1) the Si_3N_4 is made by reacting ultrapure silicon with N_2, or (2) Y_2O_3 or ZrO_2 are incorporated into the starting powders, so as to devitrify the glass in the boundaries.

14–8

RADIATION DAMAGE AND RECOVERY

Material selection is the key to successful nuclear-reactor design. First, the reactors must be "fail-safe" in case there are any operational problems. Second, radiation will alter the internal structure of many materials. Since the property changes that accompany these alterations are generally (but not always) undesired, we speak of *radiation damage.* Although we shall focus much of our attention on neutron radiation, other types of radiation — such as α-particles (He^{2+} ions), β-rays (energetic electrons), and γ-rays (high-frequency photons) — also are present.

Unlike elevated temperatures, which energize all the atoms within a solid, radiation processes may focus relatively large amounts of energy locally. As sketched in Fig. 14–8.1, a *neutron* possessing a number of electron volts may collide with an atom, dislodging it from its crystal site. Typically, the neutron has sufficient energy remaining after the first collision to ricochet to other atoms, dislodging as many as a dozen or more before being finally captured by the nucleus of an atom. In the meantime, considerable damage has occurred within the crystal structure. Slip processes are thereby inhibited, with the result that hardness and strength increase as shown in Figs. 14–8.2 and 14–8.3. There is a concurrent decrease in ductility and toughness.

The initial property changes caused by neutron irradiation of crystalline materials are similar to the changes that arise during cold work (Section 9–3). Similarly, *annealing* will soften and toughen materials to achieve *damage recovery.* Despite these comparisons, however, the structural mechanisms are basically different, because point defects are mainly responsible for radiation damage, whereas linear defects are mainly responsible for strain hardening. This difference is reflected in the fact that annealing temperatures required to remove radiation damage are lower than those required to remove strain hardening.

Scission (chain splitting) is common within polymeric materials when the latter are exposed to radiation, either of the particulate type (neutrons, α-rays, etc.) or of the electromagnetic type (γ-rays, x-rays, etc.). After scission, the broken ends of the chains and the exposed side radicals possess reactive sites with unsatisfied bonding. In the simplest case, the reactive sites rejoin. More commonly, however, there are sufficient displacements that other types of reactions can occur; for example, a chain half can be grafted onto another chain to produce *branching,* with consequent changes in properties (Fig. 4–3.8).

By noting several items from earlier chapters, we should expect that *electrical properties* will be sensitive to radiation. Recall first that disordered structures have

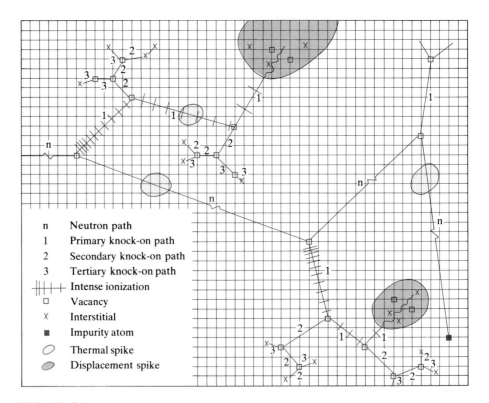

FIG. 14–8.1

**Radiation Effects in a Crystal Lattice Arising from the Neutron Entering at the
Left.** (C. O. Smith, *Nuclear Reactor Materials,* Addison-Wesley, Reading, Mass., with
permission.)

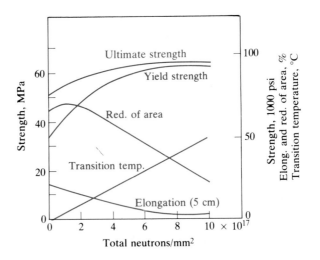

FIG. 14–8.2

**Radiation Damage to Steel
(ASTM A–212–B Carbon–
Silicon Steel).** (Adapted from
C. O. Smith, ORSORT, Oak
Ridge, Tenn.)

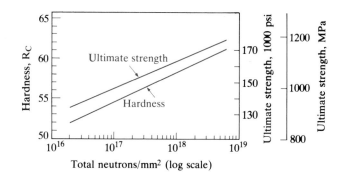

FIG. 14-8.3

Radiation Hardening (Type 347 Stainless Steel). Neutrons dislodge atoms, and therefore restrict slip in metals. Since the neutron flux is plotted on a logarithmic scale, each cycle requires appreciably longer exposures. (Adapted from C. O. Smith, ORSORT, Oak Ridge, Tenn.)

shorter mean free paths for electron movement (Section 11–2). Thus, any radiation damage that introduces imperfections into the crystal structure of a metal increases the electrical resistivity. Also recall from Section 11–4 that photoconduction occurs when an electron is excited across the energy gap. Therefore, irradiation by photons in the ultraviolet–to–γ-ray range can increase the number of intrinsic carriers (both electrons and holes) by several orders of magnitude. Hence, the resistivity of semiconductors is decreased. Finally, vacancies in some ionic solids may produce color centers, which modify dielectric and optical behaviors. Table 14–8.1 presents examples of property changes that are attributable to neutron radiation.

Radiation damage may be erased by appropriate annealing at elevated temperatures. The mechanism of damage removal is somewhat analogous to the mechanism of recrystallization (Section 9–3). However, the required temperature usually is lower, apparently because the atoms were displaced individually into positions of energy higher than that of the average position along a dislocation.

Example 14–8.1

Assume that all the energy required to produce scission in a polyethylene molecule comes from a photon (and that none of the energy is thermal).

(a) What is the maximum wavelength that can be used?

(b) How many eV are involved?

Procedure The energy of the C—C bond is listed in Table 2–2.2 as 370,000 J/(0.6 × 10^{24}). This energy must be related to the photon frequency. From introductory chemistry or physics, $E = h\nu = hc/\lambda$.

Answer

(a) $\lambda = \dfrac{hc}{E} = \dfrac{(0.66 \times 10^{-33} \text{ J·s})(3 \times 10^8 \text{ m/s})}{370{,}000 \text{ J}/0.6 \times 10^{24}}$

$\qquad = 0.32 \times 10^{-6}$ m　　　　　　　　　　　　　　(or 320 nm)

(b) $(370{,}000 \text{ J}/0.6 \times 10^{24})/(0.16 \times 10^{-18} \text{ J/eV}) = 3.8$ eV

TABLE 14–8.1 Effects of Radiation on Various Materials*

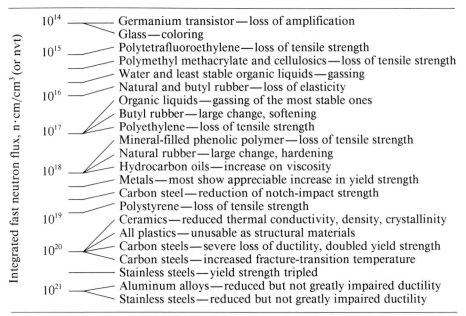

* Indicated exposure levels are approximate. Indicated changes are at least 10 percent. From C. O. Smith, *Nuclear Reactor Materials.* Addison-Wesley, Reading, Mass. with permission.

Comment This is in the ultraviolet (UV) range; thus, photons of visible light will not cause scission alone. However, a visible photon can cause scission, if it hits a bond that momentarily possesses high thermal energy. Laboratory test cells use UV light to accelerate stability evaluations.

SUMMARY

1. A material has failed if it is *no longer capable of fulfilling its intended purpose.* The causes of failure include (i) improper design judgments, (ii) unexpected service conditions, (iii) deficient maintenance, (iv) misuses, and (v) insufficient quality control during the manufacturing process. Failure during service commonly arises from an alteration of the composition or internal structure that the material undergoes during service. These changes may lead to *corrosion* (by chemical reactions); *fatigue* (from delayed fracture); *creep,* *oxidation,* or *spalling* (at elevated temperatures), or *radiation damage.*

2. Corrosion occurs at the anode; electrons are supplied to the cathode. The most common of several cathode reactions is

$$O_2 + 2\,H_2O + 4\,e^- \rightarrow 4\,(OH)^- \qquad (14\text{--}2.5)$$

Oxygen accelerates corrosion, but corrosion occurs in the oxygen-deficient areas.

The types of galvanic cells are listed in the following table.

CELL	ANODE	CATHODE
	Base Metal	**Noble Metal**
Zn versus Fe	Zn	Fe
Fe versus H_2	Fe	H_2
H_2 versus Cu	H_2	Cu
	Higher Energy	**Lower Energy**
Boundaries	Boundaries	Grain
Stresses	Cold worked	Annealed
Stress corrosion	Stressed areas	Nonstressed areas
	Lower Concentration	**Higher Concentration**
Electrolyte	Dilute solution	Concentrated solution
Oxidation	Low O_2	High O_2
Dirt or scale	Covered areas	Clean areas

Corrosion is commonly localized; *pitting* and *crevice corrosion* are the result.

3. Corrosion is minimized by the addition of *protective coatings* and by the isolation of anodes from cathodes.

Corrosion is restricted by surface *passivation*. Al_2O_3 forms on aluminum; chromium forms monolayers of oxide on stainless steels; and rust inhibitors provide adsorbed chromate surfaces on metals. Thus, although oxygen accelerates corrosion by its cathode reaction (Eq. 14-2.5), it can also restrict corrosion by isolating the anode.

Corrosion is reversed by *impressed currents* and by *sacrificial anodes,* which are expendable.

Corrosive environments reduce the fracture toughness of metals. *Stress corrosion* proceeds through metals as nonductile cracks.

4. *Fatigue* is a time-delayed fracture that is commonly associated with cyclic stressing. Repetitious loading reduces the allowable design stress. Many (but not all) materials have an *endurance limit,* below which unlimited cycling is tolerable. For materials without an endurance limit, the design stresses must be based on a *fatigue strength,* which is the stress corresponding to the number of expected cycles during service.

Stress corrosion and *static fatigue* are delayed fractures that occur when a material is stressed in a corrosive environment. (Such environments include moist air for glass.)

5. It is common for materials to lose strength at elevated temperatures. Reasons include easier atom movements, and microstructural coarsening that produces *overaging* and *overtempering.* Although *creep* occurs slowly, it must be considered in engineering design. *Oxidation* commonly produces a scale; however, since carbon oxidizes to a gas, *decarburization* produces a soft, nearly pure ferrite surface on steel.

6. Elevated temperatures lead to softening and distortion of many polymer products, thus subjecting these products to *creep* and *stress relaxation.* Polymers also are subject to degradation, oxidation, and swelling during service.

7. Ceramic materials generally have high-temperature stability. However, they can fail by *fluxing* and by *spalling.* The latter is thermal cracking that results from stresses developed by sharp thermal gradients. Ceramic nitrides and carbides are newly developed materials that have potential uses in "heat engines" and in other high-temperature, dynamic applications.

8. Radiation exposure can produce *radiation damage* if the radiation knocks the atoms or electrons out of their normal locations. These imperfections harden and embrittle metals, generally weaken polymers, and alter the conductivity of both metals and semiconductors. *Damage recovery* may be secured by heating. The resulting atom movements restore the former undamaged structures.

KEY WORDS

Anode
Anode reaction
Anodized
Cathode
Cathode reactions
Cell, composition
Cell, concentration
Cell, galvanic
Cell, oxidation
Cell, stress
Corrosion
Creep
Creep rate
Crevice corrosion
Decarburization
Degradation

Electrode potential (ϕ)
Electrolyte
Endurance limit
Fatigue
Fatigue strength
Galvanic protection
Galvanic series
Galvanized steel
Half-cell reaction
Hydrogen electrode
Impressed current
Inhibitor
Nernst equation
Oxidation (general)
Passivation
Pit corrosion

Radiation damage
Reduction
Refractories
Relaxation time (τ)
Sacrificial anode
Scale
Scission
$S-N$ curve
Spalling
Stainless steel
Static fatigue
Stress corrosion
Stress relaxation
Stress rupture

PRACTICE PROBLEMS

14–P21 How many coulombs ($A \cdot s$) are required to plate 1 g of nickel from an Ni^{2+} electrolyte?

14–P22 A metallic reflector that has 1.27 m² of surface is being chrome plated. The current is 100 A. (a) How many g of Cr^{2+} must be added to the electrolyte per hr of plating? (b) What thickness will accumulate per hr?

14–P23 The electrodes of a standard galvanic cell are nickel and magnesium. What potential difference will be established?

14–P24 Why do we use the hydrogen electrode as the reference electrode, rather than one of the metals (e.g., lead)?

14–P25 Distinguish between *anode* and *cathode* on the basis of electron movements. How does the cathode-ray tube fit into the definitions of Section 14–2?

14–P26 What copper concentration (g/l) is required in an electrolyte for it to have an electrode potential of 0.32 V (with respect to hydrogen)?

14–P27 In Fig. 14–2.2, copper is replaced by gold and zinc by tin. Based on the half-cell reactions of Table 14–2.1, will the weight of tin corroded per hr be greater or less than the weight of the gold that is plated?

14–P28 What is the source of the electrons in an ordinary, old-fashioned dry cell? (The electrolyte is a gelatinous paste containing NH_4Cl. The cathode reaction changes Mn^{4+} ions in MnO_2 to Mn^{2+}.)

14–P29 Consider Eqs. (14–2.2b, 14–2.4, 14–2.5, and 14–2.6), and the corrosion of iron. Determine which, if any, of these, provides the prevalent *cathode* reaction if (a) iron is in the water along a seashore, (b) iron is in a copper sulfate solution, (c) iron is in the can containing a carbonated beverage, (d) iron is the blade of a rusty spade, (e) iron is a car fender that has been scratched.

14–P31 With a current density of 0.1 A/m^2, how long will it take to corrode an average of 0.1 mm from the surface of aluminum?

14–P32 Pit corrosion punctured the hull of an aluminum boat after 12 months in seawater. The aluminum sheet was 1.1 mm thick. The average diameter of the hole was 0.2 mm. (a) How many atoms were removed from the pit per second? (b) What was the corrosion current density?

14–P33 A discarded tin-coated can is left on the shoreline of a local freshwater lake. As rusting proceeds, Eq. (14–2.1$_{Fe}$) is obviously the anodic reaction. (a) What is the probable cathodic reaction? (b) Why is the answer to part (a) *not* ($Sn^{2+} + 2 e^- \rightarrow Sn$)?

14–P34 Explain why metal in an oxygen-deficient electrolyte becomes anodic to metal in an oxygen-enriched electrolyte.

14–P35 How do barnacles accelerate corrosion on a ship hull?

14–P36 An enterprising mechanic suggests the use of a magnesium drain plug for the crankcase of a car as a means of avoiding engine corrosion—specifically of the bearings and cylinder walls. Discuss this proposal.

14–P37 Undercoatings of various types are applied to new cars. Under what conditions are they helpful? Under what conditions are they detrimental?

14–P41 Sometimes, people assume that, if failure does not occur in a fatigue test in 10^8 cycles, the stress is below the endurance limit. The test machine is connected directly to a 1740-rpm motor. How long will it take to log that number of cycles?

14–P42 Examine the crankshaft of a car. Point out specification and design features that alter the resistance to fatigue.

14–P51 (a) Why do aluminum kitchen utensils corrode less readily than do iron ones? (b) Why does chromium make steel "stainless"?

14–P52 The following data were obtained in a creep-rupture test of Inconel "X" at 815° C (1500° F): 1 percent strain after 10 hr, 2 percent strain after 200 hr, 4 percent strain after 2000 hr, 6 percent strain after 4000 hr, "neck-down" started at 5000 hr, and rupture occurred at 5500 hr. Sketch the *e–t* curve. What is the second-stage creep rate?

14–P61 A stress relaxes from a 0.7 MPa to 0.5 MPa in 123 days. (a) What is the relaxation time? (b) How long would it take to relax to 0.3 MPa?

14–P62 An initial stress of 10.4 MPa (1500 psi) is required to strain a piece of rubber 50 percent. After the strain has been maintained constant for 40 days, the stress required is only 5.2 MPa (750 psi). What would be the stress required to maintain the strain after 80 days? Solve this problem without using a calculator.

14–P63 Raw polyisoprene (i.e., nonvulcanized natural rubber) gains 2.3 w/o by becoming cross-linked by oxygen in the air. What fraction of the possible cross-links are established? (In this case, assume that the O_2 cleaves and that each cross-link involves a single oxygen.)

14–P81 The average energy of a C—Cl bond is 340 kJ/mol according to Table 2–2.2. Will visible light [400 nm (violet) to 700 nm (red)] have enough energy to break one of these bonds?

14–P82 (a) What frequency and wavelength must a photon have to supply the energy necessary to break the average C–H bond in polyethylene? (b) Why can some bonds be broken with slightly longer electromagnetic waves?

TEST PROBLEMS

1421 A galvanic couple includes iron in an 0.002 molar solution of Fe^{2+} ions, and magnesium in an 0.3 molar solution of Mg^{2+} ions. What is the electrode potential difference?

1422 The electrode potential difference between copper and cobalt in 1-molar (1-M) solutions is -0.62 V. What is the electrode potential of cobalt with respect to zinc within 0.1-M Co^{2+} and 0.1-M Zn^{2+} solutions?

1423 What is the ratio of Zn^{2+} concentrations required in electrolytes to introduce a potential difference of 35 mV?

1424 In a stagnant solution, the copper-ion concentration is 0.012 M at point A of a copper surface; it is 0.002 M at point B. (a) What potential difference develops for possible corrosion? (b) At which point will corrosion proceed?

1425 What current density, i, is required to plate 20 μm chromium onto a surface in 20-min, from a Cr^{2+} solution?

1426 A dry cell with a zinc anode (cell wall) provides 15 A of current from its 175 cm^2 of surface. How long will it take for 15 percent of the 0.5-mm thick wall to corrode? (Although you may assume uniform corrosion for this problem, local pitting corrosion will be the rule.)

1427 It is necessary to set up a small laboratory electroplating bath with a maximum capacity of 1 lb (0.45 kg) of nickel per day. What is the minimum dc amperage requirement?

1428 Local corrosion produces a 0.1-mm diameter "pinhole" through a 0.5-mm thick aluminum pan in 30 days. Assume that the hole is cylindrical. What was the current density at this site?

1429 Two pieces of metal, one copper and the other zinc, are connected with a copper wire and are immersed into seawater. Indicate the galvanic cell by writing the half-cell reactions (a) for the anode, and (b) for the cathode. Also, indicate (c) the direction of electron flow in the wire, and (d) the direction of "current" flow in the electrolyte. (e) What metal might be used in place of copper so that the zinc changes polarity?

1431 An average current density of 100 A/m^2 is used to build up an anodized coating on aluminum. The Al_2O_3 is to be 1-μm thick. What time is required?

1432 When copper tubing and steel pipe are coupled in home plumbing, the building codes call for a polymeric insulator such as PTFE to be placed between the two metals. Explain the basis for this requirement.

1433 Electrical codes call for a "jumper" to be placed around the insulated joint described in Problem 1432. Does this requirement defeat the purpose of the plumbing code? Explain your answer.

1434 Explain why metal in a dilute electrolyte becomes anodic to metal in a more concentrated electrolyte.

1435 Given that oxygen accelerates corrosion (Eqs. 14–2.5 and 14–2.6), why does corrosion occur in the oxygen-lean areas?

1436 Why is corrosion accelerated when the anode is smaller than the cathode, but not when the situation is reversed?

1437 A flashlight with old dry cells will dim during use. However, if left unused for a period of time, it will shine more brightly (for a short time). Explain what is happening.

1451 Design considerations would permit a pressure tube in a steam generator to have 3 l/o strain during 1 yr of high-temperature service. What maximum creep rate is tolerable when reported in the normal percent/hr figures?

1452 One cm^3 of magnesium ($\rho = 1.74$ g/cm^3) is oxidized to MgO ($\rho = 3.65$ g/cm^3). (a) What is the volume change in the oxidation process? (b) Account for the smaller volume.

1453 Refer to Example 14–3.1. How do the closest Ni-to-Ni distances compare in NiO and in metallic nickel? (Nickel is an fcc metal; the oxide has the NaCl structure.)

1461 The relaxation time for a polymer is known to be 60 days, and the modulus of elasticity is 70 MPa (both at 100° C). The polymer is compressed 5 percent and is held at 100° C without further change in dimension. Calculate the stress (a) initially, (b) after 1 day, (c) after 1 month, and (d) after 1 yr.

1462 The relaxation time for a nylon thread is reduced from 4000 to 3000 min if the temperature is increased from 15 to 35° C. (a) Determine the activation energy for relaxation. (b) At what temperature is the relaxation time 2000 min?

1463 Lucite (PMMA of Fig. 10–1.8) is loaded at 50° C for 1 hr. (a) By how much would the stress have to be increased to give the same strain in 36 sec? (b) Repeat for 100° C.

1464 A chloroprene rubber gains 3 w/o by oxidation.

Assume oxygen-produced cross-linking. Also assume that one-half of the cross-links are single oxygens and that one-half involve a pair of oxygen atoms. (See Fig. 4–3.10.) What fraction of the possible anchor points (cross-linkages) are connected?

1481 Photons of blue light ($\lambda = 480$ nm) lead to extensive scission (bond breaking) in a polymer. How much energy is involved with each photon?

1482 Polyformaldehyde has $\begin{smallmatrix} H \\ | \\ -C-O- \\ | \\ H \end{smallmatrix}$ as mers. A

photon must have what wavelength of radiation to be the sole source of energy for scission of the molecular chain?

APPENDIXES

APPENDIX A Constants and Conversions

Constants*

Acceleration of gravity, g	9.80 . . . m/s^2
Atomic mass unit, amu	1.66 . . . $\times 10^{-24}$ g
Avogadro's number, N	0.6022 . . . $\times 10^{24}$ mol^{-1}
Boltzmann's constant, k	86.1 . . . $\times 10^{-6}$ eV/K
	13.8 . . . $\times 10^{-24}$ J/K
Electron charge, q	0.1602 . . . $\times 10^{-18}$ C
Electron moment, β	9.27 . . . $\times 10^{-24}$ A·m^2
Electron volt, eV	0.160 . . . $\times 10^{-18}$ J
Faraday, $\mathcal{F}$	96.5 . . . $\times 10^3$ C
Fe–Fe$_3$C eutectoid composition	0.77 w/o carbon
Fe–Fe$_3$C eutectoid temperature	727° C (1340° F)
Gas constant, R	8.31 . . . J/mol·K
	1.987 . . . cal/mol·K
Gas volume (STP)	22.4 . . . $\times 10^{-3}$ m^3/mol
Permittivity (vacuum), ϵ	8.85 . . . $\times 10^{-12}$ C/V·m
Planck's constant, h	0.662 . . . $\times 10^{-33}$ J·s
Velocity of light, c	0.299 . . . $\times 10^9$ m/s

Conversions*

1 ampere	=	1 C/s
1 angstrom	=	10^{-10} m
	=	10^{-8} cm
	=	0.1 nm
	=	3.937×10^{-9} in.
1 amu	=	1.66 . . . $\times 10^{-24}$ g
1 Btu	=	1.055 . . . $\times 10^3$ J
1 Btu/° F	=	1.899 . . . $\times 10^3$ J/° C
1 [Btu/(ft^2·s)]/[° F/in.]	=	0.519 . . . $\times 10^3$ [J/(m^2·s)]/[° C/m]
	=	0.519 . . . $\times 10^3$ (W/m^2)/(° C/m)
1 Btu·ft^2	=	11.3 . . . $\times 10^3$ J/m^2
1 calorie	=	4.18 . . . J
1 centimeter	=	10^{-2} m
	=	0.3937 in.
1 coulomb	=	1 A·s
1 cubic centimeter	=	0.0610 . . . in.^{3}
1 cubic inch	=	16.3 . . . $\times 10^{-6}$ m^3
1° C difference	=	1.8° F
1 electron volt	=	0.160 . . . $\times 10^{-18}$ J

* All irrational values are rounded downward.

1° F difference	=	0.555 . . . ° C
1 foot	=	0.3048 . . . m
1 foot·pound$_f$	=	1.355 . . . J
1 gallon (U.S. liq.)	=	3.78 . . . × 10^{-3} m^3
1 gram	=	0.602 . . . × 10^{24} amu
	=	2.20 . . . × 10^{-3} lb$_m$
1 gram/centimeter^3	=	62.4 . . . lb$_m$/ft^3
	=	1000 kg/m^3
	=	1 Mg/m^3
1 inch	=	0.0254 . . . m
1 joule	=	0.947 . . . × 10^{-3} Btu
	=	0.239 . . . cal
	=	6.24 . . . × 10^{18} eV
	=	0.737 . . . ft·lb$_f$
	=	1 watt·sec
1 joule/meter^2	=	8.80 . . . × 10^{-5} Btu/ft^2
1 [joule/(m^2·s)]/[° C/m]	=	1.92 . . . × 10^{-3} [Btu/(ft^2·s)]/[° F/in.]
1 kilogram	=	2.20 . . . lb$_m$
1 megagram/meter^3	=	1 g/cm^3
	=	10^6 g/m^3
	=	1000 kg/m^3
1 meter	=	10^{10} Å
	=	10^9 nm
	=	3.28 . . . ft
	=	39.37 in.
1 micrometer	=	10^{-6} m
1 nanometer	=	10^{-9} m
1 newton	=	0.224 . . . lb$_f$
1 ohm·inch	=	0.0254 . . . Ω·m
1 ohm·meter	=	39.37 Ω·in.
1 pascal	=	0.145 . . . × 10^{-3} lb$_f$/in.^{2}
1 poise	=	0.1 Pa·s
1 pound (force)	=	4.44 . . . newtons
1 pound (mass)	=	0.453 . . . kg
1 pound/foot^3	=	16.0 . . . kg/m^3
1 pound/inch^2	=	6.89 . . . × 10^{-3} MPa
1 watt	=	1 J/s
1 (watt/m^2)/(° C/m)	=	1.92 . . . × 10^{-3} [Btu/(ft^2·s)]/[° F/in.]

SI prefixes

giga	G	10^9	kilo	k	10^3	micro	μ	10^{-6}
mega	M	10^6	milli	m	10^{-3}	nano	n	10^{-9}

APPENDIX B Table of Selected Elements*

ELEMENT	SYMBOL	ATOMIC NUMBER	ATOMIC MASS, amu	ORBITALS				MELTING POINT, °C	DENSITY (SOLID), Mg/m³ (=g/cm³)	CRYSTAL STRUCTURE, 20° C	APPROX. ATOMIC RADIUS nm†	VALENCE (MOST COMMON)	APPROX. IONIC RADIUS, nm‡
				1s									
Hydrogen	H	1	1.0078	1				−259.14	—	—	0.046	1+	very small
Helium	He	2	4.003	2				−272.2	—	—	0.176	inert	—
				2s	*2p*								
Lithium	Li	3	6.94	He +	1			180.7	0.534	bcc	0.1519	1+	0.068
Beryllium	Be	4	9.01	He +	2			1290	1.85	hcp	0.114	2+	0.035
Boron	B	5	10.81	He +	2	1		2300	2.3	—	0.046	3+	~0.025
Carbon	C	6	12.011	He +	2	2		>3500	2.25	hex	0.077	—	—
Nitrogen	N	7	14.007	He +	2	3		−210	—	—	0.071	3−	—
Oxygen	O	8	15.999	He +	2	4		−218.4	—	—	0.060	2−	0.140
Fluorine	F	9	19.00	He +	2	5		−220	—	—	0.06	1−	0.133
Neon	Ne	10	20.18	He +	2	6		−248.7	—	fcc	0.160	inert	—
				3s	*3p*								
Sodium	Na	11	22.99	Ne +	1			97.8	0.97	bcc	0.1857	1+	0.097
Magnesium	Mg	12	24.31	Ne +	2			649	1.74	hcp	0.161	2+	0.066
Aluminum	Al	13	26.98	Ne +	2	1		660.4	2.699	fcc	0.14315	3+	0.051
Silicon	Si	14	28.09	Ne +	2	2		1410	2.33	§	0.1176	4+	0.042
Phosphorus	P	15	30.97	Ne +	2	3		44	1.8	—	0.11	5+	~0.035
Sulfur	S	16	32.06	Ne +	2	4		112.8	2.07	—	0.106	2−	0.184
Chlorine	Cl	17	35.45	Ne +	2	5		−101	—	—	0.101	1−	0.181
Argon	Ar	18	39.95	Ne +	2	6		−189.2	—	fcc	0.192	inert	—
				3d	*4s*	*4p*							
Potassium	K	19	39.1	Ar +		1		63	0.86	bcc	0.231	1+	0.133
Calcium	Ca	20	40.08	Ar +		2		839	1.55	fcc	0.197	2+	0.099
Titanium	Ti	22	47.90	Ar +	2	2		1668	4.51	hcp	0.146	4+	0.068
Chromium	Cr	24	52.00	Ar +	5	1		1875	7.19	bcc	0.1249	3+	0.063
Manganese	Mn	25	54.94	Ar +	5	2		1244	7.47	—	0.112	2+	0.080

ELEMENT	SYMBOL	ATOMIC NUMBER	ATOMIC MASS, amu	ORBITALS			MELTING POINT, ° C	DENSITY (SOLID), Mg/m³ (=g/cm³)	CRYSTAL STRUC-TURE, 20° C	APPROX. ATOMIC RADIUS nm†	VALENCE (MOST COMMON)	APPROX. IONIC RADIUS, nm‡
				$3d$	$4s$	$4p$						
Iron	Fe	26	55.85	Ar + 6	2		1538	7.87	bcc	0.1241	2+	0.074
									fcc	0.1269	3+	0.064
Cobalt	Co	27	58.93	Ar + 7	2		1495	8.83	hcp	0.125	2+	0.072
Nickel	Ni	28	58.71	Ar + 8	2		1455	8.90	fcc	0.1246	2+	0.069
Copper	Cu	29	63.54	Ar + 10	1		1084	8.93	fcc	0.1278	1+	0.096
Zinc	Zn	30	65.38	Ar + 10	2		420	7.13	hcp	0.139	2+	0.074
Germanium	Ge	32	72.59	Ar + 10	2	2	937	5.32	§	0.1224	4+	—
Arsenic	As	33	74.92	Ar + 10	2	3	816	5.78	—	0.125	3+	—
Krypton	Kr	36	83.80	Ar + 10	2	6	−157	—	fcc	0.201	inert	—
				$4d$	$5s$	$5p$						
Silver	Ag	47	107.87	Kr + 10	1		961.9	10.5	fcc	0.1444	1+	0.126
Tin	Sn	50	118.69	Kr + 10	2	2	232	7.17	bct	0.1509	4+	0.071
Antimony	Sb	51	121.75	Kr + 10	2	3	630.7	6.7	—	0.1452	5+	—
Iodine	I	53	126.9	Kr + 10	2	5	114	4.93	ortho	0.135	1−	0.220
Xenon	Xe	54	131.3	Kr + 10	2	6	−112	2.7	fcc	0.221	inert	—
				$4f$	$5d$	$6s$						
Cesium	Cs	55	132.9	Xe +		1	28.6	1.9	bcc	0.265	1+	0.167
Tungsten	W	74	183.9	Xe + 14	4	2	3410	19.25	bcc	0.1367	4+	0.070
Gold	Au	79	197.0	Xe + 14	10	1	1064.4	19.3	fcc	0.1441	1+	0.137
Mercury	Hg	80	200.6	Xe + 14	10	2	−38.86	—	—	0.155	2+	0.110
Lead	Pb	82	207.2	Hg + $6p^2$			327.4	11.38	fcc	0.1750	2+	0.120
Uranium	U	92	238.0	Rn + $5f^3$	$6d$	$7s^2$	1133	19.05	—	0.138	4+	0.097

* From various sources, including the ASM *Handbooks*.

† One half of the closest approach of two atoms in the elemental solid. For noncubic structures, the average interatomic distance is given; e.g., in hcp, the atom is slightly ellipsoidal.

‡ Radii for CN = 6; otherwise, 0.97 $R_{CN=8} \approx R_{CN=6} \approx 1.1\ R_{CN=4}$. Patterned after Ahrens.

§ Diamond cubic.

APPENDIX C Properties of Selected Engineering Materials (20° C)*

MATERIAL	DENSITY Mg/m³ (=g/cm³)	THERMAL CONDUCTIVITY, $\left(\dfrac{\text{watts}}{\text{mm}^2}\right)\Big/\left(\dfrac{°\text{C}}{\text{mm}}\right)$†	LINEAR EXPAN-SION, °C^{-1}‡	ELECTRICAL RESISTIVITY, ρ ohm·m§	AVERAGE MODULUS OF ELASTICITY, $\bar{E}$	
					GPa	psi
Metals (Annealed)						
Aluminum (99.9+)	2.7	0.22	22.5×10^{-6}	29×10^{-9}	70	10×10^6
Aluminum alloys	2.7(+)	0.16	22×10^{-6}	$\sim45 \times 10^{-9}$	70	10×10^6
Brass (70 Cu – 30 Zn)	8.5	0.12	20×10^{-6}	62×10^{-9}	110	16×10^6
Bronze (95 Cu – 5 Sn)	8.8	0.08	18×10^{-6}	$\sim130 \times 10^{-9}$	110	16×10^6
Cast iron (gray)	7.15	—	10×10^{-6}	—	140(±)	$20 \times 10^6\pm$
Cast iron (white)	7.7	—	9×10^{-6}	660×10^{-9}	205	30×10^6
Copper (99.9+)	8.9	0.40	17×10^{-6}	17×10^{-9}	110	16×10^6
Iron (99.9+)	7.88	0.072	11.7×10^{-6}	98×10^{-9}	205	30×10^6
Lead (99+)	11.34	0.033	29×10^{-6}	206×10^{-9}	14	2×10^6
Magnesium (99+)	1.74	0.16	25×10^{-6}	45×10^{-9}	45	6.5×10^6
Monel (70 Ni – 30 Cu)	8.8	0.025	15×10^{-6}	482×10^{-9}	180	26×10^6
Silver (sterling)	10.4	0.41	18×10^{-6}	18×10^{-9}	75	11×10^6
Steel (1020)	7.86	0.050	11.7×10^{-6}	169×10^{-9}	205	30×10^6
Steel (1040)	7.85	0.048	11.3×10^{-6}	171×10^{-9}	205	30×10^6
Steel (1080)	7.84	0.046	10.8×10^{-6}	180×10^{-9}	205	30×10^6
Steel (18 Cr – 8 Ni stainless)	7.93	0.015	9×10^{-6}	700×10^{-9}	205	30×10^6
Ceramics						
Al_2O_3	3.8	0.029	9×10^{-6}	$>10^{12}$	350	50×10^6
Brick						
building	2.3(±)	0.0006	9×10^{-6}	—	—	—
fireclay	2.1	0.0008	4.5×10^{-6}	1.4×10^6	—	—
graphite	1.5	—	5×10^{-6}	—	—	—
paving	2.5	—	4×10^{-6}	—	—	—
silica	1.75	0.0008	—	1.2×10^6	—	—
Concrete	2.4(±)	0.0010	13×10^{-6}	—	14	2×10^6

MATERIAL	DENSITY Mg/m³ (=g/cm³)	THERMAL CONDUCTIVITY, $\left(\dfrac{\text{watts}}{\text{mm}^2}\right)\Big/\left(\dfrac{^\circ C}{\text{mm}}\right)$[†]	LINEAR EXPAN-SION, $^\circ C^{-1}$[‡]	ELECTRICAL RESISTIVITY, ρ ohm·m[§]	AVERAGE MODULUS OF ELASTICITY, $\bar{E}$	
					GPa	psi
Glass						
plate	2.5	0.00075	9×10^{-6}	10^{12}	70	10×10^{6}
borosilicate	2.4	0.0010	2.7×10^{-6}	$> 10^{15}$	70	10×10^{6}
silica	2.2	0.0012	0.5×10^{-6}	10^{18}	70	10×10^{6}
vycor	2.2	0.0012	0.6×10^{-6}	—	—	—
wool	0.05	0.00025	—	—	—	—
Graphite (bulk)	1.9	—	5×10^{-6}	10^{-5}	7	1×10^{6}
MgO	3.6	—	9×10^{-6}	10^{3} (1100° C)	205	30×10^{6}
Quartz (SiO_2)	2.65	0.012	—	10^{12}	310	45×10^{6}
SiC	3.17	0.012	4.5×10^{-6}	0.025 (1100° C)	—	—
TiC	4.5	0.030	7×10^{-6}	50×10^{-8}	350	50×10^{6}
Polymers						
Melamine-formaldehyde	1.5	0.00030	27×10^{-6}	10^{11}	9	1.3×10^{6}
Phenol-formaldehyde	1.3	0.00016	72×10^{-6}	10^{10}	3.5	0.5×10^{6}
Urea-formaldehyde	1.5	0.00030	27×10^{-6}	10^{10}	10.3	1.5×10^{6}
Rubbers (synthetic)	1.5	0.00012	—	—	~0.05	600–11,000
Rubber (vulcanized)	1.2	0.00012	81×10^{-6}	10^{12}	3.5	0.5×10^{6}
Polyethylene (L.D.)	0.92	0.00034	180×10^{-6}	10^{13}–10^{16}	0.1–0.35	14,000–50,000
Polyethylene (H.D.)	0.96	0.00052	120×10^{-6}	10^{12}–10^{16}	0.35–1.25	50,000–180,000
Polystyrene	1.05	0.00008	63×10^{-6}	10^{16}	2.8	0.4×10^{6}
Polyvinylidene chloride	1.7	0.00012	190×10^{-6}	10^{11}	0.35	0.05×10^{6}
Polytetrafluoroethylene	2.2	0.00020	100×10^{-6}	10^{14}	0.35–0.7	50,000–100,000
Polymethyl methacrylate	1.2	0.00020	90×10^{-6}	10^{14}	3.5	0.5×10^{6}
Nylon	1.15	0.00025	100×10^{-6}	10^{12}	2.8	0.4×10^{6}

* Data in this table were taken from numerous sources.

† Alternatively, W/mm·K. Multiply by 1.92 to get Btu/(ft²·s)/(° F/in.).

‡ Or, K^{-1}; divide by 1.8 to get $^\circ F^{-1}$.

§ Multiply ohm·m by 39 to get ohm·in.

ANSWERS TO PRACTICE PROBLEMS

Please refer to the *Study Guide* for the solutions to these problems.

Chapter 1

1–P21 37 MPa (5300 psi)

1–P22 69,000 MPa (9,800,000 psi)

1–P23 32,000 psi (220 MPa)

1–P24 Problems 1–P21 and 1–P22 use diameters with only one sig. fig. (significant figures). Therefore, the results are accurate to 1 sig. fig. only. However, we commonly include one additional digit to indicate which way the figure "tilts;" e.g., in Problem 1–P21, > 5000 psi, rather than < 5000 psi. (Do *not* give 5252.11312 psi, or 37.433243 MPa answers. It would show that you are *not* thinking!) Problem 1–P23 is somewhat ambiguous, since 70 GPa can be 70 ± 0.5, or 70 ± 5. Use a second digit in this case.

1–P25 (a) 0.0012 ohm   (b) 16×10^6 ohm$^{-1} \cdot$m^{-1}

1–P26 Strength: stress to failure
Ductility: strain to fracture
Toughness: energy for fracture

1–P27

Resistance:	ohm	Function of material and shape
Resistivity:	ohm·m	Function of the material, independent of shape
Conductivity:	ohm$^{-1} \cdot$m^{-1}	Reciprocal of resistivity

1–P41

Metal:	Oven box	Easy shaping by "deep drawing"
	Resistor wire	Selected electrical resistivity
	Element jacket	Oxidation resistance
Ceramic:	Insulation	Isolate resistance wire from element jacket
	Oven box coating	Easy cleaning, oxidation resistance, color
Polymer:	Dial knobs	Easy shaping, color, cost
	Power wire coating	Insulation, flexibility

Chapter 2

2–P11 (a) 40.4×10^{-24} g/Mg
(b) 43×10^{21} Mg/cm^3

2–P12 (a) 84.5×10^{18} Cu/mm^3
(b) 8.92 g/cm^3 (8.92 Mg/m^3)

2–P13 (a) 5.58×10^{21} Ag atoms/g
(b) 0.017 cm^3 (17 mm^3)

2–P14 (a) 87 mg (0.087 g)   (b) 1.07 A

2–P15 (a) Atomic number is the number of electrons (and protons) per neutral atom.   (b) Atomic weight is the mass per atom (amu), or per mole (grams).

2–P21 (a) 32   (b) 46   (c) 17   (d) 78   (e) 94
(f) 30   (g) 58   (h) 60   (i) 28   (j) 62.5

2–P22 2×10^{20} bonds

2–P31 (a) 62.5 amu   (b) 560 mers/molecule

2–P32 -24.5 kJ

2–P33 (a) 65,000 g/mol (65,000 amu/molecule)
(b) 630 mers/molecule

2–P34 (a) $C{=}C \rightarrow \cdots C{-}C{-}$
(b) -1 double bond per mer; $+2$ single bonds
per mer

2–P35 -10^{-19} J/mer

2–P36 $C{=}C \rightarrow \cdots C{-}C{-}$

2–P41 (a) 0.121 nm   (b) 0.137 nm

2–P42 0.399 nm   Also, $d_{Cs^+-Cs^+} = 0.4607$ nm

2–P43 (a) Ca^{2+}, Mn^{2+}, Hg^{2+}, Pb^{2+}
(b) Hg^{2+}, Pb^{2+} (possibly Ca^{2+})

2–P44 The interatomic (and interionic) distances are
governed by the local electric fields around the atoms.
Neutrons carry no charges, so they can move through
these fields unaffected. Positive and negative ions with
protons $\neq$ electrons are not free to penetrate these
fields without interference.

2–P45 $R_{Sr^{2+}} \approx 0.13$ nm (actually, 0.127 nm)

2–P61 $\alpha_L \approx 6 \times 10^{-6}/°$ C (actually, $5.5 \times 10^{-6}/°$ C)

2–P62 $E_{Mo} \approx 300$ GPa   (43,000,000 psi) [actually,
330 GPa (47,000,000 psi)]

Chapter 3

3–P11 2.00 Fe/u.c.

3–P12 0.0353 nm^3

3–P13 (a) 8.1×10^{20} unit cells
(b) 1.8×10^{-22} g (107.9 amu)

3–P21 0.68

3–P22 0.1249 nm (App. B: 0.1249 nm)

3–P23 See *Study Guide.*

3–P24 (a) 15.3×10^{18} Ba/mm^3 (b) 0.68
(c) bcc

3–P25 (a) 0.684   (b) 0.73

3–P26 5.90 Mg/m^3 (5.90 g/cm^3)

3–P27 4 Mg^{2+} and 4 O^{2-}

3–P28 (a) 0.71   (b) 0.79

3–P29 0.405 nm (vs. 0.402 nm)

3–P31 0.049 nm^3

3–P32 3/1

3–P41 $+56$ v/o

3–P42 $+1.4\%$

3–P43 -14 v/o

3–P44 $\rho_d > \rho_g$

3–P45 See *Study Guide.*

3–P51 0, 0, 0 and $\frac{1}{2}, \frac{1}{2}, \frac{1}{2}$

3–P52 0, 0, 0 $\frac{1}{2}, \frac{1}{2}, 0$ $\frac{1}{2}, 0, \frac{1}{2}$ $0, \frac{1}{2}, \frac{1}{2}$

3–P53 0.4427 nm

3–P54 (a) 0.375 nm   (b) 0.53 nm

3–P55 $\frac{3}{4}, \frac{3}{4}, \frac{3}{4}$ $\frac{1}{4}, \frac{1}{4}, \frac{3}{4}$ $\frac{1}{4}, \frac{3}{4}, \frac{1}{4}$

3–P56 $\frac{1}{4}, \frac{1}{4}, \frac{1}{4}$ $\frac{3}{4}, \frac{3}{4}, \frac{1}{4}$ $\frac{3}{4}, \frac{1}{4}, \frac{3}{4}$

3–P57 (a) CN = 6; "dia." = $a - 2R$
(b) CN = 4; "dia." = $(a\sqrt{\tfrac{3}{4}} - R)2$

3–P61 (a) 1, 1, $\frac{1}{2}$ (b) 0, 0, $\frac{1}{2}$

3–P62 1, $\frac{1}{2}$, 1   0, $-\frac{1}{2}$, 0 (plus others)

3–P63 (a) 0.71   (b) 0.71

3–P64 (a) 64.8°   (b) 0.75

3-P65 (a) 0.255 nm   (b) 0.406 nm

3-P66 (a) 0.442 nm   (b) 0.703 nm

3-P67 [012] [021] [102] [201] [210] [120]
         [01$\bar{2}$] [02$\bar{1}$] [10$\bar{2}$] [20$\bar{1}$] [2$\bar{1}$0] [1$\bar{2}$0]

3-P70 (320)

3-P71 (212)

3-P72 (6$\bar{4}$9)

3-P73 (a) 15.3 × 10^{12}/mm^2
         (b) 10.8 × 10^{12}/mm^2
         (c) 17.7 × 10^{12}/mm^2

3-P74 (a) [110]   (b) The same.

3-P75 (a) [0$\bar{2}$1]   (b) [11$\bar{1}$]

3-P76 [11$\bar{2}$2]

3-P77 (a) (100) (010), and their negatives ($\bar{1}$00) (0$\bar{1}$0)
         (b) (001), and its negative (00$\bar{1}$)

3-P78 (a) [101], [011], and [1$\bar{1}$0]
         (b) [011], [110], and [10$\bar{1}$]

3-P79 (a) [11$\bar{1}$], and [$\bar{1}$11]   (b) [111], and [11$\bar{1}$]

3-P81 (a) 0.1750 nm   (b) 0.2858 nm
         (c) 0.2425 nm

3-P82 (a) 0.154 nm   (b) 10.2°

Chapter 4

4-P11 Decreases

4-P12 0.06 nm

4-P13 (a) CN = 6   (b) NaCl

4-P14 8.920 g/cm^3 (vs. 8.94 g/cm^3, theoretically)

4-P15 The charges must remain balanced.

4-P16 1 percent

4-P17 Point: vacancies, interstitials
         Linear: dislocations, screw and edge
         2-D: surfaces, grain boundaries

4-P18 (a) Parallel   (b) Perpendicular

4-P21 0.65

4-P22 10,000 J/mol

4-P23 Below T_g: Expansion from increased vibrational energy only.
         Above T_g: Expansion from (1) increased vibrational energy, and (2) the continuous rearrangements of atoms (or molecules).

4-P24 Metals crystallize rapidly (ms or μs), because each atom moves individually. The ordering of molecules into a crystalline array involves the simultaneous rearrangements of large polyatomic groups.

4-P25 See *Study Guide.*

4-P30 Greater

4-P31 Dextrose: 0.014 H_2O: 0.986

4-P32 19,200 g/mol

4-P33 (a) 31.7 amu   (b) 21.6 amu

4-P34 (a) 4.65 nm   (b) 7125 amu/molecule

4-P35 3.7 nm

4-P36 (a) 48%   (b) 42%

4-P37 0.14

4-P38 See *Study Guide.*

4-P39 See *Study Guide.*

4-P41 Brass: Zn in Cu
         Stainless steel: Cr (and Ni) in Fe.
         12 carat gold: 50% Au (bal., commonly Cu)
         Bronze: Sn in Cu
         Sterling silver: Cu in Ag
         Monel: Cu in Ni (App. C)

4-P42 91.4 a/o Cu, 8.6 a/o Sn

4-P43 95.5 w/o Al, 4.5 w/o Mg

4-P44 1.3 w/o C

4-P45 (a) 41.5 w/o Cu   (b) 12.8 w/o Cu

4−P46 (a) 4 sites/u. c.   (b) 6−f sites

4−P47 Nickel

4−P48 Pure Cu

4−P51 (a) 51 w/o FeO   (b) 39.8 w/o Fe^{2+}
(c) 30.8 w/o O^{2-}

4−P52 48 w/o MgO

4−P53 (a) Mg^{2+}, 24 w/o; Li^+, 16 w/o; O^{2-}, 16 w/o; F^-, 44 w/o
(b) 3.02 g/cm³

4−P61 4.8 to 1

4−P62 $f_{PVC} = 0.76$ $f_{PVDC} = 0.24$

Chapter 5†

5−P11† 138 g added sugar

5−P12† 26.2 g Sn

5−P13† 4.6 T ice

5−P21 L − from 1100 → 1010° C
$(L + \alpha)$ − from 1010 → 830° C
α − from 830 → 340° C
$(\alpha + \epsilon)$ − from 340 → 20° C

5−P22 300 to 700° C − α; 800†, 900° C − $(\alpha + \beta)$; 1000, 1100° C − Liquid

5−P23 (a) 100°C, $(\alpha + \beta)$; 200° C, α; 300° C, $(\alpha + L)$
(b) Only α, 150 to 270° C; only liquid, above 305°C

5−P24 Immediately below all of the $(x + L)$ fields with horizontal lines at 903, 835, 700, 598, and 424° C.

5−P25† 700° C (%Al): α, (0 > 8%); $\alpha + \beta$, (8 > 11%); β, (11 > 12.5); $\beta + \gamma$, (12.5 > 15); γ, (15 > 21); $\gamma + \epsilon_2$, (21 > 23); ϵ_2, (23 > 24); $\epsilon_2 + L$, (24 > 36); L, (36 > 100%)

450° C: α, (0 > 9%); $\alpha + \gamma_2$, (9 > 16); γ_2, (16 > 20), $\gamma_2 + \delta$, (20 > 21), δ, (21−21.5); $\delta + \zeta$, (21.5 > 24.5); ζ, (24.5 > 26); $\zeta + \eta_2$, (26 > 28); η_2, (28 > 29); $\eta_2 + \theta$,

(29 > 46.5); θ, (46.5 > 47); $\theta + \kappa$, (47 > 97.5); κ, 97.5 > 100)

900° C: α, (0 > 7.5); $\alpha + \beta$, (7.5 > 10); β, (10 > 14); $\beta + \gamma_1$, (14 > 15.5); γ_1, (15.5 > 18); $\gamma_1 + \epsilon_1$, (18 > 21); ϵ_1, (21 > 22); $\epsilon_1 + $ Liq., (22 > 25.5); Liq., (25.5 > 100)

5−P26 The solidus temperature is the upper limit for hot-working a metal, if the presence of a liquid is to be avoided. Liquid has the potential for introducing "hot-shortness" into the metal — meaning rupturing during deformation.

5−P27† 300° C (%Mg): α, (0 > 6%); $\alpha + \beta$, (6 > 35); β, (35 > 37); $\beta + \beta'$, (37 > 40⁺); β', (40⁺ > 42); $\beta' + \gamma$, (42 > 49); γ, (49 > 57); $\gamma + \epsilon$, (57 > 94); ϵ, (94 > 100%).

400° C: α, (0 > 11%); $\alpha + \beta$, (11 > 35); β, (35 > 37); $\beta + \gamma$, (37 > 46); γ, (46 > 59); $\gamma + \epsilon$, (59 > 91); ϵ, (91 > 100%)

5−P31 (a) 11% Mg   (b) 76% Mg
(c) Liquid: 40 Al−60 Mg (not saturated)
(d) Liq., 24 Al−76 Mg; ϵ, 7.5 Al−92.5 Mg

5−P32 (a) 78% Al_2O_3
(b) 6% Al_2O_3
(c) α: 98 Al_2O_3-2 ZrO_2; β: 4 Al_2O_3-96 ZrO_2
(d) Liquid: 42 Al_2O_3-58 ZrO_2; β: 6 Al_2O_3-94 ZrO_2

5−P33 300, 400, 500, 600, 700° C: α (65 Cu − 35 Zn); 800° C: α (66 Cu − 34 Zn), and β (61 Cu − 39 Zn); 900° C: α (67.5 Cu − 32.5 Zn), and β (63 Cu − 37 Zn); 1000 and 1100° C: Liquid, (65 Cu − 35 Zn)

5−P34 (Fig. 5−6.2)

Temp. ° C	α: Cu−Sn	ϵ: Cu−Sn	Liq.: Cu−Sn
1100	–	–	90−10
1000	98−2	–	89−11
900	95−5	–	81−19
800−400	90−10	–	–
300	94−6	62.8−37.2	–
200	98−2	62.8−37.2	–
100	~100−0	62.8−37.2	–

5−P41 (a) 620° C   (b) 600° C
(c) 590° C   (d) 520° C

---

† Color displacement during printing may produce a slight variance (±0.5%).

5–P42 (a) 600° C (b) 575° C
(c) 560° C (d) 465° C

5–P43 (a) 26 Ag–74 Cu (b) 23 Ag–77 Cu

5–P44 (a) 75 Al_2O_3–25 ZrO_2
(b) 40 Al_2O_3–60 ZrO_2

5–P45 (a) 1370° C (2500° F)
(b) Liquid: 67 Ni–33 Cu, α: 76 Ni–24 Cu

5–P46 1100° C no solid; 1000° C 170 g; 900° C 1000 g; 800–400° C 1500 g; 300° C 1350 g; 200° C 1150 g; 100° C 1100 g

5–P47 3.2 g

5–P48 300 → 700° C, 200 g; 800° C, ~180 g; 900° C, ~100 g; 1000° C, none

5–P49 780° C. Three phases coexist at this invariant temperature (Eq. 5–2.2).

5–P51 β (24% $CaTiO_3$) $\xrightarrow[105°C]{}$ α (20% $CaTiO_3$) + γ (91% $CaTiO_3$)

5–P52 β (25% Sn) $\xrightarrow[586°C]{}$ α (15.8% Sn) + γ (25.5% Sn)
γ (27% Sn) $\xrightarrow[520°C]{}$ α (15.8% Sn) + δ (32% Sn)
δ (32.6% Sn) $\xrightarrow[350°C]{}$ α (11% Sn) + ϵ (37% Sn)
ζ (34% Sn) $\xrightarrow[580°C]{}$ δ (33% Sn) + ϵ (37% Sn)

5–P53 L (21 w/o An) $\xrightarrow[15°C]{}$ α (0 w/o An) + β (49.7 w/o An)
L (90 w/o An) $\xrightarrow[-12°C]{}$ β (49.7 w/o An) + γ (100 w/o An)

5–P54 See *Study Guide.*

5–P61 (a) 780°C (b) 750^{+}° C (c) 800^{+}° C

5–P62 (a) 5.6 lb (b) 5.6 lb (c) 5.3 lb

Chapter 6

6–P11 It takes energy (+) to break the C═C double bonds, before the C—C— single bonds release energy (−).

6–P12 Rock candy, candy part of peanut brittle. See *Study Guide.*

6–P13 7.5 w/o Cu was at the practical limit for processing a single-phase alloy.

6–P14 See *Study Guide.*

6–P21 See *Study Guide.*

6–P22 (a) 24 Al–76 Mg (b) 3.2 kg liquid
(c) 437°C (d) 32 Al–68 Mg

6–P23 The solidifying Si is almost pure. The Al remains in the liquid. See *Study Guide.*

6–P31 $r_c \approx 1.25$ nm, or approximately 1000 atoms, rather than 60–70

6–P32 Foreign particles, external surfaces, grain boundaries, dislocations, solute atoms

6–P41 31 MPa (4500 psi)

6–P42 (a) ~18 × 10^{-6}/° C (b) α increases with temperature

6–P43 Mean > median energy

6–P44 1/(3.6 × 10^6 atoms)

6–P45 928° C

6–P51 4 × 10^{17}/m²·s

6–P52 (a) 200/ (b) 150,000/1 (c) 6000/1
(d_a) $PF_{bcc} < PF_{fcc}$ (d_b) $R_C < R_{Ni}$
(d_c) thermal energy

6–P53 −3.3 × 10^{29}/m³/m

6–P54 (a) 2 × 10^{-15} m²/s
(b) $\log_{10} = (-15 + 0.3)$, OK

6–P55 $J_{Zn\,in\,Cu}/J_{Al\,in\,Cu} = 1.6$

6–P56 1293° C

6–P57 (a) $PF_{bcc} < PF_{fcc}$ (b) $r_C < R_{Fe}$
(c) $PF_{boundary} < PF_{xtal}$ (d) Cf. T_m

6–P58 (a) 3 × 10^{-23} m²/s (b) 300/m²·s

6–P59 (a) 500
(b) O^{2-} ions are larger, and $\frac{1}{3}$ of the cation sites are open

Chapter 7

7–P11 ~80 mm²/mm³

7–P12 (a) ~60 mm²/mm³
(b) ~15 mm²/mm³ (380 in.²/in.³)

7–P13 2.8

7–P14 The grains continue to grow but at a much *reduced* rate.

7–P21 (a) Grain size, grain shape, and grain orientation
(b) (Size, shape, orientation) plus phase fraction, and phase distribution

7–P22 (a) 10 μm   (b) 4 μm

7–P23 (a) 0.05 v/o   (b) 0.2

7–P24 2.74 g/cm³

7–P25 See *Study Guide.*

7–P31 β (24 CaTiO₃) $\xrightarrow{105°\,C}$ α (20 CaTiO₃) + γ (91 CaTiO₃)

7–P32 β (11.5 Al) $\xrightarrow{565°\,C}$ α (9 Al) + γ_2 (15 Al)

γ_1 (15 Al) $\xrightarrow{780°\,C}$ β (13 Al) + γ_2 (15.5 Al)

χ (15.5 Al) $\xrightarrow{965°\,C}$ β (14.5 Al) + γ_1 (16 Al)

7–P33 (a) α and γ (b) α (99.99⁺ Fe)
γ (99.7 Fe–0.3 C)   (c) 66γ–34α

7–P34 See *Study Guide.*

7–P35 540 unit cells

7–P36 0.160 nm³

7–P37 R.T. $\xrightarrow{\alpha+\overline{C}}$ 727° C; ($\alpha + \overline{C} + \gamma$) at 727° C;
727° C $\xrightarrow{\alpha+\gamma}$ 770° C; 770° C $\xrightarrow{\gamma}$ 1200° C

7–P38 α: 0% at 20° C to 0.02% at 727° C to 0% at 912° C
$\overline{C}$: Fe₃C has 6.7% C at all temperatures of interest
γ: ~0.8% at 727⁺° C; dropping to 0.45% at 770° C; 0.45% above 770° C
Also, label the ($\alpha + \overline{C}$), the ($\alpha + \gamma$), and the ($\gamma + \overline{C}$) fields.

7–P39 See *Study Guide.*

Chapter 8

8–P11 0.84 mm (0.033 in.)

8–P12 +0.05 v/o

8–P13 (a) 140 GPa (20,500,000 psi)
(b) 40 GPa (5.8 Mpsi)

8–P14 284 MPa (41,000 psi)

8–P15 (a) −0.35%   (b) −0.7%

8–P16 0.87 mm (0.034 in.)

8–P17 0.38°

8–P18

Young's modulus:	Axial stress/axial strain	$E = s/e$
Shear modulus:	Shear stress/shear strain	$G = \tau/\gamma = \tau/\tan\alpha$
Bulk modulus:	Hydrostaticpressure/ volumechange or [compressibility]⁻¹	$K = P_h/(\Delta V/V)$ $K = 1/\beta$

8–P21 See *Study Guide.*

8–P22 0.005 (0.005)

8–P23 (a) 0.43   (b) 1.54

8–P31
[11$\bar{1}$](101)	[1$\bar{1}$1](110)	[11$\bar{1}$](011)
[$\bar{1}$11](101)	[$\bar{1}$11](110)	[1$\bar{1}$1](011)
[1$\bar{1}$1](10$\bar{1}$)	[11$\bar{1}$]($\bar{1}$10)	[111](01$\bar{1}$)
[111](10$\bar{1}$)	[111]($\bar{1}$10)	[$\bar{1}$11](01$\bar{1}$)

8–P32 (a) See *Study Guide.*
(b) [1$\bar{1}$0](110)   [101](10$\bar{1}$)
[110]($\bar{1}$10)   [01$\bar{1}$](011)
[10$\bar{1}$](101)   [011](0$\bar{1}$1)
(c) 0.373 nm

8–P33 *Resolved shear stress:* Axial stress, *s*, resolved into the shear component, τ
Critical shear stress: The shear stress on the slip plane that initiates slip

8–P34 See *Study Guide.*

8–P35 Both metals are hexagonal with very few slip systems

8–P36 $b_{111} > b_{100}$

8–P41 76°

8–P42 14 mm (0.56 in.)

8–P43 148.3 MPa·$\sqrt{m}$

8–P44 110 J

8–P45 See *Study Guide.*

Chapter 9

9–P11 4.5 v/o

9–P12 338 ft

9–P13 15.0 mm

9–P14 0.08

9–P15 1.25 v/o

9–P16 (a) 9.7 v/o   (b) 2.13 g/cm³
(c) 2.36 g/cm³

9–P21 See *Study Guide.*

9–P22 $(S_y)_{Be} = 2.3(S_y)_{Sn}$

9–P23 65 Cu–35 Zn

9–P31 (a) 24.3 m
(b) More ductile—undeformed;
higher yield strength—deformed

9–P32 (a) 18%   (b) 280 MPa, 22%

9–P33 330 MPa

9–P34 See *Study Guide.*

9–P35 0.95 mm

9–P36 Choice 1: Hot work from 2.5 → 1.1 mm; cold work 17% to 1.0 mm
Choice 2: Cold work-anneal cycles → 1.1 mm; anneal; cold work 17% to 1.0 mm

9–P37 *Step 1* 30% CW; *Step 2* CW 64%; *Step 3* Anneal > 300° C, 1 hr; *Step 4* CW 30%

9–P38 (a) 12,000 hr   (b) 11,500 hr

9–P39 415° C

9–P41 Age hardening is not possible without supersaturation and precipitation.

9–P42 260° C (or 500° F)

9–P43 6×10^7 s (~2 yrs)

9–P44 Solution treat

9–P51 100% increase

9–P52 50% decrease

9–P61 (a) ~850° C (~1575° F)
(b) ~780°C (~1440° F)
(c) ~855° C (~1580° F)

9–P62 At 800° C: all γ; at −60° C: all M; at 300° C, plus 10 s: M

9–P63 (a) only γ (b) $\gamma + (\alpha + \overline{C})$ (c) $(\alpha + \overline{C})$

9–P64 See *Study Guide.*

9–P65 (a) γ (b) M + some retained γ

9–P66 At 730° C, at 1045 steel contains a mixture of α and γ (0.8% C)

9–P67 8 s

9–P68 See *Terms and Concepts* and *Study Guide.*

9–P69 C diffuses more rapidly than Mo. See *Study Guide.*

9–P71 *Hardness:* Resistance to penetration or to scratching
Hardenability: "Ability" to produce maximum hardness

9–P72 (a) 17° C/s   (b) 13 mm
(c) 50° C/s; ~5 mm

9–P73 See *Study Guide.*

9–P74 (a) 25 R_c (b) 28 R_c

9–P75 (a) 24 R_c (b) 26 R_c

9–P76 (a) $D_{qe} = 1.5$ mm; 200°/s; $D_{qe} = 8$ mm; 30°/s
(b)

	S	$\frac{3}{4} R$	$M-R$	C
D_{qe}	1.5 mm	4+	6⁻	9
R_c	55	38	31	27

(c) See *Study Guide*.

9–P77 See *Study Guide*.

9–P78

	S	$\frac{3}{4} R$	$M-R$	C
D_{qe}	11 mm	18.5	22	27
R_c	56	53	52	51

9–P79

Location	D_{qe}	(a) Carburized		(b) Non-carburized	
Surface	6.5 mm	0.62% C	34 R_c	0.20% C	23⁻ R_c
2-mm	7	0.35	28	"	
$\frac{3}{4}$ radius	7.5	0.20	22+	"	22+
$M-R$	8.5	"	22	"	22
Center	10	"	21	"	21

9–P81 See *Study Guide*.

9–P82 $s_p \approx +25$ MPa (+3500 psi);
$s_b \approx -55$ MPa (−8000 psi)

9–P83 $s_p \approx +55$ MPa (+8000 psi);
$s_b \approx -25$ MPa (−3500 psi)

Chapter 10

10–P11 14×10^6 Pa·s

10–P12 (a) 24×10^{-21} J (b) 0.0016 Pa·s

10–P21 0.20

10–P22 −47,300 lb$_f$ (−210,000 N)

10–P41 $f_{St} = 0.49$; $f_{Al} = 0.51$

10–P42 See *Study Guide*.

10–P43 1.4 g/cm³

10–P44 0.00033 W/mm·°C

10–P45 (a) 94% (b) 0.08 ohm/m

10–P46 4.6×10^{-6}/°C

10–P51 162 amu

10–P52 (a) 8.2 g (b) 1.38 g/cm³

Chapter 11

11–P11 (a) 0.15 amp (b) 0.225 W
(c) 810 J (0.225 W·h)

11–P12 (a) 5×10^6 A/m² (b) 22.5 W
(c) 0.5 mm

11–P13 Al, Al alloys, brass, Cu, Mg, sterling silver

11–P14 32 m/s

11–P15 4×10^{21}/m³

11–P16 9×10^{20}/m³

11–P17 Electrons— Negative, delocalized.
Anions— Atoms with excess electrons, negative.
Cations— Atoms with missing electrons, positive.
Electron holes— Positive, missing delocalized electrons.

11–P21 60 ohm·nm

11–P22 390×10^{-9} ohm·m

11–P23 Only the alloys

11–P24 (0.16 ohm·nm)/(a/o Cu); Cu is the solvent in Table 11–2.2,

11–P25 Pure metals and low alloys have a W–F ratio of $\sim 7 \times 10^{-6}$ W·ohm/m. See *Study Guide*.

11–P26 380 W·ohm/m

11–P27 (a) 9 ↔ 22% Zn (b) Use 20%

11–P28 (a) 1.7 mm (b) 1.4 mm (c) 1.1 m

11–P29 Brass: none. Cu–Ni: 12 ↔ 30%, and >80% Ni. Use 85 Cu–15 Ni.

11–P31 (a) 0.17 (b) 0.23 (c) 0.30 (d) 0.37

11–P32
$E_f + 0.15$ eV = 0.003 $E_f - 0.15$ eV = 0.997

$E_f + 0.10$ eV $= 0.019$ $E_f - 0.10$ eV $= 0.981$
$E_f + 0.05$ eV $= 0.121$ $E_f - 0.05$ eV $= 0.879$
$E_f = 0.500$

11–P33 See *Study Guide.*

11–P34 See *Study Guide.*

11–P41 (a) 0.043   (b) 0.96

11–P42 (a) 62.5 $m^2/V \cdot s$ (b) 0.39 $m^2/V \cdot s$
(c) 1.95 m/s, and 0.195 m/s

11–P43 $\sigma_{InP} > \sigma_{InAs}$

11–P44 $1/10^{13}$

11–P45 (a) 0.0074 V   (b) 6.6×10^{20} m^3
(c) 24.5 $(ohm^{-1} \cdot nm^{-1})$

11–P46 10^{-3} $ohm^{-1} \cdot m^{-1}$

11–P47 (a) 0.68 eV   (b) 640 $ohm^{-1} \cdot m^{-1}$

11–P51 (a) 5×10^{28} Si/m^3 (b) $3.3 \times 10^{21}/m^3$

11–P52 1.7×10^6

11–P53 Greater
$(3.4$ $ohm^{-1} \cdot m^{-1} > 5 \times 10^{-4}$ $ohm^{-1} \cdot m^{-1})$

11–P54 Accept electrons from VB; donate electrons to the CB

11–P55 90° C

11–P56 See *Study Guide.*

11–P57 See *Study Guide.*

11–P58 (a) p-type   (b) $7.8 \times 10^{18}/mm^3$

11–P59 (a) p-type   (b) 42.5 $ohm^{-1} \cdot m^{-1}$

11–P61 $+0.03°$ C

11–P62 4° C

11–P63 6.8

Chapter 12

12–P11 $\sim 33,000$ J/m^3

12–P21 *Strain-hardened* steels contain dislocations. *Annealing* removes dislocations and permits grain growth.

12–P22 (a) Low remenance, low coecive force, low energy product
(b) Few dislocations; minimum boundaries

12–P23 ~ 0.07 J

12–P31 0.25×10^6 A/m

12–P32 (a) 34 β /u.c.   (b) 32 β /u.c.

12–P33 0.2×10^6 A/m

12–P41 (a) 1.57×10^{-23} $A \cdot m^2/Co$ (b) 1.7 β/Co

12–P42 See *Terms and Concepts* and *Study Guide.*

12–P43 120,000 A/m

Chapter 13

13–P21 10 V

13–P22 250×10^9 el/m^2

13–P23 2.5

13–P31 See *Study Guide.*

13–P32 1200 cm^2

13–P41 -10^5 psi $(-700$ MPa$)$

13–P42 0.0107 nm

13–P43 (a) $PbZrO_3$ (b and c) See *Study Guide.*

13–P44 See *Study Guide.*

13–P45 See *Study Guide.*

13–P51 (a) 1.524   (b) 1.97×10^8 m/s

13–P52 -3.7 db/km

13–P53 Glass 1: 12°43′   Glass 2: 12°51′

13–P61 (a) 267 min   (b) 764 min

13–P62 125

Chapter 14

14–P21 3300 C/g

14–P22 (a) 97.2 g/hr   (b) 0.01 mm

14–P23 -2.1 V (with Mg anodic)

14–P24 Convenient. See *Study Guide.*

14–P25 *Anodes* provide electrons to the external circuit. *Cathodes* receive electrons from the external circuit.

14–P26 13.4 g/l

14–P27 $m_{Au} > m_{Zn}$ (1.1)

14–P28 $Zn \rightarrow Zn^{2+} + 2$ e$^-$ supplies the electrons

14–P29 (a) $O_2 + 2 H_2O + 4$ e$^- \rightarrow 4(OH)^-$
(b) $Cu^{2+} + 2$ e$^- \rightarrow Cu$
(c)(closed): $2 H^+ + 2$ e$^- \rightarrow H_2\uparrow$
  (open): $O_2 + 2 H^+ + 2$ e$^- \rightarrow 2 H_2O$
(d) $O_2 + 2 H_2O + 4$ e$^- \rightarrow 4(OH)^-$
(e) $O_2 + 2 H_2O + 4$ e$^- \rightarrow 4(OH)^-$

14–P31 ~ 11 mo

14–P32 (a) 6.6×10^{10} Al/s
(b) 10^{-6} A/mm^2

14–P33 (a) $O_2 + 2 H_2O + 4$ e$^- \rightarrow 4(OH)^-$
(b) No tin in the water

14–P34 See *Study Guide.*

14–P35 Oxygen deficient area under the barnacle

14–P36 See *Study Guide.*

14–P37 See *Study Guide.*

14–P41 40 days

14–P42 See *Study Guide.*

14–P51 See *Study Guide.*

14–P52 (a) See *Study Guide.*   (b) 0.001 %/hr

14–P61 (a) 366 da
(b) 187 *additional* days (310 total)

14–P62 2.6 MPa

14–P63 10%

14–P81 Not enough energy

14–P82 (a) 1.09×10^{15}/s
(b) Thermal energy is also present.

TERMS AND CONCEPTS

Acceptor An impurity that accepts electrons from the valence band and, therefore, produces electron holes in the valence band.

Acceptor saturation Filling of acceptor sites. As a consequence, additional thermal activation does not increase the number of intrinsic carriers.

Activation energy (E or Q) Energy barrier that must be met before a reaction can occur.

Additive (plastics) Minor addition for modifying properties or performance.

Age hardening See *precipitation hardening.*

Alloy A metal containing two or more elements.

Amorphous Noncrystalline, and without long-range order.

Anion Negative ion.

Anisotropic Having different properties in different directions.

Annealing point (glass) Stress-relief temperature ($\sim T_g$). This temperature provides a viscosity of $\sim 10^{12}$ Pa·s ($\sim 10^{13}$ poises).

Annealing (strain-hardened metal) Heating a cold-worked metal to produce recrystallization and therefore softening.

Annealing (steels) For a *full* anneal, austenite is formed, then the steel is cooled slowly enough to form pearlite. (Annealing to remove strain hardening is called *process* anneal. In that process, the steel is heated to just below the eutectoid temperature.)

Anode The electrode that supplies electrons to the external circuit. The electrode that undergoes corrosion. The negative electrode.

Anode reactions Electrochemical oxidation reactions.

Anodized Surface coated with an oxide layer; achieved by making the part an anode in an electrolytic bath.

Antiferromagnetism Magnetic coupling with balanced alignment of oppositely oriented spins; therefore, the net magnetism is zero.

Arrhenius equation Thermal activation relationship. (See comments with Example 6–4.2.)

ASTM grain size (G.S.#) Standardized grain counts. (See Eq. 7–1.3).

Atactic Lack of long-range repetition in a linear polymer (as contrasted to isotactic).

Atomic mass unit (amu) One-twelfth of the mass of C^{12}; also, g/(0.602 . . . $\times 10^{24}$).

Atomic number The number of electrons possessed by an uncharged atom. The number of protons per atom.

Atomic radius (R) One-half of the interatomic distance of like atoms.

Atomic weight Atomic mass expressed in atomic mass units, or in g per mole.

Attenuation (optical) Reduction of radiation intensity with distance.

Austempering Transformation of $\gamma \rightarrow (\alpha + \overline{C})$ below the "knee" of the I-T curve to form bainite, which is a dispersion of carbide in a ferrite matrix.

Austenite (γ) Face-centered cubic (fcc) iron, or an iron-rich, fcc solid solution.

Austenite decomposition Eutectoid reaction that changes austenite to (α + carbide).

Austenization Heat treatment to dissolve carbon into fcc iron, thereby forming austenite.

Bainite Microstructure of ($\alpha + \overline{C}$) formed isothermally below the "knee" of the I-T curve.

Band, conduction (CB) Band above the energy gap. Electrons become carriers when they are activated to this band.

Band, valence (VB) Filled energy band below the energy gap. Conduction in this band requires electron holes.

***BH* product** Energy required for demagnetization. The maximum product of the second quadrant usually is used.

Bifunctional (molecule) Molecule with two reaction sites for joining with adjacent molecules.

Birefringence Difference in index of refraction with orientation.

Blow molding Processing by expanding a parison into a mold by air pressure.

Body-centered cubic (bcc) The center of the cube is identical to the cube corners. All translations of $\pm a/2, \pm b/2, \pm c/2$ from any location produce positions that are identical in every respect.

Boltzmann's constant (k) Thermal energy coefficient ($13.8 \ldots \times 10^{-24}$ J/K, or 86.1×10^{-6} eV/K).

Bond angle Angle between stereospecific bonds in molecules, or in covalent solids.

Bond energy Energy required to separate two chemically bonded atoms. Generally expressed as energy per mole of 0.6×10^{24} bonds.

Bond length Interatomic distance between centers of covalently bonded atoms.

Boundary stresses Shear stresses between phases that arise from differential dimensional changes.

Bragg's law Diffraction law for periodic structures (Eq. 3–8.2).

Branching Bifurcation of a polymer chain.

Brass An alloy of copper (>60 percent) plus zinc.

Bronze An alloy of copper and tin (unless otherwise specified; e.g., an aluminum bronze is an alloy of copper and aluminum).

Bulk modulus (K) Hydrostatic pressure per unit volume strain.

Butadiene-type compound Prototype for several rubbers based on $C{=}C{-}C{=}C$. (See Table 4–3.1.)

Carbide ($\overline{C}$) Compound of metal and carbon. (Unless specifically stated otherwise, refers to the iron-base carbide (Fe_3C) in this text, and is labeled $\overline{C}$.)

Carburize Introduction of carbon through the surface of a steel by diffusion. The purpose is to harden the surface.

Casting Shaping by solidifying a liquid or suspension.

Cathode The electrode that receives electrons from the external circuit. The electrode on which electroplating is deposited. The positive electrode.

Cathode reactions Electrochemical reduction reactions. (See Eqs. 14–2.2b, 14–2.4, 14–2.5, and 14–2.6.)

Cation Positive ion.

Cell, composition An electrochemical cell between electrodes of different compositions.

Cell, concentration An electrochemical cell arising from nonequal electrolyte concentrations. (The more dilute solution establishes the anode.)

Cell, galvanic An electrochemical cell containing two dissimilar metals in an electrolyte.

Cell, oxidation An electrochemical cell arising from unequal oxygen potentials. (The oxygen-deficient area becomes the anode.)

Cell, stress An electrochemical cell in which a plastically strained area produces an anode.

Ceramic A material that is a compound of metallic and nonmetallic elements.

Charge carriers Electrons in the conduction band provide n-type (negative) carriers. Electron holes in the valence band provide p-type (positive) carriers.

Charge density (D) Charge per unit area (e.g., on an electrode).

Cis Prefix denoting unsaturated positions on the same side of the polymer chain. (Also see *trans.*)

Coalescence Growth of dispersed particles within a microstructural matrix.

Coercive field (electric, $-\mathscr{E}_c$) Electric field required to remove remanent polarization.

Coercive field (magnetic, $-H_c$) Magnetic field required to remove remanent magnetic induction.

Cold work, percent Amount of cold working, calculated from the change in cross-sectional area, $100(A_0 - A_f)/A_0$.

Cold working Plastic deformation below the recrystallization temperature.

Component (phases) One of the basic chemical substances required to create a chemical mixture or solution.

Composite Unified combinations of two (or more) distinct materials.

Compounds, III–V Semiconducting compounds of Group III and Group V elements.

Compound A phase composed of two or more elements in a given ratio.

Concentration gradient (dC/dx) Change in concentration with distance. Concentration is expressed in number per unit volume.

Conductivity Transfer of thermal or electrical energy along a potential gradient.

Conductivity (electrical) Product of carrier density, carrier charge, and charge mobility. Reciprocal of resistivity. Charge flux per unit voltage gradient.

Coordination number (CN) Number of closest ionic or atomic neighbors.

Copolymer Polymers with more than one type of mer.

Copolymer, block Copolymer with clustering of like mers along the chain.

Copolymer, graft Polymeric molecule with branches of a second type of polymer.

Coring Segregation during solidification (occurs because the initial solid does not equilibrate with the succeeding solid).

Corrosion Surface deterioration by electrochemical reaction.

Coulombic forces Forces between charged particles, particularly ions.

Covalent bond A pair of shared electrons that produces a bond between two adjacent atoms.

Creep Time-dependent strain that occurs as a result of mechanical stresses.

Creep rate Creep strain per unit time.

Crevice corrosion Localized corrosion in an oxidation cell.

Critical shear stress Minimum shear stress to initiate slip on a crystal plane.

Cross-linking The tying together of adjacent polymer chains.

Crystal A solid with a long-range repetitive pattern of atoms in the three coordinate directions.

Crystal direction [*uvw***]** A line from an arbitrary origin through a selected unit-cell location. The indices are the lattice coefficients of that location.

Crystal plane (*hkl***)** A two-dimensional array of ordered atoms. (See also *Miller indices.*)

Crystal pulling Method of growing single crystals by slowly pulling a seed crystal away from a molten pool.

Crystal system Categorization of unit cells by axial and dimensional symmetry.

Curie point (electric) Transition temperature between a symmetric crystal and polar crystal.

Curie temperature (magnetic, T_c) Transition temperature for magnetic-domain formation.

Debonding Uncoupling of reinforcement fibers from the matrix of a composite.

Decarburization Removal of carbon from the surface zone of steel by elevated temperature oxidation.

Defect semiconductor Nonstoichiometric compounds of elements that have more than one valence.

Defect structure Nonstoichiometric compounds that contain either vacancies or interstitials within the structure.

Deformation, elastic Reversible deformation without permanent atomic (or molecular) displacements.

Deformation, plastic Permanent deformation arising from the displacement of atoms (or molecules) to new surroundings.

Deformation, viscoelastic Concurrent viscous flow and elastic deformation.

Degradation Destruction of polymers to smaller molecules.

Degree of polymerization, *n* Mers per average molecule. Also, molecular mass/mer mass.

Delocalized electrons Valence electrons not bound to specific atoms.

Depleted zone The region adjacent to an $n-p$ junction that lacks charge carriers. (A reverse voltage bias increases the width of this depleted zone.)

Devitrification Crystallization of glass. Process for producing "glass-ceramics."

Diamagnetism Repulsion of magnetic flux because the permeability is less than that of a vacuum. (Zero for superconductors.)

Diamond-cubic structure The fcc structure of an element with CN = 4. (See Fig. 3–2.6a.)

Dielectric An insulator. A material that can be placed between two electrodes without conduction.

Dielectric constant, relative (κ) Ratio of charge density arising from an electric field (1) with, and (2) without, the material present.

Dielectric strength Electrical breakdown potential of an insulator per unit thickness.

Diffraction line The diffracted beam from a crystal surface. (Detected photographically or with a Geiger-type counter.)

Diffraction (x-ray) Deviation of an x-ray beam by a periodic structure.

Diffusion The movement of atoms or molecules in a material.

Diffusion flux (J) Transport per unit area and time.

Diffusivity (D) Diffusion flux per unit concentration gradient.

Dipole An electrical couple with positively and negatively charged ends.

Dipole moment (p_e) Product of electric charge and charge separation distance.

Dislocation, edge ($\perp$) Linear defect at the edge of an extra crystal plane. The slip vector is perpendicular to the defect line.

Dislocation, screw ($\S$) Linear defect with slip vector parallel to the defect line.

Dispersion Difference in index of refraction with wavelength.

Domain (electric) Microstructural region with coordinated alignment of electric dipoles.

Domain (magnetic) Microstructural region of coordinated magnetic alignments.

Domain boundary (Bloch wall) Transition zone between magnetic domains; moves during magnetization and demagnetization.

Domain boundary (ferroelectric) Transition zone between electrical domains.

Donor An impurity that donates carriers to the conduction band.

Donor exhaustion Depletion of donor electrons. When it occurs, additional thermal activation does not increase the number of extrinsic carriers.

Drawing Mechanical forming by tension through, or in, a die (e.g., wire drawing (Fig. 9–1.3), or sheet drawing). This process usually is carried out at temperatures below the recrystallization temperature.

Drift velocity ($\bar{v}$) Net velocity of electrons in an electric field.

Ductility Total permanent (plastic) strain prior to fracture; measured as elongation, or as reduction of area.

Ductility-transition temperature (T_{dt}) Temperature that separates the regime of brittle fracture from the higher temperature range of ductile fracture.

Earlywood First wood growth during the growing season. The part of the wood-growth ring that has larger, thinner walled cells. (Also called spring wood.)

Elastic modulus (E) Stress per unit of elastic strain.

Elastic strain (e_{el}) Strain that is recoverable when the load is removed.

Elastomer Polymer with a large elastic strain. This strain arises from the unkinking of the polymer chains.

Electric field ($\mathscr{E}$) Potential gradient, V/m.

Electrode potential (ϕ) Voltage developed at an electrode (referenced to a standard electrode).

Electrolyte Conductive ionic solution (liquid or solid).

Electron charge (q) The charge of 0.16×10^{-18} coul (or 0.16×10^{-18} A·s) carried by each electron.

Electron hole (p) Electron vacancy in the valence band that serves as a positive-charge carrier.

Electron–hole pair An electron in the conduction band and an accompanying electron hole in the valence band, which results when the electron jumps the energy gap in an intrinsic semiconductor.

Electronic repulsion Repelling force of too many electrons in the same vicinity.

Elongation (El.) Total plastic strain before fracture, measured as the percent of axial strain. (A gage length must be stated.)

End-quench test (Jominy bar) Standardized test, performed by quenching from one end only, for determining hardenability.

Endurance limit The maximum stress allowable for unlimited stress cycling.

Energy distribution Spectrum of energy levels arising from thermal activation.

Energy band Permissible range of energy levels for delocalized electrons in a material.

Energy gap (E_g) Forbidden energies between the valence band and the conduction band.

Energy well Potential energy minimum between two atoms.

Engineering The adaptation of materials and/or energy for society's needs. (Other definitions are possible.)

Equilibrium The state at which net reaction ceases (because the minimum free energy has been reached).

Eutectic, induced A eutectic reaction resulting from segregation during solidification.

Eutectic composition Composition of the liquid-solution phase that possesses a minimum melting temperature (at the intersection of two solubility curves).

Eutectic reaction $L_2 \rightleftharpoons S_1 + S_3$.

Eutectic temperature Temperature of the eutectic reaction at the intersection of two solubility curves.

Eutectoid composition Composition of the solid-solution phase that possesses a minimum decomposition temperature (at the intersection of two solid solubility curves).

Eutectoid reaction $S_2 \rightleftharpoons S_1 + S_3$.

Eutectoid temperature Temperature of the eutectoid reaction at the intersection of two solid solubility curves.

Extrusion Shaping by pushing the material through a die (Fig. 9–1.2c).

Face-centered cubic (fcc) A structure in which the centers of the cube faces are identical to one another and to the cube corners. All translations of $\pm a/2, 0, \pm c/2$ (and their permutations) from any location produce positions that are identical in every respect.

Family of directions $\langle uvw \rangle$ Crystal directions that are identical except for our arbitrary choice of axes.

Family of planes $\{hkl\}$ Crystal planes that are identical except for our arbitrary choice of axes.

Fatigue Time-delayed fracture; commonly occurs under cyclic loading or in a reactive environment.

Fatigue strength Design stress during the expected lifetime of a plastic.

Fermi distribution, $F(E)$ Probability of electron occupancy over the range of energy levels (Eq. 11–3.2).

Fermi energy (E_f) Energy level with a 50-percent probability of occupancy. Maximum filled energy level of a metal at $0°$ K.

Ferrimagnetism Net magnetism arising from unbalanced alignment of magnetic ions within a crystal.

Ferrite (α) Body-centered cubic iron; or an iron-rich, bcc solid solution.

Ferrites Compounds containing trivalent iron; commonly magnetic.

Ferroelectrics Materials with spontaneous electric-dipole alignment.

Ferromagnetism Metallic materials with spontaneous magnetic alignment.

Fiber debonding Loss of cohesion between fiber and matrix caused by shear stresses.

Fiber-reinforced plastic (FRP) A product with a polymer matrix and a fiber reinforcement (commonly glass).

Fick's first law Proportionality between diffusion flux and concentration gradient.

Filler Particulate or fibrous additive for reinforcement and dilution.

Floating zone Method of growing single crystals, which melts an isolated zone within a bar. The material solidifies as a single crystal on the bottom side of the rising molten zone.

Fluorescence Luminescence that occurs almost immediately after excitation.

Forward bias Potential direction that moves the charge carriers across the n–p junction.

Fracture, brittle Fracture with negligible plastic deformation and minimum energy absorption.

Fracture, ductile Fracture accompanied by plastic deformation and, therefore, by energy absorption.

Fracture toughness The critical stress intensity factor, K_{Ic}, for fracture propagation.

Galvanic protection Corrosion protection achieved by making the material cathodic.

Galvanic series Sequence (cathodic to anodic) of corrosion susceptability for common metals. Varies with passivity and with electrolyte composition. (See also *electrode potentials,* which are for standard electrolytes.)

Galvanized steel Zinc-coated steel. The zinc serves as a sacrificial anode.

Glass An amorphous solid below its transition temperature. A glass lacks long-range crystalline order, but normally has short-range order.

Glass-transition temperature (T_g) Transition temperature between a supercooled liquid and its rigid glassy solid.

Grain Individual crystal in a polycrystalline microstructure.

Grain (wood) Seasonal growth pattern, alternating between earlywood and latewood.

Grain boundary The zone of crystalline mismatch between adjacent grains.

Grain boundary area (S_v) Intergranular area/unit volume; for example, in.^{2}/in.^{3} or mm^2/mm^3.

Half-cell reaction See Table 14–2.1.

Hall effect Cross-voltage induced by a current moving at $90°$ to a magnetic field.

Hardenability The ability to develop maximum hardness by avoiding the $\gamma \rightarrow (\alpha + \overline{C})$ reaction.

Hardenability curve Hardness profile of end-quench test bar.

Hardness Index of resistance to penetration (or to scratching if the indentor is moved). There are several common test procedures — Brinnell, Rockwell, and so on.

Hardness traverse Profile of hardness values.

Heat of fusion (H_f) Energy per mole (or other stated unit) to melt a material.

Hexagonal close-packed metal (hcp) A hexagonal metal with CN = 12, and PF = 0.74.

High-stiffness composite Composite with a high elastic modulus–to-density ratio, E/ρ.

Homogenization (soaking) Heat treatment to equalize composition by diffusion.

Hot-working Deformation that is performed above the recrystallization temperature, so that annealing occurs concurrently.

Hydrogen bond Secondary bonds in which the hydrogen atom (proton) is attracted to electrons of neighboring atoms.

Hydrogen electrode Standard reference electrode with the following half-cell reaction:

$$H_2 \rightarrow 2H^+ + 2\ e^-$$

Hypereutectoid A composition with *more* solute than is found in the eutectoid composition.

Hypoeutectoid A composition with *less* solute than is found in the eutectoid composition.

Immiscibility Mutually insoluble phases.

Imperfection (crystal) Defect in a crystal; may be a point, a line, or a boundary.

Impressed current Direct current applied to make a metal cathodic during service.

Index of refraction (n) Ratio of light velocity in a vacuum to that in the material.

Induced compression Prestressing of the surface region of a material, commonly by differential thermal contraction.

Induction (magnetic, B) Flux density in a magnetic field.

Induction (remanent, B_r) Flux density remaining after the external magnetic field has been removed.

Induction hardening Hardening by heating the surface with high-frequencing induced currents.

Inhibitor An additive to an electrolyte that promotes passivation.

Injection molding Process of pressure molding a material in a closed die. For thermoplasts, the die must be cooled. For thermosets, the die is maintained at the curing temperature of the plastic.

Insulator (electrical) Material with a filled valence band and a large energy gap.

Interfacial stress Shear stresses that transfer the load between the matrix and the reinforcement.

Intergranular precipitation Nucleation and growth of a second phase along the boundaries of preexisting grains.

Internal reflection Surface reflection of light within the higher-index phase.

Interplanar spacing (d_{hkl}) Perpendicular distance between two adjacent planes with the same index.

Interrupted quench Two-stage quenching of steel that involves heating to form austenite, initial quenching to a temperature above the start of martensite formation, followed by a second (slower) cooling to room temperature.

Interstice Unoccupied space between atoms or ions.

Intragranular precipitation Nucleation and growth of a particulate phase within preexisting grains.

Ion An atom that possesses a charge because it has added or removed electrons.

Ion vacancy ($\square$) Unoccupied ion site within a crystal structure. The charge of the missing ion must be compensated appropriately.

Ionic bond Atomic bonding through coulombic attraction of unlike ions.

Ionic radius (r, R) Semiarbitrary radius assigned to ions. The sum of the radii of two adjacent ions is equal to their interatomic distance. Varies with coordination number. (See the footnote with Appendix B.)

Isotactic Long-range repetition in a polymer chain (in contrast to atactic).

Isotherm Line of constant temperature.

Isothermal precipitation Precipitation from supersaturation at constant temperature.

Isothermal Transformation (I-T) Transformation, with time, by holding at a constant temperature.

I-T curve Plot of reaction progress for isothermal transformation.

Jominy distance (D_{qe}) Cooling rate indexed to a distance from the quenched end of a Jominy bar.

Junction (n–p) Interface between an n-type and a p-type semiconductor.

Laser Light amplifications stimulated by emission of radiation.

Latewood The part of the wood grain that has smaller, thicker-walled cells. Formed late in the growing season. (Also called summer wood.)

Lattice A space arrangement with extended periodicity.

Lattice constants (*a*, *b*, *c*) Edge dimensions of a unit cell.

Lever rule (inverse) Equation for interpolation to determine the quantity of phases in an equilibrated mixture.

Light-emitting diode (LED) An $n-p$ junction device designed to produce photons by recombination.

Linear density Items (e.g., atoms) per unit length.

Liquidus The locus of temperatures above which there is only liquid.

Lone pair Electron pairs in a nonconnecting sp^3 orbital.

Long-range order A repetitive pattern over many atomic distances.

Luminescence Light emitted by the energy released as conduction electrons recombine with electron holes.

Macromolecules Molecules made up of hundreds to thousands of atoms.

Magnet, permanent (hard) Magnet with a large $(-BH)$ energy product, which therefore maintains domain alignment.

Magnet, soft Magnet that requires negligible energy for domain randomization.

Magnetization (*M*) Magnetic-moment density; that is, magnetic moment per unit volume.

Magneton, Bohr (*β*) Magnetic moment of an electron $(9.27 \times 10^{-24} \text{ A} \cdot \text{m}^2)$.

Martensite (M) A phase arising from a diffusionless, shearlike phase transformation. In steels containing >0.15 w/o carbon, martensite is a hard, brittle, body-centered tetragonal phase that is supersaturated with carbon.

Materials (engineering) Substances used for technical products. (These include metals, ceramics, polymers, semiconductors, glasses, and natural substances such as wood and stone, but generally exclude food, drugs, and related substances.)

Materials balance Mathematical calculation of "the whole is equal to the sum of its parts."

Matrix Principal material of a composite that envelopes the reinforcement.

Mean chord length, $\overline{L}$ A measure of grain size, based on the average boundary intercept distance.

Mer, $\rightarrow$ $\rightarrow$ The smallest repetitive unit in a polymer.

Metallic bond Interatomic bonds in metals characterized by delocalized electrons in the energy bands.

Metal (Chapter 1) Material characterized by its high electrical and thermal conductivities. (Chapter 2) Element located in the left and lower portions of the periodic table, as normally presented. (Chapter 11) Material with only a partially filled valence band.

Microfibrils Discrete filamentary units of the wood-cell wall.

Microcracking A multiplicity of cracks on a microscopic scale in the region of failure.

Microstructure Structure of grains and phases. Generally requires magnification for observation.

Miller index (*hkl*) Index relating a plane to the references axes of a crystal. *Reciprocals* of axial intercept, cleared of fractions and of common multipliers.

Mixture A combination of two, or more, phases.

Mobility (*μ*) The drift velocity of an electric charge per unit electric field, $\bar{v}/\mathscr{E} = (\text{m/s})(\text{V/m})$. Alternatively, the diffusion coefficient of a charge for a potential difference, $D/E = (\text{m}^2/\text{s})/\text{V}$.

Modulus of elasticity (elastic modulus, Young's modulus) Stress per unit elastic strain.

Mole Mass equal to the molecular weight of a material. Also, $0.602 \ldots \times 10^{24}$ molecules.

Molecular crystal Crystals with molecules as basic units (as contrasted to atoms).

Molecular length ($\overline{L}$) End-to-end, root-mean length.

Molecular orientation Collective alignment of macromolecules, usually by an applied axial stress.

Molecular weight Mass of one molecule (expressed in amu), or the mass of $0.602 \ldots \times 10^{24}$ molecules (expressed in grams). Mass of one formula weight.

Molecular weight (mass-average), $\overline{M}_\text{m}$ Average molecular size based on the mass fraction.

Molecular weight (number-average), $\overline{M}_\text{n}$ Average molecular size based on the number fraction.

Molecule Finite group of atoms bonded by strong attractive forces (primary bonds). Bonding between atoms is by weak, secondary bonds.

Monomer A molecule with a single mer.

NaCl structure The fcc structure of an AX compound with CN = 6. (See Figs. 3–1.1 and 3–2.5.)

Nernst equation Electrode potential as a function of electrolyte concentration and temperature (Eq. 14–2.3).

Network structure Molecular structure with primary bonds in three dimensions.

Nonstoichiometric compounds Compounds with noninteger atom (or ion) ratios.

Normalizing Heating of steel into the austenite range (~50° C, or 100° F) so that it will contain a uniform, fine-grained microstructure.

Nucleation The start of the growth of a new phase.

Nucleation, heterogeneous Nucleation on a preexisting surface (or by introduction of "seeds").

Nucleation, homogeneous Unaided nucleation within a preexisting phase.

Orbital Wave probabilities of atomic or molecular electrons.

Orthorhombic A crystal with three unequal, but mutually perpendicular, axes.

Overaging Aging continued until softening occurs.

Oxidation (general) The raising of the valence state of an element.

Paramagnetism Magnetic structure with the magnetic moments that are uncoupled, except in the presence of an applied magnetic field.

Passivation The condition in which normal corrosion is impeded by an absorbed surface film.

Pearlite (P) A lamellar mixture of ferrite and carbide formed by decomposing austenite of eutectoid composition.

Performance (materials) The behavior (and alteration) of materials during manufacture and or service.

Periodic table See Fig. 2–1.1.

Permeability (magnetic, μ) Ratio of induction to magnetic field, B/H.

Permittivity (ϵ) Ratio of charge density to electric field in a vacuum, 8.85×10^{-12} C/V · m.

Phase A physically homogeneous part of a materials system. (See Section 5–1.)

Phase, transition Metastable phase that forms as an intermediate step in a reaction.

Phase boundary Compositional or structural discontinuity between two phases.

Phase diagram Graph of phase-stability areas with composition and environment (usually temperature) as the coordinates.

Phase diagram, isothermal cut A constant-temperature section of a phase diagram.

Phase diagram, one-phase area Part of a phase diagram possessing a single unsaturated solution.

Phase diagram, two-phase area Part of a phase diagram beyond the solubility limit curves, such that a second phase is necessary. Temperature and phase compositions *cannot* be varied independently.

Phase diagram, three-phase temperature Invariant temperature in a binary system at which three phases can coexist.

Phases, chemical compositions of Phase compositions expressed in terms of chemical components.

Phases, quantities of Material compositions expressed in terms of phase fractions.

Phosphorescence Luminescence that is delayed by extended relaxation times.

Photoconduction Conduction arising from photon activation of electrons across the energy gap.

Photomultiplier Device that uses a photon to trigger an electron avalanche in a semiconductor. Thus, weak light signals can be amplified.

Photon A quantum of light.

Photovoltaic Production of an electrical potential by incident light.

Piezoelectric Dielectric materials with structures that are asymmetric, such that their centers of positive and negative charges are not coincident. As a result, the polarity is sensitive to pressures that change the dipole distance and therefore the polarization.

Pit corrosion Localized corrosion that penetrates a metal nonuniformly.

Planar density Items (e.g., atoms) per unit area.

Plastic Material composed primarily of polymers.

Plastic constraint Prevention of plastic deformation in a ductile material caused by the presence of an adjacent rigid material.

Plastic strain (e_{pl}) Permanent strain. It is not recoverable, since the atoms have moved to new neighbors.

Plasticizer An additive of small-molecular-weight molecules to a polymeric mix for the purpose of reducing the viscosity.

Point defect Crystal imperfections involving one (or a very few) atoms.

Poisson's ratio (v) Absolute ratio of lateral to axial strain.

Polar group A local electric dipole within a molecule.

Polarization, dielectric ($\mathscr{P}$) Electric dipole density; that is, dipole moment per unit volume.

Polarization, electronic (p_e) Dipole moment from electronic displacements.

Polarization, ionic (p_i) Dipole moment from ionic displacements.

Polarization, molecular (p_m) Dipole moment from molecular orientation.

Polarization, remanent ($\mathscr{P}_r$) Polarization remaining after the external electric field has been removed.

Polyblend A mixture of two or more polymeric phases.

Polycrystalline Material with multiple crystals and accompanying grain boundaries.

Polydispersity index, PDI The $\overline{M}_m/\overline{M}_n$ ratio. (Used as a measure of molecular-size distribution).

Polyfunctional Molecule with multiple (>2) reaction sites.

Polymer Nonmetallic material consisting of (large) macromolecules composed of many repeating units; the technical term for a plastic.

Polymorphism A composition with more than one crystal structure (allotrophic).

Precipitation Separation from a supersaturated solution.

Precipitation hardening Hardening by the formation of clusters prior to precipitation (also called age hardening).

Preferred orientation A nonrandom alignment of crystals or molecules.

Prestressing Introduced compressive stresses, usually obtained by tension rods or similar mechanical means.

Primary bond Strong (>200 kJ/mol) interatomic bonds of the covalent, ionic, and metallic types.

Proeutectoid A phase that separates from a solid solution *before* the latter decomposes into the eutectoid product.

Property Characteristic quality of a material. (Quantitative values are preferable.)

Proportional limit The limit of the proportional range of the stress–strain curve. (Elastic strain increases beyond this "limit," but is masked by plastic strain.)

Quench Process of accelerated cooling, usually in agitated water or oil.

Radiation damage Structural defects arising from exposure to radiation.

Rate ($R = t^{-1}$) Reaction per unit time.

Recombination Annihilation of electron–hole pairs.

Recovery Loss of resistivity, or related behaviors that originated during strain hardening, by annealing out point defects. (See also *recrystallization*.)

Recrystallization The formation of new annealed grains from previously strain-hardened grains.

Recrystallization temperature Temperature above which recrystallization is spontaneous.

Rectifier Electric "valve" that permits forward current and prevents reverse current.

Reduction Removal of oxygen from an oxide. Also, the lowering of the valence level of an element.

Reduction of area (R of A) Total plastic strain before fracture, measured as the percent decrease in cross-sectional area at the fracture point.

Refractory A material capable of withstanding extremely high temperatures.

Reinforcement Component of composites with a high elastic modulus and high strength.

Relaxation time (τ) Time required for exponential decay to $1/e$ of the initial value.

Repetition distance Translation vector between two identical lattice points.

Resistivity (ρ) Reciprocal of conductivity (usually expressed as ohm·m).

Resistivity coefficient, solutioin (y_x) Coefficient of resistivity versus solid solution content (atom fraction basis), $d\rho/dX$.

Resistivity coefficient, thermal (y_T) Coefficient of resistivity versus temperature, $d\rho/dT$.

Rolling Mechanical working through the use of two rotating rolls (Fig. 9–1.2).

Root mean square length ($\overline{L}$) Statistical end-to-end length of linear molecules.

Sacrificial anode Expendable metal that is anodic to the product that is to be protected.

Scale Surface layer of oxidized metal.

Scission Fragmentation of a polymer by radiation.

Secondary bonds Weak (<40 kJ/mol) interatomic

bonds arising from dipoles within the atoms or molecules.

Segregation Compositional nonuniformity due to delayed reactions.

Semiconductor A material with controllable conductivities, intermediate between an insulator and a conductor.

Semiconductor, extrinsic Semiconduction from impurity sources.

Semiconductor, intrinsic Semiconduction of a pure material. The electrons are excited across the energy gap.

Semiconductor, *n*-type Semiconductor in which impurities provide donor electrons to the conduction band. Electrons are the majority carriers.

Semiconductor, *p*-type Semiconductor in which impurities provide acceptor sites for electrons from the valence band. Electron holes are the majority carriers.

Shear modulus (*G*) Shear stress per unit shear strain.

Shear strain (*γ*) Tangent of shear angle, α, developed from shear stress.

Shear stress (*τ*) Shear force per unit area.

Sheet molding Thermal forming of plastic products from starting sheets of fiber-reinforced plastics.

Short-range order Specific first-neighbor arrangement of atoms, but random long-range arrangements.

Simple cubic (sc) A cubic unit cell with lattice points at the corners only. (It may contain atoms at other locations, which are not identical.)

Sintering Bonding by thermal means.

Slip direction Crystal direction of the displacement vector on which slip takes place.

Slip plane Crystal plane along which slip occurs.

Slip system Combination of slip directions on slip planes that have low critical shear stresses.

Slip vector (b) Displacement distance of a dislocation. It is parallel to a screw dislocation and perpendicular to an edge dislocation.

S–N curve Plot of fatigue stress versus number of stress cycles.

Solder Metals that melt below approximately 425° C (800° F) and are used for joining. Commonly, these are Pb–Sn alloys; however, other alloys (even glass) may be used.

Solid solution A homogenous crystalline phase with more than one chemical component.

Solid solution, interstitial Crystals that contain a second component in their interstices. The basic structure is unaltered.

Solid solution, ordered A substitutional solid solution with a preference by each of the components for specific lattice sites.

Solid solution, substitutional Crystals with a second component substituted for solvent atoms in the basic structure.

Solidus The locus of temperatures below which only solids are stable.

Solubility limit Maximum solute addition without supersaturation (See Section 5–1).

Solute The minor component of a solution.

Solution hardening Increased strength present in solid solutions (from pinning of dislocation by solute atoms).

Solution treatment Heating to induce solid solutions.

Solvent The major component of a solution.

Space charge Polarization from conductive particles in a dielectric.

Spalling Cracking from thermal stresses.

Spheroidite A two-phase microstructure of spherelike carbides in a ferrite matrix.

Spheroidization Process of making spheroidite, generally by extensive overtempering to develop spherelike carbides.

Spinning (polymers) Fiber-making process by filament extrusion.

Stainless steel High-alloy steel (usually containing Cr, or Cr + Ni) designed for resistance to corrosion and or oxidation.

Static fatigue Delayed fracture arising from stress corrosion.

Steel, low-alloy Steel containing up to 5 percent alloying elements other than carbon. Phase equilibria are related to the Fe–Fe$_3$C diagram.

Steel, plain-carbon Basically Fe–C alloys with minimal alloy content.

Stereoisomer Isomeric molecules differing in mer configurations along the molecular chain.

Stereospecific Covalent bonding (and hydrogen bridges) between specific atom pairs (in contrast to the omnidirectional coulombic attractions).

Sterling silver An alloy of 92.5 Ag and 7.5 Cu. (This corresponds to nearly the maximum solubility of copper in silver.)

Stoichiometric compounds Compounds with integer atom (or ion) ratios.

Strain (*e*) Unit deformation. Elastic strain is recoverable; plastic strain is permanent.

Strain hardening Increased hardness (and strength) arising from plastic deformation below the recrystallization temperature.

Strength (*S*) Critical stress to produce failure. Yield strength, S_y is the stress to initiate the first plastic deformation; ultimate tensile strength, S_u, is the maximum calculated stress on the basis of the original area.

Stress (*s*) Force per unit area. Nominal stress is based on the original design area; true stress is based on the actual area.

Stress corrosion Corrosion accentuated by stresses; or fracture accelerated by corrosion.

Stress intensity factor (K_I) Stress intensity (tension) at the root of a crack, $Pa \cdot m^{-1/2}$.

Stress intensity factor, critical (K_{Ic}) A value (for a given material) that relates the depth of a propagating crack to the design limit for the nominal stress (Eq. 8–4.1).

Stress relaxation Decay of stress at a constant strain (by molecular rearrangement).

Stress relief Removal of residual stresses by heating.

Stress rupture Time-dependent rupture resulting from constant load (usually at elevated temperatures).

Stress–strain diagram Plot of stress as a function of strain. (Stress normally is plotted on the basis of original area.)

Superconductivity Property of zero resistivity at temperatures approaching absolute zero.

Supercooled ( ) A phase cooled beyond its range of full equilibrium. In this text, we identify a supercooled phase with parentheses; for example, (γ).

Surface Boundary between a condensed phase and gas.

Tempered glass Glass with surface compressive stresses induced by heat treatment.

Tempered martensite A two-phase microstructure of ferrite and carbide obtained by decomposing martensite.

Tempering A toughening process in which martensite is heated to initiate a ferrite-plus-carbide microstructure.

Tetragonal Two of three axes equal; all three at right angles.

Thermal expansion coefficient (α) (Change in dimension)/(change in temperature).

Thermistor Semiconductor device with a high dependence on temperature. It may be calibrated as a thermometer.

Thermoplastic Softens with increased temperature, thus becoming moldable; rehardens on cooling.

Thermosetting Hardens with heating. Plastics that undergo further (three-dimensional) polymerization with heating.

Tie-line Isotherm across a two-phase field connecting the solubility limits of two equilibrated phases.

Toughness A measure of the energy required for mechanical failure.

***Trans* (polymers)** Prefix denoting unsaturated positions on the *opposite* side of the polymer chain. (See also *cis*.)

Transducer A material or device that converts energy from one form to another, specifically electrical energy to, or from, mechanical energy.

Transistor Semiconductor device for the amplification of current. There are two principal types: field effect and junction.

Translation Vector displacement between lattice points.

True stress (σ) Stress based on the actual area.

True strain (ϵ) Strain based on the actual area at the point of fracture.

T-T-T curve Time-Temperature-Transformation curve (See *I-T curves*.)

Twin (crystal) Two parts of a crystal that possess a mirror image relationship.

Ultimate strength (S_u) Maximum stress, based on nominal area.

Unit cell A small (commonly the smallest) repetitive volume that possesses the maximum symmetry of a crystal lattice.

Vacancy ($\square$) A normally occupied lattice site that is vacant.

Valence electrons Removable electrons from the outer shell(s) of an atom.

Vinyl compounds (C_2H_3R), where **R** is one of several atoms or radicals.

Viscoelastic deformation Combined deformation by viscous flow and elastic strain.

Viscoelastic modulus (M_{ve}) Ratio of shear stress to the sum of elastic deformation, γ_{el}, and viscous flow, γ_f.

Viscoelasticity Combination of viscous flow and elastic behavior.

Viscosity (η) Ratio of shear stress to the velocity gradient of flow. Reciprocal of fluidity.

Viscous flow Time-dependent, irreversible flow.

Vulcanization Treatment of rubber with sulfur to cross-link elastomer chains.

Yield strength (S_y) Resistance to initial plastic deformation.

Young's modulus (E) Modulus of elasticity (axial). Stress per unit elastic strain.

Zener diode An n–p junction that has a controlled breakdown voltage with a reverse bias.

Zone refining Purification performed by passing a molten zone along a bar of material. The liquid retains the solute; the solid crystallizes at the solidus composition.

BIBLIOGRAPHY

T. Alfrey and E. F. Gurnee, *Organic Polymers,* Prentice-Hall, Englewood Cliffs, N.J., p. 95, 1967.

E. B. Allison and P. Murray, *Acta Metallurgica,* Vol. 2, p. 487, 1954.

I. A. Aksay and J. A. Pask, *J. Amer. Ceram. Soc.,* Vol. 58, p. 507, 1975.

Aluminum, Vol. 1, ASM-International, pp. 98, 102, 167, 1967.

E. C. Bain, *Alloying Elements in Steel,* ASM-International, p. 39, 1939.

R. M. Bozorth, *Ferromagnetism,* Litton Educational Publishing Co., Inc., Division of Van Nostrand Reinhold, New York, 1956.

W. L. Bragg and J. F. Nye, *Proc. Roy. Soc. (London),* Vol. A190, p. 474, 1947.

L. Broutman, *Modern Composite Materials,* Addison-Wesley, Reading, Mass., p. 373, 1967.

A. H. Cottrell, *The Mechanical Properties of Matter,* J. F. Wiley, New York, p. 358, 1964.

A. H. Cottrell and D. Hull, *Proc. Roy. Soc. (London),* Vol. A242, 1950.

B. D. Cullity, *Elements of X-ray Diffraction,* 2nd Ed., Addison-Wesley, Reading, Mass., pp. 98, 99, 1977.

B. D. Cullity, *Introduction to Magnetic Materials,* Addison-Wesley, Reading, Mass., 1972.

Constance Elam (Tipper), *Distortion of Metal Crystals,* Clarendon Press, Oxford, England, 1935.

Electron Microstructures of Steel, Amer. Soc. Test. and Materials, pp. 39–41, 1950.

Encyclopedia of Materials Science and Engineering, M.I.T. Press, Cambridge, Mass., Vol. 6, p. 4760, 1986.

R. A. Flinn, *Fundamentals of Metal Casting.* Addison-Wesley, Reading, Mass., p. 8, 1963.

R. S. French and W. H. Hibbard, *Trans. A.I.M.E.,* Vol. 188, p. 53, 1950.

R. E. Gardner and G. W. Robinson, Jr., *J. Amer. Ceram. Soc.,* Vol. 45, p. 46, 1962.

M. F. Garwood, H. H. Zurburg and M. A. Erickson, *Interpretation of Tests and Correlation with Service,* ASM-International, p. 10, 1951.

S. D. Gehman, *Chemical Reviews,* Vol. 26, p. 203, 1940.

M. Gell, D. H. Duhl and A. F. Giamei. *Superalloys,* ASM-International, p. 206, 1980.

M. Gordon, *High Polymers,* Illife, London, and Addison-Wesley, Reading, Mass., p. 90, 1963.

A. G. Guy and J. J. Hren, *Elements of Physical Metallurgy,* 3rd Ed. Addison-Wesley, Reading, Mass., pp. 63, 78, 444, 486, 1974.

Journal of Metals, Vol. 39, No. 4, pp. 20, 23, 1987.

H. K. Hardy and T. J. Heal, *Progress in Metal Physics* 5, London: Pergamon Press, 1954, p. 256.

W. C. Leslie, R. L. Rickett and W. D. Lafferty, *Trans. AIME.,* Vol. 218, p. 703, 1960.

C. W. Mason, *Introduction to Physical Metallurgy,* ASM-International, p. 33, 1947.

F. A. McClintock and A. S. Argon, *Mechanical Behavior of Metals,* Addison-Wesley, Reading, Mass., 1966.

Metals Handbook, 8th Ed., Vol. 8, ASM-International, 1973.

W. J. Moore, *Physical Chemistry,* Prentice-Hall, Englewood Cliffs, N.J., pp. 140, 149, 1955.

National Academy of Science, *Materials and Man's Needs,* Committee on the Survey of Materials Science and Engineering, p. 10, 1974.

C. J. Nuese, *J. Materials Education,* Vol. 2, p. 140, 1980.

R. E. Peterson, *Materials Res. and Stand.,* ASTM, p. 124, 1963.

M. N. Riddell, et al, *Polymer Science and Engineering,* Vol. 6, p. 363, 1966.

F. N. Rhines, *Metals Progress,* Vol. 112, p. 60, 1977.

B. Rodgers, *The Nature of Metals,* 2nd Ed., ASM-International, pp. 20, 34, 36, 44, 1964.

A. Schmidt and C. A. Marlies, *Principles of High Polymer Theory and Practice,* McGraw-Hill, New York, pp. 87, 573, 1948.

E. Schmid and W. Boas, *Plasticity in Crystals,* Chapman Hall, London, p. 191, 1968

C. O. Smith, *Nuclear Reactor Materials,* Addison-Wesley, Reading, Mass., p. 67, 1967.

F. N. Speller, *Corrosion: Causes and Prevention,* McGraw-Hill, New York, p. 84, 1951.

L. H. Van Vlack, *Materials for Engineers: Concepts and Applications,* Addison-Wesley, Reading, Mass., 1982.

L. H. Van Vlack, *Physical Ceramics for Engineers,* Addison-Wesley, Reading, Mass., p. 100, 1964.

L. H. Van Vlack, *A Textbook of Materials Technology,* Addison-Wesley, Reading, Mass., p. 3, 1973.

D. Walsh, et al., eds., *Polymer Blends and Mixtures,* Martinus Nijhoff Publishers, Dordrecht, The Netherlands, pp. 430, 449, 1985.

H. Wiedersich, *Journal of Metals,* Vol. 16, p. 425, 1964.

O. H. Wyatt and D. Dew-Hughes, *Metals, Ceramics, and Polymers,* Cambridge University Press, London, p. 290, 1974.

C. A. Zapffe, *Stainless Steels,* ASM-International, p. 60, 1949

INDEX

A

α-iron 228
Absorption, water, 383
AC Spark Plug, 480
Acceleration of gravity, 552
Acceptor, 419, 442, **569**
Acceptor level, 419
Acceptor saturation, 420, **569**
Acetone, 25
Acta Metallurgia, 287, 581
Activation energy, 199, 201, 350, **569**
Addition polymerization, 32
Additives (plastics), 356, **569**
Age hardening, 304, **569**
 and corrosion, 511
 mechanism, 306
Aging, over-, 307, 341, 545, **576**
Ahrens, L.H., 50
AISI number 318
Aksay, I.S., 158
Alcoa, 510
Alfrey, T., 524, 581
Allison, E.B., 287, 581
Allegheny-Ludlum, 504
Allotrope, 78
Alloy, **569**
 galvanic series, 518
 single-phase, 216
 supersaturated, 224
Alloy retardation, 317
Alloying Elements in Steel, 581
Alnico, 457
Alper, A., 157
Aluminum, 297, 303, 307, 581
Aluminum Research
 Laboratories, 166
Amax Corp., 319
American Ceramic Society, 157, 581
American Forest Institute, 381

American Information
 Technologies, 491
American Journal of Science, 165
American Optical Co., 487
American Society for Testing and
 Materials, 219, 520, 522, 582
Ammonia, 25, 29
Amoco Performance Products, 2
Amorphous, 114, **569**
Amu, 20
Analog Devices, 11
Analysis of phases, 161
Angle, axial, 62
 bond, 28
 diffraction, 96
Angles between directions, 84
Aniline, 181
Anion, 22, 393, 442, **569**
Anisotropic, 254, 373, **569**
 magnetic, 463
 optical, 489
 wood, 380
Annealing, 192, 292, 296, 313, 314
 and cold work, 301, 340
 full, 314, 340
 process, 314, 340
 steels, 314, **569**
 strain-hardened metal, 296, **569**
Annealing point (glass), **569**
Anode, 505, 506, **569**
 reactions, 505, **569**
 sacrificial, 519, 545
Anodize, 517, **569**
Antiferromagnetism, 461, **569**
Apparent density, 288, 462
Apparent porosity, 288, 462
Apparent strain, 338, 376
Apparent volume, 288, 462
Area, grain-boundary, 219
 phase-boundary, 236
 reduction of, 259
Argon, A.S., 529, 582

Aramid fibers, 372
Archimedes principle, 462
Arrhenius equation, 200, 212, **569**
Askill, J., 208
ASM International, 154, 155, 157, 159, 166, 175, 176, 177, 178, 297, 303, 307, 531, 581
ASTM grain size, 219, 220, **569**
Atactic, 124, **569**
Atomic bonding, 19
Atomic coordination, 19
Atomic diffusion, 200
Atomic disorder, 107
Atomic mass, 20, 554
Atomic mass unit (amu), 20, 54, 552, **569**
Atomic movements, 200, 212
Atomic number, 20, 554, **569**
Atomic order, 59
Atomic packing factor, 66
Atomic radius, 49, 50, 54, 554, **569**
Atomic vibrations, 197
Atomic weight, 20, 54, **569**
Atoms, 20
Attenuation (optical), 493, **569**
Attractions, coulombic, 45
Auger, 360
Austempering, 313, **569**
Austenite, 158, 228, 248, **569**
 decomposition of, 229, **569**
 isothermal transformation, 233
Austenization, **569**
Avalanche, 429
Average molecular weight, 119, 121
Avogadro's number, 20, 552
AX structures, 68, 69
 CsCl, 69
 NaCl, 61, 68, 100
Axis (crystal), 62

Boldface page numbers refer to items that are defined in *Terms and Concepts.*

585